Environmental Sedimentology

Environmental Sedimentology

Edited by Chris Perry and Kevin Taylor

Department of Environmental and Geographical Sciences,
Manchester Metropolitan University

 **Blackwell** Publishing

BLACKWELL PUBLISHING
350 Main Street, Malden, MA 02148-5020, USA
9600 Garsington Road, Oxford OX4 2DQ, UK
550 Swanston Street, Carlton, Victoria 3053, Australia

First published 2007 by Blackwell Publishing Ltd

Library of Congress Cataloging-in-Publication Data

Environmental sedimentology / edited by Chris Perry and Kevin Taylor.
p. cm.
Includes bibliographical references and index.
ISBN-13: 978-1-4051-1515-5 (pbk.: acid-free paper)
ISBN-10: 1-4051-1515-7 (pbk.: alk. paper)
1. Sedimentology.   2. Geology–Environmental aspects.
I. Perry, Chris (Chris T.)   II. Taylor, Kevin (Kevin G.)

QE33.E58 2007
551.3–dc22
2006023456

A catalogue record for this title is available from the British Library.

Set in 10/12.5pt Sabon
by Graphicraft Limited, Hong Kong

For further information on
Blackwell Publishing, visit our website:
www.blackwellpublishing.com

Contents

Case studies

Contributors

Andrew Cooper   School of Biological and Environmental Sciences, University of Ulster, Coleraine Campus, Coleraine, Co. Londonderry BT52 1SA, UK

Peter French   Department of Geography, Royal Holloway, University of London, Egham, Surrey TW20 0EX, UK

Lars Håkanson   Department of Earth Sciences, Uppsala University, Villav. 16, 752 36 Uppsala, Sweden

Karen Hudson-Edwards   School of Earth Sciences, Birkbeck College, University of London, Malet Street, London WC1E 7HX, UK

Piers Larcombe   Centre for Environment, Fisheries and Aquaculture (CEFAS), Pakefield Road, Lowestoft, Suffolk NR33 OHT, UK

Anne Mather   Department of Geography, University of Plymouth, Plymouth, Devon PL4 8AA, UK

Chris Perry   Department of Environmental and Geographical Sciences, Manchester Metropolitan University, John Dalton Building, Chester Street, Manchester M1 5GD, UK

Kevin Taylor   Department of Environmental and Geographical Sciences, Manchester Metropolitan University, John Dalton Building, Chester Street, Manchester M1 5GD, UK

Jeff Warburton   Department of Geography, University of Durham, Durham DH1 3LE, UK

Preface

It will be clearly apparent to students of the earth sciences that a number of environmental issues have gained in importance over the past few years, and that sedimentology has a role to play in informing the debate about their impacts. These issues include recent and future climate change, and a wide range of anthropogenic pressures that influence earth surface systems from a range of physical, chemical and ecological perspectives. Many sediment systems act as excellent archives of past environmental change, allowing us a window into the recent past, as well as providing tools for monitoring change within active sedimentary environments. These in turn can be used to inform management strategies.

Sedimentology text books traditionally do a good job of introducing the concepts of sedimentology and dealing in detail with the principles of sedimentology. These text books, however, tend to focus primarily on facies analysis and basin processes of sediment accumulation, and on the interpretation of ancient sedimentary environments. Although these cover many aspects of modern depositional environments, the emphasis is primarily on using modern systems to interpret the geological record. More applied sedimentology text books effectively focus on the application of sedimentology to the oil and gas industry. It is increasingly being recognized that a more environmental approach to sedimentology is needed. In this context Environmental Sedimentology represents a rapidly expanding research field, which draws upon both the traditional aspects of sedimentology and the more applied areas of the discipline, and also interfaces with the fields of hydrology, geomorphology, engineering, biogeochemistry and ecology.

Essentially it is concerned with understanding the development of recent sedimentary systems, and examining their response to both natural and anthropogenically induced disturbance events.

The aims of this text book are:
1 to outline the boundaries of the field of Environmental Sedimentology;
2 to allow those students with a prior grounding of sedimentological principles to develop their knowledge in the environmental aspects of the subject;
3 to allow environmental scientists and physical geographers access to the field of environmental sedimentology;
4 to allow more specialist groups (e.g. civil engineers, legislators) to gain information on this topic.

The target audiences for this book are earth scientists, environmental scientists, physical geographers, hydrologists and civil engineers. Although we assume that earth scientists will have a grounding in the principles of sedimentology, we recognize that some of the other groups may not have. Therefore, the introductory chapter not only summarizes the principles of environmental sedimentology, but also provides an overview of the basics of sedimentology. We encourage readers of this text to make free reference to the available text books on fundamental sedimentology where appropriate.

The book is structured such that it outlines processes and issues associated with a range of terrestrial (i.e. upland, fluvial, lake, arid and urban) and coastal and shallow-marine sedimentary environments. We do not incorporate aspects of groundwater hydrology, which, although often recently described as an environmental sedimentological topic, essentially deals with water, rather than sediments. We also do not intend to specifically address the topic of soils, which are not strictly sediments, but the initial products of bedrock and organic material degradation. Each chapter has a similar structure and, for each specified environment, examines aspects of: sediment sources and sediment accumulation; the processes and impacts of natural disturbance events; the processes and impacts of anthropogenic activities; sediment system management and remediation; and issues likely to be of concern in the future, such as short to medium term (< 100 years) climatic change, sea-level rise and increased anthropogenic influence. Each chapter is authored by a leader in their particular field and by the very nature of this approach, each chapter should not be considered to be uniform in its format. Each author has presented the major environmental processes and products in a manner in which they feel most appropriate, inevitably reflecting that author's strengths. The reader will, therefore, find some chapters that focus on physical processes, others on the chemical aspects, and yet others with a more numerical eye on environmental sedimentology. This reflects the wide range of disciplines that inform the subject.

Each chapter provides a wealth of up-to-date references to allow the student to follow up on the processes and products discussed in the chapters in more detail. Most, but not all, of these references are in primary scientific journals and reports, but where it is felt advantageous, reference is also made to text books in the field. An important part of each chapter is the inclusion of a number of case studies, which act as self-contained examples to illustrate some of the key concepts discussed in the respective chapters. Each case study, geographically specific, has additional references pertinent to that example.

Numerous individuals have provided inputs into the development of this book, but in particular we would like to thank the following for their comments and reviews at various stages: Helene Burningham, Sue Charlesworth, Ian Drew, Simon Haslett, Piers Larcombe, David Nash, Phil Owens, Laura Shotbolt and Chris Vivian. We would also like to thank Delia Sandford, Ian Francis and Rosie Hayden at Blackwell for help and advice throughout this project.

Chris Perry and Kevin Taylor
Manchester Metropolitan University

1

Environmental sedimentology: introduction

Chris Perry and Kevin Taylor

1.1 ENVIRONMENTAL SEDIMENTOLOGY: DEFINITION AND SCOPE OF CHAPTERS

Environmental sedimentology represents a relatively new subdiscipline of the earth sciences and, as such, the boundaries of the field are not clearly defined. Herein we define environmental sedimentology as . . . *'the study of the effects of both man and environmental change upon active surface sedimentary systems'*. Consequently, environmental sedimentology can be regarded as the study of how both natural and anthropogenic inputs and events modify the production and accumulation of the physical and biogenic constituents of recent sedimentary deposits. The field of environmental sedimentology has evolved gradually over the past two decades, largely owing to an increased recognition of the influence that anthropogenic activities are exerting upon sediment production and cycling. Studies in these areas reflect a need to address issues of sedimentological change driven by environmental or land-use modification or contamination. This, in turn, has promoted increasingly integrated approaches to examining the dynamics of, and interlinkages between, sedimentary environments, the nature of which can be illustrated by one example, namely studies that link catchment processes (and anthropogenically induced changes in catchment sediment yields) with sediment supply to (and through) the coastal zone. Here, disciplines such as slope geomorphology, fluvial sedimentology, hydrology, coastal and marine sedimentology, and coastal management combine to assess interlinked issues of sediment production, transport and accumulation.

This book is divided into nine main chapters, each of which deals with a distinct sedimentary system. These are delineated as follows; mountains, fluvial, arid, lacustrine, urban, temperate intertidal (estuarine and deltaic), temperate coastal (beach, barrier island and dune), tropical coastal (coral reef and mangrove) and continental shelf. Although this structure provides a convenient approach for distinguishing and describing individual sedimentary environments, such subdivisions are to some extent arbitrary and, in light of the sediment exchanges that occur between environments, not necessarily ideal. As a result, the linkages that exist between environments are highlighted where appropriate. In addition, the book does not set out to review every type of sedimentary environment, but rather to provide a framework for the subject in the context of a number of key sedimentary systems and settings. Given these caveats, each of the chapters examines aspects of sediment supply and accumulation, the response of the individual sedimentary systems to natural and anthropogenic change, and issues of sediment management and remediation, and reviews the potential responses of these sedimentary systems to issues such as climatic and environmental change.

The chapters in this book essentially deal with sedimentological processes and geomorphological changes that have been operating over short-to-medium (< 100 yr) time-scales (*biological* time-scales of Spencer 1995; *secular* time-scales of Udvardy 1981), although many of

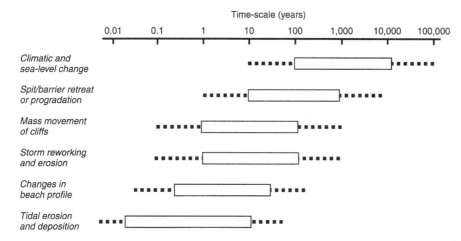

Fig. 1.1 Time-scales over which different geomorphological processes operate within a hypothetical coastal environment. (Adapted from Woodroffe 1992.)

the sedimentary landforms discussed also inevitably represent the products of longer-term sediment accumulation. Hence, in each chapter these may encompass the daily processes of sediment transport and reworking, through to the effects of rapid climatic and sea-level change (Fig. 1.1). In this context, short-term changes (< 1 yr) within a coastal sedimentary environment may, for example, include the processes of tidal erosion and deposition, seasonal changes in beach profiles or morphological change due to storm events; in the short-to-medium term (< 10 yr), mass movement of unstable cliffs, spit progradation, or barrier breaching; and in the medium-term (up to 100 yr), coastal landform progradation, changes in delta morphology, or coastal retreat. Superimposed upon these processes may be a range of anthropogenic activities (e.g. sea-wall construction, sand dredging or contaminant inputs), which may influence not only the dynamics of the sedimentary system, but also the associated floral and faunal components of the system. Many of these biological components (e.g. dune plants, mangroves or corals) are often of sedimentological significance in their own right, either as sources of sediment or as agents of sediment trapping and stabilization.

This introductory chapter reviews the fundamentals of sediment production and the principles of sediment transport and deposition. In addition, consideration is given to the types of issues that are discussed in the respective chapters, and which help to define the discipline of environmental sedimentology. This chapter also outlines the magnitude and frequency of predicted shifts in global climatic and environmental conditions that have relevance to the development of both terrestrial and marine sedimentary systems over the coming century. These include issues such as sea-level change, global warming and changes in temperature, precipitation and storm frequency, and thus provide a framework for discussion within the respective chapters.

1.2 SEDIMENT PRODUCTION AND SUPPLY WITHIN SEDIMENTARY ENVIRONMENTS

The composition of sediments that accumulate within individual sedimentary environments is primarily a reflection of three main factors:
1 the sediment source;
2 the processes of sediment transport and deposition, which determine whether sediment is retained or transported through a specific environment (these mechanisms are outlined in section 1.3);
3 the chemical processes operating within the sediment or water column, for example, carbonate and evaporite precipitation, chemical diagenesis.
In terms of the initial supply of sediment into a sedimentary system, three basic sediment types can be delineated. These are: (i) detrital minerals, (ii) biogenic or organic sediments and (iii) anthropogenic particles and compounds.

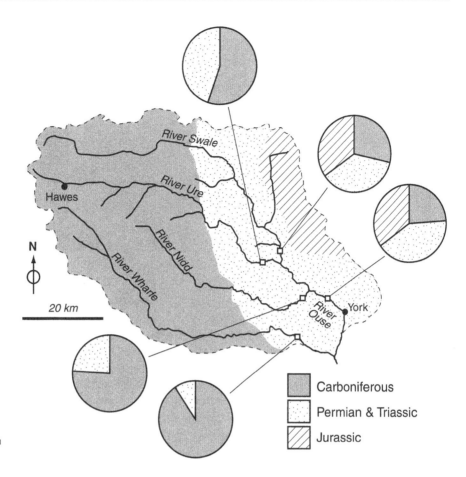

Fig. 1.2 Composition of suspended sediment samples recovered from the River Ouse (UK) and four of its tributaries that have different underlying geologies. (Adapted from Walling et al. 1999.)

1.2.1 Detrital minerals

Detrital minerals, such as quartz and feldspar, along with heavy minerals, form a primary component of many terrestrial and marine sediments. These minerals are initially released by weathering processes and are progressively eroded and transported into, and through, a range of sedimentary environments. As a result, initial mineralogical composition of the bedrock often influences the relative abundance of the individual minerals that are released. This control is clearly illustrated in studies of suspended sediment compositions within river catchments where individual tributaries are underlain by bedrock of differing geological compositions. In the River Ouse catchment (north-east England), for example, individual tributaries drain areas of differing geology (Carboniferous, Permian/Triassic and Jurassic). Suspended sediments in

the different tributaries have distinct mineralogical and magnetic signatures that demonstrate variations in the relative importance of different rock units as sources for fluvial sediment (Fig. 1.2). Local variations are attributed to variations in the rates of erosion and sediment supply from the different geological units (Walling et al. 1999).

In reality, these detrital minerals rarely undergo a simple source to sink transport route, but instead are subject to numerous phases of weathering, transport, deposition, storage, lithification, reworking and redeposition. For example, detrital sands in the Orinoco drainage-basin of South America are derived from similar bedrock material, but the nature of relief and chemical weathering markedly alter the grain composition (Johnsson et al. 1991). Material derived from steep, orogenically-active terrains undergoes limited chemical weathering, whereas

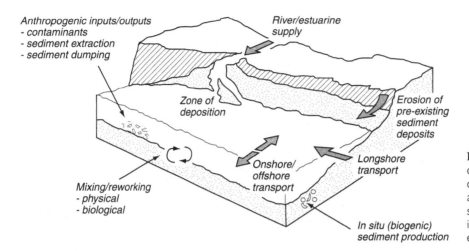

Fig. 1.3 Schematic diagram illustrating the different potential natural and anthropogenic sources of sediment into a nearshore marine environment.

in low-relief parts of the catchment thick soils accumulate and the detrital grains become highly altered chemically. Furthermore, sediment deposited in low-lying alluvial floodplains undergoes additional chemical weathering and alteration, further modifying their composition. Consequently, within any individual environment sediments tend to be derived from a range of source areas and weathering regimes, and these both contribute to, and influence, the composition of the accumulating sediment (Fig. 1.3). Over the time-scales considered in this book, the release of sediment from previously deposited sedimentary sequences should, therefore, also be regarded as a key sediment source.

1.2.2 Biogenic and organic sediments

In addition to detrital minerals, significant amounts of sediment are derived from the remains of skeletal carbonate-secreting organisms. These form across a wide range of marine environments (Schlager 2003), although marked latitudinal variations occur both in the types and rates of biogenic sediment production (Lees 1975; Carannante et al. 1988). Such production peaks in the low latitudes and, in particular, in the vicinity of coral reefs, where carbonate sediments often represent the primary sediment constituents (see Chapter 9). Even in these environments, however, marked variations in the composition of sediment assemblages are

evident across individual reef or platform environments and reflect subtle spatial variations in marine environmental conditions (e.g. light and wave energy). These influence the composition of the reef community and, hence, the composition of the sediment substrate. In high-latitude settings, biogenic carbonate production may remain important but in many settings is volumetrically 'swamped' by terrestrial inputs of detrital minerals.

Carbonate deposits are also relatively common within a range of intertidal, terrestrial and freshwater settings, and are associated primarily with physico-chemically induced carbonate deposition. On a localized scale these represent important sources of carbonate to the sediment record. For example, within intertidal and supratidal settings that are characterized by high aridity and high evaporation rates, the precipitation of evaporite minerals such as gypsum ($CaSO_4.2H_2O$) and anhydrite ($CaSO_4$) is common. At present, extensive evaporite-rich sediments occur in the south-east Arabian Gulf and an excellent review of their occurrence is given by Alsharhan & Kendall (2003). The hydrology of such depositional settings has been reviewed by Yechieli & Wood (2002).

In some freshwater fluvial and lacustrine settings carbonate deposition also occurs and can lead to the development of significant carbonate bodies (in some cases these have been described as forming 'freshwater reefs'; Pedley 1992).

Deposition is driven by a range of physico-chemical and biologically mediated processes, the latter in association with microbial mats that facilitate local reductions in CO_2 and thus carbonate precipitation (Pedley 2000). These deposits are termed *tufa* where deposition occurs under ambient environmental conditions, or *travertine* where the deposits are associated with thermal activity (Ford & Pedley 1996) and have been shown to have potential as recorders of palaeo-climatic information (Andrews et al. 1994).

Organic inputs, derived from plant material, can also contribute abundant material to sediment substrates. This is particularly the case in salt marshes (see Chapter 7) and mangroves (see Chapter 9). Along mangrove-colonized shorelines, where external inputs of sediment (siliciclastic or carbonate) are minimal, this organic material can be the main substrate contributor and leads to the development of mangrove peat (Woodroffe 1983). Biogenic, but non-carbonate sediment, contributors such as diatoms are also important within, for example, lake environments (Chapter 4). The progressive accumulation of such microfossils within lake sediments has proved to be an effective long-term recorder of a range of environmental parameters such as effective moisture, i.e. lake water levels, and of temperature (Battarbee 2000) and have thus been widely used as proxy records of climatic and hydrological change (e.g. Bradbury 1997).

1.2.3 Anthropogenic particles and compounds

Increasingly important in many sedimentary systems are inputs of anthropogenically sourced sediments. These include both sediment grains that come from material that is anthropogenic in origin (e.g. building material, industry) and sedimentary materials that have been heavily impacted by anthropogenic activity. A good example of anthropogenic-derived sediment is that present within urban environments (see Chapter 6). In this environment, as well as soil and vegetation sources, sediment is sourced from vehicle wear, building material, combustion particles and industrial material. All of this material has chemical and mineralogical

properties distinct from natural sediment grains and, as a consequence, interacts with the environment in a different manner.

Another significant component of modern sediment, mostly absent from pre-industrial age sediments, are contaminants. A *contaminant* is commonly defined as a substance released into the environment without a known impact (Farmer 1997), or the presence of elevated concentrations of substances in water, sediments or organisms (GESAMP 1982). In neither of these definitions is the potential to cause environmental harm attributed to a contaminant. This is in contrast to a *pollutant*, which is more specifically defined as a substance that either causes harm to the environment or exceeds an environmental standard. Contaminants in sediments take a variety of forms, including metals, inorganic elements, nutrients, organic compounds and radionuclides, and the major sources of these contaminants are highlighted in Chapter 3 (Table 3.2). It is important to be aware that many of these contaminants can be sourced from natural processes as well as anthropogenic activities, although in most cases anthropogenic inputs tend to dominate.

Contaminant sources to sediments may be of particulate, dissolved or gaseous form, but for most contaminants the particulate form is dominant (Horowitz 1991). Although contaminant sources are predominantly particulate, there are important exceptions. Contaminants from sewage treatment works (e.g. Zn and P) can be predominantly in solute form, and metals from acid mine drainage are also in dissolved form at source owing to the low pH of the waters. These dissolved contaminants, however, commonly become associated with the sediment phase, via mineral precipitation or surface adsorption, as solution concentration, pH and Eh change through mixing with dilute river water (Boult et al. 1994).

Contaminant sources may take one of two general forms, point source and diffuse (non-point) source. Point sources of pollution originate from a single location and include mines and mine waste, landfill sites, factories, waste water treatment works and bedrock mineralization. Diffuse sources of contaminants originate from

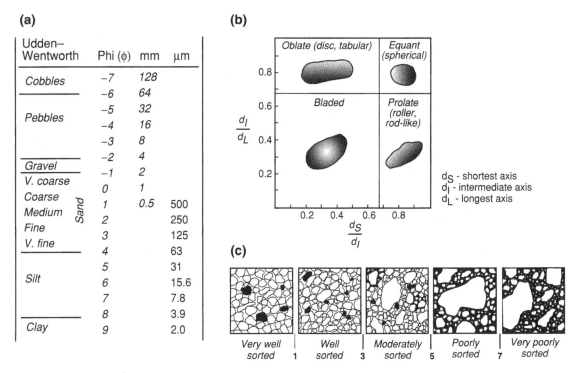

Fig. 1.4 (a) The Udden–Wentworth scheme, which is widely used to describe grain-size categories. (b) Classification scheme used to describe particle form based on the ratio between the short, intermediate and longest grain axes. (Adapted from Graham 1982.) (c) Visual comparison chart used for describing the degree of sediment sorting within a sediment deposit. (Adapted from Graham 1982.)

a wide area and can be defined as 'pollution arising from land-use activities (urban and rural) that are dispersed across a catchment or subcatchment, and do not arise as a process of industrial effluent, municipal sewage effluent, deep mine or farm effluent discharge' (Novotny 2003). Examples include direct atmospheric deposition, urban runoff and sediments (i.e. from the road network), agricultural runoff and sediments (i.e. from soil erosion), the reworking of floodplain sediments (i.e. by bank erosion) and background geology.

1.2.4 Particle description and classification

Regardless of origin, individual sedimentary particles are typically described in terms of their grain size and shape. Grain size is an important parameter both from a descriptive perspective and in relation to understanding sediment transport and deposition (see section 1.3). For larger particles, measurements of three orthogonal axes

are typically made and are used to calculate a mean diameter. For smaller particles, grain size is typically determined by grading the samples through a set of sieves (see McManus 1988). A number of schemes have been devised to describe and measure grain size, but one of the most widely used is the Udden–Wentworth scheme (Fig. 1.4a).

Descriptions of sediment shape are somewhat more complex and may be taken to comprise elements of a particle's form, roundness and texture. Roundness is usually described on the basis of comparisons with visual identification charts. Form is also usually quantified by describing grains in terms of one of four standard classes: oblate, equant, bladed or prolate, which reflect the relationship between the short, intermediate and long axes of grains (Fig. 1.4b). Other useful schemes combine elements of both roundness and sphericity in visual comparison charts (e.g. Powers 1982). Particle sorting describes the range of grain sizes that occur within a sedimentary deposit (Fig. 1.4c) and can be calculated by measuring the dispersion of grain size around the mean. This is again a useful parameter as it can be used, along with grain size data, to infer information about the environments of sediment deposition and the history of sediment reworking (e.g. McManus 1988).

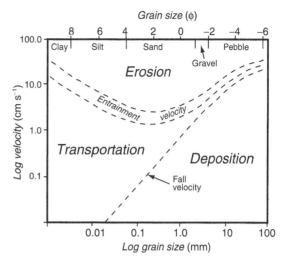

Fig. 1.5 Hjulström's (1935) graph showing the relationship between flow velocities and sediment grain size and the corresponding fields in which erosion, transport and deposition occur.

1.3 MECHANISMS OF SEDIMENT TRANSPORT AND ACCUMULATION

The transport and deposition of sediment within and through different sedimentary environments may occur within a variety of mediums (water, wind or ice), and the thresholds for sediment entrainment and transport represent a fundamental control on both the character and development of specific sedimentary deposits, as well as their response to fluctuating energy regimes. The classic work of Hjulström (1935) demonstrated the relationship that exists between the velocity of fluid flow and the size (diameter) of sediment that can be moved within a fluid. At its most simplistic this demonstrates that sediment will be deposited when flow rates drop below the fall velocity for a particle of a given size. However, the relationship between these two parameters is non-linear so that, for example, much higher flow velocities are required to entrain highly cohesive fine clay and silt-rich sediments (Fig. 1.5). Although in reality these entrainment/transport thresholds vary from this model depending upon sediment substrate and individual grain characteristics (e.g. shape, structure, density) as well as the flow characteristics of the fluid medium, a basic grain-size–flow-velocity relationship is demonstrated that can be broadly applied within both fluvial and marine settings. This section outlines some of the key physical parameters that control sediment entrainment, transport and settling (deposition), and highlights the main sedimentary processes that operate within the different sedimentary environments discussed in this book. For further details about the physics of sediment transport and deposition reference should be made to texts such as Allen (1985) or Leeder (1999).

1.3.1 Sediment entrainment

The entrainment of sediment by a fluid (most commonly water or wind), and thus the potential for sediment transport, is determined by the relationship between (i) fluid density (which is the weight per unit volume of a fluid – usually expressed as 'specific gravity'), (ii) fluid viscosity (the resistance of a fluid to deformation or flow – this is measured as a ratio between the shear stress and the rate of deformation) and (iii) the velocity of fluid flow. These parameters exert an influence on the nature of the flow regime within a fluid medium and, in particular, determine whether flow that occurs immediately above the sediment substrate (within the boundary layer) is laminar or turbulent. In situations characterized by laminar flow, flow streamlines run parallel to the substrate, flow velocity is low and viscosity is high (Fig. 1.6a). Under turbulent flow, the streamlines move in a series of random eddies, flow velocity is high and viscosity is low (Fig. 1.6a). The threshold between these two flow states is expressed by the Reynolds number (R), which describes the ratio between mean velocity over a defined distance or depth, and

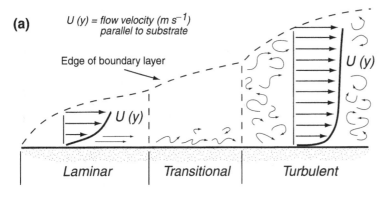

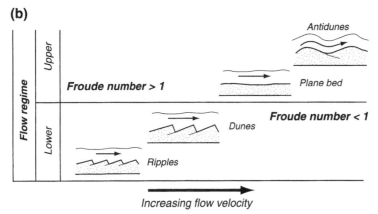

Fig. 1.6 (a) The nature of flow regimes within a fluid medium. During laminar flow, the flow streamlines run parallel to the substrate, flow velocity is low and viscosity high. Under turbulent flow, the streamlines move in a series of random eddies, flow velocity is high and viscosity low. Intermediate flow is described as being transitional. (Adapted from Allen 1985.) (b) The relationship between flow regime and sediment bedform development. As flow velocity increases from lower to upper flow, the amount and size of sediment that can be entrained and transported increase, leading to a change in sediment bedform structure. (Adapted from Selley 1994.)

fluid viscosity, and is determined by the following equation:

$$R = Udp/\mu$$

where U is the particle velocity, d is the particle diameter, p is the particle density and μ is the fluid viscosity. In the context of particles moving in a fluid, a low Reynolds number (< 500) describes fluid flow occurring in a laminar fashion, whereas at high Reynolds numbers (> 2000), fluid flow is turbulent. Between these two values flow is described as transitional (Allen 1985). Flow turbulence increases both proportionally with velocity and as bed surface roughness increases.

Another important coefficient in terms of fluid dynamics (the Froude number) explains the ratio between the force required to stop a moving particle and the force of gravity. Within open channels the Froude number (F) is determined as:

$$F = U/\sqrt{gD}$$

where U is the average current velocity, g is acceleration due to gravity and D is the depth of the channel. As flow velocity increases the Froude number approaches 1, a value that separates lower (< 1) flow regimes from higher (> 1) flow regimes (Allen 1985). Flow velocities increase from lower to upper flow and associated with this is an increase in the amount and size of sediment that can be entrained and transported. This, in turn, will influence the structure of the sedimentary bedforms that develop (Fig. 1.6b and section 1.4.1).

A number of forces act upon a sediment lying on a substrate surface (Fig. 1.7a), and these influence the potential for entrainment. The key force here is bed shear stress, which is related to the velocity of flow. This represents the force acting per unit area parallel to the bed and exerts a fluid drag across the grain. If this drag exceeds the frictional and gravitational forces acting on the grain, then lift and entrainment will occur. Sediment entrainment thresholds thus

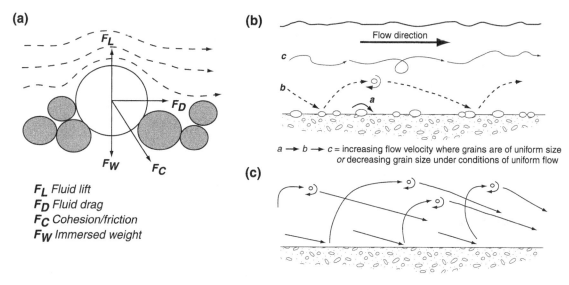

Fig. 1.7 (a) Schematic diagram illustrating the main forces acting on a sediment particle within a moving fluid medium. (Adapted from Allen 1985.) (b) The processes of sediment movement within flowing water; *a*, rolling; *b*, saltation; *c*, suspension. (c) Grain transport due to saltation under conditions of aeolian transport. Grains typically exhibit steeper and longer trajectories during aeolian transport.

occur at a critical shear velocity, the value of which varies with sediment grain size (although this is complicated by substrate specific variations in grain size, particle density, sediment packing and grain imbrication). The interrelationship between these variables is highlighted in the Hjulström graph (Fig. 1.5), which illustrates a general (and fairly intuitive) rule whereby increasing flow velocities are required to entrain increasingly larger sized particles. This rule, however, breaks down where the substrate is dominated by very fine sands, silts and clays because of the cohesive nature of such material. In such cases, much higher flow velocities are required to entrain particles and this helps to explain why fine silts and clays can accumulate within tidally influenced estuarine and deltaic environments (see Chapter 7). Velocity–entrainment relationships are also complicated as fluid moves across a sediment substrate because grains protruding from the substrate cause flow to be constricted. This causes the streamlines above the grain to accelerate and thus to exert a fluid lift (Fig. 1.7a). Within aeolian environments, wind velocities that exceed the critical shear stress for specific sediment grain sizes are also required for entrain-

ment, although they occur at higher velocities than in water.

1.3.2 Sediment transport

Once entrained, sediment movement occurs in three ways:
1 as bedload material that is too heavy to be lifted up into the water and moves by *rolling* along the substrate (Fig. 1.7b);
2 via the process of *saltation* whereby lighter grains are temporarily lifted into the fluid and then settle out (in water, saltating grains typically exhibit short, flat trajectories due to the cushioning effects of fluid viscosity, whereas in air the trajectories tend to be steeper and longer; Fig. 1.7b);
3 in *suspension* where the lightest particles are held within the fluid and moved, often in an erratic path (Fig. 1.7b).

For sediments of a given grain size these transport mechanisms occur along a gradient of increasing flow velocity. Consequently, for a sediment deposit comprising a mix of grain sizes it follows that the occurrence of different sediment size fractions can be attributed to different transport mechanisms. In aqueous environments, solute transport can sometimes also be an important medium for the transport of contaminants (see Chapters 3 and 6). These transport mechanisms, which operate in a range of fluid mediums and in different environments, produce very different types of sedimentary deposits. These are discussed below in the context of (i) aqueous environments, (ii) aeolian environments and (iii) glacial environments. Consideration is also given to the movement of material by gravitational processes.

1.3.3 Sediment settling

As with sediment entrainment and transport, the major controls on sediment deposition relate to grain size and flow velocity. The rate at which sediment settles (the settling velocity W) within a fluid of a given density is determined by Stokes law:

$$W = [(P_1 - P)g/18\mu]\,d^2$$

where $(P_1 - P)$ is the density difference between the fluid and particle, g is the acceleration due to gravity, μ is the fluid viscosity and d is the grain diameter. The law states that the settling velocity of a spherical particle is related both to its diameter and to the difference between the density of the particle and that of the surrounding fluid. In simple terms this means that larger sediment grains will settle faster than smaller grains providing they are of equal density. Within most sedimentary systems this process is complicated, however, by three factors:

1 the fact that few grains are completely spherical;

2 the fact that grains are often continually in contact within one another and hence disrupt settling;

3 the fact that different minerals have different densities (e.g. quartz 2.65 g cm^{-3}, feldspars 2.55–2.76 g cm^{-3}, biotite 2.80–3.40 g cm^{-3}; Allen 1985).

Hence, in the case of terrigenous sands, which are commonly dominated by quartz, but with variable amounts of other detrital and heavy minerals, the different particles have different settling velocities and thus different transport and settling thresholds. Differences in settling velocities are even more complex in systems dominated by skeletal carbonates because the grains not only have very different internal skeletal structures (and hence densities), but also very different morphologies (see Chapter 9).

1.4 SEDIMENT TRANSPORT IN DIFFERENT SEDIMENTARY ENVIRONMENTS

1.4.1 Sediment transport in aqueous environments

Sediment transport within aqueous environments occurs primarily in association with either traction or turbidity currents. Within traction currents the primary mechanisms of sediment movement are rolling and saltation (Fig. 1.7b), whereas within density currents transport is associated both with traction and suspension. Consideration is given first to traction currents, which are important within both fluvial and shallow marine environments. Within fluvial systems, current flow is nearly always unidirectional and progressive reworking of fine sediment commonly leads to the finest material being transported furthest downstream. Hence, fluvial systems are typically characterized by downstream reductions in mean grain size (see Chapter 3). In contrast, within nearshore settings and in the marine-influenced lower reaches of rivers, currents tend to be bi-directional (owing to the variable flood- and ebb-tide influence) and hence sediments will be reworked both on- and offshore during the tidal cycle (see Chapters 7 and 8).

Studies of flow regimes under unidirectional conditions have highlighted clear changes in sediment transport mechanisms and sedimentary structures associated with different current velocities. As flow increases, the critical velocities required to entrain sediment particles are reached and at this stage sediment starts to move by rolling and saltation. This leads to the development of ripples and, at slightly higher velocities, dunes (Fig. 1.6b). Such structures are associated with lower flow regimes (Froude numbers < 1).

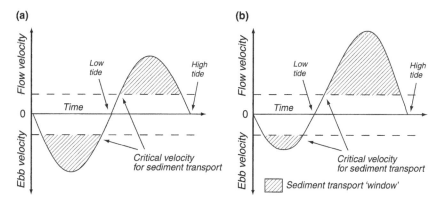

Fig. 1.8 Graphs showing changes in ebb- and flood-tide velocities within (a) symmetrical and (b) asymmetrical tide-cycle settings. Within each phase of the tidal cycle sediment transport occurs only when velocities exceed the critical thresholds for transport. In symmetrical settings, sediment is moved back and forth but there is no net sediment transport direction. In asymmetrical settings, a stronger ebb- or flood-tide phase may result in a net direction of sediment transport.

As flow velocities increase, upper flow regimes (Froude number > 1), characterized by turbulent flow, are reached and under these conditions the sediment bedforms are initially smoothed out to form planar beds and eventually antidunes which may migrate upstream (Fig. 1.6b). As flow reduces, a reverse sequence is followed. Hence through cycles of river flooding, the mechanisms and processes of sediment movement change with flow velocity (see Chapter 3).

In contrast, shallow marine environments are characterized by bi-directional flow, although the magnitude and frequency of the flood- versus ebb-tidal phase varies depending upon local tidal regime and nearshore geomorphology. The potential for sediment transport changes through each tidal cycle as flow velocity increases through either the ebb- or flood-tide phase and then decreases approaching either low or high tide ('slack' water). In settings where the tide cycle is symmetrical (Fig. 1.8a) sediment will be reworked first seaward and then landward, but there will be no net transport in either direction. It is more common, however, for tidal cycles to be asymmetrical (Fig. 1.8b), and under these conditions there will be a net sediment transport direction. This situation is common in many estuaries where fluvial outflow exerts an influence on the tidal cycle (see Chapter 7), or in mangrove settings where strong ebb-tide flows in the mangrove creeks can occur owing to the

frictional effects of high vegetation cover on the mangrove flats (see Chapter 9).

Sediment transport is also initiated in nearshore (marine or lacustrine) environments where wave-generated water motion interacts with the shoreline substrate. Waves are generated by the frictional effects of wind and this initiates water particle motion within the upper part of the water column. The orbital particle motion decreases with depth (Fig. 1.9a) until it reaches effective wave base (defined as half the wavelength), below which there is no wave-induced water motion. In open, deeper water, water motion therefore exerts no influence on sea-floor substrates, but as the water shallows nearshore the oscillating water particles start to interact with the sea-bed (Fig. 1.9b). As this occurs, the water particles move in an increasingly ellipsoidal fashion and initiate on- and offshore movement of sediment.

Transport in aqueous fluids may also occur in density currents. These are associated both with traction and suspension transport, and occur due to density differences between two fluid bodies. Within aqueous environments, density differences commonly result from variations in temperature, salinity or suspended sediment load where two bodies of water meet. Where the fluid entering a body of water has a higher density, for example where sediment-laden water enters a lake, the denser fluid will flow beneath the less dense fluid

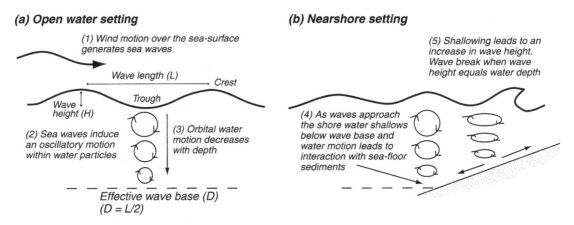

(a) Open water setting

(1) Wind motion over the sea-surface generates sea waves

Wave length (L)

Crest

Trough

Wave height (H)

(2) Sea waves induce an oscillatory motion within water particles

(3) Orbital water motion decreases with depth

Effective wave base (D)
(D = L/2)

(b) Nearshore setting

(5) Shallowing leads to an increase in wave height. Wave break when wave height equals water depth

(4) As waves approach the shore water shallows below wave base and water motion leads to interaction with sea-floor sediments

Fig. 1.9 Schematic diagram illustrating wave motion in (a) open-water settings, and (b) as waves approach the shoreline.

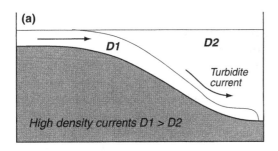

(a)

D1

D2

Turbidite current

High density currents D1 > D2

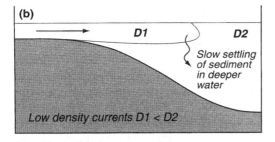

(b)

D1

D2

Slow settling of sediment in deeper water

Low density currents D1 < D2

Fig. 1.10 Schematic diagrams illustrating differences in density flows as fluid enters a standing body of water. (a) High density flows occur where the entering fluid has a higher density than the standing water body. (b) Low density flows occur when the situation is reversed. (Adapted from Selley 1994.)

and create a density current (Fig. 1.10a; see Chapter 4). Where the fluid entering the body of water has a lower density, for example where freshwater enters the sea, flow will typically occur as a plume across the water surface (Fig. 1.10b). In the former case, a specific type of density current, known as a turbidity current, is commonly generated. These are capable of transporting very large volumes of sediment across even very

low slope angles and are thought to be a major cause of sediment distribution across continental shelf (see Chapter 10) and slope settings. In low-density flows, most of the fine sediment is transported in suspension. This mechanism of transport is common at the distal ends of turbidite deposits where the finest sediment has remained in suspension, and along the distal margins of deltas where fine suspended sediment settles out along the delta front (see Chapter 7). Settlement of fine grained suspended sediment is enhanced where mixing of fresh and saltwater occurs. Under these conditions, even slight increases in salinity (> 1) will promote the aggregation of fine clay particles. This process is known as flocculation and leads to an increase in grain size and thus in grain settling velocity. Flocculation is a common process in estuarine environments (see Chapters 7 and 9) and in salt marshes, and may lead to the development of zones of high turbidity and fine sediment deposition.

The largest proportion of the contaminant load in sediment systems is transported by the particulate matter. For example, Gibbs (1977) suggested that up to 90% of the metal load is transported by sediments in rivers, but this can vary from metal to metal. Similar observations have been made for organic contaminants, such as chlorinated organic compounds. This particulate portion of the contaminant load comprises contaminant-rich grains (e.g. metal sulphide grains from tailings effluent) or contaminant element-bearing Fe and Mn oxide coatings on other particles. Some metals, however, especially under low pH conditions, can be transported in solution (see Chapter 3). This dissolved portion encompasses contaminants that are either truly dissolved, or in colloid form. The partitioning of contaminants between the dissolved and particulate load in aquatic systems depends on both physical and chemical factors, including pH, redox, sediment mineralogy, sediment texture, suspended sediment concentration and sediment grain size. Grain size is possibly the most significant factor controlling the concentration and retention of contaminants in both suspended and bottom sediment. Metals in particular have been shown to be enriched in the fine silt and clay fractions of sediments, as a result of their large surface area, organic and clay contents, surface charge and cation exchange capacity (see Chapters 3 and 6).

1.4.2 Sediment transport in aeolian environments

Sediment transport within aeolian settings occurs primarily associated with either traction carpets or in suspension and is common in three main environments:

1 arid deserts;
2 associated with shoreward areas of beaches and barrier islands;
3 developed around ice caps.

Sediment transport in deserts (Chapter 5) and coastal dunes (Chapter 8) is associated primarily with rolling and saltation of grains in the traction carpet. As in aqueous environments, the critical velocities required to entrain sediment increase with grain size, and high velocities are required to entrain fine silt and clay-sized material. Once entrained, however, such sediment may be transported long distances as dust clouds, a mechanism that is known, for example, to transport large volumes of Saharan dust across the Atlantic and into the eastern Caribbean (Prospero et al. 1970). Controls on the development of aeolian sediment bedforms are discussed in Chapter 5, but are strongly influenced by wind direction and its variability, and by the rate of sediment supply. Aeolian sediment deposition around ice caps is thought to be primarily associated with suspension transport and these silica-rich sands are commonly termed 'loess'.

1.4.3 Sediment transport in glacial environments

Sediment transport within glacial environments occurs associated with a range of transport processes, which include suspension, aqueous suspension and aqueous traction currents. These form different types of deposits and are associated with different glacial environments. Suspension transport, as outlined above, results in the deposition of loess deposits in glacial marginal areas. Aqueous suspension is associated with the deposition of fine, laminated clay sequences (varves), whereas aqueous traction currents are responsible for extensive fluvioglacial sand and gravel transport, and the development of extensive outwash plains (see Chapter 2). Glacial ice also acts as an important sediment transport medium and, although the movement of ice is slow, it is responsible for significant erosion of underlying bedrock and sediment. The resultant debris is transported under and within the ice, and is deposited either along the flanks of glaciers or at the terminal end after the ice starts to melt. These deposits are typically structureless and comprised of poorly sorted boulder to clay-sized material. The descriptive term for the sediment is diamict and when deposited directly by glacier ice is termed till. Consequently till is a major component of glacial landforms such as moraines and drumlins. High-magnitude sediment transport events in glacial environments can occur associated with jökulhlaups – a flood caused by the sudden drainage of a subglacial or ice-dammed lake, commonly triggered by a volcanic eruption (see Chapter 2). These events can transport huge volumes of sediment, and result in extensive deposition of outwash deposits.

1.4.4 Sediment transport associated with gravitational processes

Within each of the settings described, an additional agent of sediment transport is gravity and three main categories of gravitational sediment

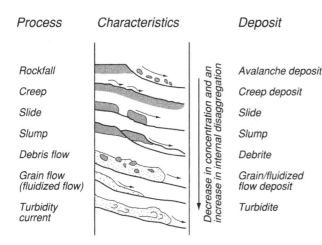

Fig. 1.11 Processes and deposits associated with rock and sediment movement along a continuum of decreased concentration and increased internal disaggregation. (Adapted from Stow 1986.)

transport are recognized: (i) rockfalls, (ii) slides and slumps, and (iii) mass flows. These are recognized to occur along a continuum whereby there is an increase in the degree of internal disaggregation and a reduction in the concentration of the sedimentary material (Fig. 1.11). Rock falls are defined as the collapse of rock or sediment primarily along a vertical plane. They may be caused by tectonic movement or by weathering in upland settings and typically produce scree deposits. Slides and slumps occur over lower angled slopes and involve transport along an inclined shear plane. They are thus characterized by movement over both vertical and horizontal displacement planes. In slides, the sediment generally remains undisturbed, whereas in slumps the original sedimentary structures are normally disrupted or destroyed. The presence of water along a shear plane acts as a medium to initiate both slumping and sliding. At higher water contents the process of slumping grades into that of mass flows, a term used to encompass a spectrum of transport processes including debris flows and grain flows. Debris flows involve the transport of rock and fine sediment that 'flows' downslope as a chaotic mass and these occur in a range of environments from deserts to continental slopes. They typically require the presence of unconsolidated sediment and steep slopes and, on land, low vegetation cover and heavy rainfall to initiate movement (Chapter 5). Grain flows occur within finer sediments and require

steep slopes and a confined channel margin. They occur most commonly on the continental slopes and form graded deposits.

1.5 POST-DEPOSITIONAL PROCESSES

Processes acting internally and externally upon a sediment after deposition can be physical, chemical or biological. Physical processes include compaction, resuspension, erosion or dredging of sediment. Chemical and biological processes include the series of early diagenetic, bacterially mediated redox reactions, which result in the oxidation of carbon species (organic matter) and the reduction of an oxidized species. Although post-depositional processes acting upon sediments are varied and have a range of impacts, of most importance in the context of environmental sedimentology is the chemical remobilization of nutrients and contaminants during early diagenesis, and the release of contaminants from floodplains.

1.5.1 Early diagenesis in aquatic sediments

Upon the consumption of O_2, a series of anaerobic bacterial reactions are favoured, utilizing oxygen in species such as nitrate (NO_3^{2-}), iron oxide (FeOOH), manganese oxide (MnO_2) and sulphate (SO_4^{2-}). These anaerobic early diagenetic reactions are many and complex. The most significant reactions are nitrate reduction, Mn(IV) reduction, Fe(III) reduction, sulphate reduction and meth-anogenesis. All of these reactions break down organic matter and, therefore, lead to an overall decrease in organic matter content as sediments are buried. Many of these reactions can only utilize simple organic molecules, such as acetate and hydrogen, as the reductant. However, some bacterial communities, particularly iron-reducing bacteria, have been shown to possess the ability to utilize complex organic molecules (Lovley & Anderson 2000). Therefore, such diagenetic reactions may act to break down persistent organic contaminants in aquatic sediments. Bacteria can also directly mediate the reduction of some contaminant metals, for example Cr, U, Se, Hg and Tc (e.g. Lovley 1993).

Early diagenetic reactions have an impact upon the short- and long-term fate of contaminants in sediments through two principal mechanisms: release of contaminants into sediment porewaters; and the uptake of contaminants into authigenic mineral precipitates. The oxidation of organic matter and the reduction of iron and manganese oxides result in the release of contaminants associated with these mineral phases to sediment porewaters (Rae & Allen 1993). These increased porewater contaminant concentrations can result in the molecular diffusion of contaminants into the overlying water column (commonly termed a 'benthic flux'). There is a growing awareness that benthic contaminant fluxes to intertidal environments can be as significant as riverine input and may act as a major long-term input of contamination into water bodies. Rivera-Duarte & Flegal (1997a; b) documented that benthic fluxes of Co and Zn from sediments in the San Francisco Bay were of the same magnitude as riverine inputs. Similarly, Shine et al. (1998) showed that the flux of Cd and Zn from coastal sediments in Massachusetts, USA was of a similar magnitude to that within the water column itself.

In marine and brackish intertidal sediment-ary environments, sulphate reduction is a major pathway for organic-matter oxidation and as a result sulphide is released into porewaters. Sulphide forms a highly stable complex with most metals (Cooper & Morse 1998) and consequently metals released by Fe(III) and Mn(IV) reduction will be precipitated out as sulphides. These precipitates are predominantly in the form of iron monosulphides, and metals may be adsorbed onto the sulphide surfaces, or incorporated into the sulphide structure (Parkman et al. 1996). Early diagenetic metal sulphides have also been documented in mining-impacted estuarine sediments, acting as a long-term sink for contaminants in these sediments (Pirrie et al. 2000; see Chapter 7). In contrast to natural sediments, early diagenetic mineral precipitates within contaminated sediments can be varied and unique. For example, Pirrie et al. (2000) described the occurrence of early diagenetic simonkolleite (a Zn–Cl mineral) from metal-contaminated estuarine sediments. Early diagenetic minerals (e.g. vivianite – iron phosphate) can also be important in non-marine sediments (see Chapter 6).

1.5.2 Remobilization from floodplains

Contaminants may also be remobilized from river floodplains. Floodplains are sites of sediment accumulation within river basins and, therefore, are classically considered to be con-taminant sinks, thereby preserving good temporal records of contaminant input (e.g. Smol 2002; Chapter 3). These sinks of contaminants, however, can also become sources as a result of post-depositional processes, both chemical and physical. For example, Hudson-Edwards et al. (1998) demonstrated that remobilization of Pb, Zn, Cd and Cu within overbank sediments of the River Tyne, England, occurred as a result of changes in water-table levels and the break-down of organic matter above the water table (see Chapter 3).

Contaminants stored on floodplains also may be remobilized through physical erosion, and this may take place long after the primary con-taminating activity (e.g. mining) has ceased. For example, Macklin (1992) showed that the prim-ary source of Pb and Zn to the contemporary River Tyne, northern England, was remobilized floodplain alluvium originally deposited during eighteenth and nineteenth century metal mining.

Contaminant remobilization is often triggered by natural (climate) or anthropogenic (land use) changes that cause modifications, first in sediment load and delivery, and eventually in erosion and deposition. Macklin (1996) warned that flood-plain contaminant remobilization is increasing as a result of the hydrological changes associated with global warming, and stressed that the long-term stability of contaminant metals with respect to changes in physical (river bank and bed erosion, land drainage and development) and chemical conditions (redox and pH) is poorly understood. The remobilization of contaminants through the physical erosion of contaminated saltmarsh sediment can also be a significant source of contaminants to estuaries and coastal waters (see Chapter 7).

1.6 SEDIMENTARY RESPONSES TO ENVIRONMENTAL CHANGE

1.6.1 Sedimentary responses to natural disturbance events

Although daily or ongoing processes of fluvial or tidal flow and water/wind velocity influence background levels of sediment transport and accumulation, the amount of sediment transport that occurs during 'normal' conditions is relatively low. Most sediment transport and, as a result, much of the morphological change that occurs within sedimentary environments takes place during low-frequency but high-magnitude events. These may be associated with storms or high (seasonal) rainfall episodes, or with episodic high-energy events such as cyclones and tsunami. At these times, sediment transport rates can dramatically increase and hence a high proportion of annual sediment movement may occur over a period of only a few days. This is particularly the case in many arid and semi-arid environments, which are characterized by highly 'flashy' discharge events and where short-lived but high-intensity rainfall events lead to very high-energy flows (see Chapter 5). High-magnitude discharge events also characterize many seasonally influenced fluvial systems. In the Burdekin River catchment, North Queensland, tropical cyclones

dramatically increase discharge rates through the catchment (Fig. 1.12a), resulting in increased suspended (Fig. 1.12b) and bedload sediment transport (Amos et al. 2004). These high-energy events also influence spatial variability in sediment transport and storage within fluvial catchments. For example, in the Rajang River Delta of Sarawak, eastern Malaysia, sediments are stored on the delta plain during the 'dry' season, but undergo rapid offshore transport during the 'wet' season when discharge rates increase (Staub et al. 2000).

In coastal environments, storm waves are directly responsible for extensive reworking of unconsolidated sedimentary deposits and this is manifested in changes in beach profiles and the breaching of coastal barriers (see Chapter 8), and the on- and offshore transport of sediment and rubble (see Chapter 9). In the tropics, high wind speeds and storm waves associated with cyclones can lead to tree damage and mortality within mangroves. This, in turn, can facilitate substrate destabilization and erosion of intertidal sediment substrates. These events can thus result in significant localized ecological damage and often marked short-term changes in patterns of nearshore sediment accumulation. However, cyclones and other high-energy episodic events are also important controls on the longer term distribution and development of sedimentary environments. On the Great Barrier Reef shelf, for example, cyclones generate northward flowing alongshore currents, which result not only in significant along-shelf sediment transport, but also a marked partitioning of sediment across the shelf (Larcombe & Carter 2004; see Chapter 9).

1.6.2 Anthropogenic modifications of sediments and sedimentary systems

Although seasonal and natural (i.e. storm-induced) changes in energy levels can lead to changes in sediment dynamics and accumulation rates, anthropogenic activities also have potential to modify rates of sediment input, sediment transport pathways and the composition of the accumulating sedimentary materials. Clear links between anthropogenic activity and sedimentary

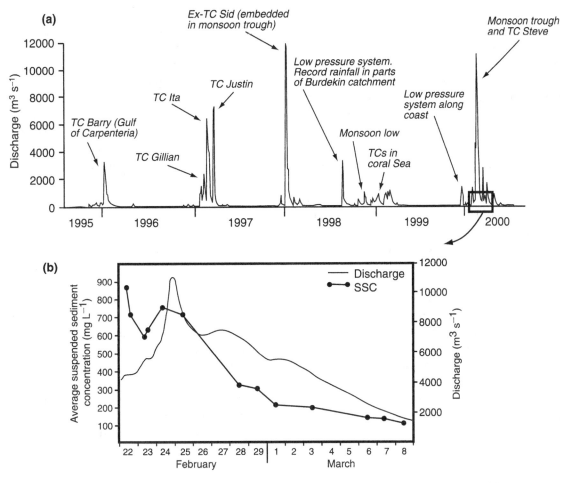

Fig. 1.12 (a) Data from hydrographs recording fluvial discharge between 1995 and 2000 in the lower part of the Burdekin River catchment. (b) Detail of discharge and suspended sediment concentrations (SSC) between February and March 2000. Note that SSC generally decreased through the discharge event and that, in this catchment, SSC levels appear to be supply limited. (Adapted from Amos et al. 2004.)

system response occur, for example, in areas where construction or resource extraction activities in the upstream sectors of catchments result in downstream sediment starvation and/or erosion. At a global scale, the transport of sediment through river systems represents a major pathway of sediment movement from upland 'source' to marine 'sink'. The transport of such sedimentary material is, however, highly sensitive to a range of anthropogenic influences, including reservoir construction, land-use change, soil and water conservation activities, and sediment control programmes (Walling & Fang 2003). Some of these activities lead to increased sediment loads

but others, and in particular reservoir construction, lead to reduced sediment transport. The impacts vary between catchments, but in some regions reduced sediment supply has resulted in marked changes in the behaviour and geomorphology of fluvial systems (see Chapters 2 and 3). In the Alpine region of Europe, for example, sediment deficits have been recorded in many rivers over the past 30–40 years. Such reductions have resulted from excessive gravel extraction from rivers and the retention of sediment behind dams. The result, on many upland rivers, has been widespread erosion and entrenchment (Descroix & Gautier 2002).

Similar links between sediment sources and sinks are evident along many coastlines where sediment is supplied either from fluvial sources or from one area of coastline to another, often via longshore transport. Reductions in sediment supply often lead to increased rates of erosion along the coastal sectors that are deprived of sediment. This occurs, for example, where sediments are trapped behind dams located on the rivers that feed the coastal sector. In California, reduced fluvial sediment supply due to damming has led to increased rates of cliff erosion (see Chapter 8), and extensive subsidence and shoreline erosion of the Nile delta is attributed to significantly reduced sediment supply to the delta front (see Chapter 7). In the northern Gulf of California fundamental changes in the sources and rates of sediment supply, and in the composition of accumulating sediment, have also been directly linked to the effects of dam construction and have resulted in a 95% reduction in sediment supply from the Colorado River (see Case Study 1.1). Coastal retreat may also occur where 'upstream' sediment supply has been restricted by sea-defence construction. In such cases, seawall or groyne systems either prevent or restrict sediment throughput, leading to increased rates of downstream beach erosion and shoreline change (see Chapter 8). The recognition of these important sedimentary links has been a key driver in the development of integrated catchment and coastal management schemes.

Anthropogenic-related modifications to sediment source and transport pathways may be significantly exacerbated by the effects of urbanization. In relation to sedimentary systems, the most important influence occurs where constructional activities occur within the sedimentologically active zone. Problems arise either because of construction in areas where episodic sedimentological and geomorphological changes can be expected, or where deliberate constructional activities have a consequent effect upon pathways of sediment transport and zones of sediment accumulation. Coastal dune and barrier island sequences, for example, form part of active sedimentary systems that will respond to high-energy storm events. Hence, roads and houses built in such zones become susceptible to storm damage (see Chapters 8 and 9). Similarly, construction of seawalls or other landward constraints may lead to 'coastal squeeze' as landward migration

---

Case study 1.1 Anthropogenic modifications to sediment supply, northern Gulf of California

The Gulf of California is a narrow epicontinental sea, 1500 km long, that has formed from tectonic activity along the Californian coast (Case Fig. 1.1). The Northern Gulf of California (NGC) receives sediment from four areas, the Colorado River, the batholith of the Baja California, the Sierra Madre Occidental, and the deserts of north-west Mexico. Each of these source areas has a distinctive mineralogical signature enabling the provenance of accumulating marine sediments in the NGC to be determined. On this basis four distinct sediment provinces have been identified, these being, (i) the Colorado River Delta Province, (ii) the Concepción River Province, (iii) the Transitional Province and (iv) the Baja-Sonora Province (Case Fig. 1.1).

Historically, the Colorado River has been the primary source of sediment into the NGC, with an estimated annual sediment discharge of 160×10^6 t. Fluvial sediment supply has, however, been dramatically reduced over the past 100 years following the construction of a series of dams along the river. Of particular significance have been the Hoover Dam (built in 1934) and the Glenn Canyon Dam (built in 1952). The result of this extensive water flow regulation has been to reduce fluvial sediment supply by around 95%, resulting in sediment starvation to both estuarine and deltaic environments of the Colorado River mouth, and in the northern areas of the Gulf.

In the vicinity of the Colorado River, major changes in sediment supply and transport have been identified and, as a result, oceanic (rather than fluvial) hydrodynamic forces now exert the major influence on sediment dynamics within the estuary and delta (Carriquiry & Sánchez 1999). Rather than a predominant north to south (fluvial to basinal) transfer of sediment, sediment is now transported from south-east to north-west along the eastern side of the NGC into the estuarine basin, and then reworked southwards along the western sides of the NGC (Case Fig. 1.1). Despite significant reductions in fluvial sediment input from the Colorado River, however, average sedimentation rates in this NGC are reported to have remained relatively constant over the past 100 years. This is attributed to a transition in the source areas that supply sediment to the Gulf (Carriquiry et al. 2001). In particular, a high proportion of sediment is now supplied from resuspension and reworking of the Colorado Delta sediments and from the shallower part of the NGC shelf. These sediments form the Colorado River Delta Province and dominate the central areas of the NGC marine basin (Case Fig. 1.1). Additional sources of sediment are derived from the desert areas of north-west Mexico

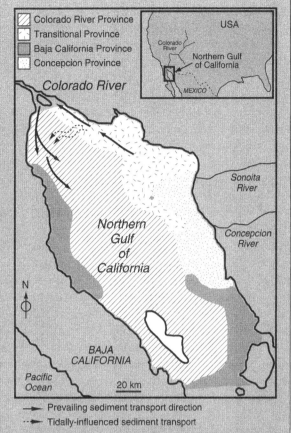

Case Fig. 1.1 Distribution of the main sediment provinces in the Northern Gulf of California. The main sediment transport pathways in the vicinity of the Colorado River Delta are also shown. (Adapted from Carriquiry & Sánchez 1999; Carriquiry et al. 2001.)

and south-west USA. These form the Concepción River Province and feed into the basin via the Sonoita and Concepción Rivers (Case Fig. 1.1). In addition, intense desert winds from the Sonora Desert represent an important transport medium for aeolian sediment transport into the Gulf. These sediments are rich in zircon and garnet, and contribute primarily to the Transitional Sediment Province. The area therefore emphasizes the effects of anthropogenically influenced reductions in sediment supply through fluvial systems, and the consequent 'downstream' impacts on both sediment transport pathways and on the composition of the accumulating marine sediments.

Relevant reading

Carriquiry, J.D. & Sánchez, A. (1999) Sedimentation in the Colorado River delta and Upper Gulf of California after nearly a century of discharge loss. *Marine Geology* **158**, 125–45.

Carriquiry, J.D., Sánchez, A. & Camacho-Ibar, V.F. (2001) Sedimentation in the northern Gulf of California after cessation of the Colorado River discharge. *Sedimentary Geology* **144**, 37–62.

of coastal sedimentary environments is restricted. Such interactions with sedimentary systems or a restriction in the way sediment systems respond to increased energy levels often brings with it a management or remediation 'cost'. Hence in many cases, the need for management is often driven not so much by the actual event, but as a result of (increasing) human occupation and modification of the environment, i.e. urbanization of environments that will naturally respond to changes in the energy inputs associated with storm or flood events. The influence of urbanization of the coastal fringe is seen particularly clearly in relation to estuarine environments where large areas of intertidal land have been claimed over a period of several centuries (see Chapter 7). The result is often a fundamental change in the character and extent of intertidal land, and a suppression of an estuary's ability to respond to changes in nearshore energy regimes or sea-level state.

Urbanization also has major impacts upon the hydrology of catchments and river basins, which in turn influences the nature of sediment movement and accumulation (see Chapter 6). The increase in runoff rate in urban systems leads to enhanced flooding pressures in river systems, and this is often exacerbated by the past removal of floodplains and river culverting, which inhibits the accumulation of sediment. These increases in runoff rate also have marked impacts upon sediment transport in urbanized river basins, with large storm events accounting for the majority of suspended sediment transport flux over short periods of time (Goodwin et al. 2003; Old et al. 2003).

Another highly significant anthropogenic impact on sediments is that of sediment composition and quality. The increase in contaminant loading in sediments has been extensively documented for virtually all sedimentary systems globally, including those that are generally assumed to be pristine. These have been documented both through monitoring programmes on sediment composition, and sedimentary archives of contaminant accumulation, such as salt marshes, lakes, reservoirs and floodplains. The former approach allows for short-term data

sets only, as monitoring programmes on sediment composition have not been in place for long. Longer temporal records of sediment composition may be recorded by sediments accumulating in depositional environments (see Case Study 1.2). The nature and length of the sediment record will depend on a number of factors, including sediment accumulation rates, extent of sediment disturbance and post-depositional changes. Examples of temporal records of sediment compositional change for lakes and river basins can be found in Chapter 4 and in Smol (2002). Such compositional changes, and associated records of environmental pollution have also been clearly documented for saltmarsh sediments (see Chapter 7).

1.6.3 Response of sedimentary systems to climatic and environmental change

Given the influence that climate exerts on the development of sedimentary environments, ongoing and projected climatic and environmental changes are potentially significant in relation to the dynamics and functioning of most sediment systems. Climate exerts, for example, an important influence on weathering regimes, the hydrological cycle (including seasonality of rainfall) and the frequency and magnitude of high-energy (storm) events, all of which are important in determining rates of sediment supply and transport (e.g. Chapters 2–4). In the marine environment, climatic conditions also influence environmental factors such as levels of dissolved CO_2 and sea-surface temperatures. These are primary controls on the distribution and development of biogenic sedimentary deposits (Chapter 9). Sea-level itself is a major control both on the distribution and extent of coastal sedimentary environments (Chapters 7 and 8) and, because it influences base level, a major forcing factor with fluvial systems (Chapter 3). Hence many of the predicted changes in global climatic conditions need brief consideration here. These include changes in atmospheric CO_2 concentrations, increased atmospheric and sea-surface temperatures, increased UV radiation, changes to patterns of storm frequency and

Case study 1.2 Temporal changes in sediment composition as a result of pollution: archives of lead pollution

Sediments that have accumulated in aquatic systems may be good records of the external inputs affecting sediment composition, and can be used as both indicators of pollution impacts upon sediment composition, and as archives of local and global pollution changes. Extensive studies on lake, reservoir and river basins have been undertaken. This example focuses on lead in lake and reservoir sediments. Lead (Pb) is a natural element, which is supplied to sediments through geological weathering of rocks and mineral deposits. However, Pb has also been sourced to sediments through two anthropogenic processes: the mining and smelting of Pb ores, and the combustion of fossil fuels, especially vehicle fuel with added Pb. Levels of Pb in the environment are a concern because Pb can act as a powerful neurotoxin.

Studies of sediment composition in lakes in northern Scandanavia have documented long-term (over 2000 yr) records of sediment-Pb composition (Brännvall et al. 1999). Sediment composition in these studies (Case Fig. 1.2) shows consistent changes in Pb deposition to these sediments over a wide area. An early peak at around 2000 yr BP was related to lead smelting during the Greek and Roman cultures. This was followed by peaks in Pb deposition around AD 1200 and 1530, again related to lead smelting and coinciding with known peaks in metal production in Europe. The presence of significant Pb in sediments of these ages indicates that sediment composition was being markedly altered prior to the Industrial Revolution, and that pollution levels in the environment are not all a recent phenomenon. These lake sediments also show a clear input of Pb in the latter half of the twentieth century (Case Fig. 1.2), related to the burning of fossil fuels and the use of Pb in petrol. This late twentieth century impact upon sediments has also been documented by numerous other studies from a wide range of sedimentary environments (e.g. Renberg et al. 1994; Shotyk et al. 1998; and see Chapter 6). The use of stable isotopes of Pb ($^{206}Pb/^{207}Pb$) further clarifies the increases of anthropogenic Pb pollution from natural Pb inputs from soil weathering (Case Fig. 1.2a).

Shorter time periods of Pb-pollution impacts on sediment composition have been studied in reservoir sediments, as these systems commonly display faster sediment accumulation rates, thereby allowing increased temporal resolution. For example, Callender & Van Metre (1997) studied sediment cores from water reservoirs in the south-west and Midwest USA (Case Fig. 1.2b). This study clearly documented a high Pb peak in sediment between 1970 and 1980, linked to atmospheric Pb input from leaded-fuel combustion. More recently deposited sediments display a clear, and rapid, reduction in Pb levels, as a result of the Clean Air Act of 1970 in the USA, and also the phasing out of leaded-fuel.

Similar records of Pb pollution have also been documented from Scottish freshwater lakes (Eades et al. 2002). Sediments in Loch Lomond show an increase in Pb past 1800 as a result of industry and fossil-fuel burning (Case Fig. 1.2c). A large increase in the latter half of the twentieth century, coupled with a significant drop in the $^{206}Pb/^{207}Pb$ of this Pb was a result of vehicle combustion of fuel with Pb additives. A significant fall in Pb levels since the 1980s was a result of the increasing use of unleaded petrol in vehicles.

Relevant reading

Brännvall, M.-L., Bindler, R., Emeteryd, O., et al. (1997) Stable isotope and concentration records of atmospheric lead pollution in peat and lake sediments in Sweden. *Water, Air and Soil Pollution* **100**, 243–52.

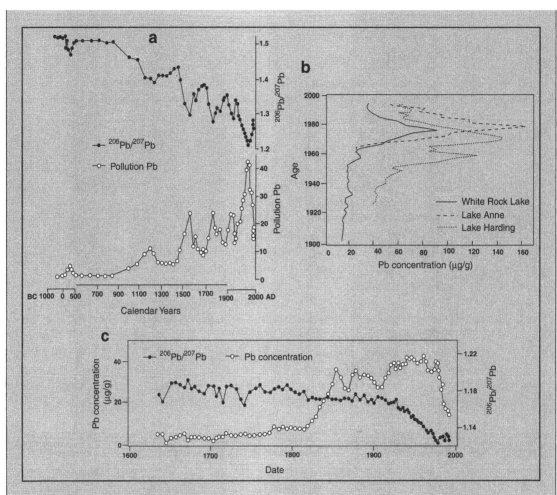

Case Fig. 1.2 (a) Lead pollution and lead isotope profile preserved in an annually laminated lake, Lake Koltjärn, Sweden. (After Brännvall et al. 1999.) (b) Historical records of lead inputs preserved in urban and suburban lakes in Georgia, USA. (After Smol 2002; Callender & Van Metre 1997.) (c) Records of lead pollution and lead isotope signature in a Scottish freshwater lake, Loch Lomond. (After Eades et al. 2002.)

Brännvall, M.-L., Bindler, R., Renberg, I., et al. (1999) The Medieval metal industry was the cradle of modern large-scale atmospheric lead pollution in Northern Europe. *Environmental Science and Technology* **33**, 4391–5.

Callender, E. & Van Metre, P.C. (1997) Reservoir sediment cores show U.S. lead declines. *Environmental Science and Technology* **31**, 424A–8A.

Eades, L.J., Farmer, J.G., MacKenzie, A.B., et al. (2002) Stable lead isotopic characterisation of the historical record of environmental lead contamination in dated freshwater lake sediment cores from northern and central Scotland. *The Science of the Total Environment* **292**, 55–67.

Renberg, I., Wik Persson, M. & Emeteryd, O. (1994) Pre-industrial atmospheric lead contamination detected in Swedish lake sediments. *Nature* **368**, 323–6.

Shotyk, W., Weiss, D., Appleby, P.G., et al. (1998) History of atmospheric lead deposition since 12,370 ^{14}C yr BP from a peat bog, Jura Mountains, Switzerland. *Science* **281**, 1635–40.

intensity, and increased sea-levels. Depending upon the magnitude of these changes, both positive and negative responses may occur within different sedimentary environments.

The long-term links between climate and sediment system response are clearly illustrated in studies that have examined the response of large fluvial systems to Quaternary or Holocene climatic change. The Ganges river system, for example, represents one of the largest sediment dispersal systems in the world. It extends for a distance of over 3000 km, presently discharging around 300×10^9 m^3 of water and 520×10^6 t of sediment annually (Goodbred 2003). Despite its vast size, strong sediment linkages occur between the source areas in the Himalaya, through the catchment basins on the Ganges Plain, to the depocentres in the Bengal Basin. This is believed to reflect the strong seasonal control that exists on river flow, with around 80% of fluvial discharge and 95% of sediment load delivered over the 4 month summer monsoon period. As a result, the system is highly susceptible to changes in atmospheric circulation patterns and, in particular, any change in the strength of the summer monsoon. This has altered several times over the past 150 kyr, resulting in changes in rates and patterns of fluvial sediment production, erosion, transport and accumulation, and is manifested by shifts from periods of upstream erosion and entrenchment, to periods of sediment accumulation on the delta plain and within the Bengal Basin.

Large-scale studies thus establish clear links between sediment dispersal system behaviour and climatic regime. Predicting future changes in sediment system dynamics is, however, complicated by the uncertainty that exists in relation to the magnitude and regional variability of future climate change. Future climate projections are based, in part, upon constructed scenarios of future human behaviour (in relation to issues such as greenhouse gas emissions and land-use practice). They are also dependent, however, upon atmospheric circulation patterns and atmospheric interactions with large-scale ocean currents and land features (including albedo, vegetation cover and soil moisture content).

Among the more complex models in use are the coupled atmosphere–ocean general circulation models (AOGCMs). Examples of these include the recent HadCM3 model developed at the UK's Hadley Centre (see Hadley Centre 2004), the projections from which are discussed below. Such models assume that future greenhouse gas emissions will follow the widely used IS92a scenario (Houghton et al. 1992), whereby CO_2 levels will double during the twenty-first century, there will be mid-range economic growth, and no measures made to reduce greenhouse gas emissions. The dynamics of, and interactions between, the complex variables involved in such models means, however, that although climate projection models are progressively improving, there remain many unknowns and significant uncertainties about the feedbacks that occur between different parameters. Hence, a range of climate simulation models have been developed that are based on different input data and account for different future scenarios (Hadley Centre 2004; McCarthy et al. 2001).

Although future changes in variables such as temperature, rainfall and sea-level are clearly significant from a sedimentological perspective (outlined below), it is important that these projections are placed in the context of past environmental change. The Quaternary has, for example, been characterized by marked climatic shifts associated with glacial and interglacial phases. At the global scale, these oscillations resulted in marked changes in climate, but the actual effects were spatially very variable. Hence, although previously temperate regions in the northerly latitudes cooled and were subject to glacial or periglacial conditions, more arid regions to the south became more temperate in character. Consequently the nature of environmental change and, as a result, the processes influencing sediment transport and accumulation differed between regions. Such spatial variability is likely to be a key factor in the context of any future climatic change. It is also relevant to note that the magnitudes of projected future change in, for example, temperature and sea-level are both above and below those experienced in the recent past. During the early Holocene, for example, global temperatures were rising at rates of around 1–2°C/1000 years. This is below the rates of most projections through to 2100 (see below). By contrast, sea-level has risen by around 120 m over the last 17,000 years, and most of this rise occurred prior to 6000 ybp. Hence rates of rise were in the region of 10 mm yr^{-1}, considerably above current projections of change (Jones 1993a).

Although future projections are subject to a high degree of uncertainty, what is evident from recent data is that marked increases in atmospheric concentrations of greenhouse gases have occurred over the past 200 years. The following increases have, for example, been reported for the period 1750–2000: CO_2 (280 ppm to 368 ppm), CH_4 (700 ppb to 1750 ppb) and N_2O (270 ppb to 316 ppb) (McCarthy et al. 2001). These are all predicted to increase further through to 2100 (e.g. atmospheric CO_2 concentrations are predicted to increase to between 540 and 970 ppm). Linked to these greenhouse gas increases are projected changes in global mean surface temperatures, which will have potential consequences for sea-levels, sea-surface temperatures and climate circulation systems. Global mean surface temperatures have, for example, increased by 0.6 ± 0.2°C over the twentieth century, although these increases have been more significant over land areas than the oceans. There has also been an increase in the number of hot days and a reduction in the number of days experiencing frosts. In addition, Arctic sea-ice has thinned by around 40% in the past few decades and decreased in extent by 10–15% since the 1950s (McCarthy et al. 2001).

Based on outputs from the HadCM3 models, which show differences between current climate (defined as the period 1960–1990) and climate at the end of the twenty-first century (2070–2100), mean surface air temperatures are predicted to increase by 0.3°C (range 0.6–9.2°C; see Hadley Centre 2004). Increases are evident across much of the globe with the exception of the southern Pacific Ocean. Annual precipitation rates are also projected to increase by an average of 0.2 mm day^{-1} (range − 3.7–8.9 mm day^{-1}; see Hadley Centre 2004). These projections are more

variable regionally, but show increases over the northern mid-latitudes, tropical Africa and southeast Asia. Decreases are predicted in Australia, central America and southern Africa.

Such changes in temperature and rainfall are likely to have an impact upon the functioning of many terrestrial sedimentary systems. As patterns of river channel erosion and sedimentation are influenced by streamflow over time and, especially, by flood frequency, any changes in the hydrological cycle will significantly influence fluvial sediment transport and depositional processes (see Chapter 3). Changes in precipitation may result in modified drainage densities and either higher or lower sediment yields depending on the regional effect. In either case there would be significant change in the downstream depositional environments. Lakes are also highly susceptible to changes in air temperature and rainfall because these influence rates of evaporation, lake-level and hydrochemical and hydrobiological regimes (see Chapter 4). Under extreme conditions, lakes may disappear entirely. Responses are also likely to vary between open (exorheic) and closed (endorheic) lakes. The latter are dependent upon rates of fluvial input and evaporation and are thus highly sensitive to changes in both. Hence lake sediments are good sources of information about past climatic and environmental conditions.

Climate changes are also likely to have an impact upon rates of sea-level rise and on the functioning of large-scale ocean–atmosphere circulation systems. Over the past 100 years, global mean sea-level has increased at an average rate of 1–2 mm yr^{-1}. Predictions from the HadCM3 models suggest that mean sea-level will rise by 0.38 m by the end of the twenty-first century (range 0.09–0.74 m; see Hadley Centre 2004). This will occur primarily because of thermal expansion of the oceans and the melting of glaciers and ice caps. Significant regional variations in the magnitude of these rises are likely, however, superimposed upon which will be spatial variations in rates of isostatic change. Even relatively small increases in sea-level will, however, exert a significant influence on most coastal sedimentary systems (e.g. Nicholls 2004) and increase rates of coastal cliff failure (Jones 1993b). In large part, this will occur because increasing sea-levels raise the plane of activity over which wave influence is exerted. Along temperate sediment-dominated coasts, this is likely to result in a landward migration of beaches, dunes and barrier islands, although the potential for landward retreat will depend upon the nature of the backshore environment and on the presence or absence of coastal infrastructure. Where roads and urban conurbations exist, landward migration may be prevented by coastal defence structures and thus, in such cases, progressive loss of coastal sedimentary environments ('coastal squeeze') may occur (Chapters 7 and 8). Rates of change will be exacerbated along low-lying coasts and in areas that are undergoing active subsidence (see Pirazzoli 1996, fig. 117).

Predicting the response of individual coastal systems is further complicated by local sediment dynamics and rates of sediment supply. Coastal dune systems represent an interesting example where these factors are highly site-specific and thus shoreline response modes highly variable. Dunes are sites of temporary sediment storage on the coast and significant sediment exchange occurs with adjacent environments. This may be a response to changes in energy levels during storm events, but will also occur during periods of rising sea-level as the plane of wave influence is reset (see Chapter 8). Hence dune responses to sea-level change, as with most coastal systems, can be considered realistically only on a site-by-site basis. In the tropics, sea-level rise may bring the benefit of renewed phases of growth on coral reefs that have reached sea-level and ceased to accrete during the late Holocene. The potential for such expansion will, however, depend upon reef community status and in severely degraded coral reefs partial submergence of the reef structures may occur. This, in turn, may facilitate increased wave-overtopping and adjacent shoreline erosion (see Chapter 9). Along mangrove-dominated coasts, shoreline response to sea-level rise will depend, in part, upon whether sediment supply is sufficient to maintain the seaward fringe. As along temperate coastlines, landward migration of mangroves may occur if the backshore area is unimpeded. Hence environmental responses are likely to be highly site-specific. Predicting future responses to sea-level rise along many coastal fringes is further complicated by the fact that sea-level is rising from a position of pre-existing sea-level highstand. Hence many coastal sedimentary environments will be migrating across areas of relatively low-lying land and thus the conditions under which the environments transgress will be very different to those that were submerged during the Holocene sea-level rise. This will complicate attempts to model future landform migration on the basis of changes that occurred during the early Holocene.

Climate change models also predict an increase in the frequency of severe weather conditions (McCarthy et al. 2001). These include not only storms and cyclones, but also shifts in large scale, ocean-driven climate oscillations such as the El Niño–Southern Oscillation (ENSO) and the North Atlantic Oscillation (NAO). There is, for example, some evidence to suggest that El Niño events have become more frequent and intense during the past 20–30 years and that these events may increase in frequency and intensity through to 2100 in tropical Pacific areas. Such changes may result in modified rainfall patterns, in rainfall intensity and the frequency of drought conditions. Predicted increases in high-intensity precipitation events may result in increased flooding, landslides and mudslides, increased rates of soil erosion and increased flood runoff, with obvious implications both for upland and fluvial sedimentary systems. Past records of increased slope failures in Central and South America have, for example, been linked to interannual variations in the magnitude and intensity of rainfall events between El Niño and La Niña years (Trauth et al. 2000). Any change in storm frequency and intensity will also have significant implications for many coastal sedimentary systems due, in large part, to likely increases in storm-wave surges. Under these conditions, increased rates of coastal erosion (Chapter 8) and, in the tropics, increased damage to coral reefs and mangroves can be expected (Chapter 9).

1.7 ASSESSMENT AND MANAGEMENT OF SEDIMENTS

Sediments are managed on many scales (both spatially and temporally) for a wide range of reasons, and different environments have unique management challenges and solutions (see individual chapters). In many cases it can be recognized that there is a problem from the sediment perspective which needs to be addressed. For example, sediment may be having an adverse impact upon the ecological functioning of a river basin, or excess sediment accumulation may be having a negative impact upon the economic functioning of a port or other navigable waterway. In most cases, these issues will be about either sediment quantity or sediment quality, although in many cases they are interlinked (see Case Study 1.3).

Case study 1.3 Sediment dredging and treatment in the Port of Hamburg, Germany

Hamburg harbour is the largest port in Germany, and one of the ten largest in the world. It is near the mouth of the River Elbe, approximately 100 km from the North Sea, which drains central Europe (Case Fig. 1.3A). As it is an economic port, water depths need to be maintained to allow shipping access into the port. Sediment is supplied to the port both from the upstream catchment of the River Elbe, but also by tidal movement of sediment from the North Sea. Sedimentation rates in some parts of the port are in the order of several metres per year, and this sediment accumulation has a major impact upon shipping access. As a result, there is a need for sediment to be dredged from the port; approximately 3 to 4 million cubic metres of sediment from the Elbe each year.

This creates an issue of how to dispose of this dredged sediment. Before the 1970s this dredged material was either placed on agricultural land or disposed of further down the system. However, high levels of contamination in this sediment, together with stricter environmental

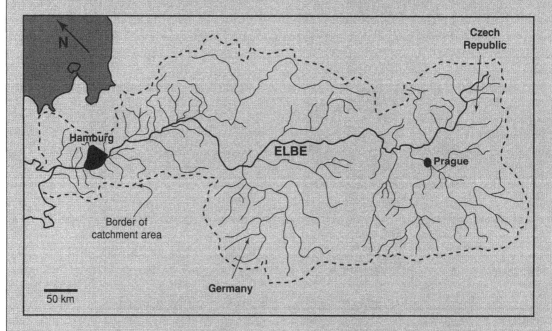

Case Fig. 1.3A The drainage basin of the River Elbe, which drains central Europe and has near its mouth the harbour of Hamburg, the largest port in Germany.

controls, have meant that this dredged sediment can no longer be disposed of in this manner.

Typical levels of metal and organic contaminants in the dredged sediment are shown in Case Table 1.3. The largest portions of the contaminants present in the sediments deposited in the port have been sourced from discharges in the upper reaches of the River Elbe. The upper Elbe drains industrial regions in the Czech Republic, and it is these regions that contribute the largest contaminant load. The levels of contaminant inputs into the port have decreased in recent years, partly as a result of environmental legislation on sources, but also due to economic decline and decreases in industrial activity.

Case Table 1.3 Levels of contamination in sediment dredged from the harbour of Hamburg. (From Netzband et al. 2002.)

Contaminant	Levels in Hamburg Port dredged material ($\mu g\ g^{-1}$)
Arsenic	50–150
Lead	150–300
Cadmium	5–25
Chromium	150–300
Copper	250–600
Nickel	50–100
Mercury	5–20
Zinc	1000–2500
Mineral Oil	Up to 3000
PCBs	Up to 1.5
PAHs	Up to 15

Given the problem of high contamination levels, the Port of Hamburg has developed a dredged sediment treatment process, whereby dredged sediment is treated, cleaned and the waste material minimized (Case Fig. 1.3B). Contaminants in the dredged sediment are concentrated in the fine fraction (< 63 µm), with the sand-sized fraction (> 63 µm) having much lower levels of contaminants. Therefore, this treatment separates these two size fractions. The

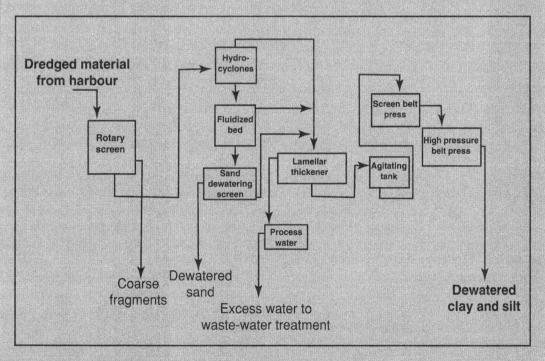

Case Fig. 1.3B The stages in the treatment of contaminated dredged sediment taken from the harbour of Hamburg. (Adapted from Kroning 1990.)

steps in the treatment process are shown in Case Fig. 1.3B. The first stage of the process is the removal from the wet sediment of most of the coarse fraction by centrifugation. Any remaining fine grains with the sand fraction are then removed in a fluidized bed under flowing water conditions. Sand-sized material is dewatered and can be used in the construction industry. The fine-grained material is dewatered through a series of flocculation and filtration procedures. The resulting contaminated fine sediment is disposed of in a purpose-built disposal facility. Annually, 1.2 to 1.4 million cubic metres of sediment are treated in this way, with 50% of this volume being contaminated fine sediment placed in the disposal facility.

The removal of contaminated sediments via this dredging, sediment treatment and disposal has removed about 30% of the heavy metal contaminants from the River Elbe and a similar percentage of the organic contaminants. Therefore, this dredging acts as a pollutant filter to the North Sea. This is not a sustainable solution, however, and it is now widely recognized that minimization of contaminants at source (i.e. in the upper reaches of rivers) is the most effective sediment pollution strategy.

Relevant reading

Adams, M.-S., Ballin, U., Gaumert, T., et al. (2001) Monitoring selected indicators of ecological change in the River Elbe since the fall of the Iron Curtain. *Environmental Conservation* 28, 333–44.
Kroning, H. (1990) Separation and dewatering of silt from the Port of Hamburg. *Aufbereitungs-Technik* 4, 205–14.
Netzband, A., Reincke, H. & Bergemann, M. (2002) The River Elbe: a case study for the ecological and economical chain of sediments. *Journal of Soils and Sediments* 2, 112–16.

Sediment management approaches generally take one of two forms.

1 Those that address issues resulting from the presence of sediment (either a quality or quantity issue) and where sediment requires removing or remediating. This type of management is most common in engineered or contaminated environments, e.g. urban environments (Chapter 6) and estuaries (Chapter 7).

2 Those that use management strategies to trap or retain sediments in a system to maximize the ecological or environmental functioning of that system. Such management practices are particularly common in coastal environments (see Chapters 7–9).

1.7.1 Risk assessment and sediment guidelines for contaminated sediments

A first step in addressing these issues is to assess the environmental risk of the sediments, and take appropriate action. In order to determine if sediment is contaminated, baseline and threshold-effect information is needed. In the case of artificial compounds (e.g. pesticides, PCBs, some radionuclides) the baseline value is zero, and contamination assessment is relatively straightforward. Pre-industrial historical values need to be established for elements with both natural and anthropogenic sources (e.g. metals, phosphorus). In rare cases this can be accomplished through analysis of monitoring data or archived samples, but in most cases values are determined through the analysis of sediments that have accumulated through time (Smol 2002). Threshold-effect values are often determined through an assessment of the physical, chemical or biological nature of the sediment, or a combination of these. Increasingly, the end result is a series of threshold-effect values that collectively form Sediment Quality Guidelines, or other similar measures (e.g. critical level, critical load, etc.).

Unlike water quality, clear Sediment Quality Guidelines are generally lacking for sediments, although many countries have developed their own guidelines independently. These guidelines can take chemical, biological or integrated chemical and biological forms. Simple chemical guidelines involve either analysing for a contaminant and setting acceptable levels based on natural background levels (e.g. Ingersoll et al. 1996), or assessing the potential chemical bioavailability of contaminants in a sediment (e.g. Ankley et al. 1996). Although such guidelines are simple and relatively cheap to implement, they are not based on an ecological response and, therefore, do not provide realistic information on the risk associated with that contaminant level. Much better information may be acquired through determining the ecological response to the sediment. The most appropriate way to do this uses ecotoxicity tests. Such tests use organisms (either benthic or water column) to determine the toxicity of a sediment, the benefit of these being that they provide information on ecological risk. However, the test does not provide information on which contaminant (or combination of contaminants) is responsible for this risk. Recently, a triad approach has been gaining favour (Sediment Quality Triad; Chapman 1986). This assessment consists of three components:
1 identification and quantification of all contaminants in a sediment;
2 measurement of toxicity based on a sediment toxicity test;
3 evaluation of in situ biological effects.
The benefit of this measure is that it integrates laboratory, field, biological and chemical data. The disadvantages are that it has not been fully accepted and is expensive.

1.7.2 Managing issues of sediment quantity and quality

Once a sediment has been determined to be a problem, either as a result of quantity or quality issues, there are a number of remediation strategies that can be put in place to address this. These can be placed into three main categories: physical removal of material (dredging), phys-

ical isolation of the material (containment) and chemical treatment of the sediment. Dredging of sediment can be carried out for reasons of quality, but by far the most important reason is for quantity reasons: for example, the need to dredge sediment to maintain a minimum draft for shipping in docks, channels and canals (see Case Study 1.3). Once such sediment has been dredged, however, there is commonly a secondary quality issue as the sediment may be too contaminated to dispose of it in a normal manner. Physical containment is used in cases of sediment quality issues and involves covering the contaminated sediment, either with clean sediment or concrete. This procedure effectively isolates the contaminated sediment from the overlying water column and ecosystem, and may often be the cheapest alternative. Chemical treatment of sediment can take place *ex situ* or *in situ*. *Ex situ* treatment is generally used to produce sediment of low enough contamination levels to be disposed of safely or reused (see Case Study 1.3). *In situ* treatment of sediments is a novel application and, in a similar manner to biochemical remediation of contaminated land, is often site-specific, depending on the problem.

1.7.3 Managing for sediment retention and stabilization

A number of sediment management strategies focus on either retaining or trapping sediment for the purpose of limiting change within the sedimentary system. Most commonly this is done in response to either one-off erosional events, such as can occur following cyclone or tsunami damage, or in response to the progressive sediment loss that occurs when the volume of sediment being supplied to a particular environment is reduced, i.e. there is a shift to a negative sediment budget. Such problems are particularly common in coastal environments and reflect a need to protect coastal infrastructure. In many cases the need for shoreline protection occurs due to inappropriate siting of properties or roads in areas that should be expected to be periodically influenced by marine processes. The techniques used are varied, but include:

1 the construction of shore normal groynes that facilitate the trapping of sediment being moved alongshore;

2 the artificial emplacement of sediments onto beaches to maintain their morphological integrity – a process known as beach recharge;

3 the artificial stabilization of coastal dunes in an attempt to retain sediment within the dune system – numerous approaches including the use of fences to trap windblown sand and biodegradable fabrics to stabilize sediments have been used (see Chapter 8).

Many of these techniques are of short-term benefit only and may bring with them either a continual maintenance cost (e.g. in the case of ongoing beach recharge), or may have downdrift sedimentological consequences (e.g. in the case of groyne construction).

1.8 FUTURE ISSUES AND RESEARCH

As discussed at the start of this chapter, environmental sedimentology represents a rapidly expanding field of research. It is increasingly clear that most active sediment systems are, to varying extents, influenced by human activity and thus significant attention is being targeted at trying to understand the impacts and effects of human activity on contemporary sediment systems. One of the aims of this introductory chapter has been to provide an overview of the nature of such influences. These include, at a generic level, the modification of sediment transfer and accumulation pathways, the contamination of actively accumulating sediments, and modifications to sediment chemistry and sediment diagenesis. Examples of each are included in individual chapters, but in many cases there is a need for much higher resolution monitoring of sedimentological and geochemical processes in order to enhance understanding of the temporal and spatial dynamics of individual sedimentary systems. An excellent example of the advances that can be made by adopting more integrated sedimentological, geochemical and ecological approaches, as well as airborne sensing and satellite imagery, is seen in the recent developments

that have been made in understanding the dynamics of the Amazon–Guianas coast in South America (see Baltzer et al. 2004 and references therein). Such bodies of work demonstrate the benefits of small-scale, localized studies that are placed in the context of an improved understanding of regional sediment dynamics. In many sedimentary environments there are also complexities emerging in terms of distinguishing between natural (background) sediment dynamics and those associated with anthropogenic activities. Addressing such issues, as well as understanding the interactions between natural and anthropogenically induced change, represent important areas for future research.

Considerable research attention is also currently being focused on issues of sediment management and remediation, both in relation to addressing issues of sediment pollution, as well as sediment abundance. There is a growing awareness that a physical and chemical understanding of sediments needs to integrate water and ecological information to better inform habitat monitoring and management (e.g. see Chapter 10). It is also important that the full range of scales on which sedimentary processes operate is considered in sediment management. For example, within river basins a number of models have recently been produced to manage sediment at the catchment scale (Apitz & White 2003; Owens et al. 2004). Such approaches are important for compliance with, and implementation of, legislations and guidelines; for example, the European Community Water Framework Directive (2000/60/EC). Sediment is also increasingly being considered as an economic, ecological and social resource, and as such needs to be managed sustainably.

Overriding many of these themes are the potential changes that will be associated with climate-change-induced shifts in, for example, atmospheric and oceanic temperatures, rainfall patterns, sea-level, oceanic CO_2 concentrations and storm frequencies. Understanding or predicting the likely response of sediment systems (both terrestrial and marine) to such changes is, in part, reliant upon improved climate models. There is a need, however, for much better understanding of likely sediment system responses at a range of spatial scales. Large-scale studies, such as those of Goodbred (2003) that are outlined in section 1.6.3, are useful in terms of providing an insight into the way in which climate can control large-scale sediment system evolution. However, even modelling more localized sediment responses is fraught with difficulties and is constrained by the limitations that exist in terms of the data that can be reliably placed within appropriate modelling software.

In the immediate future the most pressing research needs are likely to be driven by the fact that many human-induced as well as environmental or climatic-induced changes to sediment systems bring with them a 'human cost'. This will increasingly occur where change within any individual sediment system progressively impinges upon human usage of the environment. This may occur either because of increased or decreased sediment accumulation, or because contaminant accumulation is having an impact on the ecological and/or sedimentary functioning of the environment. Understanding, mitigating and managing these changes represent major research challenges for the near future.

2

Mountain environments

Jeff Warburton

2.1.1 Introduction

Mountain environments account for approximately one-third of the Earth's surface and it is estimated that about 10% of the world's population live in mountain and upland regions (Gerrard 1990). Geomorphologically these are amongst the most active areas on Earth, often being characterized by some of the highest recorded rates of erosion and sedimentation (Walling & Webb 1983; Jansson 1988). Mountain regions generally have steep slopes and large relative relief. These are factors that make mountain environments sensitive to natural and anthropogenic activity such as extreme climate events, seismic perturbations, deforestation and land-use change. The dynamic nature of mountain environments, however, must not be overemphasized at the risk of ignoring slow, continually acting processes, the cumulative effect of which can be highly significant (Messerli 1983). Furthermore, sedimentary activity in mountain environments varies enormously between different topographic settings (Milliman & Sivitski 1992) and even within the same general setting; differences caused by variations in the pressures posed both by natural and anthropogenic agents produce differing sedimentary responses (Dedkov & Moszherin 1992).

An understanding of the environmental sedimentology of mountain environments is important because mountains provide essential resources such as water supply, sustainable energy (hydroelectric power), recreation and tourism,

ecological refuge and specialist agricultural niches. Their significance varies from country to country and is generally proportional to the degree of mountain cover, for example in Europe the proportion of area greater than 1000 m of altitude varies markedly: 5% in the UK, 44% in Norway and 100% in Andorra. Mountain areas, in common with many other environments, are increasingly endangered by socio-economic changes, increased recreation and traffic, and changing land-use, often leading to environmental degradation. These changes often can be related to direct impacts such as building and development works, or may be the result of subtle environmental changes such as changing precipitation patterns, shifts in habitats, changes in runoff rates, water and soil pollution and changes in the ground thermal regime. Environmental degradation of this type is often manifest in changes in sedimentary processes acting in mountain areas and may result in slope instabilities and enhanced fluvial erosion and sedimentation. An understanding of the environmental sedimentology of mountain areas provides a useful framework for studying the effects of humans and environmental change on active surface sedimentary systems. Such an approach is particularly pertinent to mountain environments where, due to high relief and steeper slopes, geomorphological and sedimentary processes often operate at greater rates than in lowland environments and the extreme physical conditions make mountain environments susceptible to even slight changes in climate and land-use.

Given the large range in mountain environments it is impossible in this short chapter to fully characterize the environmental sedimentology of all mountain areas. The aim is, therefore, to highlight the general functioning of active sedimentary processes in mountain environments and, using case studies and examples, show how these processes are affected by climate change and human actions and how adjustments to sediment systems have an impact on human use of mountain regions. The chapter begins by considering the main characteristics of mountain environments (section 2.1.2) that are most relevant to environmental sedimentology. This forms a basis for classifying mountain environments (section 2.1.3). The bulk of the chapter that follows is divided into five main sections: sediment sources and transfer processes (section 2.2); processes and impacts – events related to natural disturbance (section 2.3); processes and impacts – anthropogenic influences (section 2.4); management and remediation (section 2.5); and future issues (section 2.6). Inevitably some topics, such as climate change, depositional processes (i.e. alluvial-fan and glacial-lake sedimentation) and some slow geomorphological processes (e.g. creep), are not covered in great depth.

2.1.2 Characteristics of mountain environments relevant to environmental sedimentology

There are seven general themes that need to be considered when dealing with the environmental sedimentology of mountainous areas.

1 Mountains are generally regions of abundant sediment supply and high erosion potential. Typically erosion rates and sediment yields are globally some of the highest recorded (Milliman & Syvitski 1992).

2 High rates of sediment production translate into elevated rates of sediment transfer and increased sediment deposition. However, sediment delivery in tectonically stable and tectonically active regions differs markedly depending on the nature and rate of the different geomorphological processes that operate (Marutani et al. 2001).

3 The importance of steep slopes is fundamental to many processes operating in mountain envir-

onments. High levels of gravitational stress result in rapid sediment movements both in terms of slope instability and channel sediment transport (Jones 1992).

4 There is considerable variability in the spatial and temporal rates of sediment transfer and this has important environmental and social consequences for mountain environment development (Butler et al. 2003).

5 Mountain environments are sensitive to disturbance both from climate change and anthropogenic impacts (Ives & Messerli 1989).

6 The incidence of geomorphological hazards tends to be high in mountainous, high-energy environments where narrow valley floors are juxtaposed with steep unstable side slopes. Infrastructure and population are always at risk and this risk increases as expansion of roads and settlements continue. These issues are often greatest in mountainous terrain where population and infrastructure have developed along upland river valleys. There is, however, growing recognition that heavy engineering approaches designed to manage such active geomorphological settings are unsustainable. A greater understanding of catchment-wide sediment budget dynamics can provide the necessary knowledge to better manage such systems (Gerrard 1990).

7 Mountain sediment systems are often only a part of a larger drainage basin structure. Therefore, sediment delivery from the headwaters will have an impact downstream on floodplain processes. The degree of coupling needs to be established so that floodplains can be managed effectively (Brizga & Finlayson 1994; Piégay et al. 2004) and hazards at the mountain front reduced (White et al. 1997).

2.1.3 Definition and classification of mountain environments

Mountains occur in virtually all geographical and climate settings on Earth. Mountain areas vary significantly from small isolated mountains to huge mountain chains stretching for many hundreds of kilometres across continents.

Fig. 2.1 Two contrasting mountain types: (a) typical hilly terrain of Dooncarton Mountain in County Mayo, Ireland, 260 m a.s.l.; (b) high-mountain relief of Hunza Peaks in the Karakoram, Pakistan, 7000 m a.s.l. (Photographs: J. Warburton.)

Classifying different mountain systems is problematic and several schemes have been proposed to account for the variability (Price 1981; Gerrard 1990). As a result, the definition of mountain regions is largely arbitrary because multiple criteria can be used to define such areas, for example relative relief, the presence of particular geomorphological features, a threshold altitude (1000 m), etc. Whatever definition is used, however, mountain environments have several common features, namely: the presence of steep slopes, and vertical differentiation of climate and vegetation cover (Barry 1992). In very general terms as altitude increases relief is greater, vegetation cover diminishes and the climate becomes more extreme in terms of precipitation, wind and temperature. These characteristic elements (greater relief, diminished vegetation, extreme climate) are highly significant in terms of the environmental sedimentology of mountain regions because of their potential impact on erosion, i.e. the climatic control on weathering and therefore sediment production, the high energy associated with steep slopes and the transport and removal of sediment, and the diminished vegetation which decreases resistance to erosion. In this chapter mountains are defined as areas of steep relative relief where sedimentation and erosion are actively conditioned by hillslope and/or channel processes. This geomorphological definition avoids the problem of specifying a minimum altitude for mountain relief and can include contrasting examples of the type shown in Fig. 2.1: (a) Dooncarton Mountain in Co. Mayo, Ireland (260 m of relief, maritime temperate climate) and (b) the Hunza Peaks in the Karakoram, Pakistan (3000 m of vertical relief, arid continental interior). Processes operating in the two environments are markedly different.

Globally, mountain environments cover all climate regimes and, as seen from the examples just quoted, can range from small isolated coastal peaks to immense mountain ranges containing the world's highest summits. Louis (1975), as part of an assessment of global relief, provides one of the few estimates of the area of mountain and high plateaux (Table 2.1). Assuming the total land surface is approximately 149 million km^2 (oceanic islands cover 2 million km^2), mountains occupy about 20% of this area. The distinction between mountain and plateau relief is somewhat arbitrary so the total extent of world mountain area is open to some debate. However, the main point is that mountains form a significant proportion (approximately one-fifth) of the global land area. This global estimate is interesting but in terms of mountain sediment systems it is the different relief regimes that are most important in determining the processes operating within a particular environment. Barsch & Caine (1984) distinguish four categories of mountain relief (Table 2.2), varying from subdued hilly terrain to high mountain systems (e.g. Fig. 2.1). Figure 2.2 shows maps of (a) the major relief elements of the Earth surface (Dongus 1980; as cited in Barsch & Caine 1984) and (b) the global distribution of suspended sediment yield (Walling & Webb 1983). Although some exceptions exist, such as the low suspended sediment yields from areas of old tablelands and continental shields, the correspondence between areas of high relief and active tectonics and seismicity, and elevated suspended sediment delivery at the global scale is clearly apparent.

Table 2.1 Global estimates of areas of mountain and plateaux. (Source: after Louis 1975; Barry 1992.)

Altitude range (m)	Mountains (10⁶ km²)	Plateaux (10⁶ km²)	Mountain cover as a proportion of total land surface (%)
> 3000	6		4.0
2000–3000	4	6	2.7
1000–2000	5	19	3.4
0–1000	15	92	10.2
Total	30	117	20.3

Type	Altitudinal difference (m) (over 5 km distance)	Relative relief (m km^{-2})
High mountain system	> 1000	500
Mountain system	500–1000	200
Mountainous terrain	100–500	100
Hilly terrain	50–100	50

Table 2.2 Contrasting relief regimes from different mountain systems. (Source: Barsch & Caine (1984). Reproduced with permission *Mountain Research and Development*.)

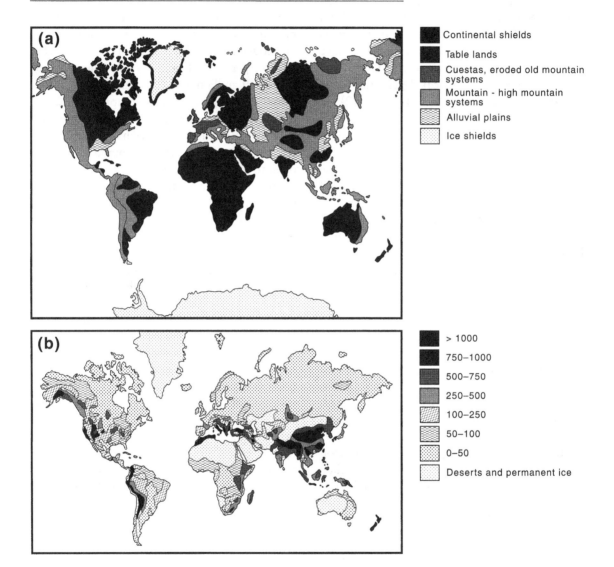

Fig. 2.2 (a) Global map of the major relief elements of the earth surface (Dongus 1980; as cited in Barsch & Caine 1984). (Reproduced with permission, *Mountain Research and Development*.) (b) Global distribution of suspended sediment yield. (From Walling & Webb 1983, *Background to Palaeohydrology*. (Ed.) Gregory, K.J. (1983). © John Wiley & Sons Limited. Reproduced with permission.)

Research themes in mountain geomorphology have been discussed by Barsch & Caine (1984), who make the important distinction (Fig. 2.3) between studies focusing on mountain landforms and those that study land-forming processes in mountain environments. In terms of research into the environmental sedimentology of mountain systems it is the morphodynamics of mountains

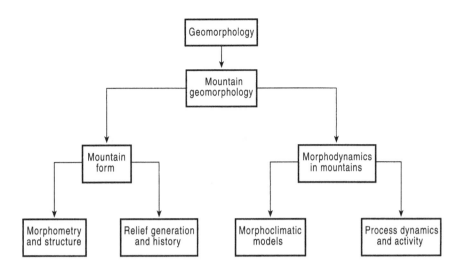

Fig. 2.3 Research themes in mountain geomorphology (Barsch & Caine 1984). Environmental sedimentology is largely focused on the morphodynamics of mountain environments but has important links to mountain form. (Reproduced with permission *Mountain Research and Development*.)

that are of greatest interest and in particular the process dynamics and activity, because this is what is most closely related to sedimentary processes. Process dynamics, however, can never be viewed in isolation because these are closely controlled by morphoclimatic factors as well as morphometry, structure and relief. Environmental sedimentology, therefore, embodies the full range of research themes in mountain geomorphology and in the case studies that follow each of these themes is covered.

2.2 SEDIMENT SOURCES AND TRANSFER PROCESSES

2.2.1 The mountain sediment cascade

An important conceptual model in understanding the environmental sedimentology of mountain environments is the notion of the mountain sediment cascade (Caine 1974). In simple terms this depicts the mountain sediment system as a series of sediment stores linked by a series of transfer processes (Fig. 2.4). Recognizing and quantifying significant sediment stores and the processes that link them is the basis of the sediment budget approach. The seminal work by Rapp (1960) working in the Kärkevagge catchment in Sweden used this methodological approach and inspired others to follow a similar approach in identifying the significance of sediment storage in regulating the sediment yield

from mountain catchments (Church & Ryder 1972). Caine (1974) provided an early overview of alpine geomorphological processes and described these as a series of sediment flux cascades. This consisted of two dynamic geomorphological subsystems, namely the slopes and stream channels (Fig. 2.4). Caine (1984) subsequently developed this concept by superimposing three sediment subsystems: the geochemical, the fine sediment and the coarse detritus. This basic description was later modified by Barsch & Caine (1984) into a four-way classification of mountain processes each with different controls, responses and rates of activity.

Barsch & Caine (1984) distinguished:
• valley glacier sediment system
• coarse debris system
• fine sediment system
• geochemical system

All four systems interact and material is transferred between the different systems and hence it is often convenient to couple some of these systems together. For example, the valley glacier and coarse sediment subsystems are most characteristic of mountain areas of greatest elevation and relief. Table 2.3 provides examples of these subsystems and the main morphological units which typify them.

Interactions between the coarse sediment system and fine sediment system in many mountain environments are difficult to separate as zones of activity are closely coupled. For example, there

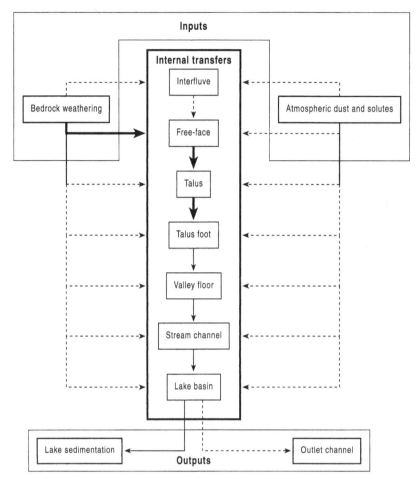

Fig. 2.4 The alpine sediment cascade process system proposed by Caine (1974).

Table 2.3 Examples of mountain geomorphological process subsystems and typical geomorphological units as described by Barsch & Caine (1984).

Sediment system	Morphological units	Transfer processes	Typical mountain environment	Case study
Glacia	Glacierized valleys and terrain; moraine	Glacial transport	Icelandic glaciers Ggjkull and Kvrjkull	Spedding (2000)
Coarse debris	Steep bedrock slopes and talus	Rock fall, avalanches; debris flows; rock slides; talus creep	Randa rock slide, Valais, Switzerland	Gtz & Zimmermann (1993)
Fine sediment	Waste mantled slopes	Solifluction; soil creep; slopewash	Colorado Front Range, USA	Benedict (1970)
Fluvial and geochemical	Stream channels; valley floors; fans and lakes	Fluvial transport; solute transport; lake sedimentation	Kärkevagge, northern Sweden	Rapp (1960)

is significant slope–channel coupling in most mountain environments because, given the high relative relief and steep slopes, valley sediment storage is often small and sediment runout from mountain slopes enters stream channels directly. The usefulness of Fig. 2.4 is that it provides an overall framework for evaluating sediment fluxes under natural and disturbed conditions. Greater

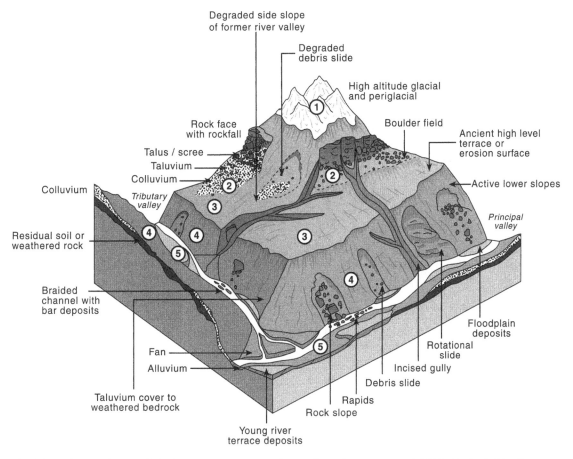

Fig. 2.5 Land-system diagram of a high-mountain environment showing five major terrain zones: (1) high-altitude glacial and periglacial; (2) free rock faces and debris slopes; (3) degraded middle slopes and ancient valley floors; (4) active lower slopes; and (5) valley floors. (Redrawn from *Engineering Geology*, Fookes, P.G., Sweeney, H., Manby, C.N.D. & Martin, R.P. (1985) Geological and geotechnical engineering aspects of low cost roads in mountainous terrain, **21**, 1–152, with permission from Elsevier.)

understanding can be achieved if the landform associations in mountainous terrain can be shown in a land-systems model. Fookes et al. (1985) have developed such a model based on east Nepal, but as Gerrard (1990) argues, this model is more generally applicable to most extratropical high mountains that have been extensively glaciated. Figure 2.5 shows this model and the interrelationship between five main zones: high-altitude glacial and periglacial; free rock faces and debris slopes; degraded midslopes and ancient valley floors; active lower slopes; and valley floors. The model is most useful in demonstrating the coupling between the slope sediment systems and valley sediment systems that actively occur in high mountain environments and which form the

basis of conceptual models of sediment delivery (e.g. Fig. 2.4; Caine 1974; Barsch & Caine 1984). Given these different geomorphological subsystems there is an enormous variety of processes operating in mountain environments. However, in terms of environmental sedimentology the most significant are usually the destructive mass movements which generally occur in mountain torrents or on steep unstable bedrock slopes.

In terms of the fluvial system as a whole headwater mountain catchments are viewed as sediment production zones feeding bedload and suspended sediment downstream. A simple schematic model describing the linkages between mountain catchments and variations in fluvial form along the river profile from the headwaters

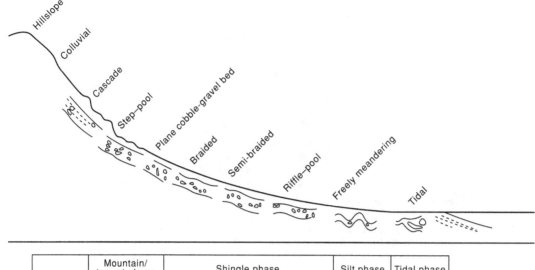

Fig. 2.6 Schematic diagram showing transitions in the fluvial system along a river profile. (Source: Mosley & Schumm 2001; reproduced with permission *New Zealand Hydrological Society*.)

to the coastal lowlands was developed by Nevin in 1965 (Mosley & Schumm 2001; Fig. 2.6). This shows the transition from the headwater catchments, which are dominated by steep channels such as cascades, step–pool systems and coarse braided rivers, through to meandering lowland channels. Generally the sediment load in the headwater channels is dominated by coarse bedload (50–90% of the total load), whereas downstream fine suspended load dominates (often exceeding 80–90% of the total annual load) (Fig. 2.6). It is well known that in the absence of significant tributaries grain-size systematically decreases downstream. Bed material size at any point in the drainage basin is a function of sediment supply and the combined action of sediment sorting and abrasion. This is often expressed as an exponential decline in grain size with distance downstream:

$$D = D_0\, e^{-aL}$$

where D is the particle size, D_o is the initial particle size, L is distance downstream and a is

a coefficient representing the combined action of sorting and abrasion. Such simple patterns do not hold for all rivers especially where there are variations in different sediment source rock types or where tributary inputs of sediment become significant (Rice & Church 1998).

2.2.2 Sediment yields from different mountain environments

The geomorphological activity in a particular mountain environment can be estimated by measuring the sediment yield from the catchments draining such areas. There have been several attempts to interpret global patterns of fluvial sediment yield (net erosion) in terms of the factors controlling sediment delivery (Walling & Webb 1983; Jansson 1988; Milliman & Syvitski 1992; Summerfield & Hulton 1994). The main factors considered have been either climate and runoff, or relief and tectonics. Clearly both are closely related (see Fig. 2.2) and the sediment discharge from a particular drainage basin is dependent on the combination of these controls. A study by

Table 2.4 Pearson correlation coefficients for mechanical denudation versus morphometric, hydrological and climatic variables. (Source: Summerfield & Hulton 1994.)

Variable		Log. mechanical denudation rate
Morphometric	Area	−0.11
	Mean trunk channel gradient	0.67
	Basin relief	0.80
	Relief ratio	0.78
	Mean model elevation	0.66
	Mean local relief	0.68
	Hypsometric integral	−0.03
Hydrological	Mean annual runoff	0.45
	Runoff variability	−0.04
Climatic	Mean annual temperature	0.41
	Mean annual precipitation	0.42

Jansson (1988) is typical of this approach in that the research poses the question, 'How does the magnitude of erosion vary globally under present human influence?' The methodological approach used in this study is a statistical analysis of sediment yield based on climatic conditions. No attempt is made to classify rivers in terms of topography. Alternatively, some studies have explicitly examined relief as a controlling variable on sediment discharge (e.g. Milliman & Syvitski 1992), whereas others have used a multivariate approach incorporating morphometric, hydrological and climatic variables (Summerfield & Hulton 1994). The importance of drainage-basin topography in influencing mechanical denudation rates is clearly demonstrated in the study of Summerfield & Hulton (1994). Results (Table 2.4) show a relatively strong statistical association between basin relief and mechanical denudation, albeit partly a function of other factors related to relief such as seismicity and weak rock structure.

Dedkov & Moszherin (1992) using suspended sediment yield data from 1872 mountain rivers assessed variations in erosion intensity and attempted to determine the significance of human impacts on mountain sediment systems. Their data (Fig. 2.7) show an interesting pattern that suggests all mountain areas are affected to some degree by human activity but it is the areas of lower relief that are most greatly impacted. Factors that promote erosion, such as forest removal, overgrazing, cultivation of slopes and road construction, occur in virtually all mountain regions

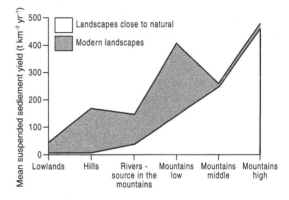

Fig. 2.7 Diagram showing the dependence of suspended sediment yield on relief and the significance of human interference in enhancing erosion (Source: Dedkov & Moszherin 1992; reproduced with permission of IAHS Press, from Dedkov, A.P. & Moszherin, V.I. (1992) Erosion and sediment yield in mountain regions of the world. In Walling, D.E., Davies, T.R. & Hasholt, B. (Eds.) *Erosion, Debris Flows and Environment in Mountain Regions.* IAHS Publication 209, 29–36.)

but the intensity of these activities varies with the degree of economic development and population pressure. However, it is in the lower relief mountains where these pressures are greatest.

Milliman & Syvitski (1992) in their analysis of 280 rivers discharging to the ocean found that sediment loads/yields were a log-linear function of basin area and maximum elevation of the river basin. Other factors controlling sediment discharge, such as climate and runoff, were of secondary importance. In particular, sediment fluxes from small mountainous rivers have been greatly underestimated in previous global sediment budgets, possibly by as much as a factor of three. Figure 2.8 shows the subdivision of river

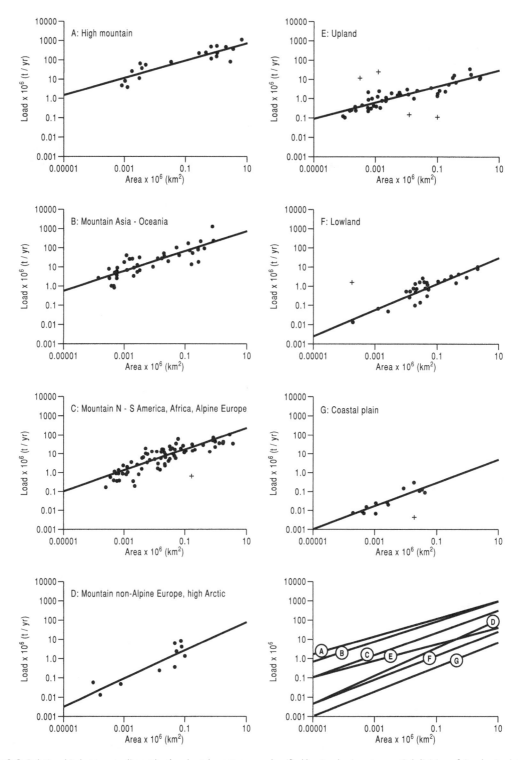

Fig. 2.8 Relationship between sediment load and catchment area as classified by river basin category. Subdivision of river basins into five categories based on maximum elevation within the hinterland: high mountain (headwaters at elevations > 3000 m), mountain (1000–3000 m), upland (500–1000 m), lowland (100–500 m) and coastal plain (< 100 m). The + symbols denote data points not included in regression calculations. (Source: Milliman & Syvitski 1992; reproduced with permission from *The Journal of Geology*, Milliman, J.D. & Syvitski, P.M. (1992) Geomorphic/tectonic control of sediment discharge to the ocean: the importance of small mountainous terrain, **100**, 525–44.)

basins into five categories based on maximum elevation within the hinterland: high mountain (headwaters at elevations > 3000 m), mountain (1000–3000 m), upland (500–1000 m), lowland (100–500 m) and coastal plain (< 100 m). Mountain rivers were further subdivided into: Asia and Oceania (generally very high sediment yield), the high Arctic and non-Alpine Europe (low sediment yields) and the rest of the World. The correlations between load and basin area for the various topographic categories generally range between 0.70 and 0.82. There is a distinct pattern to the data dependent on topographic setting. Mountain rivers have the greatest loads followed by the uplands, lowlands and coastal plains. There is some overlap in these general relationships owing to exceptions within each of the categories. For example, mountainous rivers draining South Asia and Oceania have much greater yields than (two to three times) other mountainous areas and are generally an order of magnitude greater than high Arctic and non-Alpine European mountains. Although these studies show some clear general patterns, such data should be interpreted with caution owing to the inherent errors in data collection and the incommensurate nature of the measurements (Harbor & Warburton 1993).

2.2.3 Sediment budget models of mountain sediment systems

Sediment budgets have been used as a tool for understanding sedimentary processes and

sediment fluxes for 50 years (Jäckli 1957, Table 2.5). A sediment budget accounts for the sources, transfers and storage of sediment within a landscape unit. Constructing a contemporary sediment budget for a mountain catchment is a time-consuming and labour intensive endeavour and, therefore, most budgets tend to be measured over short periods (typically 1–3 yr). Figure 2.9 shows a sediment budget framework applied to a small glacier basin in southern Switzerland in order to evaluate the significance of the proglacial zone in contributing sediment to a glacier-fed stream (Warburton 1990). A measurement framework was set up to determine the rates of sediment transport of slope and channel processes and changes in storage within the sediment system (Fig. 2.9a). Results (Fig. 2.9b) clearly demonstrate the importance of the glacier stream in terms of sediment flux but there is still a significant additional load: approximately 23% of the total catchment sediment yield is added by proglacial sources (Warburton 1990). The overwhelming proportion of the proglacial sediment (95%) was eroded from the valley floor during a brief meltwater flood.

This methodological approach is widely applied in the study of mountain sediment systems (Schlyter et al. 1993; Slaymaker et al. 2003). Two further examples of mountain sediment budget models are shown in Table 2.5. These are from the Upper Rhine (area 4307 km^2, relief 2800 m; Jäckli 1957) and Kärkevagge in northern Sweden (area 15 km^2, relief 930 m; Rapp 1960). In terms

Table 2.5 Sediment fluxes from two high mountain environments: Upper Rhine (area 4307 km^2, relief 2800 m; Jäckli 1957) and Kärkevagge in northern Sweden (area 15 km^2, relief 930 m; Rapp 1960). Units are in 10^6 J km^{-2} yr^{-1}. A joule (J) is the unit of *work* (E), which is generally defined by $E = m\,g\,(h_1 - h_2)$, where m is mass, g is acceleration due to gravity and h is elevation, with $(h_1 - h_2)$ being the change in height between two points 1 and 2. (Source: Barsch & Caine 1984.)

Catchment details	Upper Rhine	Kärkevagge
Area (km^2)	4307	15
Relief (m)	2800	930
Coarse debris bedrock slopes	729.2 (4.2%)	15.7 (58.4%)
Soil–fine sediment mantled slopes	53.5 (0.3%)	7.93 (29.5%)
Channel transport–lake sedimentation	13,798 (79.5%)	Not measured
Solute flux (output)	2781 (16.0%)	3.24 (12.1%)
Total	17,362	26.87

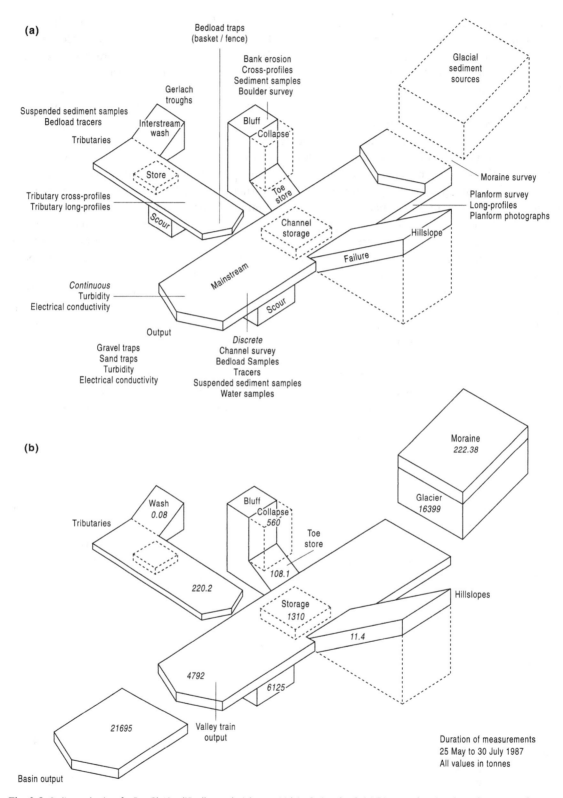

Fig. 2.9 Sediment budget for Bas Glacier d'Arolla proglacial zone, Valais, Switzerland. (a) Diagram showing the sediment transfer processes and storages, and measured sediment budget components. (b) Summary of main sediment fluxes and sediment storages in the proglacial fluvial sediment budget May–July 1987 (values in tonnes). (Source: Warburton 1990.)

Table 2.6 Summary of slope denudation estimates and dominant sedimentary process characteristics in the Kärkevagge catchment 1952–1960. (Based on: Rapp 1960.)

Process	Ton-metres (vertical)*	Sedimentary characteristics
Transportation of salts in running water	136,500	Stream and lake solute loads derived from catchment-wide chemical weathering
Earth slides and mudflows	96,375	Sources include poorly sorted till deposits and talus material. Deposition in bouldery mudflow levees, alluvial fans and sheet deposits (sorted gravel and sand)
Dirty avalanches	21,850	Slush avalanches. Transported material ranging in size from fine silt to boulders up to 5 m in length
Rock falls	19,565	Includes: pebble falls, small boulder falls and large boulder falls (varying in size from 10 to 100 m³). Rock-fall debris dominantly 20–50 cm, largest boulders up to 5 m in length
Solifluction	5300	Stony soil with occasional boulders. Vegetated or partly vegetated surfaces. Develops solifluction lobe movements 4 cm yr⁻¹ to depths of 50 cm
Talus creep	2700	Coarse, clast-supported openwork surface with fine content increasing with depth. Poorly sorted surface sediments varying in size from small pebbles to boulders. Surficial rates up to 10 cm yr⁻¹

*The tons-metres vertical concept was introduced by Jäckli (1957) and is calculated by multiplying the mass of sediment moved by the vertical component over which the sediment is transferred.

of sediment fluxes, regardless of the contrasts in scale, the coarse debris system (rock fall, snow avalanches and debris flows) is significant, and fluvial and channel processes are particularly important (although this was not measured at Kärkevagge). Slow mass movements tend to be of far less significance. Drainage-basin scale is important in determining the relative contribution of such processes because as basin size increases valley floor and channel processes become more significant (Church & Ryder 1972; Church & Slaymaker 1989).

The sediment transfer processes operating in the Kärkevagge catchment are similar to those operating in many mountain areas (Barsch & Caine 1984) (Table 2.6). The sedimentary characteristics of these processes reflect the source sediments within the catchment and the process dynamics that generate the subsequent sedimentation patterns. The overall picture is one of poorly sorted source sediments distributed across a heterogeneous landscape. Differentiation of these deposits predominantly occurs as a result of gravitational sorting by rock fall, frost sorting of susceptible soil and selective transport by fluvial processes (Table 2.6).

When evaluating models of this kind it should be kept in mind that Tables 2.5 and 2.6 and Fig. 2.9b show average values only for sediment fluxes and neglect the inherent interannual variability in such sediment systems and the longer term dynamics of change that have an impact on mountain environments. In order to evalaute such changes longer-term sediment budget models need to be developed. Such models can be developed over historical time-scales using archival evidence (Piégay et al. 2004); the late Holocene using detailed lake sediment records and geochemical analysis (Slaymaker et al. 2003) and soil geomorphology (pedology and weathering) (Birkeland et al. 2003). Furthermore, previous sediment-budget studies, such as the two examples provided in Table 2.5, can provide important baseline studies for later comparison with further sediment budgets or predictions of future change. Such an approach has been developed for Rapp's (1960) work in Kärkevagge in northern Sweden by Schlyter et al. (1993).

Although the sediment budget framework is important for establishing how mountain sediment systems operate and the relative importance of different sedimentological processes,

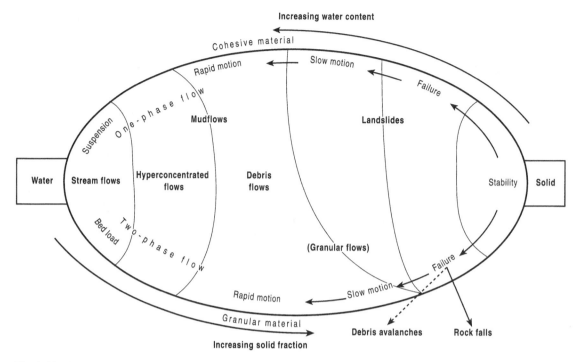

Fig. 2.10 Classification of mass movements and flows on steep slopes as a function of solid debris fraction and material type. (Source: Coussot & Meunier 1996; redrawn from *Earth Science Reviews*, Cousset, P. & Meunier, M. (1996) Recognition, classification and mechanical description of debris flows, **40**, 209–27, with permission from Elsevier.)

its usefulness for understanding environmental sedimentology is limited because most of the important events are large-scale mass movements which are highly episodic and inevitably site specific. Examples include localized flash floods, rock slides and volcanically triggered debris flows.

2.2.4 Sediment transfer processes

The magnitude and frequency of different sediment transfer processes operating in mountain environments vary according to the local or regional tectonic and geological setting. It is beyond the scope of this chapter to provide a detailed review of the environmental sedimentology of volcanic, seismic and tectonic studies. However, the coincidence between these variables and mountain areas is undeniable. For example, case studies later in this chapter clearly demonstrate links between seismic activity and debris flows (Case Study 2.1 Huascaran, Peru); volcanic

activity and sedimentation (Case Study 2.2 Mount St Helens); and sediment delivery in tectonically unstable regions (Case Study 2.3 Waipaoa, East Coast, New Zealand). Furthermore, these types of events have common characteristics in terms of very high rates of sediment production often coupled to catastrophic or rapid releases of water or runoff. This leads to a consideration of sediment–water flows because once a large mass of water, ice, snow or sediment is released on a slope it will immediately begin to flow (Pierson 1988). Such flows can rapidly entrain further material along their paths.

The behaviour of the flowing mass depends on the ratio of sediment to water. A very broad spectrum of flows is possible ranging from relatively dry (non-liquefied) granular flows of rock debris (debris avalanches) to large flood flows involving mostly water (Fig. 2.10). Eisbacher & Clague (1984) have classified destructive mass movements of this kind into five groups (Table 2.7):

Table 2.7 Examples of different types of sediment water flows and destructive mass movements in mountain environments.

Mass movement type	Location, date	Origin	Volume (m^3)	Velocity ($m\ s^{-1}$)	Travel distance (km)	Source
Debris flows from superficial deposits	Gamahara Torrent, Japan, 1996	Landslide–debris flow	5 to 10×10^4	16	4.6	Marui et al. (1997)
Debris flows from bedrock failure	Monument Creek, Grand Canyon, USA, 1984	Debris flow			4.5	Webb et al. (1988)
Mass movements on volcanoes	Mount St Helens, USA, 1980	Landslide	2.5 to 2.8×10^9	70 (maximum)	25	Pierson (1988)
Glacier-related mass movements	Kautz Creek, Mount Rainier, Washington, USA, 1947	Debris flow (glacier outburst flood)	38×10^6	4.5	9+	Driedger (1988)
Rock falls and rock avalanches	Elm, Switzerland, 1881	Rock avalanche	10×10^6	50	2	Heim (1882)
Flood	Skeiðarársandur jökulhlaup, Iceland, 1996	Glacier outburst	3.5×10^9	5–10	35+	Magilligan et al. 2002

- debris flows from superficial deposits
- debris flows from bedrock failure
- mass movements on volcanoes
- glacier-related mass movements
- rock falls and rock avalanches

Recognizing the dominant flow type that occurs during a mountain flood event is an important task. Differentiation of the types of water and sediment flows, whether they are water floods, hyperconcentrated flows or debris flows, can be based on the degree of sorting, composition, texture and sedimentary structures (Table 2.8, Costa 1988). Water floods involve turbulent Newtonian fluid flow with non-uniform sediment concentration profiles with sediment concentrations less than 40% by weight. Debris flows on the other hand are rheologically very different. They are non-Newtonian having laminar flow and uniform sediment concentration gradients, with sediment concentrations varying from 70 to 90% by weight. Hyperconcentrated flows are transitional between these two extremes and, as such, retain some characteristics of both water floods and debris flows (Costa 1988; Table 2.7).

Furthermore debris flows are often considered as an intermediate phenomenon between hyper-concentrated flows and landslides, both in terms of their initiation mechanism and dynamics. Coussot & Meunier (1996) propose a simple scheme for classifying mass movements and flows that occur on natural steep slopes. This is based on material type and proportion of solid in the moving mass (Fig. 2.10). Material types vary from fine, cohesive clays to coarse, cohesionless granular materials, and solid content generally increases from water flow, to hyperconcentrated flows, to debris flows and landslides. This is a useful summary of many of the main slope and valley processes that operate in mountain environments and how they are interrelated in terms of sedimentary continua, e.g. solid:water ratio (Table 2.7).

2.3 PROCESSES AND IMPACTS – EVENTS RELATED TO NATURAL DISTURBANCE

This section will focus on natural changes in mountain sediment systems, particularly glacier and slope systems, that make up the coarse debris and fine sediment components of the sediment cascade model (Caine 1974).

Table 2.8 Geomorphological and sedimentologic characteristics of water and sediment flows in channels (Source: Costa 1988; reproduced from *Flood Geomorphology*. Baker, V.R., Kochel, R.C. & Patton, P.C. (Eds). 1988. © John Wiley, New York. This material is used by permission of John Wiley & Sons, Inc.)

Flow	Landforms and deposits	Sedimentary structures	Sediment characteristics*
Water flood	Bars, fans, sheets, splays; channels have large width:depth ratios	Horizontal or inclined stratification to massive; weak to strong imbrication; cut and fill structures; ungraded or graded	Average Trask sorting coefficient 1.8–2.7; clast-supported; normally distributed; rounded clasts; wide range of particle sizes
Hyperconcentrated flow	Similar to water flood	Weak horizontal stratification to massive; weak imbrication; thin gravel lenses; normal and reverse grading	Φ graphic sorting 1.1–1.6 (poor); clast-supported openwork structure; predominantly coarse sand
Debris flow	Marginal levees, terminal lobes, trapezoidal to U-shaped channel	No stratification; weak to no imbrication; inverse grading at base; normal grading near top	Average Trask sorting coefficient 3.6–12.3; Φ graphic sorting 3.0–5.0 (very poor to extremely poor); matrix-supported; negatively skewed; extreme range of particle sizes; may contain megaclasts

*Trask sorting coefficient: calculated by dividing the 75th percentile by the 25th percentile of the grain size distribution. Φ graphic sorting: inclusive graphic standard deviation (in ϕ units) $= [(\phi84 - \phi16)/4] + [(\phi95 - \phi5)/6.6]$.

2.3.1 Glacier systems and environmental change

A characteristic feature of many mountainous environments is the presence of glacier ice. At present the distribution is restricted to the higher mountains and polar ice sheets but in the past was considerably more extensive. The legacy of past glaciations still conditions sediment transfer in most mountainous regions. This is manifest as a direct influence in currently glacierized regions and as an indirect control in glaciated areas where the impact of glaciers still significantly alters mountain sediment systems. The concept of paraglaciation (Church & Ryder 1972) provides a useful framework for understanding contemporary mountain sediment budgets and the disequilibrium that often exists between sediment production and delivery in previously glaciated areas (Church & Slaymaker 1989). Ballantyne (2002) provides a comprehensive review of periglacial geomorphology and the important concept of glacially conditioned sediment availability. He identifies six paraglacial land-systems and recognizes the significance of these sediment stores and sinks as sources of easily eroded sediment. Sediment stores in mountain environments are fundamental in controlling basin sediment yield and may be sensitive to environmental changes in climate and/or human disturbance. In addition to these long-term influences glacial processes can have short-term effects on mountain sediment systems and in some cases pose significant hazards. Three main hazards can be identified: glacier fluctuations, glacier outburst floods (jökulhlaups) and avalanches (Cooke & Doornkamp 1990). Hazards related to ice and snow are common in most glacierized high mountain areas. Their impact on society, however, depends on the degree to which human structures and settlements are developed in those regions. The European Alpine countries are particularly affected by glacier hazards owing to the combination of steep, unstable slopes and the proximity of infrastructure and villages to the glacial environment. Of particular importance is the sensitivity of the glacier environment to small changes in temperature and precipitation, which may considerably increase the risk for communities living near them. Outburst floods (jökulhlaups), landslides, debris flows and debris avalanches can destroy property and take lives. Given the uncertainties of recent environmental change it is not clear yet whether some of these hazards are a normal part of glacier behaviour, or whether they represent an evolving new threat from a changing cryosphere.

2.3.1.1 Fluvioglacial sedimentation and glacier outburst floods

A jökulhlaup is an Icelandic term used to describe a catastrophic flood caused by the sudden drainage of a subglacial or ice-dammed lake. Lakes may develop and drain seasonally or may build up over many years before being drained. Volcanic eruptions or geothermal heating under ice caps can often be the trigger for such periodic drainage, leading to catastrophic floods. The Grímsvötn area on Vatnajökull in southern Iceland is particularly prone to such activity, which is triggered by the Katla volcano under the ice cap of Myrdalsjökull (Gerrard 1990). At Grímsvötn water is stored in a subglacial lake located in a volcanic caldera of about 30–40 km^3 in the centre of Vatnajökull ice cap. Geothermal heating causes internal drainage in the ice cap, which causes the subglacial lake to rise by about 100 m over a period of five or six years. Eventually the ice dam is breached and water flows under the ice cap to emerge at the ice margin several tens of kilometres from the source. Floods can have peak discharges of up to 100,000 m^3 s^{-1}. Two main mechanisms lead to the release of stored water. Drainage may begin at basal water pressures less than the ice overburden pressure through the slow expansion of glacial conduits by melting of ice walls through frictional or sensible heat. Alternatively high subglacial lake levels lift the glacier off the bed along the flow path resulting in extremely sudden discharges. As a consequence of the water release river levels may rise by up to 10 m and millions of cubic metres of sediment are deposited, often raising sandur (alluvial plain) levels by several metres. Sediments are transported from subglacial sources and eroded from the proximal zone of the moraine/sandur complex (Magilligan et al. 2002). Björnsson (2003) estimates the sediment load of a large jökulhlaup may be as great at 10 × 10^6 t per event, but if this occurs at

the same time as an eruption this may rise by another order of magnitude. Similarly, Maizels (1997) distinguishes three types of proglacial sedimentary (outwash) deposits in relation to jökulhlaup type:

Type I – 'normal' braided river outwash
Type II – produced by sudden drainage of ice dammed lakes
Type III – associated with drainage during subglacial geothermal activity, with deposits resulting from high sediment concentrations and hyperconcentrated flows.

Each type of activity results in a distinct set of depositional landforms and sediments. In historical times, the Skeidarsandur jökulhlaup of November 1996 stands out as unprecedented in its magnitude and duration (Magilligan et al. 2002). The event reached a peak discharge of 53,000 m^3 s^{-1} in 17 hours and was responsible for widespread incision and aggradation with 3.8 × 10^7 m^3 of sediment being deposited in the proglacial depression of the sandur. Deposits of over 9 m depth were recorded. The total volume of water released from the glacier was estimated at 3.5 km^3, and was the most rapid jökulhlaup recorded for this area. The event also eroded large stretches of the main highway around Iceland, destroyed two bridges, and caused damage estimated at US$15,000,000 (Magilligan et al. 2002).

Because glacier outburst floods are sudden discharges from water bodies dammed within or at the margins of glaciers in steep glacierized environments the downstream impacts of such events can be devastating, leaving paths of total destruction. At Mount Rainier (4364 m) in the Cascade Range, Washington State, USA glacial outburst floods occur on a relatively small scale. Mount Rainier is a strato-volcano consisting of overlapping layers of lava and tephra. The mountain is topped by a summit ice cap, with 25 glaciers extending radially in all directions from the summit (Fig. 2.11a). Most of the activity on Mount Rainier is restricted to the glaciers on south-western slopes of the mountain due to the local geographical–climatological conditions

(a)

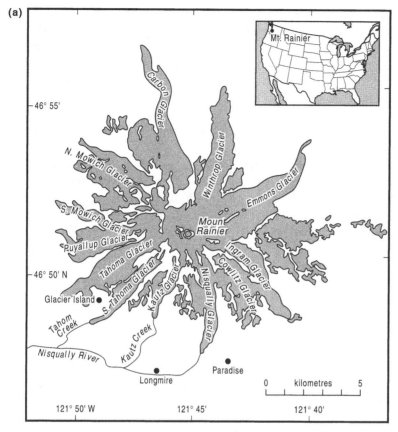

(b)

Fig. 2.11 (a) Location map of Mount Rainier, Washington Sate, USA showing principal glaciers. (b) Tree damage caused by debris flow activity in Kautz Creek (1989).

Table 2.9 Characteristics of glaciers susceptible to outburst floods – Mount Rainier, Washington, USA. (Source: Driedger & Fountain 1989.)

Glacier	Surface area (km^2)	Mean surface slope (degrees)	Number of floods in record up to 1988
Nisqually	4.6	25	9
South Tahoma	2.8	23	12*
Kautz–Success	1.8	29	5
Carbon	11.2	18	1
Winthrop	9.1	21	2

*Revised estimate (Walder & Driedger 1995) 22 floods since 1967.

favouring subglacial cavity formation: increased exposure to storms and more intense solar radiation (melt) (Driedger & Fountain 1989). At Mount Rainier water originates from snow and ice melt and liquid precipitation, and is stored within glacier cavities and at the bed of the glacier. Ice movements deform these cavities, resulting in the catastrophic release of the water. Hydrological studies of the South Tahoma glacier (Driedger & Fountain 1989) suggest that water is stored at the bed of the glacier and the flood magnitudes originating from such storage are in the order of 1×10^5 m^3. These floods threaten life and property because they occur without warning and quickly develop into rapidly moving debris flows as they entrain loose volcanic debris.

Kautz Creek (Fig. 2.11a) is the catchment affected by the largest recent event, which occurred on 2–3 October 1947. This was triggered by heavy rain and it was estimated that 38×10^6 m^3 of material was moved in the event, with some boulders up to 4 m in diameter being transported (Driedger 1988). The highway 9 km downstream from the glacier was engulfed by the flow, which deposited over 3 m of mud and debris. Since 1947 smaller significant floods have occurred in 1961, 1985 and 1986. Tahoma Creek shows clear evidence of frequent outburst floods from South Tahoma glacier. At least 22 outburst floods have been recorded since 1967, including 14 in the years 1986–1992 (Walder & Dreidger 1995) (Table 2.9). In 1967, 1971, 1973, 1986 and 1988 small floods triggered debris flows that destroyed trail bridges and campground-picnic areas. Recent activity has resulted in valley scour of 2 m in places, with boulders up to 0.5 m transported and local deposition exceeding 1 m

in places. On the Nisqually River (Fig. 2.11a), outburst flooding from the Nisqually glacier destroyed and damaged bridges in 1926, 1932, 1934 and 1955, resulting in a high-level bridge eventually being built. Since then large floods in 1968, 1970, 1972 and 1985 resulted in massive rearrangement of the stream bed but have left the bridge unscathed. Other activity has also been noted at the Winthrop glacier, and the Carbon, South Mowich and Emmons glaciers are also suspected of being susceptible to this kind of activity (Fig. 2.11a). The most recent activity reported from Mount Rainier was on 14 August 2001 and constituted a moderate debris flow in the Van Trump drainage within the Kautz Glacier area.

The outburst floods tend to occur in late summer or autumn; usually in the late afternoon or early evening, often during rainstorms. The flood waves have some common characteristics. They have been described as a noisy, churning mass of mud and rock. Local winds can develop along the flows and thick dust clouds can accompany the events. There is often a smell of freshly cut vegetation and chipped rock as boulders smash trees and collide with bedrock and other sediment (Fig. 2.11b). The flows are very rapid, often in excess of 4.5 m s^{-1}. Observers have said there is generally less than 2 minutes between hearing the flow and it passing the observer (Driedger 1988) (Table 2.7).

2.3.2 Seismically triggered slope instabilities

Landslides or rock falls occur when a surficial mass fails along a steeply dipping fracture plane. The rock mass breaks up and moves downslope, breaking into fragments and then continues as a slide or flow. Such mass movements have several common characteristics:

1 landslides occur in unstable, faulted and jointed rock masses;

2 a trigger mechanism is usually involved in setting them off;

3 flow within the landslide is complex – often involving several stages or modes of movement;

4 once deposited the mass is often still unstable and undergoes modification by other slope processes;

5 these are very rapid mass movements.

Seismically triggered slope instabilities have many of these features (see Case Study 2.1). For example, the Sherman Glacier rock avalanche in Alaska was triggered by the 1964 Alaska Good Friday earthquake (McSaveney 1978). The bedrock of this area consists of highly deformed slightly metamorphosed sedimentary sequences, which are heavily faulted and the fault planes are steeply dipping. The rock avalanche fell 600 m and then spread 5 km across the Sherman Glacier, depositing a blanket 3–6 m thick. The avalanche flowed as a large lobate mass with the particles behaving like loose aggregates, spreading as it

---

Case study 2.1 Huascarán, Yungay, Peru 1970

The greatest mountain landslide disaster recorded to date occurred in Peru following an offshore earthquake (magnitude 7.7) on 31 May 1970. The seismic event triggered a catastrophic rock and snow avalanche from the summit ice-cap of Huascarán Mountain (6654 m), the highest peak in the Peruvian Andes. The displaced mass fell vertically to the glacier below, where it generated a gigantic debris flow surge that travelled down-valley at a speed of over 70–100 m s^{-1} entraining sediment in its path (Case Fig. 2.1). On reaching the meltwater river of the Rio Sacsha the flow had changed into a mudflow of over 1 km wide, carrying gigantic boulders, some as great as 15 m^3. Eyewitnesses described the flow as a huge wave at least 80 m high (Whittow 1980). The flow travelled 15 km down to the confluence with the Rio Santa in just a few minutes.

The mountain has had a history of devastating landslides. A mudflow had swept down the Rio Sacsha in 1962 killing 4000 people and depositing 13×10^6 m^3 of material (Smith 2001). The 1970 mudflow was of much greater magnitude and completely overran the towns of Yungay and Ranrahirca, leaving 18,000 people dead or missing. It is estimated that $50–100 \times 10^6$ m^3 of material was involved, burying some settlements up to 10 m deep in sediment (Case Table 2.1).

This is not the only catastrophic mass movement event identified in this valley. A smaller avalanche also took place on 10 January 1962 and an earlier (Pre-Columbian) event, larger than the 1970 avalanche, is identified from deposits preserved on the valley floor (Case Table 2.1).

Case Table 2.1 Summary characteristics of selected data from three of the largest avalanches observed from Nevados Huascarán, Peru. The Pre-Columbian event is identified from deposits preserved outside the limits of the 1970 event. (Source: Plafker & Ericksen 1978.)

	Pre-Columbian	10 January 1962	31 May 1970
Area covered (km^2)	> 30	6	22.5
Volume (10^6 m^3)	100–200*	> 13	50–100
Average velocity (km h^{-1})	315–355†	170	280
Runup height at Rio Santa (m)	123	30	83
Velocity at Rio Santa‡ (km h^{-1})	> 140	> 60	> 120

*Estimated from the extent and nature of deposits.
†Derived by comparison with historic avalanches.
‡Calculated from runup height assuming avalanche front thicknesses of 15 m (1962), 30 m (1970) and 45 m (Pre-Columbian).

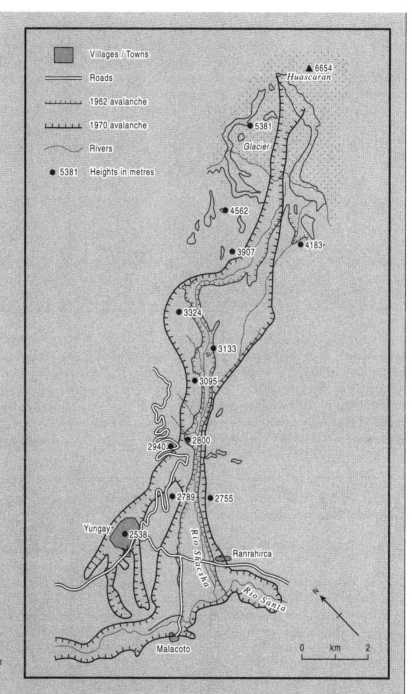

Villages / Towns

Roads

1962 avalanche

1970 avalanche

Rivers

● 5381   Heights in metres

▲ 6654
Huascaran

● 5381

Glacier

● 4562

● 3907

● 4183

● 3324

● 3133

● 3095

● 2940   ● 2800

● 2789   ● 2755

Yungay   ● 2538

Ranrahirca

Rio Shacsha

Rio Santa

Malacoto

0   km   2

Case Fig. 2.1 Map of the Rio Shacsha and Nevado Huascará area, Peru showing the runout tracks of the 1962 and 1970 avalanches. (Modified from Whittow 1980.)

Although details of the Pre-Columbian event are less well known and no estimates of loss of life are available, the 1962 event destroyed nine towns including most of Ranrahirca (Case Fig. 2.1), killing approximately 4000 people. Although the trigger mechanism for this event is not known the avalanche appears to have originated as an icefall from the same part of the west face of the North Peak of Huascarán. Based on evidence from deposits this was predominantly a rock avalanche.

Observation of the summit area of Nevados Huascarán suggests that there is still considerable potential for further destructive avalanches. Slopes in the source area remain oversteepened and show signs of recent cracks parallel to the avalanche scar. Therefore the potential for further disasters remains. Although the settlements are nearly 23 km from the source of the debris flows, given the steepness of the mountain slopes, travel times are less than four minutes (Plafker & Eriksen 1978). Even if a warning could be given evacuation to safe ground in such a short time period is not a viable option. Therefore a more viable option would be the relocation of settlements in the Santa valley to positions outside the maximum extent of the Pre-Columbian avalanche. This, however, would require a major initiative in development planning.

Relevant reading

Plafker, G. & Eriksen, G.E. (1978) Nevados Huascaran avalanches, Peru. In *Rockslides and Avalanches*, Vol. 1. *Natural Disasters* (Ed. B. Voight), pp. 48–55. Elsevier, Amsterdam.
Smith, K. (2001) *Environmental Hazards – Assessing Risk and Reducing Disaster*, 3rd edn. Routledge, London.
Whittow, J. (1980) Landslides and avalanches – avalanches. In *Disasters: the Anatomy of Environmental Hazards*, pp. 163–70. Penguin, Harmondsworth.

went. Because of the high speed of this flow and the mechanical fluidization through jostling of the clasts, flow was very plastic (high-viscosity–low-yield stress). High viscosity prevented turbulence and the flow was almost laminar. Because little energy was used up in internal deformation most energy was lost to friction at the base, thereby eroding the substrate.

Another example is the 1980 Mount St Helens debris avalanche in Washington. This was caused by the eruption of 18 May (Case Study 2.2). It began catastrophically with a large lateral blast

Case study 2.2 Impact of an extreme tectonic/volcanic event on a mountain sediment system – Mount St Helens eruption 1980

The 18 May 1980 eruption of Mount St Helens in south-west Washington was a major geological disaster claiming 57 lives and affecting several hundred thousand people. A magnitude 5+ earthquake triggered a lateral blast, which sent a huge mass failure down the northern flank of the mountain eventually exploding in a cloud of ash, rock and hot gas, sending ash up to 18 km vertically into the atmosphere. The avalanche of rock, ice and mud released in the earthquake surged 28 km down the North Fork Toutle River valley (Case Fig. 2.2a & b). A second part of the slide flowed into South Coldwater Canyon and Spirit Lake raising the lake level by 70 m and damming the outlet with debris over 100 m deep. Downstream mudflows choked the Cowlitz and Toutle rivers bringing shipping to a halt on the Columbia River (Case Fig. 2.2a).

The immediate engineering response to the Mount St Helens disaster was undertaken by the US Army Corps of Engineers, who were responsible for navigation and flood control on the Columbia River. The worst affected area was the Cowlitz River on highway 15 where efforts were concentrated on raising river levees and roads and clearing channel constrictions (Case Fig. 2.2a). Overnight the navigation channel on the Columbia River was reduced in depth from 12 to only 4.2 m. Using large dredges by 23 May the channel was partially cleared and by

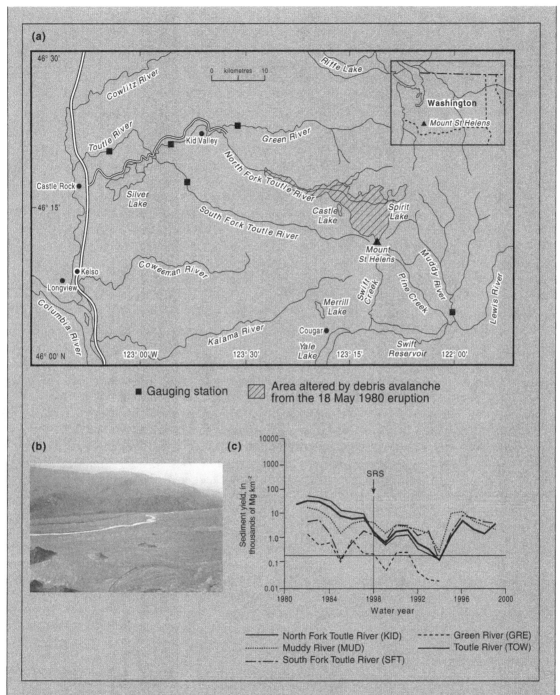

Case Fig. 2.2 Impacts of the 1980 Mount St Helens eruption. (a) Location map showing main river systems. (b) North Fork Toutle River Valley, 1989. (c) Annual suspended sediment yields in Mount St Helens rivers. The vertical line (SRS) is the location of the sediment retention structure. The horizontal line is mean sediment yield value for western Cascade Range rivers (Source: Major et al. 2000; reproduced with permission *Geological Society of America*.)

November the shipping lane was fully restored. Temporary dams and sediment storage pits were excavated in the winter of 1980 in the Toutle River to hold back sediment delivery from the upper catchment, and dredging operations were completed in May 1981. These temporary structures have now been overwhelmed by aggradation of the river (Case Fig. 2.2b).

By 1982 lake waters at Spirit Lake were rising dangerously high behind the eruption debris dam. By 19 August 1982 a State of Emergency was declared at Spirit Lake and a full-scale Federal response was initiated. In November 1982 water was being pumped from Spirit Lake to immediately reduce water levels, however, in order to permanently drain the lake to safe levels a water transfer tunnel connecting the lake to South Coldwater Creek was built and this was operational by 6 May 1985.

In order to produce a long-term solution and management strategy for the sediment problem the Mount St Helens Sediment Retention Structure (SRS) was designed and built. In 1986 work began on the construction of the SRS on the North Fork Toutle River. The purpose of the structure was to stop the advance of debris avalanche deposits from the 1980 volcanic eruption moving downstream and causing long-term navigation problems on the Toutle, Columbia and Cowlitz rivers. This was one element in a three part solution of long-term sediment management; this also involved river levee construction and a dredging programme. The US Army Corps of Engineers initiated the project in 1986 and impoundment began in November 1987. The aim of the sediment retention structure was to dam sediment not water. The structure reduces the flow to such an extent that sediment is deposited naturally in the upstream side of the structure and does not migrate downstream causing flooding problems and shipping hazard. The SRS consists of a 600 m wide embankment standing nearly 56 m above the pre-eruption stream bed. The embankment is constructed of crushed and fractured rock with an impervious core of clay. The embankment rests on gravel, and water passes underneath and into the embankment as the lake rises. The outlet works occupy a large central block, which consists of six rows of five outlet pipes through which water and fish can pass into the plunge pools and downstream outlet channels. The pipes are closed off permanently as the level of sediment rises in the impounded lake upstream. Once the conduits are fully blocked the river will flow continuously over the wide unlined spillway, which is approximately 38 m above the original stream bed. The structure is sited about 35 km downstream of the volcano.

The impounded area is estimated to have a sediment retention capacity of approximately 198×10^6 m^3 of sand and gravel and is expected to fill in 50 years. In 2000 the structure was approximately 30% full. The rate of sediment retention varies greatly from year to year. For example, in the wet winters of 1996 and 1997, the higher than average discharges resulted in 23.5×10^6 m^3 of sediment being deposited behind the structure. This was nearly four times the amount trapped in the previous two years.

Relevant reading

Major, J.J. (2003) Post-eruption hydrology and sediment transport in volcanic river systems. *Water Resources Impact* 5, 10–15.

Major, J.J., Pierson, T.C., Dinehart, R.L., et al. (2000) Sediment yield following severe volcanic disturbance – a two decade perspective from Mount St Helens. *Geology* 28, 819–22.

Major, J.J., Scott, W.E., Driedger, C., et al. (2005) Mount St. Helens erupts again: activity from September 2004 through March 2005. *U.S. Geological Survey Fact Sheet FS2005-3036.* http://vulcan.wr.usgs.gov/Volcanoes/MSH/Publications/FS2005-3036/FS2005-3036.pdf (accessed July 2005).

to the north of the mountain followed by the eruption. Erupted debris and the collapse of the north side of the mountain, together with melting of the summit icefields, resulted in pyroclastic flows off the summit cone (lateral flow of hot gases and unsorted volcanic fragments). This initial surge deflated and was translated into a ground-based debris flow, which continued as a debris avalanche down the North Fork of the Toutle River. This became a debris flow and gradually evolved into a hyperconcentrated flow as it moved downstream (Tables 2.7 & 2.8).

2.3.3 Natural and climatically induced slope failures

Mass wasting of slopes in mountain environments proceeds by a combination of small-scale processes and infrequent large-scale events (see Case Study 2.3). In the Swiss Alps debris flows are a widespread phenomenon occurring in all altitudinal belts, although they are especially common in the periglacial zone. Debris flow starting zones tend to occur in poorly consolidated debris on slopes or in gullies. The association

---

Case study 2.3 Impact of an extreme climate event on the hillslope sediment system – Cyclone Bola (New Zealand)

In 1998 Cyclone Bola had a major impact on much of the North Island, New Zealand with the worst affected area being on the east coast (Case Fig. 2.3c). Cyclone Bola was the largest storm event in the Waipaoa catchment that had occurred since European settlement in the 1830s. Between 6 and 9 March up to 900 mm of rain fell in the north-east of the catchment (Page et al. 1999). The Waipaoa is soft-rock (Cretaceous greywackes and Miocene–Pliocene siltstones and sandstones) hill country on the east coast of the North Island, New Zealand (Case Fig. 2.3b). This is a tectonically active area with an uplift of approximately 3 mm yr^{-1}. The storm was estimated to be a 100 year return period event. The main impact of the storm was extensive erosion of hillslopes and massive sediment transfer along the valley systems. This is not the only major erosional event in this area. Deforestation and conversion to pasture in the wake of European settlement initiated a major phase of instability in the catchment. Similarly large storms over the past 100 years have also had major impacts on erosion and sedimentation in the watershed (Trustrum et al. 1999).

A short-term sediment budget was constructed by Page et al. (1994) to assess the response of Cyclone Bola on the smaller Tutira catchment (3208 ha) to the intense rain storm event (Case Fig. 2.3a). The budget quantifies the total amount of sediment generated during the event and the relative contribution from different erosion processes. Sediment storage is also estimated along with the amount of sediment discharge into two lakes within the catchment. A total of 1.35 ($\pm$ 0.13) million cubic metres of sediment was moved during the storm (420 m^3 ha^{-1}). Of this total, 21% was stored on the hillslopes, 22% deposited on the valley floors, 51% was deposited in the lakes and the remaining 6% was discharged at the catchment outlet. Approximately 89% of the sediment generated during the storm was from landslide erosion on the slopes. Channel, gully and sheet erosion was only a minor component of the budget (Case Fig. 2.3d).

Because of the dominance of landslides in the Cyclone Bola event work has been undertaken to characterize the significance of this in sediment delivery (Page et al. 1999; Trustrum et al. 1999). Page et al. (1999) developed a method for assessing sediment production from landsliding in the Waipaoa catchment and applied this to the Cyclone Bola event with the aim of determining the contribution of landslides to suspended sediment output from the event (Case Fig. 2.3b). Using a geographical information system (GIS) containing the distribution of

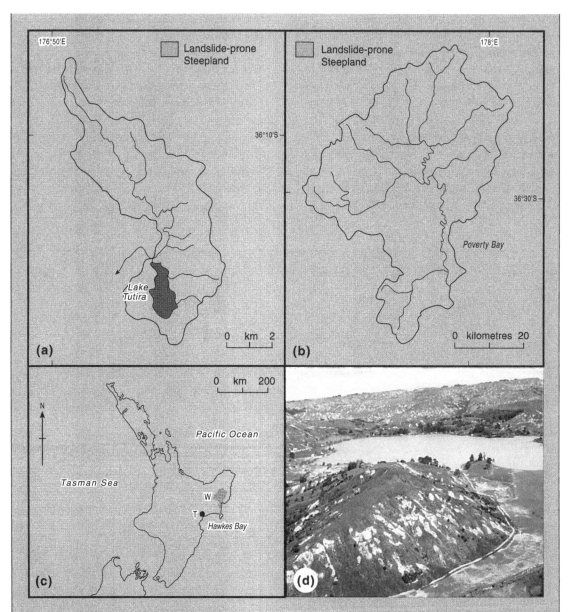

Case Fig. 2.3 (a) Tutira catchment; (b) Waipaoa River basin; (c) location of the two study areas (T, Tutira; W, Waipaoa) in North Island, New Zealand; and (d) Lake Tutira and the adjacent landslide-prone slopes, showing the 1988 Cyclone Bola storm-triggered landslides and floodplain sedimentation (photograph courtesy of N. Trustrum). (Source: Trustrum et al. 1999; reproduced with permission *Gebr. Borntraeger Verlagsbuchh.*)

landslide terrain units, vegetation and storm characteristics, relationships established between storm rainfall and landslide frequency were used to estimate landslide density across the entire 2205 km² catchment (Case Table 2.3). Each land system group has a characteristic rock type, suite of landforms and erosion processes, drainage density and channel morphology. Sixteen have been identified but only the six listed in Case Table 2.3 are prone to landslides. The Te Arai land system is most prone to landsliding, being made up of very weak bedrock, steep slopes with broken surface soil structure and steeply incising channels.

Case Table 2.3 Summary of sediment production and delivery produced from landslides in the Waipaoa catchment (North Island, New Zealand) during Cyclone Bola 1988. Estimates based on storm isohyet data. (Source: Page et al. 1999.)

Land system	Percentage area of total catchment	Percentage pasture*	Sediment generation (m³)†	Percentage of total	Percentage contribution to suspended sediment load
Te Arai	23	96	20,053,000	61	39
Wharerata	12.5	75	6,809,000	21	13
Waihora	3.5	76	577,000	2	1
Wharekopae	20.5	92	1,978,000	6	4
Makomako	6	84	2,493,000	7	5
Waingaromia	1.5	89	948,000	3	2
Total	148,480 ha		32,858,000	100	64

*As of 1988 – excludes indigenous forest, scrub and exotic forest > 8 years old.
†Adjusted for forest, scrub and soil conservation plantings.

Shallow landslides are responsible for approximately 64% of the load at the exit from the catchment. In respect of longer term suspended sediment yield, however, landslides contribute only 10–19% of the load (Page et al. 1999). Erosion of stored sediment in tributaries is an important control on longer term sediment production (Marutani et al. 1999). In the few years immediately following an event of this magnitude suspended sediment concentrations are 100% greater than in the years preceding the event owing to continued erosion of landslide scars and stored sediment (Marutani et al. 2001).

Relevant reading

Marutani, T., Brierley, G.J., Trustrum, N.A., et al. (2001) *Source-to-sink Sedimentary Cascades in Pacific Rim Geosystems*. Matsumoto Sabo Works Office, Ministry of Land, Infrastructure and Transport, Japan.

Page, M.J., Trustrum, N.A. & Dymond, J.R. (1994) Sediment budget to assess the geomorphic effect of a cyclonic storm, New Zealand. *Geomorphology* 9(3), 169–88.

Page, M.J., Reid, L.M. & Lynn, I.H. (1999) Sediment production from Cyclone Bola landslides, Waipaoa catchment. *Journal of Hydrology (New Zealand)* 38(2), 289–308.

Trustrum, N.A., Gomez, B., Page, M.J., et al. (1999) Sediment production, storage and output: the relative role of large magnitude events in steepland catchments. *Zeitschrift für Geomorphologie* N.F. Supplement 115, 71–86.

between debris flow activity and ground ice means that the formation of debris flows is susceptible to climate change, especially in marginal periglacial areas (Zimmermann & Haeberli 1992). In particular, since the Little Ice Age (AD 1450 to 1890) glacier retreat, disappearance of perennial snow patches and degradation of lowlying permafrost have resulted in the exposure of thawed debris on steep unstable slopes. Continued warming will result in an expansion of this zone. Changing climate also influences storm characteristics and changes the temporal and/or spatial pattern of precipitation and snow deposition.

An increased incidence of large-scale debris flows from within the Alpine periglacial belt poses a threat to mountain villages and infrastructure as many debris flows will be large enough to reach the valley floor.

In the summer of 1987 the Alpine countries were devastated by a number of floods. In Switzerland eight people were killed and damage exceeded 1.3 billion Swiss Francs (*c.* 1 billion US$). Heavy summer thunderstorms followed a long winter of extended snow cover and a cold/wet spring. The worst affected areas were south-east Switzerland, on 18–19 July and again on 24–25 August. Over 600 debris flows occurred during this period and because most of the precipitation fell as rain even at high altitudes, over 60% of the debris flows, including 82 large events (volumes > 1000 m^3), originated above 2300–2400 m a.s.l. in the periglacial belt; on the lower slopes conditions remained stable. Rainfall intensities were relatively low, however the period of precipitation was long and most debris flows were reported during or shortly after the maximum downpour. Starting zones were typically 28° to 33° for slope-type debris flows, but a substantial proportion (25%) were initiated in torrents, gorges and rocky ravines (Haeberli et al. 1990; Rickenmann 1990; Zimmermann & Haeberli 1992; Rickenmann & Zimmermann 1993).

The event has been investigated in detail by an interdisciplinary team of Swiss scientists examining the causes and impacts of the storms (Haeberli et al. 1990).

1 Debris flows were initiated in thick and extremely loose debris and bedrock. No sedimentary structure could be considered a factor that limited the depth of erosion. Erosion depth was more likely caused by the dynamics of the flow. The sediments affected appeared to be highly permeable and hydraulically non-homogeneous. The widespread occurrence of such material in starting zones makes these areas prone to further instabilities (Haeberli et al. 1990).

2 The regions affected by the debris flows are mainly located in crystalline bedrock covered in massive and stratified, loose glacial and talus deposits with poor sorting and low clay and limited silt contents (Roesli & Schindler 1990).

3 Most of the debris flow activity in 1987 occurred on unstable slopes or in gullies predisposed to instability. The spatial distribution of debris flows shows a distinct concentration in the periglacial zone as well as small clusters in small tributary valleys. The occurrence of the events could not be predicted from a knowledge of meteorological conditions alone (Zimmermann 1990).

Zimmermann & Haeberli (1992) did a comparison of the probability of permafrost occurrence in the 1987 debris-flow starting zones and conditions at the same locality in

Table 2.10 Comparison of the probability of permafrost occurrence in the 1987 debris flow starting zones for contemporary and Little Ice Age conditions in the Swiss Alps. (Source: Zimmermann & Haeberli 1992.)

Occurrence of permafrost in areas of 1987 debris flows	Percentage of all cases ($n = 82$)	
	1850	1987
Certain	6	6
Probable	27	9
Possible	23	21
Unlikely	44	64

the Little Ice Age. Table 2.10 indicates that the 1987 debris flows predominantly occurred in localities where permafrost is marginal or absent. Significantly in about 20% of cases conditions over the historical period changed from probable to possible permafrost occurrence and in such areas permafrost degradation and thaw instability are likely to have taken place prior to the 1987 debris flows.

Another form of mass movement which can have a large impact in developed mountain countries are massive rock-slides. The 1991 Randa rock slide in southern Valais, Switzerland caused extensive environmental damage and a heavy economic burden. The rock slide occurred in the tourist valley leading to Zermatt and the Matterhorn (Fig. 2.12). Approximately 30×10^6 m^3 of rock fell from a south-east facing rock face near the village of Randa approximately 10 km north of Zermatt. The rock fall occurred in two main parts: on 18 April 1991 and again on 9 May 1991. There were no fatalities except for loss of some livestock and 31 chalets were buried in the events (Quanterra 2003). The first event occurred without warning and caused major disruption and devastation to the valley infrastructure. The first event deposited approximately 20×10^6 m^3 of rock and initial assessments showed there was still a considerable mass of unstable rock in place. This prompted authorities to install a seismographic and geodetic warning system. As a result the second event was forecast from field evidence and geodetic and seismic surveys, and the area was successfully evacuated. Prior to the second slide geodetic displacement

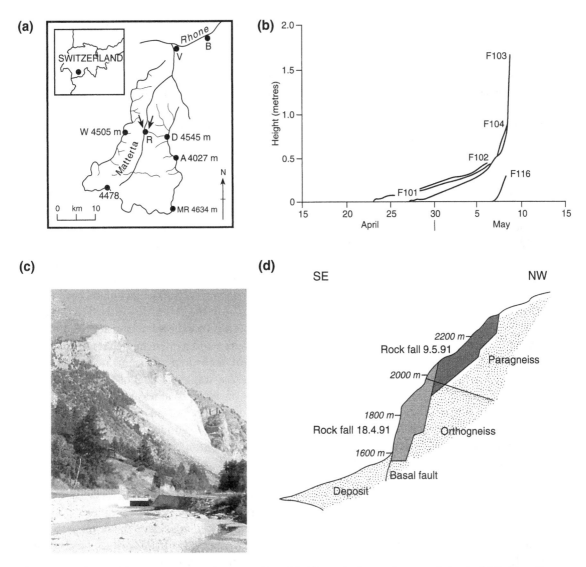

Fig. 2.12 Randa rock slide. (a) Location map showing position of the Mattertal valley, south-western Switzerland (MR, Monte Rosa; A, Allalinhorn; D, Dom; R, Randa; W, Weisshorn; V, Visp; B, Brig). (b) Records of heights of displacement based on geodetic measurements taken in the central part of the sliding rock mass (Source: Götz & Zimmermann 1993). F101, etc., refer to specific monitoring locations on the slide. (Reproduced with permission of The Japan Landslide Society.) (c) Photograph of the Randa rock slide in 2003. The entrance to the River Vispa bypass tunnel is shown in the foreground. (Photograph courtesy of R.M. Johnson.) (d) Cross-section through the rock face and rock slide deposits. (Source: Quanterra 2003.)

measurements indicated an accelerated motion of the rock mass (Götz & Zimmermann 1993) (Fig. 2.12b). During the evening of 9 May 10 × 10^6 m^3 of rock was deposited in a series of slides (generally < 1 × 10^6 m^3) over a period of 7 hours. Deposits covered approximately 0.8 km of the Zermatt–Rhône Valley railway line and about 0.2 km of the main road. The deposit caused a rock-slide dam across the Vispa River

resulting in 30 houses being flooded. Dust was deposited to a depth of 10–40 cm in a 1 km radius around the slide area and some local housing was buried in debris up to 60 m deep (Götz & Zimmermann 1993).

Formation of the rockslide-dammed lake in the valley posed two major hazards: local flooding and a sudden dam break flood downstream. Remedial measures were taken to keep lake levels as low as possible through pumping and excavation of a drainage trench. Eventually a 3.7 km long bypass tunnel for the Vispa River was completed (Fig. 2.12c). Following the event, road and rail connections were rerouted away from the site. Costs of works and surveys exceeded 110 million Swiss Francs (*c.* 88 million US$).

The rock slides were produced by structural weaknesses in the valley-side rock slopes. Relief joints parallel to the surface developed up to 200 m below the surface. At the base of the slope a steep continuous fracture traversed the slope and three joint sets divided the face into a large block (Fig. 2.12d). The face was loaded from above by pre-existing instabilities which also increased water infiltration into the rock mass, promoting weathering and build-up of porewater pressures. The rock's mass was estimated to have 7 to 15% void content (Schindler et al. 1993). Prior to the event small rock-falls had been observed with the final 1991 failure occurring during a snowmelt period. Jets of water were observed near the basal slip-plane just before and during the vent. Rock slides of this magnitude are relatively rare, and over the past 100 years in the Alps only a dozen events of similar size have occurred.

Events such as this in Switzerland, although infrequent, are not unusual. One of the most famous historic landslide events is the Elm landslide in 1881 (Heim 1882). This developed in unstable, highly deformed sedimentary rocks. Local slate mining was thought to be partly responsible for the failure. The mass movement occurred in three stages. Following failure the rock-fall debris hit a rocky ledge on its steep descent, the debris then cascaded like a waterfall on to lower slopes where the main mass continued downslope towards the village of Elm. A small part of the landslide continued under its own momentum and surged 100 m up the opposite hillslope. The landslide lasted an estimated 40 seconds and travelled about 2 km. The total volume is estimated to have been about 10 million m^3. Although termed a landslide the debris did not slide but flowed as a granular mass (sediment-gravity flow) with a dispersive stress developing as a result of grain collisions in the mobile mass. There were 115 fatalities and local communities were devastated by the event.

2.3.4 Volcanically triggered mass movements and long-term sediment delivery

Sediment delivery following major volcanic events is conditioned by the nature and extent of sediment deposited from the initial eruption and how these are reworked over time. An important element is how geomorphological processes act to determine the local dominance of sediment transfer processes. Figure 2.13 shows a schematic diagram of the changing dominance of different sediment transfer processes with distance from the Mount St Helens crater following the 1980 eruption (Scott 1988). The section shows a 60 km transect from the crater down the Toutle River. Over this distance there is a transition from pyroclastic surge and avalanche behaviour to debris flows and finally hyperconcentrated streamflow. In order to assess the impact of a mass movement event of the like seen at Mount St Helens a useful way of classifying catastrophic mass movements is in terms of the runout distances (Siebert 1984). A suitable relationship for describing the behaviour of different large-scale mass movements types is the ratio of vertical drop to travel distance (H/L). Figure 2.14 compares various types of mass movement using this simple relationship. Although there is considerable scatter in the data it is clear that volcanic landslides have greater travel distances, with some pyroclastic flows and lahars travelling great distances ($H/L < 0.02$). Hsu (1975) developed an index to describe the excessive travel distance (L_e) of large mass movements that travel distances greater than the maximum point expected by sliding of a body with a 0.62 friction coefficient (the value of 0.62 is the friction coefficient applicable to sliding of a rigid block). The relationship can be expressed as $L_e = H/0.62$. In terms of Mount St Helens the distances travelled by the pyroclastic flows and lahars were excessive, resulting in an extensive tract of devastation. This special quality of extended runout means that large-scale mass failures in volcanic mountainous terrain should be considered differently in terms of their sedimentology.

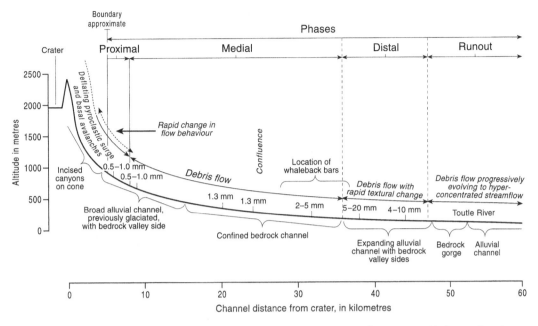

Fig. 2.13 Schematic diagram showing the changing dominance of different sediment transfer processes with distance from the crater following the 1980 eruption of Mount St Helens. Variation in dominant grain-size (mm) of the deposits with distance away from the crater is also shown. (Source: Scott 1988; Courtesy of U.S. Geological Survey.)

Fig. 2.14 Relationship between travel distance (*L*) and vertical drop (*H*) of large landslides. The value 0.62 is the friction coefficient applicable to sliding of a rigid block. (Reprinted from *Journal of Volcanology and Geothermal Research*, Siebert, L. (1984) Large volcanic debris avalanches: characteristics of source areas, deposits and associated eruptions, **22**(304), 163–97.)

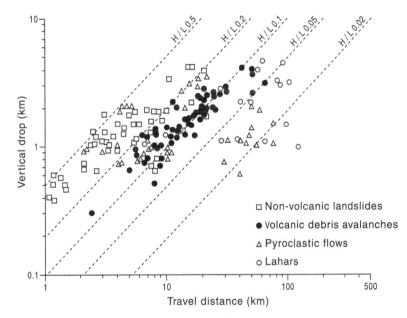

Along active continental margins earthquake and volcanic activity can result in greatly enhanced sediment loads. At Mount St Helens in the four months following the eruption the Cowlitz River (a tributary of the Columbia River, see Case Study Fig. 2.2) had a sediment load of 140×10^6 t, an amount that needs to be compared with the estimated normal annual load of the Columbia of 10×10^6 t. In the years following the eruption the Columbia had an annual sediment discharge of 35×10^6 t (Meade & Parker 1985). Dramatic post-eruption changes in channel morphology and reworking of volcanic debris resulted in sediment yields that are two orders of magnitude greater than pre-eruption levels (Major et al. 2000; Major 2003). At Mount St Helens post-1980 sediment yields declined non-linearly for a decade but increased abruptly in response to higher than normal runoff in the late 1990s (Major 2003). Even after 20 years some drainage basins still have sediment yields 10 to 100 times greater than the pre-eruption levels. Although sediment sources in small drainage basins tend to stabilize fairly rapidly (Collins & Dunne 1986), sediment stored in larger river valleys remains active and is unlikely to be stabilized for several decades (Major 2003).

2.4 PROCESSES AND IMPACTS – ANTHROPOGENIC INFLUENCES

A natural hazard can be defined as 'A physical event which makes an impact on human beings and their environment' (Alexander 1998). Mountainous environments are highly active geomorphologically and increasing human use of mountains has led to an increase in natural hazards (Hewitt 2004). In the context of fluvial hazards in mountain environments 'Mountain rivers become a hazard only when they threaten human life or property, by inundation, erosion, sediment deposition or destruction' (Davies 1991). Although a large range of fluvial hazards occur (e.g. glacial outburst floods, Alpine debris flows, etc.) they generally result from extreme temporal and spatial variability in fluvial and hydrological processes. Because of these characteristics mountain environments pose unique problems for hazard assessment, prediction and mitigation. More data are required on the frequency and magnitude of hazardous events in mountain regions.

This section considers two main examples of human interaction with mountain sediment systems. The first outlines the debate surrounding human-induced accelerated soil erosion in a steep upland environment – deforestation in Nepal. The second examines human infrastructure construction in an unstable mountain environment and discusses issues regarding the Karakoram Highway.

2.4.1 Anthropogenic impact on upland sediment systems – deforestation in Nepal and the 'Himalayan environmental degradation theory'

The so-called Himalayan environmental degradation theory (HEDT) as proposed by Ives & Messerli (1989) neatly illustrates how changes in population structure and pressure (human impacts) on a mountain environment can lead to a change in the natural sedimentary system (Fig. 2.15). What is most significant about the theory is the short time-scale over which it appears to have developed and the very large scale of the mountain area it potentially affects. This qualitative model (Fig. 2.15) is deceptively simple and can be summarized as follows (Gerrard 1990):

• Population growth. Introduction of new health care from 1950 has produced rapid population growth. This population explosion has been amplified by migration from the Indian plain and Nepal with consequent greater demands on fuel, construction materials, fodder and agricultural land.

• Deforestation. Population expansion results in massive deforestation.

• Soil erosion and landslides. Deforestation on marginal and steep mountain slopes leads to catastrophic soil erosion and landsliding with a break down in the normal hydrological cycle.

• Increased runoff and flooding. The change in slopes has resulted in increased runoff during the summer monsoon with increases in flooding and sedimentation in the Ganges and Brahamaputra rivers resulting in the extension of these great river deltas (Fig. 2.15, macro-level).

• Positive feedback – accelerated deforestation and greater soil loss. Loss of agricultural land in the mountains results in greater pressure and more deforestation and a switch to animal dung as fuel, depriving the land of much needed fertilizer.

• Degraded soil structure. Crop yields decline and soil structure is degraded leading to further soil erosion and regolith instability.

Although conceptually attractive this simple model is fraught with difficulties. These relate

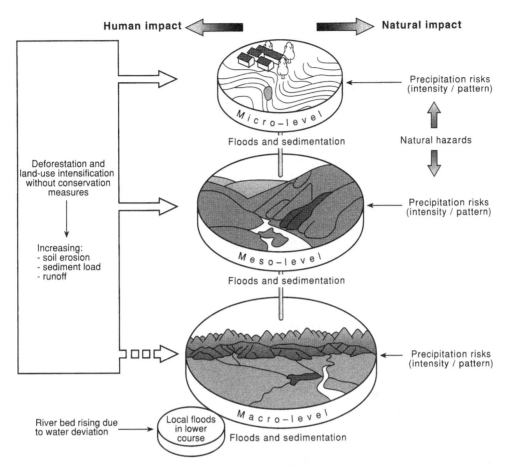

Fig. 2.15 General summary of the key elements of the Himalayan environmental degradation theory as proposed by Ives & Messerli (1989). Schematic diagram showing the relationship between human-induced and natural processes at the micro-, meso- and macro-levels. (Source: Ives & Messerli 1989; reproduced with permission from *The Himalayan Dilemma – Reconciling Development and Conservation*. Ives, J.D. & Messerli, B. (1989). Routledge, London.)

to the uncertainties attached to gathering hard evidence to test the model. Gerrard (1990) provides a useful review of the main criticisms of the theory and poses several key questions requiring new knowledge on the rates of deforestation, the extent of deforestation, the main forms of soil erosion and landsliding, the spatial extent of erosion, the nature of the river sediment load and whether the observed changes are the result of land-use change or natural variability in the sediment system.

Environmental sedimentology is ideally suited to answering such problems through better understanding of the impact of human actions on the sedimentary system. For example, Laban (1978, 1979) has shown that very high soil erosion rates

in Nepal can be tolerated within the landscape and half the landslides are from natural causes. Ironically areas least affected by landsliding are the heavily populated and terraced agricultural systems. In the lesser Himalaya evidence for accelerated erosion has been found from studies of the small Nana Kosi watershed where agriculture and deforestation have increased erosion rates by a factor of five to ten (Rawat & Rawat 1994). Such discrepancies in erosion-rate evidence from site-specific locations are hard to accommodate in a general model, however, overall the evidence generally points to rates of soil erosion that are generally less than predicted and the sediment system is fairly robust under such high rates of change.

A major part of the problem is assessing the representativeness of the available data on rates of natural and human-induced erosion in space and time, especially the relative and absolute importance of rare high-magnitude events compared with continuous slow mass movements (Ives & Messerli 1989). In general there is no substantive evidence that human activities such as deforestation, subsistence farming, irrigation and road construction have had a major impact on soil erosion on the large scale. Locally such processes may be a problem but at the large scale the overwhelming importance of geophysical and climatic controls on the delivery of sediment from the rapidly uplifting and eroding mountains to the Ganges–Brahmaputra plains is paramount (Ives & Messerli 1989). Incorrect diagnosis of the cause of the large-scale problem could result in costly but unproductive changes in land management, the longer term effects of which would be hard to predict.

2.4.2 Road construction in mountain environments – the Karakoram Highway

Increased access to mountain areas usually involves the construction of roads and highways for the passage of motor vehicles. This brings with it a suite of special engineering problems (Fookes et al. 1985). Gerrard (1990) outlines five stages in mountain road construction which usually involve: feasibility and project planning; reconnaissance of road corridors; site investigation of the road alignment; construction; and post-construction maintenance and observation. Routing a road through a steep, unstable mountain environment presents major challenges to the highway engineer at each of these stages. For example, if a road is planned across a talus slope then the alignment should be across the lower slope because these areas have the lower slope angles and larger sediment sizes, which provide greater strength during excavation. Furthermore rock-fall hazard reduces towards the toe of the slope. Alternatively if a road has to cross a mudflow zone the road should be as high on the slope as possible where mudflows are thinnest (Gerrard 1990).

Fig. 2.16 The Karakoram highway in the Hunza Valley Pakistan. Frequent road maintenance is required to keep the road open and free of debris. (Photograph C. Warburton.)

Perhaps the best known and most widely studied mountain highway is the Karakoram Highway in the Hunza valley, Pakistan (Fig. 2.16). The geographical setting makes road construction and maintenance a continuing problem owing to:
1 an overdeepened glaciated valley with very steep unconsolidated valley slopes;
2 an unstable valley floor with very varied terrain consisting of moraine ridges, outwash fans, gorge sections and sedimentation zones of till, outwash gravel and debris flow deposits;
3 a highly variable hydrological regime with runoff in excess of 900 mm yr^{-1} (peak flow > 2000 m^3 s^{-1}) and flood history caused by outburst floods from adjacent glaciers or dam-break floods from landslide and debris-flow valley dams;

4 the fact that most of the route of the road traverses unstable Holocene and Pleistocene deposits and passes through highly dynamic outwash zones such as the Batura glacier terminus.

A general assessment of the highway undertaken by Jones et al. (1983) showed the relative significance (% of length) of different terrain and material types along 129 km of the highway:

alluvial and outwash fan	40.6
rock fall and scree	20.7
rock	18.3
terrace deposits	9.4
river	0.4
till	10.6

Considering the dynamic nature of the terrain types and unstable character of the materials it is inevitable that despite best practice in designing and building the road, sections of it will be periodically destroyed during seismic events, floods and debris flows. The only action is to rebuild following these natural catastrophes (Fig. 2.16). The value of environmental sedimentology in designing such a highway is in careful planning of the alignment of the road in relation to the geomorphology and terrain types; assessment of slope instability types and likely hazards; and siting of local engineering measures to minimize risk from specific slope processes.

In more developed mountain countries road and railway construction is a highly sophisticated engineering practice involving the construction of switchbacks, blasting to achieve preferred alignments; elaborate bridges; protective structures from rock fall and avalanches; and tunnels. In the Rhaetian Alps in Switzerland the railway runs over a distance of 240 km and on this route has 376 bridges and 76 tunnels (Price 1981). In some cases tunnels pass completely through a mountain such as the 12 km Mont Blanc Tunnel (completed in 1965) between France and Italy.

2.5 MANAGEMENT AND REMEDIATION

Environmental sedimentology is an essential element in the management and remediation of sediment related hazards in mountain regions. This can be illustrated by considering examples of hazard assessment and mapping techniques and the role monitoring can play in the remediation of mountain sedimentation events.

2.5.1 Hazard assessment

In mountain environments geomorphological processes are highly active and the terrain inherently unstable (see section 2.1.2). Any move to develop such areas results in a potential hazard. Mountain hazards are on the increase due to increasing development pressures and recent environmental change. Mountain hazard mapping is therefore becoming an increasingly important component of regional and local land-use planning. Over the past few decades understanding of mountain sediment systems has advanced considerably (Owens & Slaymaker 2004), however, the problem of ranking the importance of geomorphological processes in terms of hazards is still problematic because of difficulties in:

1 recognizing and understanding causal mechanisms;
2 adequately characterizing the time-scale and frequency over which processes operate;
3 accurately mapping the occurrence of such phenomena.

Slope instability in mountainous terrain is a natural occurrence. In many regions, however, this represents an important hazard that needs to be carefully assessed. In Switzerland, there is a long history of landslide disasters and genuine concern about whether future climatic warming will lead to an increase in hazards, for example greater frequency of debris flows from the periglacial zone. This is a significant problem because more than 6% of the country is affected by hazards that are related to slope instability (Raetzo et al. 2002). Examples of specific events include the summer debris flows of 1987 and the Randa rock avalanches of 1991 (see section 2.3.3). Following the devastating events of the summer of 1987 in Switzerland revisions were made to the Federal Flood Protection Law and the Federal Forest Law, which came in to force in 1991 to protect the environment, human lives and property from damage caused by water, mass movements, snow avalanches and forest fires. It is now a requirement that canton administrative units produce registers of hazard types and maps of hazard areas for use in land-use planning. Raetzo et al. (2002) describe a three-step procedure for the hazard assessment of mass movements. This involves hazard identification, which consists of classification of landslides, mapping of landslide phenomena and a register of slope instability events. The second step involves hazard assessment, which requires the determination of the magnitude or intensity of an event over time and the expression of this on a hazard map. The hazard maps are designed to show degree of danger and by convention use colour coded danger zones (red – high danger; blue – moderate danger; yellow – low danger; and white no danger). These categories are closely related to the intensity and probability of the event occurring. The third step is risk management and land-use planning. This utilizes the hazard map, which is the basic document for land-use planning. This feeds directly into the local authority planning process. For example, standard colour coded hazard zones indicate the degree of development that can occur, for example: in red zones construction is prohibited; in blue zones construction is allowed only if certain conditions are met; and in yellow zones construction is permitted without additional restrictions.

In Austria a similar scheme for the classification and mapping of hazardous mountain events has been produced for torrents, avalanches and floods (Aulitzky 1994). This is based on work undertaken by the Austrian Forest Technical Service for Torrent and Avalanche Control and is legally bound by the 1975 Forest Law. It was realized that following a series of major mountain disasters in the 1960s in Tyrol that two-thirds of losses were related to some form of human influence. Because most of these hazards occur on alluvial cones and fans a formula was developed which delimits different degrees of risk across the fan surface in relation to the amount of debris deposited (G) (Hampel 1980). This is based on an assessment of the gradient of

the alluvial fan (J) and the mean particle size of material of the sediment transported (d_m):

$$G(\%) = \frac{(J\% - 55d_m^{1.65})^{1/(0.42-0.4d_m)}}{3.6}$$

This simple formula is very sensitive to the measured fan slope (Aulitzky 1994). Additional criteria are used for hazards in torrents (Torrent Index method) and landslides. This information is used to produce hazard maps, which show hazard zones for events with an estimated return period of less than 150 years. These include a red zone (permanent settlement and traffic prohibited); yellow zone (damage certain – some development allowed but with restrictions) and a white zone (no recognized hazard).

2.5.2 Monitoring

Hand-in-hand with hazard assessment and mapping goes hazard monitoring. In areas that have been established to be at high risk hazard, monitoring will be undertaken. Usual hazard monitoring involves direct measurements of the behaviour of a river, slope or glacier, which can be relayed in real-time or over short timescales to an observer who can plan a response, for example activate a warning and evacuation system.

The Randa rock slide discussed earlier (section 2.3.3) provides an excellent example of how both short-term (emergency) monitoring and longer term monitoring provide important information. The initial rock slide at Randa in April 1991 prompted the authorities to install a seismographic and geodetic warning system. As a result the second event in May was forecast from field evidence and geodetic and seismic surveys and the area was successfully evacuated. Prior to the second slide geodetic displacement measurements indicated an accelerated motion of the rock mass (Gotz & Zimmermann 1993) (Fig. 2.12b). Approximately 2.5×10^6 m^3 of active block are still being monitored. The rock mass is still moving towards the south-east at a maximum rate of 15 mm yr^{-1} (Ornstein et al. 2001). The sliding rock mass is bounded by shallow dipping joints and sets of steep near-vertical joints. Tension cracks on the surface behind the slope face tend to open parallel to the steeply dipping natural joints. Currently the Swiss Seismological Service have installed a network of surface seismographs to monitor local activity. This has proved useful in identifying fracturing within the underlying rock mass (Eberhardt et al. 2001). Monitoring will be extended to included borehole installation of microseismic triaxial transducers around the developing shear plane. The movements are being monitored by geodetic survey and an extensiometer gauge, which will enable the brittle fracture development of the rock mass to be assessed. Based on observations of the current instability the local authorities have been able to act in advance of another failure, i.e. to realign the road and railway to avoid the impact of potential new landslides.

Case Study 2.3, which focuses on the sediment related problems following the 1980 Mount St Helens eruption, provides additional details on remediation. It is clear from the Mount St Helens example and the examples just discussed that there are three phases in which environmental sedimentology can play a significant role:

1 hazard assessment and mapping;
2 monitoring of mountain sedimentary processes;
3 prediction of long-term sediment delivery from disturbed mountain environments and design of appropriate remediation measures to deal with the sedimentation problem.

2.6 FUTURE ISSUES

All environmental and economic indicators suggest that over the next few decades mountain environments will become more stressed and erosion and environmental degradation will become greater problems. This section considers how sedimentology can provide useful tools for remediation of geomorphological problems, thus enabling mountain life to continue in a changing environment. The two main examples include: predictions involving the evaluation of climate change scenarios and the occurrence of debris flows, and microscale hydraulic modelling of sedimentary systems.

2.6.1 Considering future climate change scenarios: debris flows in the Alps

Many mountain environments are influenced by glacier and permafrost hazards, which present a direct threat to infrastructure, settlements and human life in high-altitude and high-latitude mountain areas. Global warming is affecting the thermal stability of surface and subsurface ice, and glacier and permafrost equilibrium limits are shifting in response to this general trend (see Chapter 1). The situation is made worse as development pressures in mountain areas are forcing human settlement and activities further into the mountain cryospheric zone. For example, under the prediction of a warming scenario for the Alps the following are all likely consequences (Zimmermann & Haeberli 1992; Haeberli 1995):

1 greater frequency of outburst floods and glacier hazards;
2 permafrost degradation and slope instability – increased debris flow activity;
3 decline in material strength of foundations of high mountain buildings;
4 damage to reservoir systems (damage to infrastructure) and increased sedimentation rates (reduced storage capacity).

More generally the impact of such warming is a range of glacier-related hazards that include: glacier lake outburst floods; rock and ice avalanches; deep seated instability; and retreat of glacier and permafrost limits. Many of these have important impacts on the sedimentology of mountain environments, resulting in enhanced rates of debris supply, extensive flood deposits and large-scale mass movements. Examples include the increased frequency of rock avalanches in relation to permafrost degradation in glacier environments: the Brenva Glacier, Mont Blanc in 1997 (Bottino et al. 2002; Haeberli et al. 2002) and the September 2002 Kolka–Karmadon rock and ice avalanches and mudflows in the Caucasus Mountains of Russia (Kääb 2002; Kääb et al. 2003).

The Kolka–Karmadon rock avalanche began on the evening of 20 September 2002 high on the peak of Dzimarai-khokh. Several million cubic metres of ice and debris fell on to the Kolka glacier tongue, shearing off the front of the glacier before crossing the Maili glacier. Travelling at around 100 km h^{-1} the avalanche picked up lateral moraines and valley-bottom sediments before running out and halting approximately 18 km from the source, where approximately 80×10^6 m^3 of ice and debris were deposited and after which a mudflow approximately 300 m wide continued for another 15 km down the confined valley. A total of 120 people were killed in the catastrophe and since the event the deposits continue to influence sedimentary processes in the valley system. Rivers were dammed by the avalanche deposits, which caused the formation of lakes estimated to be over 10×10^6 m^3 in volume. These posed a considerable danger to settlements down-valley owing to the risk of dam-break flooding.

Because of the remote nature of the environments in which these high-mountain processes operate, rapid assessment of glacier hazards is needed through the use of remotely sensed data. Rapid imagery from ASTER (Advanced Spaceborne Thermal Emission and Reflection Radiometer) provides 15 m resolution imagery, which is updated regularly every 16 days but this can be shortened to 2 days where needs demand (Kääb et al. 2002). From these data digital elevation models (DEMs) can be generated, which are essential for assessing the dynamic changes in topography that characterize these rapid sedimentary events.

The above example illustrates how recent environmental change is having an impact on mountain hazards. It should be recognized, however, that the incidence of mountain hazards also varies over much longer cycles. For example, Soldati et al. (2004) examine the relationship between the temporal occurrence of landslides and climatic changes in the Italian Dolomites since the last glacial stage (Würm). Increases in landslide activity were identified corresponding to the boundaries between the Late-glacial and Holocene and the Atlantic and Sub-boreal.

Landslide type and cause also differed over time. Following retreat of the glaciers after the Last Glacial Maximum there followed a phase of increased slope instability (11,500 to 8500 cal. yr BP) involving large translational bedrock slides in the dolomitic slopes and complex movements (both rotational slides and flows) affecting the underlying pelitic formations. A second phase of enhanced activity was also identified around 5800–200 cal. yr BP, when slope processes were dominated by mainly rotational slides and flows. The mechanisms for such activity vary. In the earlier slides high groundwater levels related to increased precipitation and/or permafrost melt were probably important. In the second phase, many of the slope movements were reactivations of earlier landslides and probably were triggered by increased precipitation. Studies of this kind are valuable when considering potential impacts of future climate change because although patterns of historical landscape instability can be linked to climate, and to some extent these fit with general patterns of slope movements across Europe, non-climatic factors (e.g. geological–structural factors and changing human land-use) also have a role to play and should not be excluded from the analysis.

2.6.2 Microscale modelling

Understanding sedimentation problems in mountain environments is difficult because problems evolve over several decades or centuries and the main governing processes operate at scales that cannot be measured easily. Microscale modelling provides a valuable tool for the environmental sedimentologist to investigate large-scale sedimentation processes in mountain-valley sediment systems. Microscale loose-bed hydraulic models are very small-scale and conventionally have been used to evaluate channel designs on large river systems (Gaines & Maynord 2001). Typically model studies are carried out to determine channel response to engineering structures and identify overall flow patterns within the fluvial system. A basic model consists of a hydraulic flume, a rigid channel insert to define the valley-system boundaries and a loose bed of natural or synthetic bed material. Typical model scales are in the order of $1:10^{-3}$ or 10^{-4}. At this horizontal scale microscale models differ from traditional loose bed hydraulic models because in order to ensure that sediment transport occurs in the model, water depths are exaggerated and as a consequence vertical scale distortion is inherent. Also shallow depths lead to surface tension effects and laminar flows, which do not occur in the real river. The success of microscale models, however, is not judged on the basis of hydraulic similarity but more so on how well overall the model reproduces the gross features of the prototype river morphology (Gaines & Maynord 2001). In this sense the approach is useful in that it provides an initial assessment method for rapidly simulating river behaviour at large scales.

The technique has been used by Davies et al. (2003) to examine anthropogenic aggradation of the Waiho River, Westland, New Zealand. Long-term aggradation of the Waiho River alluvial fan over much of the past 100 years has raised the fanhead to an unprecedented high level. This now poses an unacceptable flood risk to the adjacent village of Franz Josef Glacier and the main State Highway 6, which is a key tourist corridor. The behaviour of this fan has puzzled geomorphologists for some time and has led to a number of hypotheses regarding the evolution of the fan system in relation to river control measures (Davies 1997; Davies & McSaveney 2001). The only suitable way of testing these ideas, however, was to construct a model capable of simulating river behaviour for a variety of imposed boundary conditions. A 1:3333 microscale physical hydraulic model was used to study the problem (Fig. 2.17a). In the model an alluvial fan was generated and allowed to develop to equilibrium with steady inputs of water and sediment. The fan was laterally constrained by rigid boundaries which were geometrically similar to the natural unrestricted Waiho River. The boundaries were then reset to reflect the presence of stopbanks (river control structures) and the fan allowed to evolve under the same water and sediment feed rates. The model

fanhead aggraded in a similar spatial pattern to that observed in the real river. Figure 2.17b shows a comparison between relative changes in bed level in the model and those measured in the Waiho River. The correspondence in the spatial pattern of aggradation was very similar. This implies that the reduction of the flow area at the fanhead caused by the river stopbanks is sufficient to cause the observed aggradation in the Waiho.

This example clearly demonstrates the advantages of using microscale modelling for understanding large-scale sedimentation problems in mountain environments where rates of sedimentation are relatively rapid and topography exerts a major control on the spatial patterns of cut and fill. Although there are constraints in using this type of model (e.g. hydraulic conditions cannot be truly scaled) a formal physical model would require a planform of 50 m by 50 m which is beyond the scope of most physical model investigations.

2.7 CONCLUDING COMMENTS

The aim of this chapter has been to show how the functions of active sedimentary processes are altered by human actions and recent climate/environmental change in mountain environments. It has been demonstrated that mountain environments are characterized by high relief, steep slopes and local climates that vary over altitude. These environmental constraints create a distinctive set of mountain sedimentological processes. Figure 2.18 is a simple conceptual model that summarizes the relationship between mountain sedimentation zones and dominant process regimes. The diagram distinguishes sedimentation zones in terms of a sediment-cascade continuum which links slopes and valley-floor sediment systems along a gradient of downslope/valley sediment fining. Dominant process regimes map on to the sedimentation zones but these overlap to differing degrees dependent on the extent of their spatial influences (broad boxes show the main zone of operation and the arrows show the maximum extent). The large overlap

(a)

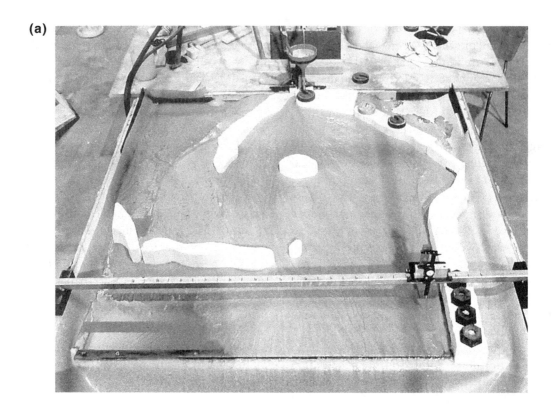

(b)

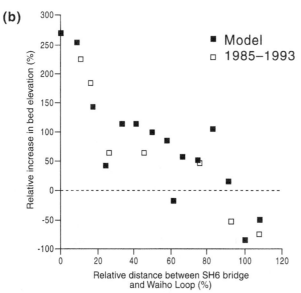

Fig. 2.17 Microscale model of the Waiho River fan, New Zealand. (a) Laboratory model showing unconstrained fan development. (Photograph courtesy of T.R. Davies.) (b) Comparison between relative changes in bed level in the model and those measured in the Waiho River. Bed-level changes are non-dimensionalized by scaling the variance in each data set to 1.0. (From Anthropogenic aggradation of the Waiho River, Westland, New Zealand: microscale modelling. Davies, T.R., McSaveney, M.J. & Clarkson, P.J. (2003) *Earth Surface Processes and Landforms*, **28**, 209–18. © John Wiley & Sons Limited. Reproduced with permission.)

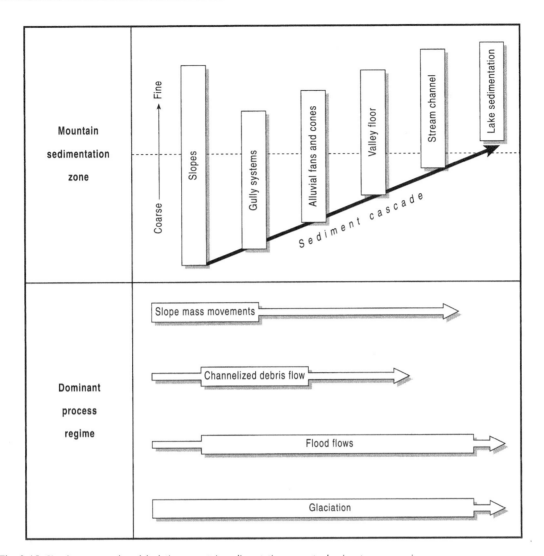

Fig. 2.18 Simple conceptual model relating mountain sedimentation zones to dominant process regimes.

between sedimentation zones and process regimes in mountain environments leads to extreme variability in space and time of mountain processes and makes understanding of such sediment systems problematic (Gerrard 1990), with the inevitable conclusion that sediment erosion and deposition in mountain environments is highly heterogeneous (Butler et al. 2003).

Nevertheless, understanding how these various sedimentological processes operate in mountain environments has been considerably aided in the last two decades by a series of methodo-

logical advances. In particular it is worth noting the importance of catchment-based sediment budget studies (Rapp 1960; Barsch & Caine 1984; Trustrum et al. 1999; Slaymaker et al. 2003). Looking to the future, similar advances in understanding are likely to be made through: the application of geophysical techniques for estimating sediment storage (Schrott et al. 2003); use of cosmogenic dating to better understand time-scales of erosion and sediment residence (Kirchner et al. 2001); use of remote sensing in hazard monitoring and mapping (Kääb et al.

2003); and use of modelling to predict future changes in mountain sediment systems (Bottino et al. 2002; Davies et al. 2003).

Finally although this chapter has been about mountain environments, it is important to stress the links between mountain regions and lowlands. Erosion in the uplands has a profound impact on downstream sedimentation (e.g. Brahmaputra and Ganges rivers draining the Himalaya) and at the junction between mountain and lowland environments sedimentological processes are often most active (e.g. alluvial fan aggradation, Waiho River, Westland, New Zealand). Sediment processes in lowland river systems are explored in Chapter 3.

3

Fluvial environments

Karen Hudson-Edwards

Rivers are arteries for the transport and storage of physically and chemically weathered material from continents, through estuaries, and ultimately to oceans; they thus play a major role in the Earth's biogeochemical cycling of materials, and influence the Earth's climate. Rivers are found in every corner of the world, and in every climatic zone. Both natural processes and anthropogenic exploitation of natural resources from prehistoric times through to the present day have, however, had a very significant impact on the hydrology, sediment regime and contamination of rivers world-wide. During the Quaternary, for example, river systems have experienced considerable natural hydrological variability due to deglaciation and sea-level change, and human activity has left very few rivers in a pristine state, except perhaps for some in Canada, Amazonia, the Congo Basin and Siberia (Meybeck 2003).

This chapter describes the nature of fluvial sedimentary environments, types and sources of sediments in these environments, processes and impacts of natural and anthropogenic disturbance events on river sediment fluxes and effects, management of fluvial sedimentary environments and, finally, issues concerning fluvial environments that need to be addressed in the future. The chapter draws on the voluminous fluvial geomorphological, sedimentological, environmental and archaeological literature. For further information, the reader is referred to this literature, and to the many excellent books and book chapters on fluvial sedimentary environments, and the background information and references contained therein (Richards 1982; Brown 1997; Thorne

et al. 1997; Benito et al. 1998; Knighton 1998; Leeder 1999; Miller & Gupta 1999; Bridge 2003). This chapter summarizes, updates and builds on this work, and, specifically, highlights the ever increasing effects of anthropogenic activity on sedimentation in rivers.

3.1.1 Definition and classification of fluvial environments, and relevance to environmental sedimentology

The word 'fluvial' pertains to a stream or river, the existence, growing or living in or about a stream or river, or something that is produced by a river (Bates & Jackson 1980). Fluvial environments occur on every continent on Earth and in every climatic zone (Fig. 3.1), and thus are often classified according to climate, with arid, semi-arid, temperate (cool), temperate (Mediterranean) and tropical types known (e.g. Jansen & Painter 1974). Rivers are also classified by their flow regimes, with perennial (flowing every year, throughout the year), intermittent (flow for only part of a year, every year, usually during or after a wet season) and ephemeral (occasional flow) types defined. Intermittent and ephemeral rivers often occur in Mediterranean or tropical environments, whereas perennial rivers occur in temperate regions.

'Fluvial sediments' are those sediments that consist of material transported by, suspended in, or laid down by a stream (Bates & Jackson 1980). They are important sources of nutrients, contaminants and other solid materials to downstream fluvial, estuarine and coastal environments. Millions of tonnes of sediment are transported annually to oceans (Table 3.1), with the

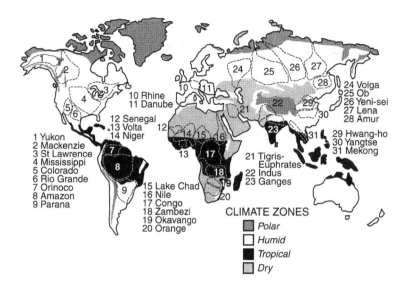

Fig. 3.1 Climatic zones and the world's largest river basins. (Based on Newson 1992, fig. 4.1. Climate zones from *The Times Atlas of the World* 1983.)

Table 3.1 Transport of sedimentary material from continents to oceans. (Based on Knighton (1998), with data from Degens et al. (1991), Meybeck (1979), Milliman & Meade (1983) and Walling (1987).)

Continent	Land area (10^6 km²)	Mean annual runoff (10^3 km³)	Total annual suspended sediment load (10^6 t yr⁻¹)
Africa	15.3	3.4	530
Asia	28.1	12.2	6433
Europe	4.6	2.8	230
North and Central America	17.5	7.8	1462
Oceania and Pacific Islands	5.2	2.4	3062
South America	17.9	11.0	1788

amounts determined by climate (precipitation), topography, vegetation cover, land-use, and the susceptibility of the underlying rocks, soils or other unconsolidated materials to physical, chemical and biological weathering.

A river system is, by definition, the system of connected river channels in a drainage (catchment) basin (Bridge 2003). The system contains a large number of features, but the two most important are its channels and floodplains (Fig. 3.2). In the upland, mountainous parts of river basins, relief is relatively high, channels are incised into bedrock or alluvium, and floodplains tend to be narrow. For a more detailed discussion of sedimentology in mountainous environments, the reader is directed to Chapter 2. In the middle to lowland portions of river systems, relief is lower, valleys are broad and larger-scale floodplains are developed. From a sedimentological

perspective, rivers are often classified according to their planform channel geometries, with four major types traditionally defined: (i) straight, (ii) meandering, (iii) braided and (iv) anastomosing (Fig. 3.3). This classification has been seen as flawed, mainly because the classes overlap. As a result, authors such as Rust (1978) and Bristow (1996) have suggested that these types should not be viewed as separate entities but, rather, as end members of a cyclical continuum (Fig. 3.4). Modifications from one end member to another can be a result of changes in sinuosity (river length divided by valley length) and the amount of braiding (channel length divided by valley length or island length divided by river length). For example, straight rivers can metamorphose into meandering rivers by an increase in sinuosity, and meandering rivers can change into braided rivers through the development of bars

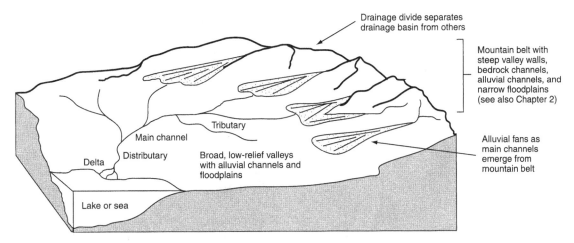

Fig. 3.2 Plan geometry of a hypothetical river system. (Based on Bridge 2003, fig. 1.1.)

(a)   **(b)**

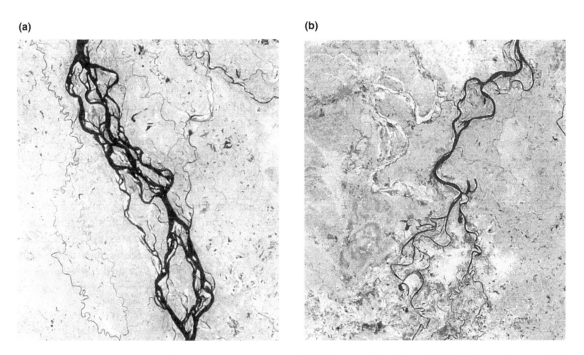

Fig. 3.3 Photographs of river channel types. (a) LANDSAT image of the Brahmaputra River in Bangladesh. This is a very large sand-bed braided river, flowing from north to south down the photograph. On the floodplain on either side are sinuous rivers that act as distributary and tributary channels draining the floodplain. (b) LANDSAT image of the anastomosed Meghna River in Bangladesh, where the sinuous river channels divide and rejoin around areas of floodplain and are separated for more than one meander wavelength. Width of photographs is approximately 60 km. (Photographs donated by A. Carter.)

or islands. Some fluvial sedimentologists group the straight and meandering types together as 'single channels', and the braided and anastomosing types together as 'multiple channels'.

The planform geometry of rivers is controlled by climate, tectonics and land-use, and more spe-

cifically by the rate of flow, flood-related sedimentary processes and the amount of sediment discharge (Schumm 1968; Knighton 1998). Little channel change occurs during low flows and low amounts of sediment transport. By contrast, the increases in discharge and sediment transport

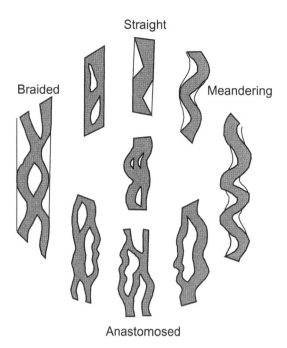

Fig. 3.4 Continuum between principal planform types of rivers, illustrating the diversity of river channel types. Between the straight, meandering, anastomosed and braided types of planform are rivers that exhibit characteristics of at least two of these end-members. (After Bristow 1996.)

that occur during floods can cause moderate to sometimes drastic channel change (Schumm & Lichty 1963). Attempts have been made to model channel changes that occur in response to changes in water and sediment discharge. Schumm (1969), for example, suggests that a decrease in streamflow and concomitant increase in sediment supply should produce a decrease in channel sinuosity and depth, and either an increase or decrease in

channel width. Channel adjustment, however, is not easy to model, because it depends on several factors including slope, and the pre-existing channel geometry, planform and bed configuration. Further examples of the types and causes of channel change are discussed later in this chapter.

Fluvial (also known as alluvial) deposits are characterized by a huge variety of bed forms and sedimentary structures, which are well-described in Collinson (1986) and Bridge (2003, chapters 4 and 5). These include longitudinal, bank-attached, linguoid, side, diagonal, point and transverse bars, islands, riffles, ripples, pebble/cobble clusters, sand ribbons, cross-stratification, dunes, sand flats, plane beds and chutes-and-pools (largely in channels), lateral deposits (in channel margins), alluvial fans (in piedmont areas), crevasse splays, levees and vertical accretion deposits (in floodplains) (Fig. 3.5). These features are used by geomorphologists to deduce the history of river sedimentation, to evaluate contamination and to record flood histories.

No study or book on environmental sedimentology would be complete without consideration of fluvial sedimentary environments. Because all humans essentially reside in river basins, they have, since pre-history, had a significant impact on them and the functioning of their sedimentary processes. In many cases, these impacts outstrip those of their natural counterparts. In other cases, natural (section 3.3) and anthropogenic processes (section 3.4) act in tandem to produce large-scale modifications to fluvial sediment transport and deposition.

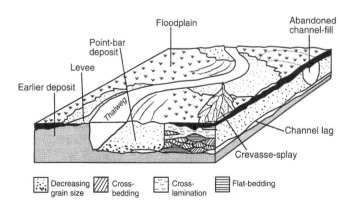

Fig. 3.5 Classic point-bar model for a meandering stream, showing various types of fluvial sedimentary deposit. (After Allen 1964, 1970; Collinson 1986.)

3.1.2 Significance of fluvial sedimentary environments

In effect, many of the issues and problems related to river basin disturbance are in turn related to differences between the Developed world, with its largely temperate rivers, and the Developing world, with its largely arid, semi-arid and tropical rivers. Issues for river development and management can be quite different for these two worlds. Many of the largest river basins in the world occur in the arid and semi-arid Developing world (Fig. 3.1), and here the major problems include drought, irrigation, poverty, salinization, pesticide contamination and desertification, all of which have an impact on river systems (Newson 1992; see also Chapter 5). By contrast, the Developed world is coming to terms with managing and restoring river systems that have been under pressure from impacts such as contamination. Both the Developed and Developing worlds will have, increasingly, to deal with the issue of climate change related to global warming.

River systems are not country-selective: they cross international boundaries. This fact raises important issues relating to sediment transport and deposition in rivers, especially with respect to contamination, erosion and flooding. Newson (1992) has called for workers in all related disciplines to consider the impact of contemporary decisions about river management in both the short- and long-term, consider basins as integrated water, sediment and contaminant entities, and for science to inform, rather than solve, problems of river basin (including sediment) management.

3.2 SEDIMENT SOURCES AND ACCUMULATION PROCESSES

3.2.1 Characteristics and provenance of fluvial sediments

3.2.1.1 Characteristics of fluvial sediment

Transported fluvial sediment can be classified as either bed load, the coarse sediment (> 0.0625 mm) carried along the river bottom, or suspended load, the finer sediment (< 0.0625 mm) moved in suspension. The suspended load forms most (generally > 90%) of a river's sediment and, because its settling velocities are generally low, is transported at the same speed as the river's flow. During floods, the size of the suspended load can increase dramatically due to increased stream power. The suspended load is supplied from the physical erosion of river banks and surface materials, and resuspension of fine channel-bed material.

Fluvial sediment also can be classified according to either its physical or chemical characteristics (Goudie 1990). Physically, sediment is classified mainly according to its size and particle shape. Estimating the relative proportions of the different size fractions in a river sedimentary unit is important because the size is related to the source and abrasion during transport, and size also plays a major role in the distribution of contaminants. Particle shape is also determined to a large part by the abrasion and corrosion that occurs during transport, but also depends on the original weathering that took place to form the grain, and on post-depositional chemical processes such as dissolution and precipitation. Particle shape is described in three ways: (i) form, (ii) sphericity and (iii) roundness (see Chapter 1). River gravels, for example, often have high to medium sphericity and are subrounded. In general, the further a river sediment particle travels and the more mechanical abrasion it undergoes, the more well-rounded, higher in sphericity and finer-grained it will become. Rounding occurs more readily in cobbles and pebbles than in sands or smaller-sized particles.

The physical and chemical characteristics of fluvial sediments are determined by the sediment provenance, that is, the origin (source) of the sediment and the physical, chemical and biological processes that have operated on it during transport to the site of deposition in the river system. Most fluvial sediment is composed of geogenic materials such as bedrock, soil and vegetation that are released to river systems through both natural (e.g. erosion, forest fire, volcanic eruption) and anthropogenic activities (damming, deforestation, mining), many of which are discussed in section 3.3. The variability of these geogenic

(a)

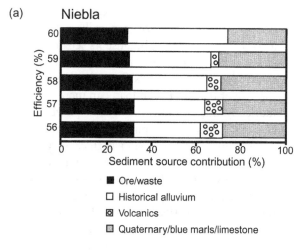

Niebla

■ Ore/waste
□ Historical alluvium
◙ Volcanics
▨ Quaternary/blue marls/limestone

(b)

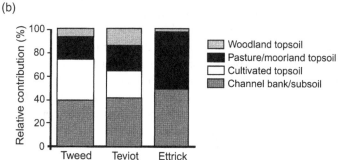

▨ Woodland topsoil
■ Pasture/moorland topsoil
□ Cultivated topsoil
▨ Channel bank/subsoil

Fig. 3.6 Examples of sediment fingerprinting studies. (a) Results of fingerprinting for sediments at Niebla, Río Tinto, Spain. Efficiency (%) is a measure of the effectiveness of the mixing model. The optimal efficiency is shown at the top, and the four nearest efficiency steps are also indicated. (Based on Schell et al. 2000, fig. 26.4b.) (b) Weighted contributions of woodland topsoil, pasture/moorland topsoil, cultivated topsoil and channel bank/subsoil in the suspended sediment load of Rivers Tweed, Teviot and Ettrick, UK, from January 1996 to February 1997. (From Owens et al. 2000, fig. 15.3a.)

materials, in turn, depends on the underlying river basin geology, the nature of soil-forming processes (climate, typography, organisms) and types of vegetation present. For example, Schell et al. (2000) were able to show that a combination of ore, mine waste, mine-derived alluvium and bedrock, derived from natural weathering, contributed to the fluvial sediment load in the mining-affected Rio Tinto basin, Spain (Fig. 3.6a), through the use of sediment fingerprinting techniques (see section 3.2.1.2).

Many anthropogenic activities affect the composition of river sediments (Table 3.2). These activities often add 'excess' amounts of substances, or 'contaminants', to river systems. The concept of 'contaminant' has been defined in Chapter 1. Contaminant concentrations of sediments are controlled by their abundance in the rocks and soils of the basin (the 'background' concentration) and by the addition, through anthropogenic activities, of excess amounts of the element or compound (the contamination). In

the case of artificial compounds (e.g. pesticides, polychlorinated biphenyls (PCBs), some radio-nuclides) the background figure is generally zero, and contamination assessment is relatively straightforward. Pre-industrial historical values need to be established for elements with both natural and anthropogenic sources, such as metals. In rare cases, this can be accomplished through analysis of archived samples, but in most cases it is determined through the analysis of sediments accumulated through time.

Table 3.2 lists the main types of sediment-bound contaminants found in river systems. These are grouped into metals and metalloids, inorganic compounds, nutrients, organic compounds and radionuclides. The partitioning of each of these contaminants will be different, depending on their geochemical properties and affinities for sorption onto the different mineral and organic phases present in suspended river sediment. This, in turn, will affect the transport and storage of contaminants in river channels and floodplains.

Table 3.2 Types and sources of natural and anthropogenic inputs to fluvial sediments.

Category	Sources
Bedrock, soil, vegetation	Physical weathering through natural erosion, forest fires, tectonic uplift, deforestation, agriculture, mining, damming and other engineering work, urbanization
Metallic and metalloid elements (Sb, As, Cd, Cu, Co, Cr, Pb, Hg, Ag, Tl, Sn, Zn)	Natural sources, industry, mining and processing, acid mine drainage, sewage treatment, agriculture, vehicle emissions and road runoff, coal combustion, atmospheric fallout
Inorganic compounds (SO_4, PO_4)	Natural sources, mining and processing, industry, acid mine drainage, acid deposition, agriculture
Nutrients (C, N, P)	Agricultural and urban land runoff (fertilizers), wastewater from sewage treatment
Organic compounds (pesticides, herbicides, petroleum hydrocarbons, viruses, bacteria)	Agriculture, industrial processes that produce dioxins, sewage, landfills
Radionuclides (^{137}Cs, ^{129}I, ^{239}Pu, ^{230}Th)	Nuclear power industry, military, natural sources

Sediment-bound contaminants enter river systems either from point (e.g. tailings effluent and other mine discharges, sewage discharges, spillages) or diffuse sources (e.g. remobilization of contaminated alluvium, agricultural runoff) (Macklin 1996; Walker et al. 1999). Point sources either operate only once (e.g. the Aznacóllar mine tailings dam failure in south-west Spain, April 1998) or repeatedly (e.g. sewage discharges; regular tailings effluent discharge into the Río Pilcomayo, Bolivia; Hudson-Edwards et al. 2001). Although diffuse contaminants enter rivers continuously, growing evidence suggests that they are mainly mobilized during storm events. For example, Kratzer (1999) suggested that runoff from infrequent winter storms would continue to deliver significant quantities of sediment-bound organochlorine pesticides to the San Joaquin River, California, even if irrigation-induced sediment transport was reduced. Muller & Wessels (1999) showed that approximately one-third of the annual inputs of total organic C, N, Cu, Pb and Zn to the River Odra, Poland, were released into the river during a major flood in 1997.

3.2.1.2 Sediment provenance and source material fingerprints

Analysis of fluvial sediment provenance can yield important information on the types of sources (e.g. topsoil or channel bank materials), relative source contributions, and on patterns of erosion,

transfer and storage. Although it is possible to directly determine fluvial sediment provenance, indirect methods have been favoured in recent decades (cf. Peart & Walling 1986). One of the main types of indirect approaches is that of sediment source 'fingerprinting', which has been used to great effect in studies of suspended and alluvial sediment provenance, over a wide range of time-scales ranging from the event-level to recent historic and Holocene. Owens et al. (2000), for example, traced contemporary sources of alluvium in the Rivers Tweed, Teviot and Ettrick, UK, using composite fingerprints (metallic element geochemistry, radionuclides, organic C, N and P) and a numerical mixing model, and showed that most of the sediment was derived from pasture/moorland topsoil and channel bank erosion (Fig. 3.6b). By contrast, Passmore & Macklin (1994) demonstrated that pre-eighteenth century alluviation in the River Tyne, UK, was related to deforestation and agricultural development during late prehistoric times. Walling et al. (2002) used ^{137}Cs measurements and sediment source fingerprinting to develop sediment budgets that showed sediment tile drains to transfer between 30 and 60% of sediment from two lowland, agricultural basins in the UK.

The scientific basis of sediment fingerprinting is that the properties of the fluvial sediment are compared with the same properties of all potential source materials (e.g. Walling et al. 1979). 'Tracers', which exhibit conservative behaviour

Table 3.3 Fingerprinting techniques used in fluvial sediment provenance studies, and selected references.

Fingerprinting technique	Method	Reference
Radionuclides (e.g. ^{7}Be, ^{137}Cs, ^{210}Pb, ^{226}Ra)	Measurement of concentration of radionuclide and comparison with a background concentration	Peart & Walling 1986
Mineral magnetism	Mineral magnetic parameters (e.g. susceptibility, frequency-dependent susceptibility, isothermal remanence magnetization) measured, statistical analysis	Walling et al. 1979
Sediment mineralogy	Mineral abundances determined by X-ray diffraction (XRD)	Woodward et al. 1992
Heavy mineralogy	Heavy minerals separated, abundances determined by point counting	Singh et al. 2004
Clay mineralogy	Clay minerals separated; abundances determined by XRD and analysis of results	Wood 1978
Major and trace element geochemistry	Determination of geochemical composition and statistical analysis (e.g. multivariate, principal components analysis)	Lewin & Wolfenden 1978; Passmore & Macklin 1994; Collins et al. 1997
Grain size and shape	Macro- or microscopic determination of grain size, statistical analysis	Knox 1987

during erosion and transport and which can, on a statistical basis, effectively differentiate between the various sources, are used for source apportionment (Foster & Walling 1994). Once suitable tracers are established, their values, or 'fingerprints' for potential source materials are compared with the corresponding tracer values for the 'unknown' sediment samples. It has been increasingly recognized that no single diagnostic tracer is capable of discriminating a range of sources, so multiple tracer studies using several fingerprints are now normally used (e.g. Walling et al. 1993). Over the past several decades, a wide range of physical and chemical fingerprinting techniques have been developed and used to great effect in determining the provenance of fine-grained suspended and alluvial sediment. The main techniques used are summarized in Table 3.3.

3.2.2 Controls on sediment supply, transport and accumulation

The natural sediment load in rivers is supplied by physical and biochemical weathering, which are in turn controlled by (i) the geochemistry, mineralogy and structure of the eroding rocks and soils, (ii) precipitation, (iii) temperature and (iv) vegetation cover (Bridge 2003). Ultimately, these

are all controlled by climate, topography and land-use. Areas with steep slopes, cold temperatures and high amounts of rainfall (such as the Himalayan mountain range) have high sediment supplies, whereas lowland areas with warmer temperatures tend to have relatively lower sediments loads (which are dominated by clay-rich materials) and higher dissolved element loads.

Sediment transport in rivers is mainly unidirectional, following the dominant flow path, but it can be influenced by turbulence and secondary flows (Sear et al. 2000). The transport of individual sediment particles occurs by entrainment, transport and deposition (Hassan & Church 1992; Sear et al. 2000). A large number of factors influence sediment transport and deposition in river systems. Among these are climate and season (Boyden et al. 1979; Macklin 1996), site geomorphology (Macklin 1996), streamflow and suspended sediment concentrations (Yeats & Bewers 1982; Kratzer 1999), water depth, flow dynamics, bed surface structure and grain size and texture (Eyre & McConchie 1993; Macklin 1996). Anthropogenic disturbances, such as channelization, damming and dredging for navigation, also play a role in sediment transport and deposition.

Although sediment can be mobilized and transported during normal flow conditions, floods

Fig. 3.7 Photograph of typical fluvial overbank sediments, with gravel base and fining upward sequence. Length of tape measure is 1.4 m.

(high flow stages) play a major role in eroding and depositing fluvial sediment, and modifying river channels and floodplains (Knighton 1998). During large floods bank erosion upstream can result in downstream overbank deposition of fine-grained sediments as thin sheets over surfaces of previously-formed terraces, or of coarser-grained gravel and sand as sheets, lobes and splays. During relatively moderate floods that do not exceed bank full conditions, coarse to fine material is transported downstream and eventually deposited as bars or overbank units on lower elevation terraces (Fig. 3.7). The accumulation and preservation of sediments within river systems also depends on the factors outlined in the previous paragraph, and particularly on the availability of space and the deposition and erosion rates (Schumm & Lichty 1963). Deposition on floodplains depends on flood characteristics such as frequency, dura-

tion and suspended sediment concentrations (James 1985). The length of time that sediment remains stored in a part of the river system is variable, ranging from 1 to 10^4 years (Fig. 3.8).

Bridge (2003) has pointed out that all discharges that result in sediment transport can affect river channel geometry. He defines *dynamic equilibrium* as adjustment of the channel to relatively consistent flood discharges over decadal time periods, and *disequilibrium* as a major and regional channel adjustment to a significant event (or events), such as an extreme flood or anthropogenic activity (e.g. mining). These are similar concepts to *passive dispersal* and *active transformation*, defined by Lewin & Macklin (1987) for mining-contaminated river systems (see Case Study 3.2). Bridge's (2003) dynamic equilibrium–disequilibrium concepts are well-illustrated in Fig. 3.9, where a system, previously at equilibrium, is affected by increase in discharge that

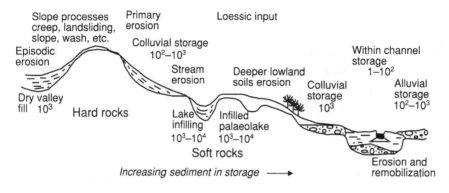

Fig. 3.8 Duration (in years) and location of long-term sediment storage in a typical humid, cool river basin. (After Brown 1997.)

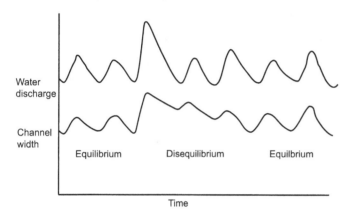

Fig. 3.9 Definition of equilibrium and disequilibrum in stream geometry. Channel width varies congruently with water discharge (i.e. equilibrium) until extreme discharge causes bank erosion and channel widening. Subsequently, the channel is in disequilibrium with water discharge and requires a certain lag time to regain, by deposition, equilibrium. (After Bridge 2003.)

causes disequilibrium between channel width and discharge. There is a lag time before the system readjusts to an equilibrium state once again.

Other authors have described *activation thresholds* for sediment transfer. In these situations, some sort of barrier must be overcome before sediment transport can occur. Rowan et al. (2000) described two large flood events that caused considerable sediment transfer in the river basin of Wyresdale Park Lake in north-west England. They argued that the activation threshold to this transfer was the accumulation of sediment within colluvial footslopes during a dry period immediately preceding the flooding. Once these threshold-crossing events occur, a river channel responds, often rapidly and dramatically, and a new set of conditions (changes in gradient, cross-sectional and planform geometry) develops. The remainder of this chapter considers many examples of responses to river sedimentation to such threshold-crossing events and their underlying environmental causes.

3.3 PROCESSES AND IMPACTS OF NATURAL DISTURBANCE EVENTS

Fluvial environments are subject to fluctuations in rates of sediment transport and accumulation that are caused by natural disturbance events including climate change, tectonic uplift, glacio-isostatic rebound, forest fires and volcanic eruptions. This section presents an overview of the effects of these processes on sedimentation in fluvial environments.

3.3.1 Climate change

Climate change can be defined principally as modifications in precipitation and temperature (and related features such as permafrost cover), and some authors also include modifications to vegetation cover in the definition (Knighton 1998). Climate change can occur on a wide range of time-scales: from 10^3 to 10^5 years (major glaciations), to 10^3 to 10^2 years (interglacial or

Climatic Optimums such as the Little Ice Age in the sixteenth to nineteenth centuries), to 10^1 years (global warming trends in the late 1990s and early 2000s). These climatic changes are related to factors such as Milankovitch cycles (Lewis et al. 2001), changes in meridional circulation (Knox 1995) and global warming, and they play a major role in influencing weathering processes and, ultimately, channel morphology and channel and floodplain sedimentation.

Many investigators have demonstrated that rivers are particularly sensitive to changes in climate, and several have sought to define the relationships between these changes and sedimentation in river systems. Early work suggested a direct relationship between sediment yield and precipitation (Fig. 3.10a) (Langbein & Schumm 1958), in that the highest yields occurred during peak periods of precipitation. Later work by Walling & Kleo (1979) showed that this relationship was more complex on a global scale (Fig. 3.10b), with a three-peak average sediment-yield–precipitation plot that reflected distinct global climatic zones. Walling & Kleo (1979) did, however, suggest that this might be too simplistic, and that other factors, such as seasonal effects on precipitation and temperature, relief, soil and rock type, and land use, may be responsible for the three peaks on Fig. 3.10b (Hooke 2000). In terms of floodplain deposition, increased discharges have been related directly to the area of floodplain that is inundated by suspended-sediment-bearing water (Hamilton 1999). The situation is not always as simple as this: Aalto et al. (2003), for example, showed that crevasse-splay deposits in the Bolivian Beni and Mamore river basins in Bolivia could be grouped temporally, and that the deposition pattern was punctuated in a stop-and-start manner. Aalto et al. (2003) related these crevasse-splay deposit groupings to the cold (La Niña) phases of the ENSO (El Niño–Southern Oscillation) cycle, in that rapidly rising floods during these phases destabilize and deposit colossal volumes of Andean sediment. Inman & Jenkins (1999) demonstrated that the wet periods of alternating, decadal-scale ENSO-induced climate changes were responsible for annual sediment fluxes that were up to 27 times greater than those in dry periods.

Flood frequency plays a major role in defining patterns of river channel erosion and deposition downstream (Rumsby & Macklin 1994). Correlations of alluvial chronologies in the UK, Europe and North America have revealed major discontinuities in the Holocene alluvial record, marked by alternating wetter and drier phases, and alternating higher and lower frequency of extreme flood events (Starkel 1983). This clustering has been linked mainly to climatically driven changes (Macklin & Lewin 1993, 2003; Rumsby & Macklin 1994; Knox 1995). The high flood-frequency clusters resulted in landslides, increased rates of deposition and lateral channel change and avulsion (Starkel 1983). Because of the intimate link between fluvial sedimentation and climate, many investigators have used the fluvial sedimentary record to reconstruct past climates (despite the sometimes sparse nature of the climate record), particularly in contemporary times when global warming is of such concern.

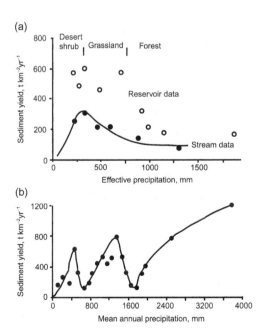

Fig. 3.10 Relationship of sediment yield to (a) effective precipitation (after Langbein & Schumm 1958; Knighton 1998) and (b) mean annual precipitation (after Walling & Kleo 1979; Knighton 1998), based on measured data.

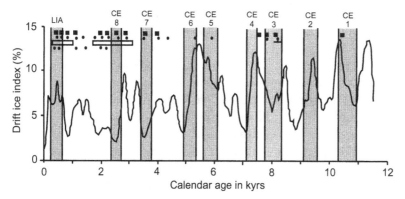

Fig. 3.11 Comparison of British Holocene flood episodes with proxy climate records for Britain, central Europe and the North Atlantic, well-date storm periods for Lake Tutira, New Zealand, and flood episodes in the upper Mississippi basin, USA: LIA, Little Ice Age; CE, central European cold-humid phases (from Haas et al. 1998). (Figure after Macklin & Lewin 2003.)

In the UK, Macklin & Lewin (2003) identified 14 major flooding phases, lying within the time periods 400–1070, 1940–3940, 7520–8100 and c. 10,420 cal. yr BP. Although these are mainly linked to variations in climate, forest clearance and subsequent agricultural activity (see section 3.4.1.1) were thought to prime the soils and sediments for erosion and subsequent redistribution by the floods, followed by alluviation (Macklin & Lewin 1993, 2003) (Fig. 3.11).

Base-level changes, which are often linked to climatic change (as well as tectonic uplift), are responsible for the accommodation space of fluvial sediments. In north-east Japan, low rates of base-level rise have been related to the development of large-scale channel fills, while rapid rates have been linked to small-scale channels (Komatsubara 2004). A drop of about 22 m in the mean level of the Dead Sea over the past 70 years resulted in a constant adjustment of the Lower Jordan River (Hassan & Klein 2002). Here, rapid drops in sea-level, particularly since the 1980s, led to deep incision, changes in channel morphology and terrace development. Stable base-levels are, in places, related to channel aggradation and only minor lateral migration.

Even though climate change operates over both long and short time periods, fluvial responses to it are often rapid within these periods. Goodbred (2003) demonstrated that the Ganges River

system in Bangladesh and India responded to $< 10^4$-year climate change in a system-wide, contemporaneous manner, with punctuated, rapid sediment transfers. Lewis et al. (2001) discussed climatic shifts within the last interglacial–glacial cycle (125–10 ka), suggesting that these were rapid and that they were recorded in alluvial sediments. They described the changes in the Thames, in which major changes in the alluvial architecture were related to climatic fluctuations at c. 70 ka and 13–11 ka (the Devensian Lateglacial climatic warm–cold–warm oscillations).

During Pleistocene cold stages, rivers in northern Europe and America were similar to contemporary cold-climate tundra rivers, with overbank deposition in large, unconfined valleys, rapid vertical aggradation of sandy material on floodplains and limited lateral channel migration due to well-established tundra vegetation and river bank cohesion (Kasse 1998). In areas with little vegetation cover, large-magnitude spring snow melts would have led to enhanced sediment erosion, transport and deposition. Thick sequences of river gravels in northwest Europe (Antoine 1993; Bridgland 2000) are thought to have aggraded in such areas, as river environments changed from incisional to depositional (Bridgland & Allen 1996). Cooling stages are often recorded in the fluvial archive by channel changes: in the Nochten area, eastern Germany, a strong cooling event at *c.* 40 ka led to a change from sandy anabranching to braided river conditions (Kasse et al. 2003). With deglaciation at the Pleistocene–Holocene transition came high runoffs and sediment loads, and considerable channel change. Thomas (1998) suggested that the change from dry to much wetter climatic conditions between *c.* 1200 and *c.* 9000 yr BP resulted in high discharges in both mountainous (e.g. Nile, Amazon) and lowland tropical rivers (e.g. Niger) and accompanying landsliding and changes in slope hydrology and processes. As vegetation re-established, slopes stabilized and erosion became more confined to bank areas.

During the Holocene, climate change has continued to significantly affect fluvial activity. Alluvial deposits in England are thought to have formed during the Late Medieval Warm Period and Late Medieval Deterioration in response to coincident heavy rain and snowmelt, whereas those from corresponding periods in Italy are related to increased cyclonic activity (Brown 1998). Episodic flooding events have been shown to play a major role in present-day sedimentation in the Eel River, California (Morehead et al. 2001), and in Scottish rivers, where rates of lateral channel shift and extent of bare gravels are related to changes in flood frequency since the mid-nineteenth century (Werritty & Leys 2001).

3.3.2 Tectonic uplift and glacio-isostatic rebound

Tectonic uplift often causes river incision and shedding of sediment from upland areas to river basins. It can result in considerable channel change, particularly in valley slope, width, discharge and, ultimately, denudation and alluviation downstream. The Bengal Basin in India and Bangladesh, for example, is rapidly aggrading due to Himalayan uplift and erosion, and sea-level changes since the last glaciation (Allison et al. 2003). Lave & Avouac (2001) used suspended sediment loads in Himalayan rivers to infer that contemporary hillslope erosion over the whole Himalayan range in Nepal was driven mainly by fluvial incision linked directly to uplift. In the Miocene, the gradual tectonic uplift of Africa has been linked to sedimenta-

tion in the Congo River arising from erosion of the edge of the uplifted Angolan margin (Lavier et al. 2001). In the Tejo river, Portugal, successive regional uplift events during increased intra-plate compression are thought to cause dramatic fluvial incision (Cunha et al. 2005). In this area, river terraces, previously thought to be related mostly to glacial–interglacial cycles superimposed upon uplift, were reinterpreted as being caused mainly by these tectonic phases, punctuated by relatively short periods of lateral erosion and some aggradation, with climatic and base-level change factors only acting as a conditioning process.

Tectonic uplift can result in a short-term or gradual change in longitudinal profile, particularly when the uplift is domal (Knighton 1998). In the eastern Carpathians, Radoane et al. (2003) demonstrated that tectonic uplift at over 6 mm per year, rather than age, was responsible for the shape of the longitudinal profile and the types of channel deposits present. Laboratory experiments carried out by Ouchi (1985) have shown meandering rivers respond to slow uplift by increasing sinuosity (in the case of a steepening valley slope), or by straightening or anastomosing (in the case of a flattening slope).

In north-west Europe, uplift as a result of glacio-isostatic rebound is thought to play a role in controlling Early Pleistocene sedimentation in rivers (Bridgland 2000; Westaway et al. 2002). The basis of this theory is that, to maintain isostatic equilibrium, a particular area will undergo uplift when adjoining areas undergo repeated and cyclical glacial surface loadings (Westaway et al. 2002). This theory has been used to explain the occurrence of thick (*c.* 100 m) sequences of gravel terraces, aggraded during glacial cycles. Although climate is the major factor responsible for these terraces, in that large-magnitude seasonal flows and reduced vegetation cover cause increased erosion and large-scale deposition (Bridgland & Allen 1996), the shedding of this sediment requires uplift during interglacial marine highstands (Bridgland 2000; Westaway et al. 2002). There may also be an element of mantle bulge or upwarp involved in response to crustal unloading as a result of terrain erosion.

Burger et al. (2001) also made a link between river morphology and glaciation-related sea-level changes. They suggested that the dendritic channel pattern (and related fluvial incision) in the Eel River Basin, California formed during glacio-eustatic lowstands and periods of continental shelf exposure, but also pointed out that local tectonic uplift may have also played a role. In fact, many other studies (e.g. Veldkamp & Van Dijke 1998) have demonstrated the complex links between tectonic, climatic and sea-level changes and their combined effects on river morphology and sedimentation. Although Harvey et al. (2003) showed that sedimentary sequences and the morphological evolution of late Quaternary alluvial fans in the Tabernas basin of south-east Spain were mainly controlled by climatic change, the depositional environment of toes of the fans was created in response to tectonic uplift. Bridgland (2000) and Bridgland & Maddy (2002) suggested that sedimentation–incision cycles of the Middle–Late Pleistocene that correspond to the low-frequency, high-amplitude 100 kyr eccentricity-driven climate cycle resulted from the overprinting of long-term fluvial responses to tectonic uplift driven by climate changes in sediment supply.

3.3.3 Forest fires

Forest fires result in the removal of vegetation, reduction of ground cover and alteration of soil properties (e.g. decreased infiltration rates, formation of water-repellent layers by vaporization of forest organic compounds). All of these processes can cause significant increases in runoff and fine sediment production in the fire-affected basins (Morris & Moses 1987; Inbar et al. 1998). In Australia, for example, hydrogeomorphological changes resulted in widespread erosion and alluvial deposition of remobilized topsoil in river systems following the Christmas 2001 bushfires near Sydney (Shakesby et al. 2003). Many other studies have demonstrated similar widespread post-fire sediment delivery to rivers (e.g. Inbar et al. 1998) (Fig. 3.12).

The fluvial response to forest fire is complex (Schumm & Lichty 1965), and depends on the extent and severity of the fire, the rate of recovery of vegetation and the fluvial geomorphology. It has been suggested that the most important factor is the percentage of burned area, because this causes a range of fluvial adjustments throughout the whole basin (Legleiter et al. 2003). Following a fire, the width, depth, planimetric geometry and longitudinal profile of the river will be modified to accommodate the elevated sediment loads arising from the increased flow magnitude and frequency. Channels are reported to respond to fire-related change by aggradation, active braiding, enlargement and lateral migration, entrenchment and narrowing (Laird & Harvey 1986; Legleiter et al. 2003). The time needed for basins to return to pre-fire surface erosion and fluvial sediment delivery rates has been reported to be relatively short, in the order of 2 to 10 years (Morris & Moses 1987; Legleiter et al. 2003). This normally follows exhaustion of available fine-grained sediment, vegetative recovery and the development of coarse surface lags (Morris & Moses 1987).

It is important to note that, although forest fires are often 'natural' events, they can equally be classified as 'anthropogenic disturbances', because many forest fires are started deliberately by humans. This is not only a contemporary phenomenon; late Mesolithic to early Neolithic charcoal finds in the North Yorkshire Moors, UK, suggest that fires were started deliberately to control vegetation or drive out game for hunting, or both (Simmons & Innes 1996).

3.3.4 Volcanic eruptions

Explosive volcanic eruptions can dramatically affect sedimentary processes in rivers, both during the eruption itself and for years or decades after the eruption has ceased. The sediment loads of rivers draining volcanoes are among the highest documented, and have been compared with those of rivers in arid climates impacted by flash floods (Hayes et al. 2002). The effects of high rainfall in volcanic areas (particularly tropical volcanic belts) compound the sediment-carrying capacity of associated rivers compared with arid rivers. During eruptions, river valleys

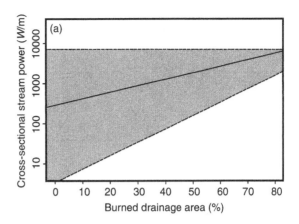

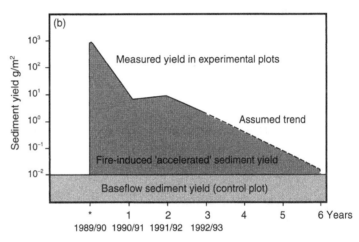

Fig. 3.12 Responses of rivers to forest fires. (a) Relationship between cross-sectional stream power and percentage of burned drainage area. The solid line represents the regression line and the dashed lines represent the minimum and maximum cross-sectional stream power values. Note the important increase in minimum stream power for more extensively burned basins. (After Legleiter et al. 2003, fig. 6.) (b) Fire-induced sediment yield against 'baseflow' yield in the Mount Carmel area, Israel. The 1992 increase in sediment yield is due to exceptionally heavy rains. (After Inbar et al. 1998, fig. 17.)

can be rapidly aggraded with great thicknesses of volcanic deposits (ash, pumice, bombs). The 1991 Mount Pinatubo eruption in the Philippines, for example, resulted in infilling of valleys with 200 m of volcanic debris (Scott et al. 1996). The most devastating effects on rivers, however, arise from lahars (Cronin et al. 1997; Hayes et al. 2002). Lahars are hyperconcentrated to concentrated volcanic debris flows, which move at high speed downslope and can reach considerable distances downstream. They are caused by pyroclastic (hot) flows combined with snow melt (Fig. 3.13), or heavy syn- or post-eruptive rainfall (Neall 1976). Lahar formation is favoured by high slope gradients, removal of vegetation due to eruptive activity, an abundance of unconsolidated volcanic ejecta and, in many cases, pre-existing channel and channel bed morphology (Hayes et al. 2002). Channel and floodplain

sediments in volcanically affected rivers are incised and reworked relatively soon (often within a matter of years) after deposition (e.g. Janda et al. 1996). This can result in the deposition of alluvial fans and in destruction of property where population centres lie downstream. For information on volcanically triggered sedimentation events see Chapter 2.

3.4 PROCESSES AND IMPACTS OF ANTHROPOGENIC ACTIVITIES

3.4.1 Anthropogenic impacts on rates and styles of sedimentation

Human exploitation of natural resources from prehistoric times through to the present day has had a very significant impact on the hydrology,

Fig. 3.13 Destruction of Armero, Columbia, by a lahar, triggered by an eruption and melting snow, which emanated from the Nevado del Ruiz volcano. More than 23,000 people died during this event. (Image and information courtesy of R.J. Janda, USGS/Cascades Volcano Observatory.)

sediment regime and ecology of many of the world's rivers. Deforestation, agriculture, land-use change, river regulation and mining are some of the anthropogenic activities that affect soil

erosion rates and sediment conveyance in river systems. Increases in flooding and flood peaks, river sediment loads and rates of valley-floor alluviation following human disturbance, or disturbance of vegetation, are now well documented (Robinson & Lambrick 1984). This section discusses several types of anthropogenic impacts on sedimentary processes in river systems.

3.4.1.1 Agriculture, deforestation and afforestation

Human settlement and cultivation have had a drastic impact on river sediment budgets, largely through deforestation and land clearance. These activities result in increased soil erosion, soil creep and landslide events, which in turn result in increased sediment input to, and sedimentation rates in, rivers (Fig. 3.14) (Knox 2001). Furthermore, the sediment composition is often altered, because progressively less weathered and less organic-rich terrigenous materials are stripped and added to the suspended load during clearance.

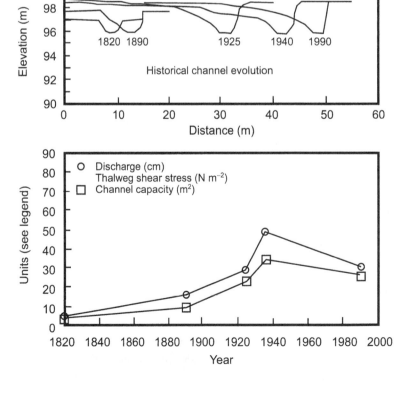

Fig. 3.14 Historical changes in channel morphology and hydraulics on the Shullsburg Branch tributary, Galena River system, USA. Accelerated overbank sedimentation from the beginning of Euro-American settlement in the 1820s increased progressively, resulting in increases of bank heights and facilitation of deeper flows with high shear stresses that led to bank erosion and channel expansion. After about 1950, overbank sedimentation decreased owing to smaller and less frequent floods as a result of improved upland land conservation practices. (After Knox 2001, fig. 9.)

Numerous studies have demonstrated increasing sediment loads to rivers since human settlement. Mei-e & Xianmo (1994) described an increase in soil erosion in the Yellow River, China, since serious cultivation began *c.* 5000 yr BP. Following another phase of clearance *c.* 1000 yr BP, Saito et al. (2001) calculated that the sediment load of the Yellow River at Sanmenxia increased by four to seven times more than its pre-1000 yr BP level of $2.7–5.0 \times 10^8$ t yr^{-1}. In the USA, sediment loads of rivers draining to the Atlantic have been estimated to have increased four to five times since European settlement and, in particular, high sedimentation rates in Chesapeake Bay and the Appalachian Piedmont have been linked to agricultural clearance in the Appalachian mountains in the nineteenth century (Kearney & Stevenson 1991). Brown (1997) discusses a plethora of examples of anthropogenically induced alluviation in the UK and rest of Europe. Among these is the Upper Thames, where increased alluviation is ascribed by Robinson & Lambrick (1984) to coupled building of large villas and hillslope agriculture during the Roman Occupation of Britain. Widespread mid- to late-medieval alluviation in Britain and a considerable part of northern Europe has also been attributed to population increases and increased ploughing during this period (Brown 1997).

Deforestation and related sediment inputs to rivers also cause significant channel change. Coltori (1997) attributed aggradation and formation of a braidplan system in the Marche basin to deforestation of the Periadriatic Basin and subsequent severe soil erosion. Post-deforestation landslides have made significant impacts on the sedimentary sequences of several depositional basins in New Zealand, and in Europe deforestation has also resulted in landslides that were triggered by high-magnitude–low-frequency climatic events, causing gully and channel erosion (Glade 2003). Changes in river channel width, sinuosity and position in basins at the foothills of the Venezuelan Andes were greatest at the most intensely deforested sites, although valley shape and other constraints on channels were important (Karwan et al. 2001).

The effects of agriculturally related land clearance are not always felt immediately. In the Chagrin River, Ohio, USA, high sediment yields matched river basin deforestation during the 1840s to 1850s, but rates declined by over 50% from the 1980s to the 1910s, despite peaks in land clearance (Evans et al. 2000). Between the 1910s and 1960s, declines in farming corresponded to increases in sediment yields. These trends were attributed to the timing of peak flows, and the resultant intrabasinal storage and delays in sediment conveyance from upstream to downstream. Walling (1999) described this delay as the 'buffering capacity' of a basin, and related it to the sediment delivery ratio, whereas Knox (2001) described the delay as a 'long-term lag response'.

Soil erosion is also promoted through agricultural practice, and currently poses a threat to the sustainability of agricultural production in many areas of the world. Remobilization of eroded soil causes siltation of irrigation networks and sedimentation of reservoirs that are used for water storage. It has been estimated that 75 billion tonnes of soil are eroded annually, and that most of this comes from agricultural areas (Myers 1993). Speth (1994) suggested that approximately 80% of the world's agricultural areas are undergoing 'moderate to severe erosion', and a further 10% from 'slight to moderate erosion'. Most of this eroded soil will channel through river basins. Rates of erosion, however, are highest in Asia, Africa and South America (1000–2000 t km^{-2} yr^{-1}) and lowest in Europe and the USA (*c.* 0–600 t km^{-2} yr^{-1}; Fournier 1960).

Agricultural and related practices also contribute to contamination of fluvial sediments, particularly with respect to fluxes of the nutrients C, N and P. Owens et al. (2001), for example, suggested that elevated particulate P concentrations in the upstream River Aire, UK, were probably related to diffuse sources of P connected with agricultural land-use. In the Amazon, progressive deforestation over the past few decades has disrupted the fluxes of humus and soil from the drainage basin, because organic-rich material that is normally retained in the soil is being progressive weathered and mobilized (Farella et al. 2001). Recently deposited alluvium is thus more humified and N-rich than older material.

Many authors (e.g. Macklin et al. 1992; Macklin & Lewin 1993; Brown 1997; Coulthard et al. 2000) have stressed that anthropogenic activity and climate change act in tandem to affect sedimentation in river basins. In a numerical (computer) modelling study, Coulthard et al. (2000) showed that, although deforestation alone increased sediment discharges by 80% in an upland basin in northern England, a change in both climate and vegetation cover resulted in a 1300% increase, confirming field-based studies in the area. River basins are therefore more sensitive to climate change when the soils are destabilized by human activities such as deforestation and agricultural practices (Coulthard et al. 2000). These activities expose relatively fresh, unweathered material, making it more susceptible to, and possibly inflating the rates of, chemical and physical weathering.

Afforestation, particularly in upland portions of basins, can also have significant effects on sedimentation in rivers, owing to soil disturbance when planting. Among the effects documented are (i) elevated rates of bank erosion (Painter et al. 1974), increased suspended sediment yields, increased bed load yields, particularly upstream (Newson 1980), and development of a flashy hydrological regime (Leeks & Roberts 1987). Mount et al. (2005) constructed sediment budgets to evaluate the impacts of upland afforestation between 1948 and 1978 in the Afon Trannon, mid-Wales. These budgets demonstrated that upland basin bed load yields of between 2 and 3 t km^{-2} yr^{-1} were equivalent to localized gravel inputs from bank erosion, and thus were probably not responsible for increased sedimentation downstream. Rather, the nature of the bank material, and increases in flood magnitude and frequency since 1988, possibly as a result of climate change, were deemed to be responsible. This study again demonstrates the interplay between different natural and anthropogenic processes in influencing river sedimentation, and the need for interdisciplinary study to understand these processes.

3.4.1.2 Mining

Present-day and historic mining activities have released large quantities of metal-bearing mineral particles into fluvial environments (Fig. 3.15a; Lewin et al. 1977, 1983; Lewin & Macklin 1987; Macklin et al. 1994; Hudson-Edwards et al. 1996, 1999b, 2001; Miller 1997). Mining-related sources of metal-bearing sediments to rivers include discharge of mine or processing waste, tailings dam failures, erosion of tailings and waste rock piles, remobilization of mining-contaminated alluvium, and mine drainage. Direct discharge of mine tailings, effluent and waste rock is one of the most common and significant sources of metal-bearing particles to river systems. This has been an important process in the past (e.g. north-east England; Hudson-Edwards et al. 1996), but is still ongoing today (e.g. Río Pilcomayo, Bolivia; Hudson-Edwards et al. 2001; Fig. 3.15b). Ore processing activities (e.g. amalgam treatment, cyanide leaching) also contaminate rivers with metals and metalloids such as Ag, Bi, Cd, Cu, Hg, Pb, Sb and Zn which, over time, can accumulate in significant amounts (Fig. 3.16).

Failures of tailings dams can release very large quantities of metal-bearing particles to river systems. The discharge of these tailings often greatly exceeds the sediment-transport capacity of a river and results in considerable channel and floodplain aggradation (James 1989). Physical remobilization of abandoned tailings or waste piles, and of channel beds and mining-contaminated floodplain alluvium (formed during historic mining activity), also provides large amounts of metal contaminants to rivers (Macklin et al. 1992; Miller 1997; Miller et al. 1998). Merrington & Alloway (1994), for example, showed that the wash load being transferred from abandoned tailings heaps in two mines in Cornwall and Wales, UK, was very high for Cu (38 kg yr^{-1}) and Pb (74 kg yr^{-1}). Erosion occurs during lateral channel migration and exposure of bank sediment (Miller et al. 1998), or channel bed, tailings, waste pile or cut-bank incision (Macklin & Lewin 1989).

(a)

(b)

Fig. 3.15 Mine waste in the Río Pilcomayo, Bolivia. (a) Braided reach, in which the entire valley is filled with mining-related sediment. The river is grey-coloured and contains sulphide mine tailings, and in places on the floodplain the sulphides have oxidized to form metal-bearing iron oxyhydroxides. Approximate width of photograph is 500 m. (b) Discharge of sulphide mine tailings directly into the river.

(a)

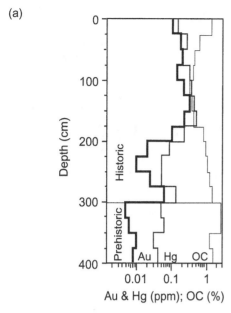

(b)

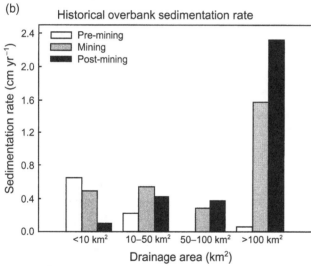

Fig. 3.16 Mining-related accumulation of metals in floodplains. (a) Vertical distribution of gold (Au), mercury (Hg) and organic carbon (OC) in floodplain alluvium, north Georgia. (After Leigh 1997.) (b) Historical overbank sedimentation rate of pre-, syn- and post-mining alluvium in the Blue River Watershed, Wisconsin. Note the spatial and temporal lag in redistribution of syn-mining alluvium in the downstream (> 100 km^2) portion of the basin. (After Lecce & Pavlowsky 2001.)

Mine drainage from both metalliferous and coal mining areas is a common source of metal contamination to river systems. The leaching of metals from exposed and buried waste is controlled by factors such as mineral dissolution, sulphide mineral oxidation and secondary mineral precipitation. These secondary minerals often precipitate or flocculate as discrete particles or as coatings on other grains; these particles co-precipitate or sorb the dissolved metals and are subsequently deposited or transported downstream. Mining operations themselves can also alter the natural fluvial geomorphology during the construction of artificial drainage networks and other structures.

3.4.1.3 River regulation and channelization

River systems are regulated for many reasons, including more efficient navigation, flow, bank erosion, drainage and flood control, stabilization of channels, provision of surface water reservoirs for drinking, irrigation or other purposes, hydroelectric power, removal of natural vegetation and waste disposal (Newson 1992). Typically, regulation is carried out by dam and reservoir building, and channelization, which can involve resectioning, sediment extraction, the use of flood walls, culverts, embankments or levees, and straightening.

Sedimentation in regulated rivers can be considerably different from that in natural rivers. In dammed rivers, reservoirs act as sediment traps, significantly affecting rivers with medium to large natural sediment loads. Downstream of dams, many rivers have been reported to supply less sediment after damming for irrigation, flood control and power generation. This reduction has resulted in catastrophic changes in downstream river reaches, and deltas and coastlines, which are often eroded as a result of increased stream power (Trenhaile 1997). In the Missouri

and Mississippi rivers, for example, the reduction of annual sediment yield from 817 million tonnes to 204 million tonnes in the twentieth century is thought to result from large-scale dam projects. This loss in sediment yield is in turn linked to rapid coastal degradation and recession in downstream Louisiana (Keown et al. 1986). Walling & Feng (2003) carried out an extensive study of long-term records of annual sediment load and runoff for 145 major rivers, and demonstrated that although c. 50% of the rivers experienced statistically significant upward or downward trends, most showed the latter. This was largely attributed to dam construction. Impacts of the Aswan High Dam on the River Nile are also discussed in Chapter 7, and the Hoover Dam in Colorado in Chapter 1.

Many studies have demonstrated effects on river sedimentation as a result of river regulation (e.g. Case Study 3.1). Arnaud-Fassetta (2003)

Case Study 3.1 Past, present and future impact of the Three Gorges Dam on the Yangtze River

The Yangtze River (Changjiang) is the longest river in Asia (> 6300 km), the third longest in the world and has the ninth largest catchment basin (1.8×10^6 km^2) in the world. Its water discharge is the largest in the western Pacific Ocean, and its sediment load (annual discharge of 480×10^6 t yr^{-1} before the 1990s) is the third largest in the world behind the Amazon and the Congo. Shanghai, the most important industrial and economic city of China, is located near the river mouth. The 39.3×10^9 m^3 Three Rivers Gorge Dam project, scheduled to be completed between 2009 and 2012, will be the largest dam in the world, with estimated power production equalling approximately 12 nuclear power plants. The dam is being built for flood management, inland navigation and hydropower.

Yang et al. (2002) reported that decadal river sediment discharge and suspended sediment concentrations (SSC) had reduced by 34% and 38%, respectively, between the 1960s and 1990s (Case Fig. 3.1a). This was in contrast to the period between the 1950s and 1960s, when annual sediment discharge and SSC increased by 10 and 12%, respectively. Yang et al. (2002) attributed these increases to deforestation, and the 1960s to 1990s decreases to a combination of dam and reservoir construction (Case Figure 3.1b). The authors used these data to predict that both river sediment discharge and SSC will probably decrease to 40% of original levels in 50 years and to 50% in 100 years, largely as a result of the dam and reservoirs. Critics have suggested that the damming will also lead to the build-up of sediment-borne contaminants, damage of turbines by sediment, increases in deforestation and soil erosion as former inhabitants relocate and erosion in the delta downstream of the dam. Wang et al. (2005) have demonstrated that, after the completion of the dam, anastomosing channels downstream of the site will cease to carry significant amounts of water and sediment, and thus disrupt the balance between fluvial discharge and

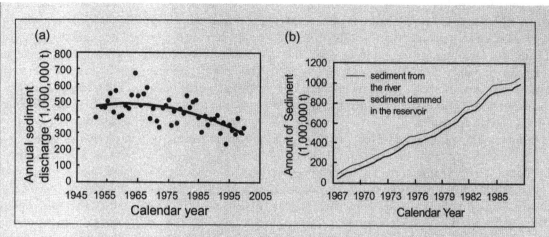

Case Fig. 3.1 (a) History of annual sediment discharge at Datong Station, suggesting a decreasing trend since the 1960s. (b) Accumulative river sediment discharge and amount of sediment dammed in Danjiang Reservoir, middle reach of the Hanjiang River, one of the major tributaries of Yichange and Datong.

basin subsidence due to the Himalayan orogeny. By contrast, others suggest that a decline in sediment supply downstream of the dam will be beneficial in that flood levels will be less. Furthermore, the Danjiangkou Dam, built on a tributary of the Yangtze, is not believed to have caused downstream incision, and concerns that coastal erosion will occur are countered by suggestions that mud flats north of the Yangtze will be supplied by tributaries downstream of the Three Rivers Gorges. The full effects of the Dam Project, and answers to these questions will not, however, be available for many decades.

Relevant reading

Fuggle, R., Smith, W.T., Hydrsult Canada Inc. & Agrodev Canada Inc. (2000) *Large Dams in Water and Energy Resource Development in the People's Republic of China (PRC)*. Country review paper prepared as an input to the World Commission on Dams, Cape Town. http://www.dams.org (accessed 25 June 2004).
Milliman, J.D. & Meade, R.H. (1983) World-wide delivery of river sediment to the oceans. *Journal of Geology* 91, 1–21.
Wang, S., Chen, Z. & Smith, D.G. (2005) Anastomosing river system along the subsiding middle Yangtze River basin, southern China. *Catena* 60, 147–63.
Wang, Y. & Zhu, D.K. (1994) *Coastal Geomorphology*. Higher Education Press, Beijing, 244 pp. (In Chinese.)
Yang, S., Zhao, Q. & Belkin, I.M. (2002) Temporal variation in the sediment load of the Yangtze river and the influences of human activities. *Journal of Hydrology* 263, 56–71.

showed that engineering of the Rhône River in France, carried out to control overbank flooding and stabilize the river planform, resulted in channel narrowing and entrenchment, and increased hazard of low-frequency, high-magnitude flood events (Fig. 3.17a). Surian & Rinaldi (2003) carried out a comprehensive review of the response of Italian rivers to sediment extraction, dams and channelization, and highlighted

channel incision, generally 3–4 m but up to 10 m, and channel narrowing of up to 50%, as the major effects (Fig. 3.17b). They did, however, point out that these effects were most pronounced immediately after the engineering works were carried out, and that the effects slowed and became asymptotic. Gaeuman et al. (2005) demonstrated that the creation of large-scale diversions and increased water withdrawal for

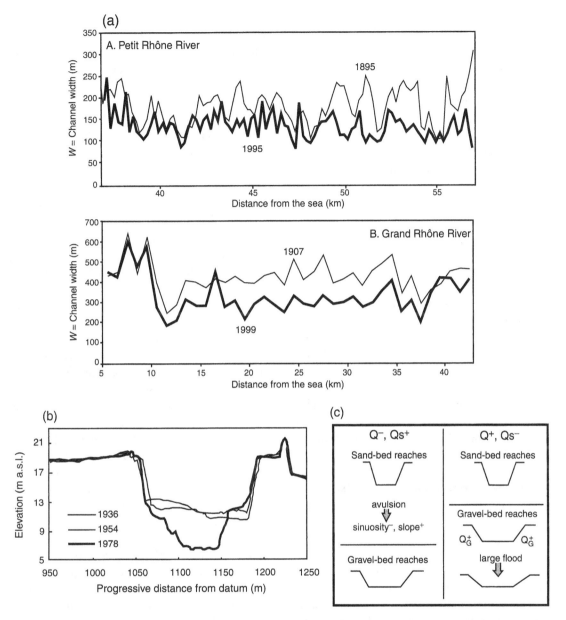

Fig. 3.17 (a) Changes in channel width in the Petit Rhône River (A) and the Grand Rhône River (B) between 1895 and 1999. Both graphs highlight the channel narrowing as a result of channelization at the end of the nineteenth century. (After Arnauld-Fassetta 2003.) (b) Channel incision along the Arno River in the Lower Valdarno–Pisa Plain, Italy, showing typical change in cross-section, with limited bed lowering from 1936 to 1954, and intense incision from 1954 to 1978. (Based on fig. 2 of Surian & Rinaldi 2003.) (c) Summary of adjustments in sand-bed and gravel-bed reaches of the lower Duchesne River, Utah, USA, owing to the creation of large-scale diversions and increased water withdrawal for irrigation. Bold arrows indicate the primary adjustment to changes in discharge (Q) and fine sediment supply (Q_s), and thin arrows indicate second-order adjustments. Q_G^+ indicates an increase in gravel-sized sediment derived from lateral erosion. (After Gaeuman et al. 2005, fig. 15.)

irrigation during the early part of the twentieth century in the Duchesne River basin, Utah, caused local gully erosion and subsequent downstream delivery of fine sediment, which in turn led to filling of secondary channels, the narrowing of the main channel and transformation of a narrow sand-bed channel into a braided gravel-bed reach (Fig. 3.17c).

3.4.1.4 Urbanization

Urban expansion can have a major effect on channel morphology and sediment supply and deposition. Activities such as building on floodplains, increased ground-water extraction, channelization, flood embankment and urban-drainage-system construction, sewage disposal and (potentially contaminated) road runoff are among those that have major effects. These effects include floods and slope failure, particularly in tropical areas (Gupta & Ahmad 1999). Often the maintenance of river channels and flood alleviation systems is extremely costly.

Sediment discharge is often highest during the initial stages of urban development, when stripping of soil and vegetation to make way for built structures is at a peak. After urbanization is complete, there is a period of adjustment as rivers respond to decreased sediment supply. The combination of the removal of soil and vegetation, creation of impermeable surfaces and installation of drainage networks in urban areas causes increased runoff during storms, with relatively short lag times and high peak discharges (Knighton 1998). More detailed information on urbanized rivers can be found in Chapter 6.

3.4.2 Sediments as sinks for contaminants: transport, deposition and remobilization

3.4.2.1 Sediment contaminants

As discussed in section 3.2.1.1, rivers are subject to contamination by a large number of metal, nutrient, organic and radionuclide elements and compounds. A large proportion of the contaminant load in fluvial systems is transported by particulate matter. Trefry & Presley (1976) and Gibbs (1977) suggested that up to 90% of the metal load is transported by sediments in many rivers, but warned that this varied from metal to metal. This was borne out by Goldstone et al. (1990), who showed that the majority of Zn released from effluent in Norwich, England, existed principally as dissolved species (95%), whereas Pb was partitioned equally between dissolved and particulate phases. The partitioning

also appears to apply to organic contaminants, as shown by Lau et al. (1989), who estimated that chlorinated organic contaminants were almost equally distributed between the soluble and particulate phases in the Detroit River, USA. The 'dissolved' portion encompasses contaminants that are truly dissolved, colloids or sulphide 'clusters'. The latter were described by Rozan et al. (2000), who showed that Fe, Cu and Zn were bound to multinuclear sulphide clusters in oxic rivers in Connecticut and Maryland, USA, which received substantial inputs of sewage effluent. The 'particulate' portion of the contaminant load comprises contaminant-rich grains (e.g. metal sulphide grains from tailings effluent) or contaminant-bearing Fe and Mn oxide coatings on other particles.

In rivers subject to acid-mine or acid-rock drainage, enormous quantities of metals can be transported downstream in the solute phase (Filipek et al. 1987). In general, however, this solute phase is considerably reduced downstream of the source as pH rises to neutral (usually as a result of tributary or ground-water inputs), and the dissolved metals are precipitated or adsorbed onto sediments (Davis & Leckie 1978). In rare cases, such as the Río Tinto in south-west Spain, acid pH is maintained along the whole course of the river and significant quantities of dissolved metals are transported to estuaries where they precipitate or are sorbed to other particulates (Hudson-Edwards et al. 1999b). In rivers of neutral or higher pH, metals and other substances are largely transported downstream in the particulate load, as discussed above (Gibbs 1973; Benjamin & Leckie 1982; Horowitz & Elrick 1987).

The partitioning of contaminants between the dissolved and particulate load in fluvial systems depends on both physical and chemical factors. The chemical factors include variations in amounts of suspended and deposited sediments (Gibbs 1977; Kratzer 1999), adsorption onto fine-grained material, co-precipitation with or sorption on hydrous Fe–Mn oxyhydroxides and carbonates, association with organic matter (Karickhoff 1981), incorporation in crystal lattices of minerals, acidification, salinity, complexing agents, biomethylation and, probably most significantly, pH and redox processes (Hem 1972; Gibbs 1977; Salomons & Förstner 1984).

3.4.2.2 Sediment-borne contaminant transport

A large number of physical factors influencing contaminant transport and the partitioning between the dissolved and particulate load have also been identified. These include sediment texture, site geomorphology (Eyre & McConchie 1993; Macklin 1996), streamflow, suspended sediment concentration (Yeats & Bewers 1982; Kratzer 1999), climate and season (Boyden et al. 1979; Macklin 1996), water depth, flow dynamics (Rust & Waslenchuk 1974; Kratzer 1999), diffusion across the sediment–water interface (McKnight & Bencala 1989) and sediment grain size (Moore et al. 1987). Gibbs (1977) and Horowitz & Elrick (1987) suggested that grain size is possibly the most significant factor controlling the concentration and retention of contaminants in both suspended and bottom sediment. Metals, in particular, have been shown to be enriched in the fine silt and clay fractions of sediments (Salomons & De Groot 1978), as a result of their large surface area, organic and clay contents, surface charge and cation exchange capacity (Horowitz & Elrick 1987).

Once sediment-bound contaminants enter river systems, they are dispersed by fluvial processes and transported downstream. Many of the significant advances in the understanding of the dispersal, storage and remobilization of sediment-bound contaminants in river systems have been made through studies of metal mining-affected rivers (see reviews by Lewin & Macklin 1987; Macklin 1996; Miller 1997). The processes described in these papers also apply to rivers contaminated by other elements and compounds (e.g. lipophilic organic compounds, Rostad et al. 1999; radionuclides, Cochran et al. 2000; nutrients, Owens et al. 2001). Work on mining-contaminated rivers has demonstrated that sediment contaminant metal concentrations tend to decrease downstream from pollution point sources in a systematic way that can be approximated using negative linear, exponential or power functions (Wolfenden & Lewin 1977; Lewin & Macklin 1987). Deviations from these models are due to floodplain storage or inputs of contaminants from diffuse or other point sources (Axtmann & Luoma 1991; Macklin 1996). The downstream decreases have been attributed to:

1 dilution of contaminated sediment by uncontaminated sediment derived from tributaries, channels and tributaries upstream of contaminant point sources and erosion of channel banks (Macklin 1996; Hudson-Edwards et al. 2001);
2 hydraulic sorting of channel-bed sediment on the basis of density, size or shape, which selectively enriches sediment near the mining source in contaminant metals (Langedal 1997a; Leigh 1997);
3 abrasion of contaminated sediment grains (Langedal 1997b);
4 storage of contaminated particles in channel and floodplain deposits (Macklin et al. 1992);
5 chemical sorption or dissolution of contaminants and/or contaminant uptake by biota (Lewin & Macklin 1987; Hudson-Edwards et al. 1996).

Geomorphological and sedimentary processes are influenced by, and control, the dispersion and transport of sediment-bound contaminants, as illustrated by the definition of two end-member responses of rivers to inputs of mine waste (Lewin & Macklin 1987): *active transformation* and *passive dispersal* (see Case Study 3.2).

3.4.2.3 Deposition of sediment-borne contaminants

Within river systems, channels, floodplains, riparian wetlands and reservoirs are the ultimate sinks for metal, nutrient and radionuclide deposition and storage (Macklin et al. 1992; Chesnokov et al. 2000). Many studies have demonstrated that enormous quantities of contaminants (up to several millions of tonnes) can be stored in these environments. Hudson-Edwards et al. (1999a), for example, established that approximately 6.2×10^8 t of Pb and 6.4×10^8 t of Zn were stored in 2000-year-old floodplains of the Yorkshire Ouse basin, UK, which had been affected by industrial activity and Pb–Zn mining.

Case Study 3.2 Fluvial responses of rivers to inputs of mine wastes: active transformation and passive dispersal

The response of rivers to inputs of both contaminated and uncontaminated particles can be viewed as a continuum, with *active transformation* at one end, and *passive dispersal* at the other. In *active transformation*, the massive input of large quantities of contaminated sediment engulfs the fluvial system, causing changes in the types, rates and/or magnitudes of fluvial erosional and depositional processes. This in turn causes the rivers to undergo significant transformations in channel form, which influences the deposition and storage of the contaminated sediment. The classic study of Gilbert (1917) demonstrated that river systems in the Sierra Nevada area, California, responded to enormous inputs of metal mining waste by aggrading their channel beds by 3–5 m during a 10–20 year period after hydraulic gold mining ceased in 1884 (Case Fig. 3.2a(i)). After all of the mining-contaminated sediment had been exhausted, the channel beds were slowly reworked and degraded (over tens to hundreds of years), eventually returning to their pre-mining elevations (Case Fig. 3.2a(i)). Gilbert (1917) modelled the downstream aggradation-degradation process as a simple, symmetrical sediment debris wave. Graves & Eliab (1977) extended the work of Gilbert (1917) and showed that the wave in fact was skewed (Case Fig. 3.2a(ii)); this reflected the temporary floodplain storage of mine waste and its subsequent remobilization.

Active transformation can also result in a channel metamorphosis from meandering to braided in response to the large sediment input, followed by incision and reversion to a single channel after mining (and the sediment supply) has ceased. This is well-illustrated by the River Nent, a tributary of the River Tyne in north-east England, which was affected by large inputs of lead-, zinc- and cadmium-bearing mining waste in the eighteenth and nineteenth centuries AD. Here, the single channel and floodplain before 1820 was transformed into a broad, aggrading valley floor due to the large influx of fine-grained, metal-rich sediment. After mining ceased in the early twentieth century, lateral reworking was initiated, and incision of this floodplain began sometime between 1948 and 1976, eventually exposing bottom gravels by 1984 (Case Figure 3.2b).

By contrast, in *passive dispersal*, the channel and floodplain are not disrupted by the influx of mine waste, because that waste is carried as part of the 'normal' sediment load. This is regarded as a system at equilibrium. 'Natural' and contaminated sediments are transported together, deposited in, and remobilized from, lateral and vertical (overbank) accretionary deposits, with little physical impact on the system.

Even though the terms *active transformation* and *passive dispersal* were defined for mining-affected river systems, they can equally be applied to rivers that are affected by other sediment-borne contaminants such as nutrients, organics and radionuclides.

Relevant reading

Gilbert, G.K. (1917) *Hydraulic Mining Debris in the Sierra Nevada.* Professional Paper 105, US Geological Survey, Washington.

Graves, W. & Eliab, P. (1977) *Sediment Study: Alternative Delta Water Facilities – Peripheral Canal Plan.* Central District, California Department of Water Resources, Sacramento.

Lewin, J. & Macklin, M.G. (1987) Metal mining and floodplain sedimentation in Britain. In: *International Geomorphology 1986*, Part 1 (Ed. V. Gardiner), pp. 1009–27. Wiley, Chichester.

Lewin, J., Bradley, S.B. & Macklin, M.G. (1983) Historical valley alluvium in mid-Wales. *Geological Journal* 18, 331–50.

Miller, J.R. (1997) The role of fluvial geomorphic processes in the dispersal of heavy metals from mine sites. *Journal of Geochemical Exploration* 58, 101–18.

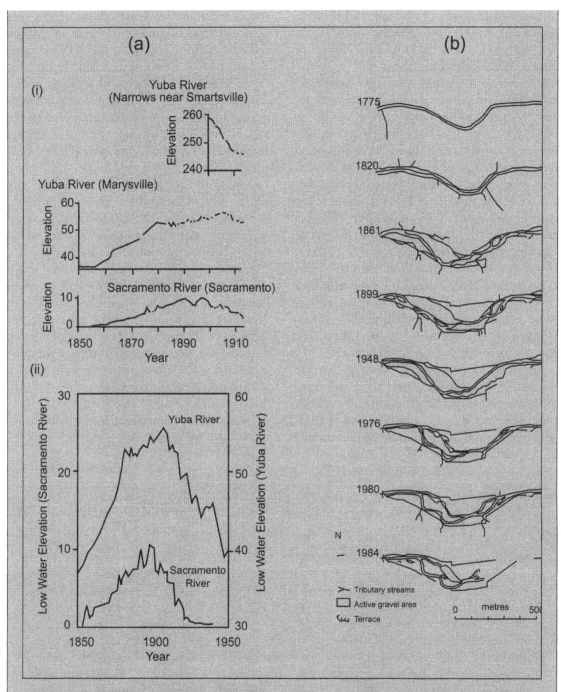

Case Fig. 3.2 (a) Changes in low-flow channel bed elevations along the Yuba and Sacramento Rivers, California: (i) Gilbert's (1917) original data (modified from Gilbert 1917); (ii) additional work by Graves & Eliab (1977) that uses and extends Gilbert's data (modified from Graves & Eliab 1977). (Figures adapted from Miller 1997.) The plots illustrate active transformation of the rivers by channel bed aggradation and subsequent degradation. (b) Channel changes (plan view) between 1775 and 1984 on the River Nent, north-east England. Active transformation has resulted in a change from a single thread to a braided channel, and deposition of fine-grained mining waste. Subsequently the latter has been incised since sometime between 1948 and 1976. (After Lewin and Macklin 1987.)

The residence times of contaminants in floodplains can be very long, ranging from 10^1 to 10^3 years. This was illustrated by Helgen & Moore (1996), who estimated that it would take several thousand years to completely remobilize metal-bearing tailings deposited on the floodplain of the Clark Fork River floodplain, Montana, USA. The length of time the contaminants remained stored in the floodplain environment generally depends on the rates of post-depositional physical, biological and chemical remobilization (see below), and on the geomorphology of the basin or reach. In stable reaches (e.g. floodplains far from the active channel, valley reaches with low stream power) or reaches where aggradation rates are high, contaminants can be stored for extremely long periods (James 1989; Miller 1997), whereas in more active reaches (e.g. near-channel floodplains, valley reaches with high stream power), the storage period may be short (James 1989).

Lewin & Macklin (1987) showed that the deposition and storage patterns of sediment-associated metals can be related directly to floodplain geomorphological processes and channel sedimentation styles. In laterally mobile braided and meandering river environments, aggradation of contaminated sediments occurs in floodplains and bars in reworked valley floors (Wolfenden & Lewin 1977; Lewin et al. 1983; Macklin & Lewin 1989). The patterns of sediment metal concentrations vary considerably across floodplains in these environments, according to the ages and quantities of contaminated sediments deposited (Lewin et al. 1983; Macklin & Lewin 1989).

Floodplain morphology often affects the nature of contaminated sediment deposition. In reaches or rivers that experience little lateral movement, contaminated sediments are added to floodplains by overbank deposition as a thin veneer over the whole floodplain surface (Lewin & Macklin 1987; Bradley & Cox 1990; Chesnokov et al. 2000) (see Case Study 3.3). River-bed incision at the end of the eighteenth century in the lower South Tyne valley, England, restricted metal-contaminated overbank sediment deposition to a relatively narrow trench adjacent to the contemporary river (Macklin 1988). In the River Aire, England, the construction of flood embankment structures at Beal confined and accelerated sedimentation in the near-channel

---

Case Study 3.3 Styles of deposition of sediment-borne radionuclide contaminants in the River Techa, Urals, Russia

The River Techa in the Chelyabinsk region of the Urals in Russia is typical of river systems affected by releases from nuclear facilities. Liquid radioactive wastes with a total activity of 10^{17} Bq, of which 12.2% was the radiogenic isotope caesium-137 (^{137}Cs), were released to the River Techa from the Mayak plutonium facility between 1949 and 1952. These waste releases coincided with large floods, resulting in widespread contamination of the Techa floodplain by ^{137}Cs. Water reservoirs and by-pass channels were constructed between 1951 and 1961 to prevent further radionuclide release to the river system, but the contaminated floodplain is currently exploited for fishing, bathing, pasture and hay collection by residents of the village of Muslomovo.

Subsequent study showed that much of the liquid waste had precipitated or sorbed onto sediment grains, with the total deposition of ^{137}Cs within a surveyed area of 2.5 km^2 estimated at 6.6 TBq. Two major styles of deposition of ^{137}Cs-contaminated sediment occurred. In many areas, deposition of highly contaminated sediment was restricted to a narrow zone near the river (Case Fig. 3.3a). This zone is confined by steep river banks (gradient of 0.0015) and has a relatively narrow floodplain (up to 75 m); maximum ^{137}Cs deposition above 7.5 MBq m^{-2} is localized in areas of bank height up to 1 m above normal water level (Case Fig. 3.3b). Further

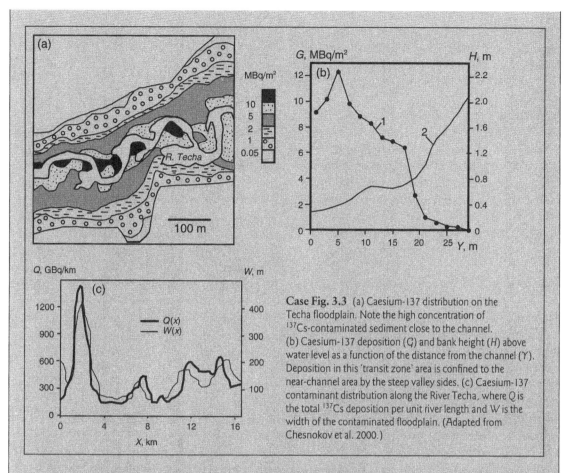

Case Fig. 3.3 (a) Caesium-137 distribution on the Techa floodplain. Note the high concentration of ^{137}Cs-contaminated sediment close to the channel. (b) Caesium-137 deposition (G) and bank height (H) above water level as a function of the distance from the channel (Y). Deposition in this 'transit zone' area is confined to the near-channel area by the steep valley sides. (c) Caesium-137 contaminant distribution along the River Techa, where Q is the total ^{137}Cs deposition per unit river length and W is the width of the contaminated floodplain. (Adapted from Chesnokov et al. 2000.)

away from the river, ^{137}Cs concentrations decline significantly, generally within 15–20 m from the river (Case Fig. 3.3c). Even though these depositional sites concentrate highly contaminated sediment, they only constitute a small percentage of the total ^{137}Cs within the surveyed area and are thus referred to as 'transit zones'.

A large proportion of ^{137}Cs-contaminated sediment is deposited in 'sedimentation zones' that comprise a well-drained floodplain with large meanders to the east of the Muslomovo and a marshy floodplain (with a gradient of 0.0003 and width of 700 m) to the west of the village. The latter contains more than 2 TBq of [137]Cs. The total deposition of ^{137}Cs-contaminated sediment per unit length of the River Techa corresponds well to the contaminated floodplain width distribution along the river (Case Fig. 3.3b), showing that the floodplain is a very effective contaminant sink, and that low-lying, wide floodplain areas such as those discussed above store large amounts of [137]Cs.

Relevant reading

Chesnokov, A.V., Govorun, A.P., Linnik, V.G., et al. (2000) ^{137}Cs contamination of the Techa river flood plain near the village of Muslumovo. *Journal of Environmental Radioactivity* 50, 179–91.

Trapeznikov, A.V., Pozolotina, V.N., Chebotina, M.Ta., et al. (1993) Radioactive contamination of the Techa river, the Urals. *Health Physics* 65, 481–8.

zone several metres above the adjacent valley floor (Hudson-Edwards et al. 1999a). Contaminants associated with fine-grained sediments also preferentially accumulate in topographically low areas on floodplains (Wolfenden & Lewin 1977; Bradley & Cox 1990). In floodplains affected by overbank deposition, sediment contaminant concentrations are either highest immediately adjacent to the active channel, dropping sharply with increasing distance from the river (Macklin 1988; Leigh 1997; Chesnokov et al. 2000; see Case Study 3.3), or are more uniform across the floodplain (Bradley & Cox 1990). Macklin (1996) suggested that these differences were due to grain-size controls on contaminant concentrations.

3.4.2.4 Uses of overbank sediment profiles

Overbank sediment profiles have many uses in contaminated river systems. Because overbank sediments record both the natural and anthropogenic geochemical evolution of floodplains, one of their main uses is as sampling and mapping media to assess metal contamination caused by mining and industrial activity (Lewin & Macklin 1987; Macklin et al. 1992, 1994; Miller 1997; Hudson-Edwards et al. 1999a). This is generally carried out by comparing near-surface overbank contaminant concentrations to those in sediments buried deeper in the sediment profile. Macklin et al. (1994) stressed that this type of assessment should be carried out only in conjunction with geomorphological mapping and dating of representative floodplain overbank profiles at a number of reaches, to take into account any lateral variations in metal concentrations and to cover as wide an age range as possible. Metal concentrations and ratios in overbank sediments also have been used as stratigraphical markers for provenancing (e.g. Passmore & Macklin 1994), dating (Davies & Lewin 1974; Lewin et al. 1977; Macklin & Lewin 1989; Macklin et al. 1992) and examining the contaminant sedimentation histories of vertically accreted fine-grained overbank deposits (Swennen et al. 1994; Hudson-Edwards et al. 1999a). This is possible because metal concentrations and ratios vary systematically with respect to the

contaminant inputs from mining or industrial areas (e.g. Knox 1987; Swennen et al. 1994).

3.4.2.5 Contaminant remobilization

Contaminants stored on floodplains may be remobilized through chemical processes (changes in redox and pH; Hudson-Edwards et al. 1998) and through physical erosion, as shown by Brunet & Astin (2000), who reported that elevated discharges of inorganic N and P were associated with increased autumn rainfall and sediment conveyance in the River Adour, south-west France. Physical remobilization may take place long after the primary contaminating activity (e.g. mining) has ceased (Miller et al. 1998). In the Carson River basin, west-central Nevada, USA, Miller et al. (1998) demonstrated that lateral instability, coupled with channel-bed incision, resulted in the exposure and erosion of mining-related, Hg-contaminated sediment from bank sediment, and suggested that the valley fill was the primary contemporary source of Hg to the river. Contaminant remobilization is often initiated by natural (climate) or anthropogenic changes (land use) that cause modifications to sediment load and delivery, and ultimately erosion and deposition (Lewin & Macklin 1987; Macklin & Lewin 1989; Macklin 1996; Miller 1997).

3.5 MANAGEMENT AND RESTORATION OF FLUVIAL SYSTEMS

River systems are continually being brought under management owing to human requirements for water, agriculture, industry and urbanization. Indeed, there are few rivers in the world today that are not regulated or managed to some degree. River management is carried out for a large number of reasons, based on perspectives that rivers are either potential hazards or benefits to anthropogenic activity (Table 3.4).

In many cases, river engineering and management has not been carried out with a full understanding of fluvial geomorphology. Gilvear (1999) has identified fundamental areas where fluvial geomorphologists can contribute to the

Table 3.4 Perspectives and objectives of river management. (Based on Knighton 1998, p. 330.)

River as hazard	River as resource
Bank protection	Aesthetics
Bridge stability	Agriculture
Deforestation	Conservation
Flood control – channelization, dams	Ecology
Floodplain zonation	Fishery
Land drainage – agricultural drains, road drainage, urban stormwater systems	Heritage
Contamination of water and sediment	Navigation
Soil erosion and sediment transport	Recreation
Gully development	Rivers as international- to district-level boundaries
	Urbanization
	Sand and gravel extraction
	Water resource – irrigation, industrial and municipal supplies, power generation

engineering of rivers and floodplains, and suggested that, ideally, river management schemes should incorporate all of these factors from the planning stage onwards. The first of these is that lateral, vertical and downstream connectivity and relationships between planform, profile and cross-section should be identified. In other words, schemes should consider the whole river basin rather than only local reaches. This premise is based on the principles that river channels are three-dimensional, with longitudinal, transverse and vertical dimensions that are modified in response to changes in water and sediment fluxes. The interconnectivity between different parts of river systems means that change in one part will eventually result in change within a contiguous part (Knighton 1998). This is clearly shown in the cases of dam construction, where trapping of the upstream sediment load leads to bed degradation and loss of habitat downstream (Kondolf 1995; section 3.4.1.3). It is thus important for engineers to view rivers from basin, rather than reach, perspectives (Knighton 1998) because by doing so they will be able to understand the underlying causes for potential problems, rather than their local (reach-based) manifestations (Sear et al. 1995).

The second of Gilvear's (1999) recommendations is that the chronology of events, landforms and sedimentary deposits in basins over a range of time-scales be determined for river systems.

James (1999) stressed this time aspect of fluvial change, because this is highly relevant to the long-term stability of engineered structures and flood-risk assessments. Gilvear (1999) suggested that time-scales of less than 1000 years are the most significant to managed rivers. Several studies have demonstrated that river channels are also extremely sensitive to environmental change and can adjust very rapidly, from days, seasons to years (Coulthard et al. 2000). Studies of fluvial landforms as indicators of stability and instability, and documentation of sedimentary histories within river basins, can be carried out by geomorphologists to aid prediction of future change (Gilvear 1999).

Recommendation three follows on from two, in that geomorphology should be linked to environmental change. By collecting field geomorphological data and combining them with other field (e.g. age dating, pollen analysis, vegetation mapping) and historical data (e.g. chronicles, weather diaries; Newson 1992), the fluvial geomorphologist can identify the processes (e.g. climate change) responsible for river landforms, and predict how rivers may respond to these processes in the immediate future. Climate change and the related frequency of extreme events such as floods and related large-scale sediment transfer can result in structural changes to rivers and damage to built structures (such as channels or dams). Baker (1994) has pointed out that such extreme events may not be predicted by using statistical analysis of hydrological data, because they are periodical on a scale outside of these relatively short-term data. Baker et al. (1983) suggested that careful, combined geomorphological and palaeohydrological studies of palaeoflood hydrology and sedimentary deposits can help to elucidate these patterns, to help predict the real flood risk and potential hazard, and to make long-term plans for river systems.

Gilvear's fourth point is that engineers should recognize the links between landforms and processes in establishing and controlling fluvial ecosystems. Pools, riffles, undercut banks and backwaters are vital habitats for many riverine species. In fact, hydrogeomorphological variables have been shown to be more important than vegetation in terms of controlling species populations (Bickerton 1995). Vegetation, in turn, is controlled by sediment type, organic matter content and stability, as well as hydraulic factors. Efforts at restoring riverine wetland ecology that focus on re-establishing pre-existing landforms by removing unnecessary sediment or adding structures to encourage the formation of habitat-friendly landforms (Brookes 1992) are generally the most successful.

As sediment transfer is a key process that links engineering and fluvial geomorphology (Sear et al. 1995), an understanding of the sources, distribution and sinks of sediment is fundamental to engineering and management practices. Having the tools to provenance suspended sediment in river systems is therefore of vital importance in controlling erosion-sensitive areas, monitoring and maintaining water and sediment quality, and developing geomorphological models (Walling 1990). Furthermore, knowledge of the influence of hydraulic factors on channel change, sediment transport processes, sediment budgets and deposition and storage of sediment, on a variety of time-scales, is necessary to improve river engineering and reduce hazards such as floods (Anthony & Julian 1999).

River management thus ideally involves the collaboration of engineers and fluvial geomorphologists, as well as end-users and managers of the river's hazards and resources. Although many schemes are focused primarily on the protection, enhancement and effective use of water supplies, they also cover the important issue of sediment management. Factors include the erosion and sedimentation, reducing contamination, restoring habitats (see below) and protecting productive floodplain zones. Because water and sediment transfer in river basins are intimately connected, many types of management involve both. Today, for example, climate change and other factors are forcing governments to adopt 'soft' engineering options for rivers, particularly in the developed world, with the creation of washlands (involving removal or addition of sediment) as a means of flood control (Newson 1992).

One branch of river management, river restoration, has seen increasing growth since its inception in the 1980s (Gore 1985). Restoration involves the creation of sustainable geomorphological features that are favourable habitats for riverine biota to recover from damage or to re-establish (Newson 1992). Methods include construction of habitats, management of riparian zones, restoration of hydrological stability and improvement of water quality (Newson 1992; Gilvear 1999). Although many schemes involve the creation of stable landforms (pools, stable riffles), increasingly, the importance of ephemeral forms such as bars, splays and scours is being recognized (Marston et al. 1995). Other methods, including revegetation, reduction of channel banks to reduce erosion and restoration of meanders, are also important (Newson 1992).

3.6 FUTURE ISSUES

A large number of issues face fluvial environments in terms of sedimentation and sediment quality. Some of the most important of these are outlined below.

3.6.1 Impacts of climate change

The predicted changes in temperature, precipitation and other global-warming-related effects on sedimentary systems are outlined in Chapter

1. During global warming the magnitude and frequency of streamflow can be significantly changed, resulting in increased frequency of high-flow events (Macklin 1996). This can result in increased flooding, especially flash flooding, hazards such as debris flows and landslides, and channel change. Rumsby & Macklin (1994) have shown that in northern England since AD 1700, channel incision is related to clustering of large floods, and lateral reworking and sediment transfer occurs during episodes of low flood frequency. They suggested that increased flooding in the future (associated with global warming) could therefore result in increased channel erosion. Longfield & Macklin (1999) attributed high-magnitude, high-frequency floods in the Yorkshire Ouse Basin, UK, during the 1980s and 1990s, to increases in the frequency and vigour of cyclonic atmospheric circulations; these also have been related to significant remobilization of mining-contaminated sediment (Dennis et al. 2003). Increases in flooding are also related to 'priming' of land surfaces by agricultural practices, which leads to increases in runoff and accelerated sediment delivery to floodplain environments (Knox 2001). Global warming could also lead to increases in the magnitude, intensity and frequency of forest fires, which may also exacerbate bank erosion, sediment transport and widespread alluviation.

Cellular computer models are increasingly being used to study the development of river basins, and the effects of climate, base level and other environmental changes on erosion and sedimentation processes (Veldkamp & Van Dijke 1998; Coulthard et al. 2000; Coulthard & Macklin 2001). These models are most effective when combined with field-based studies (e.g. Coulthard & Macklin 2001). Models such as these will, in the future, prove invaluable in assessing possible future geomorphological and sedimentological changes in river systems, and linking these to past, analogous changes. An example is the FLUVER model, constructed by Veldkamp & Van Dijke (1998) and applied to the Allier–Loire basin in France. The model takes into account tectonic uplift, changes in river longitudinal profiles and changes in effective precipitation and sea (base) level. It showed that, although changes in erosion and sedimentation were related to climatic changes in some instances, the role of tectonic uplift and changing base level had previously been underestimated.

3.6.2 Impacts of increased anthropogenic disturbance

As requirements for natural resources and land increase, river systems are, and will continue to be, put under pressure. Demand for new housing on almost all the world's continents means that floodplains are increasingly targeted as desirable building sites. This has had drastic consequences in many areas, especially with recent increases in large-scale floods associated with climate change. In the UK in the winter of 2000–2001, for example, severe winter storms resulted in the flooding of 10,000 properties in over 700 locations and £1 billion worth of damage. In terms of floodplain sediments, floods such as these cause deposition of potentially contaminated sediment in living areas, causing a potential risk to human health. The increasing urbanization of floodplains has also led to a loss of accommodation space for sediment deposition, resulting in silting of channels and reservoirs. Government organizations such as the UK Environment Agency have recommended more stringent restrictions on building on floodplains (Environment Agency 2004), in addition to more efficient land management.

Increased river regulation to provide power and surface water storage will continue to have an impact on sediment supply (as well as river and estuary ecology) both upstream and downstream of dams and reservoirs. Dam construction increased dramatically from the 1950s to the 1970s, when two to three new large dams were being commissioned per day (World Commission on Dams 2000). Even though dam building has declined since the 1970s, major large dam projects are still ongoing, particularly in China (which accounts for 46% of dams built), the USA and India (World Commission on Dams 2000). Knighton (1998) has pointed out that the long-term effects of the explosion in twentieth century river regulation projects may not be felt for at least another 100 years because river adjustment is a long-term processes (cf. Petts 1984).

Although mining in many areas of the world is now regulated and under strict environmental legislation, there are still areas where mining continues to have an impact on sediment and sediment-borne contaminant supplies to rivers. An example of this is the Ok Tedi/Fly River in Papua New Guinea, which, according to 1997 figures, received approximately 66 Mt of mining waste per year from the Ok Tedi Cu–Au mine, causing increases in the Cu- and other metal-bearing suspended sediment load by up to five to ten times natural background (Hettler et al. 1997). This is compounded by the steep relief in the area, which is linked to avalanches and land-slides in the mining area (Hearn 1995). In other areas, it is not the mining itself, but related activities that threaten river systems. The 1998 Aznalcóllar tailings dam spill in southwest Spain (Grimalt et al. 1999) and the 2000 Baie Mare tailings dam spills in Romania (Macklin et al. 2003), which released $2 \times 10^6 \, m^3$ and 40,000 t, respectively, of metal-, arsenic- and cyanide-rich tailings downstream, are notable examples of this. Events such as these overwhelm the natural sediment load of river systems, and cause immediate and long-term threats to the health of ecosystems and humans. Until safer alternatives are found for the storage of mine waste, events such as these will probably continue to occur regularly (Macklin et al. 2003).

3.6.3 Other impacts

The storage of contaminated sediments in flood-plains has been flagged as a potential serious long-term problem and, indeed, has been described as a 'chemical time bomb' (a concept defined by Stigliani 1991) by many workers in river systems (Stigliani 1991; Lacerda & Salamons 1992). Although considerable effort is being put into understanding the fate of sediment-bound con-taminants (Hudson-Edwards et al. 1998, 2003), the interplay between their physical (river bank and bed erosion, land drainage and development), geochemical (changes in nutrient loading, redox and pH) and biological remobilization (plant uptake and microbial degradation) is poorly understood (Macklin 1996). Moreover, Macklin (1996) warned that physical remobilization of these contaminants is currently increasing owing to global warming. Sediment alone has also been flagged by many government environmental bodies as a significant threat to river systems in terms of its weight and volume. This has been shown in this chapter, where processes such as increased alluviation in regulated rivers can have large secondary effects.

Finally, the present rapid rate of urbanization, particularly in the tropics (Gupta & Ahmad 1999), means that rivers are, and will continue to be, put under pressure for their resources and hazard management (cf. Knighton 1998). This requires more effective management that integrates good fluvial geomorphological and engineering practice.

4

Lake environments

Lars Håkanson

4.1 INTRODUCTION AND BACKGROUND

Lakes are important sedimentological environments. When a river enters into a lake, the section area becomes wider, hence the flow velocity of the running water is reduced and the particles suspended in the water end up in a much calmer environment where they can settle out. The coarse and heavy particles carried by the river will settle more rapidly than the smaller, less dense particles. As a result, lakes function as 'sediment traps' for materials and pollutants transported in rivers. The consequences of this are that lakes accumulate and retain pollutants, that lake sediments may be heavily contaminated and that lake sediment cores are excellent for studies of the historical development of contamination. In addition, lake sediments are good records of temporal changes in the landscape and are important systems for studying the ecosystem effects of pollutants. Lakes are also extremely important freshwater resources, important for recreation and totally dominate the landscape of certain regions. As a result, there are many reasons why lakes are target systems for research and management.

There is no generally accepted terminology as to what is a lake, a pond or an inland sea, except, of course, that ponds are small and inland seas large. From Table 4.1, however, one can note that there are 227,000 ponds, small lakes, lakes and large lakes in Sweden alone and that ponds are smaller than 1 ha and objects larger than 100 km^2 may be called large lakes. The primary focus of this chapter is to discuss the role of sediments and suspended particulate matter (SPM) in lakes. How do sediments and SPM influence the sedimentological and ecological function of lake systems? What are the sedimentological controls on the distribution and effects of environmental pollutants? These could be, and have been, discussed from many perspectives and scales. This chapter focuses on general principles and processes, many of which are analogous to those described in the chapters on riverine, coastal and marine environments. The format of the chapter will be to first discuss schemes for lake classification and controls on lake forms (section 4.1). Section 4.2 discusses sediment sources and the resulting products. Section 4.3 focuses on natural and anthropogenic disturbances on lake sedimentary systems; section 4.4 focuses on sediment

Table 4.1 The number of lakes in Sweden arranged by size (1 km^2 = 100 ha). (Data from Monitor 1986; Håkanson & Peters 1995.)

Area (km^2)	Category	Number	Cumlative number	Per cent
> 1000	Large lakes	3		0.0013
100–1000	Large lakes	19	19	0.0084
10–100	Lakes	362	381	0.16
1–10	Lakes	3987	4368	1.9
0.1–1	Small lakes	19,374	23,742	23.27
0.01–0.1	Small lakes	59,500	83,242	36.6
0.001–0.01	Ponds	144,000	227,000	100

dating and sediment records; section 4.5 discusses briefly the role of future climate change from a sedimentological perspective; and section 4.6 concerns lake management issues.

4.1.1 Lake types and classification

A fundamental question in lake management is: 'if there is a change in an important lake variable, such as the phosphorus concentration, the concentration of suspended particulate matter (SPM) and/or lake pH (which can be related to acid rain), will there also be changes in key functional groups of organisms and changes in ecosystem function and structure?' Is it possible to quantify and predict such changes? Lake classification systems may not in themselves provide answers to such questions, but they can provide a scientific framework for such analyses. Different types of lakes have different types of sediments. Such relationships will be discussed in this chapter. To start that discussion, this section presents different approaches to classify lakes.

Table 4.2 gives a compilation (from Hutchinson 1957) of all existing lake types on Earth, as classified according to form-creating processes. Most lakes are of glacial origin, i.e. they were formed by erosional or depositional glacial activities. In many parts of the world, for example

northern Africa and central Australia, lakes are rare, whereas in other areas they dominate the landscape, for example Sweden (Table 4.1), other parts of Scandinavia and the northern parts of North America and Russia. It also can be seen from Table 4.1 that there exists a clear relationship between number and size; there are many more small lakes than large lakes in the glacial landscape of Sweden.

Many schemes for classifying lakes have been put forward over the past 50 years (e.g. OECD 1982; Nürnberg 1996; Nürnberg & Shaw 1998). In addition to the criteria discussed below, other criteria to classify lakes have also been used, including the oxygen content in the water and the species composition of the flora and fauna (Håkanson & Boulion 2002). The aim here is to extract some key aspects of lake classification schemes in order to provide a framework for the following sections of this chapter.

A trophic level classification system of lakes is given in Table 4.3. This scheme recognizes four trophic levels: oligotrophic, mesotrophic, eutrophic and hypertrophic lakes. The trophic status of a lake is usually estimated by mean values of primary production measured for the growing season. Note that there is a significant overlap between the different trophic categories in Table 4.3. For example, in low-productive

Table 4.2 Major lake types on earth according to Hutchinson (1957).

Type	Subtype	Example
Tectonic lakes		Basins in faults (like Lake Bajkal and Lake Tanganyika)
Volcanic lakes		Maars, caldera lakes and lakes formed by damming of lava flows
Landslide lakes		Lakes held by rockslides, mudflows and screes
Glacial lakes	Lakes in direct contact with ice	Lakes on or in ice and lakes dammed by ice
	Glacial rock basins	Cirque lakes and fjord lakes
	Morainic and outwash lakes	Lakes created by terminal, recessional or lateral moraines
	Drift basins	Kettle lakes and thermokarst lakes
Solution lakes		Lakes formed in caves by solution
Fluvial lakes	Plunge-pool lakes	
	Fluviatile dams	Strath lakes, lateral lakes, delta lakes and meres
	Meander lakes	Oxbow lakes and cresentic levee lakes
Aeolian lakes		Basins dammed by wind-blown sand and deflation basins
Shoreline lakes		Tombolo lakes and spit lakes
Organic lakes		Phytogenic dams and coral lakes
Anthropogenic lakes		Dams and excavations made by man
Meteorite lakes		Meteorite craters

Table 4.3 Characteristic features in lakes of different trophic categories (Modified from OECD 1982; Håkanson & Jansson 1983.)

Trophic level	Primary production (g C m⁻² yr⁻¹)	Secchi (m)	Chl-a (mg m⁻³)	Algal volume* (g m⁻³)	Total P† (mg m⁻³)	Total N† (mg m⁻³)	Dominant fish
Oligotrophic	< 30	> 5	< 2.5	< 0.8	< 10	< 350	Trout, whitefish
Mesotrophic	25–60	3–6	2–8	0.5–1.9	8–25	300–500	Whitefish, perch
Eutrophic	40–200	1–4	6–35	1.2–2.5	20–100	350–600	Perch, roach
Hypertrophic	130–600	0–2	30–400	2.1–20	> 80	> 600	Roach, bream

*Mean value for the growing period (May–October)
†Mean value for the spring circulation.

(oligotrophic) lakes, the concentrations of total phosphorus (TP; generally the key nutrient regulating primary production in lakes) may vary within a year from very low to high values.

A similar classification system is shown in Fig. 4.1. It uses values of chlorophyll a (µg L⁻¹) as the basic criteria to classify lake trophic level because chlorophyll (Chl) is relatively easy and inexpensive to measure and it is a generally used biological measure of phytoplankton biomass. In oligotrophic systems the mean summer Chl values generally vary from 0.1 to 1 µg L⁻¹, in mesotrophic systems from 1 up to 10, in eutrophic systems from 10 to 100, and in hypertrophic systems the Chl values are generally higher than 100 µg L⁻¹. Thus, the boundary Chl values for every trophic class can be set equal to a power with the base 10.

A Trophic State Index (TSI) is defined accordingly:

$$TSI = 25 \cdot (\log(Chl) + 1)$$

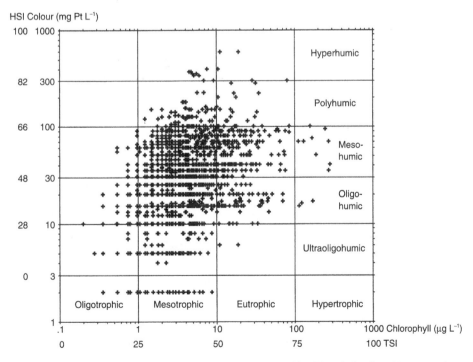

Fig. 4.1 The relationship between lake colour and humic state on the y axis and lake chlorophyll and trophic state on the x axis and data from 936 lakes: HSI, Humic State Index; TSI, Trophic State Index. (Modified from Håkanson & Boulion 2002.)

The TSI varies from 0 to 100. A TSI value of 100 relates to a Chl value of 1000 µg L^{-1} (the limit for hypertrophic lakes). A TSI value of 75 relates to a Chl value of 100 µg L^{-1}, the limit between eutrophic and hypertrophic lakes, etc. At Chl values < 0.1, then the TSI should be set to 0.

In a similar manner to trophic status, lake humic status can be determined using lake colour values as a criterion for humic state. There are several reasons to base the humic state index on colour. Coloured substances such as humic and fulvic acids reduce light transmission through water, affect bioavailability and ecological effects of nutrients and contaminants, affect plankton metabolism and may inhibit secondary production (Wetzel 2001). The colour value is generally determined by the comparative method using coloured discs and expressed in mg Pt L^{-1} (Håkanson et al. 1990). In this way, a Humic State Index (HSI) may be defined on a scale from HSI = 100, hyperhumic, to HSI = 0, ultraoligohumic, where: HSI = (100/3)log(colour − 3). For example, for HSI = 100 and colour = 300 (mg Pt L^{-1}) gives HSI = 82, etc. (Note that if colour < 3 (mg Pt L^{-1}), then HSI should be set to 0; but colour values < 3 and > 1000 mg Pt L^{-1} are very rare.)

The lake classification scheme in Fig. 4.1 uses categories of primary production (autotrophy) on the x axis and categories of catchment influence from soil, bedrock, vegetation and land-use activities (allotrophy) on the y axis. This system to classify lakes has much in common with the classification scheme for lake sediments (described later in Fig. 4.3). Lakes also can be classified according to their thermal properties, where the lake water is separated into surface water (epilimnetic water) and deep water (hypolimnetic water) by a thermocline (temperature stratification), i.e. warmer, lighter waters are found on top of colder, denser waters. Temperature stratifications occur in dimictic lakes (i.e. lakes that mix twice a year, generally during spring and autumn). Lakes also can be monomictic (such lakes circulate once a year) or polymictic (which mix many times per year; see Wetzel 2001).

4.1.2 Controls on lake form

The size and form of lakes regulate many general transport processes, such as sedimentation, resuspension, diffusion, mixing, burial and outflow (Fig. 4.2), which in turn regulate many abiotic state variables, such as concentrations of phosphorus, suspended particulate matter, pH and many water chemical variables, colour and water clarity. These in turn regulate primary production, which in turn regulates secondary production, for example of zooplankton and fish (see Håkanson 2004). The morphometry of a lake depends on the nature of the surrounding land area, drainage basin characteristics and the origin of the lake (see Table 4.2).

From the bathymetric map of a lake, one can define a set of morphometric parameters

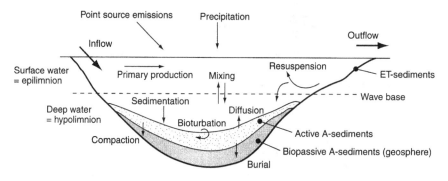

Fig. 4.2 Illustration of general and fundamental transport processes to, within and from lakes. The major sources of suspended particulate matter are the inflow from tributaries, primary production in lakes, point-source emissions and direct deposition of particles on the lake surface. The internal processes (sedimentation, resuspension, mixing, mineralization, biouptake and burial) redistribute the suspended particles and regulate the retention and outflow of particles from the lake. See section 4.2.2.2 for an explanation of A- and ET-sediments.

describing lake size, form and specific features. Among these parameters, three main groups can be identified.

1 Size parameters: (i) different parameters in length units, such as maximum length, water depth, shoreline length, maximum depth and maximum breadth, (ii) parameters expressed in area units, e.g. two-dimensional water surface area and three-dimensional bottom area, and (iii) parameters expressed in volume units, such as water volume and surface-water volume (epilimnetic volume).

2 Form parameters: based on size parameters, for example mean depth and shore development (= shoreline irregularity).

3 Special parameters: such as the dynamic ratio ($DR = \sqrt{A}/D_m$, where A is lake area in square kilometres and D_m is the mean depth in metres) and effective fetch (a measure in kilometres of the free water surface over which the winds can influence the waves and hence also the sediments).

4.1.3 Lakes as sedimentary environments

The main sedimentary environments that occur within lakes are:

1 Beach and nearshore areas dominated by wave processes, on- and offshore movement of sediments and nearshore currents (see Teller 2001). The sediments in these areas are generally coarse-grained (sand, gravel, etc.).

2 Deltaic areas in lakes with relatively large tributaries transporting coarse silt and sand.

3 Shallow areas above the wave base, where winds and waves may exert an influence on sediments (the surface-water area approximates to the area above the mean thermocline). Such sediments are generally well oxygenated and may be resuspended by storms.

4 Deep-water areas occur beneath the wave base (below the thermocline). These areas provide a relatively calm, less well oxygenated sedimentological environment for fine materials to settle.

5 On slopes inclined at more than 4 m per 100 m length, slope processes dominate the sedimentological conditions on lake bottoms. In such areas turbidites may form and they can influence the sediment record over large areas.

This is an important process with relevance to studies of sediments as historical archives, because turbidity currents can cause older, previously deposited material to settle on top of more recently deposited sediment.

4.2 SEDIMENT SOURCES AND SEDIMENT ACCUMULATION PROCESSES

4.2.1 Sources and characteristics of lake sediments

Five major sources of matter forming lake sediments can be recognized:

1 Allochthonous materials (sometimes called lithogenous materials) – particles and aggregates transported from land to lakes by rivers (e.g. sand, silt and clay).

2 Autochthonous materials (or biogenous materials) – sediments produced in the lake by organisms living in the lake (e.g. silica frustules or dead phytoplankton).

3 Hydrogenous sediments – sediments that are also produced in the lake but which emanate from materials precipitated out of solution, for example, carbonates (e.g. Pedley 1990; Pedley et al. 1996) and evaporites (e.g. Trichet et al. 2001).

4 Wet and dry deposition of matter on the lake surface – this is generally a relatively small contributor to the total amount of suspended matter found in lakes.

5 Direct point source emissions of matter, for example from urban areas and industries.

The relative contributions of each source can be assessed for a lake by means of mass-balance calculations (Fig. 4.2), where the flux (which is equivalent to the flow or transport) from each source is calculated. This can be calculated on a monthly basis (g month^{-1}) to obtain seasonal variations, or annually for overall budgets in order to rank the relative importance of each flux. The relative contribution of different sources differs between lakes as a result of variations in characteristics, lake morphometry and climatological regime.

The sediment types within lakes can be classified in a number of ways. One system, widely

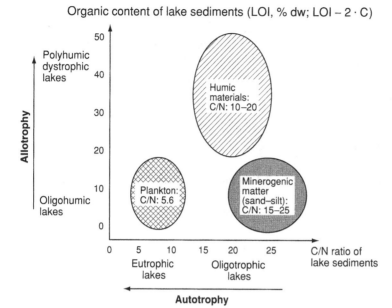

Organic content of lake sediments (LOI, % dw; LOI – 2 · C)

Fig. 4.3 Lake classification from the relationship between the C/N ratio and the loss on ignition (LOI) of surficial sediments. (Modified from Håkanson 1995.)

used, is based on the relationship between the organic content (estimated by determining loss on ignition) and the ratio between the carbon content and the nitrogen content of lake sediments (Fig. 4.3). This sedimentological classification system has much in common with the information given in Fig. 4.1 for lake classifications. Three main types of matter in lake sediments can be recognized and each has different chemical characteristics (see Hansen 1961; Håkanson & Jansson 1983).

1 The average composition of planktonic materials (autochthonous organic matter) is $C_{106}N_{16}P$, which gives a C/N ratio of about 5.6, and the loss on ignition (LOI) is usually less than 20% dry weight (dw). This group also includes hydrogenous materials precipitated out of solution and biogenous materials (from organisms).

2 Allochthonous organic matter (e.g. humus) generally (see Gjessing 1976; Thurman 1985) has a C/N ratio of 10–20 and a LOI of > 20% dw.

3 Minerogenic (inorganic) matter (such as sand and silt) generally has a C/N ratio of 15–25 and a LOI of < 20% dw.

Further classification schemes related to lake sediments include those based on (i) grain size (gravel, sand, silt or clay; see e.g. Friedman & Sanders 1978), (ii) form-creating processes (such

as shore sediments, glacial deposits, turbidites from turbidity currents, delta sediments, mixed bioturbated sediments or annually layered, laminated sediments; see e.g. Sly 1978), (iii) systems based on sediment colour and/or specific characteristics (such as black sediments as a reflection of anoxic conditions, brown sediments reflecting oxic conditions, light grey calcareous deposits, greenish grey siliceous deposits, grey sediments rich in fibre, very soft brownish sediments, such as gyttja (sedimentary peat consisting mainly of plant and animal residues precipitated from standing water) and dy (finely divided, partly decomposed organic material), and black manganese nodules; see e.g. Håkanson & Jansson 1983), (iv) systems based on chemical properties (see Table 4.4), or (v) systems that focus on the geological rather than the geographical origin of the sediments and distinguish between, for example, glacial, fluvial, post-glacial or aeolian deposits.

4.2.1.1 Suspended particulate matter (SPM) in lakes

The total amount of any substance X in the water is often separated into a particulate phase, the only phase subject to gravitational sedimentation, and a dissolved phase, generally the most

Table 4.4 A geochemical classification (Berner 1981) of sedimentary environments. C is the concentration (moles L^{-1}). H_2S is total sulphide.

Environment	Characteristic phases
I. Oxic ($CO_2 \geq 10^{-6}$)	Haematite, goethite, MnO_2-type minerals; no organic matter
II. Anoxic ($CO_2 < 10^{-6}$)	
A. Sulphidic ($CH_2S \geq 10^{-6}$)	Pyrite, marcasite, rhodochrosite, alabandite; organic matter
B. Non-sulphidic ($CH_2S < 10^{-6}$)	Glauconite and other Fe^{2+}–Fe^{3+} silicates (also siderite, vivianite, rhodochrosite);
1. Post-oxic	no sulphide minerals; minor organic matter
2. Methanic	Siderite, vivianite, rhodochrosite; earlier formed sulphide minerals; organic matter

important phase for direct biological uptake. Operationally, the limit between the particulate phase and the dissolved phase is generally determined by means of filtration, using a pore size of 0.45 μm. Evidently, this is an operational approach, and many colloidal particles will pass through such filters.

A general outline of particles in lake water, their origin, standard abbreviations and a classification scheme are given in Fig. 4.4. As stressed,

Particles in lake water

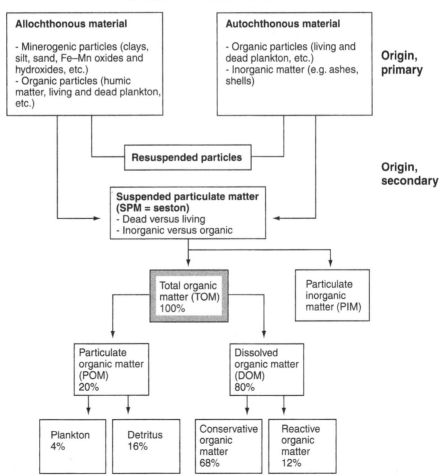

Fig. 4.4 Classification scheme and nomenclature for particles in lake water (see also Dubko 1985; Ostapenia 1985).

the first differentiation concerns the origin of the particles: allochthonous particles are transported to the lake from outside the lake, generally from tributaries or precipitation onto the lake surface – such particles may be categorized into mineral (or minerogenic) and organic (or organogenic) – and autochthonous particles, which are produced in the lake. These particles can settle on to the lake bed and some of those particles may be resuspended. This means that the total amount of suspended particulate matter (SPM) found in lakes is generally a complex mix of substances of different origins with different properties (size, form, density, specific surface, capacity to bind pollutants, etc.).

The SPM may be divided into an organic fraction (POM) and an inorganic one (PIM, particulate inorganic materials; see Fig. 4.4). Total organic matter (TOM) is generally divided into particulate (POM) and dissolved (DOM) fractions. Normally, POM is about 20% of TOM, but this certainly varies between lakes and within lakes during the year. Normally, about 4% of POM is living matter and the rest is dead organic matter (detritus). About 80% of TOM is generally in a dissolved phase, and of this about 70% is conservative in the sense that it does not change in response to chemical and biological reactions in the water mass.

4.2.2 Controls on lake sediment transport and accumulation

The sedimentological conditions in a lake will influence almost all processes in the aquatic ecosystem. For example, resuspension, especially in large and shallow lakes, controls the concentration of suspended particles in the water, influencing water clarity and hence the depth of the photic zone, often operationally defined as the Secchi depth, i.e. the depth where a black and white disc is lost from eye sight on lowering through the water (Håkanson & Peters 1995). Resuspension, therefore, has an impact on primary and secondary production. The production of zoobenthos is also controlled by sedimentological conditions, with high production in oxic sediments and low production in anoxic sediments. The macrophytes

are rooted in the sediments (Fig. 4.5) and, therefore, sediment conditions control their production. If, finally, the sediments are contaminated, this could increase the concentrations of harmful substances in zoobenthos, and, hence, also in fish that eat zoobenthos (benthivores).

Within a lake several processes regulate and control the pathways of both sediments and contaminants (see Fig. 4.2):

1 sedimentation – the transport of matter from water to sediments;

2 resuspension – the transport of matter from sediments back to water;

3 diffusion – the transport of dissolved substances from sediments back to water;

4 mineralization – the bacterial decomposition of organic matter;

5 mixing – the upward and downward transport of matter;

6 bioturbation – the mixing of the deposited materials from the movement of the bottom fauna (= zoobenthos – from their eating, digging and foraging activities);

7 compaction – the vertical change in sediment water content and sediment density due to the weight of overlying sediments;

8 burial, i.e. the transport from biologically active sediments to biopassive (geological) sediments. This latter is transport from the biosphere to the geosphere of substances that may emanate from the technosphere. These processes are discussed in more detail below.

4.2.2.1 Transport, sedimentation and resuspension

The processes of sedimentation, burial and resuspension are interlinked and to understand them requires an understanding of bottom dynamic conditions within a lake. In defining the bottom dynamic conditions (erosion, transportation and accumulation), the following definitions are often used (from Håkanson 1977).

1 Areas of erosion (E) prevail where there is no apparent deposition of fine materials but rather a removal of such materials, for example in shallow areas or on slopes; E areas are generally hard and consist of sand, gravel, consolidated clays and/or rocks.

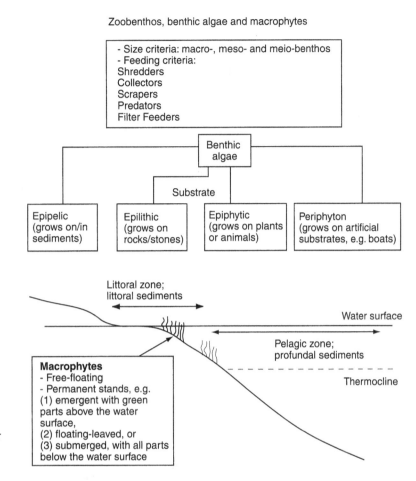

Fig. 4.5 Compilation of concepts related to sediment-living organisms (zoobenthos, benthic algae and macrophytes) (see also Vollenweider 1968, 1976; Cummings 1973; Brinkhurst 1974; Wetzel 2001).

2 Areas of transportation (T) prevail where fine materials are deposited periodically (areas of mixed sediments). This bottom type generally dominates where wind/wave action regulates the bottom dynamic conditions (see Fig. 4.6). It is sometimes difficult in practice to separate areas of erosion from areas of transportation.

3 Areas of accumulation (A) prevail where the fine materials are deposited continuously (soft bottom areas). Owing to their fine-grained nature these are the areas (the 'end stations') where high concentrations of pollutants may appear (see Table 4.5).

The water content, grain size and/or the composition of the material are often used as criteria to distinguish different sediment types (see Table 4.5; or Sly 1978). From the basic Stokes' equation for settling particles (see Chapter 1), as well as for convenience, the limit between

coarse and fine materials can be set at a particle size of medium silt (0.06 mm). The generally sandy sediments within the areas of erosion and transport (ET) often have a low water content, low organic content and low concentrations of nutrients, low benthic biomass and few contaminants (see Table 4.5 and Fig. 4.7). The conditions within the T areas are, for natural reasons, variable, especially for the most mobile substances, such as phosphorus, manganese and iron, which may react rapidly to alterations in the sediment chemical 'climate' (as given e.g. by the redox potential). Fine materials may be deposited for long periods during stagnant weather conditions. In connection with a storm or a mass movement on a slope, this material may be resuspended and transported, generally in the direction towards the A areas in the deeper parts, where continuous deposition occurs. Thus,

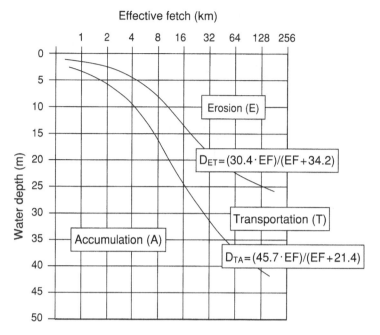

Effective fetch (km)

$$D_{ET} = (30.4 \cdot EF)/(EF + 34.2)$$

$$D_{TA} = (45.7 \cdot EF)/(EF + 21.4)$$

Fig. 4.6 The erosion, transport and accumulation (ETA) diagram showing the relationship between the effective fetch (EF, i.e. the free water surface over which winds influence waves), the water depth and the potential bottom dynamic conditions for individual sites in lakes. D_{TA} is the wave base (WB), i.e. the water depth separating T and A areas. D_{TA} can be predicted from the given equation. The mean EF value for an entire lake can be approximated by √Area. (Modified from Håkanson 1999.)

Table 4.5 The relationship between bottom dynamic conditions (erosion, transportation and accumulation) and the physical, chemical and biological character of the surficial sediments of Lake Lilla Ullevi Bay (in Lake Mälaren), Sweden. Mean values and coefficients of variation (CV) in parentheses: n = number of analyses. (Raw data from Ryding & Borg 1973.)

Category	Characteristic	Erosion ($n = 15$)	Transportation ($n = 10$)	Accumulation ($n = 14$)
Physical parameters	Water depth (m)	13.0 (0.41)	17.5 (0.31)	31.6 (0.25)
	Water content (% ww)	32.6 (0.28)	67.4 (0.14)	94.1 (0.024)
	Bulk density (g cm^{-3})	1.71 (0.087)	1.26 (0.079)	1.03 (0.019)
	Organic content (loss on ignition, % dw)	4.6 (0.48)	10.7 (0.43)	24.3 (0.10)
Nutrients (mg g^{-1} dw)	Nitrogen	0.6 (0.67)	3.4 (0.35)	10.7 (0.14)
	Phosphorus	0.8 (0.50)	2.8 (0.75)	1.6 (0.31)
	Carbon	0.5 (1.0)	22.7 (0.74)	10.4 (0.16)
Benthic biomass (mg ww m^{-2})		1000–2000	3000–4000	6000–7000
Chemically mobile elements (see also P) (mg g^{-1} dw)	Iron	24.6 (0.42)	53.5 (0.27)	41.3 (0.077)
	Manganese	0.8 (1.0)	3.5 (0.74)	2.5 (0.60)
Metals (µg g^{-1} dw)	Zinc	41 (0.46)	111 (0.24)	189 (0.090)
	Copper	18 (0.50)	31 (0.42)	59 (0.10)
	Nickel	23 (0.35)	40 (0.20)	57 (0.18)

resuspension is a natural phenomenon on T areas. It should be stressed that fine materials are rarely deposited as a result of simple vertical settling in natural aquatic environments. The horizontal velocity component in lake water is generally at least ten times larger, sometimes up to 10,000 times larger, than the vertical component for fine materials or flocs which settle according to Stokes' law (see Bloesch & Burns 1980; Bloesch & Uehlinger 1986).

Resuspension is the physical (advective) transport of matter from sediments back to water and mixing is the upward and downward transport of dissolved and suspended particulate matter across the thermocline (the thermocline is the zone in the water that separates the warmer, lighter

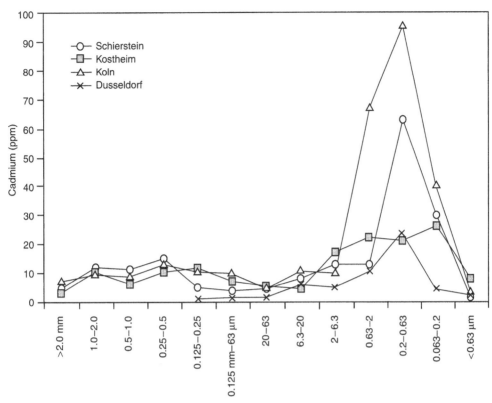

Fig. 4.7 Grain-size dependencies of cadmium concentrations in sediment samples from German rivers. (Modified from Förstner & Salomons 1981.)

surface water from the colder, heavier deep water). Some lakes are constantly mixed and do not develop a thermocline, but most lakes, and certainly those in northern and boreal landscapes, develop a thermocline in the summer. The surface water (the epilimnion) is fundamental for the primary production of matter. Many important sedimentological processes take place in the deep-water zone (= hypolimnion).

There are some basic rules regulating sedimentation in lakes (e.g. Thomas et al. 1976; Golterman et al. 1983; Håkanson & Jansson 1983; Colman et al. 2000; and Fig. 4.8).

1 River action dominates the sedimentological properties in river-mouth areas, where deltas may be formed if the amount of sandy materials carried by the tributaries is large enough. Within these areas, sedimentation rates generally decrease with distance from the mouth, and so does the grain size of the settling particles.

2 In open-water areas, dominated by wind/wave action, sedimentation rates generally increase from the wave base to the deepest parts of the lakes. The coarsest materials (sand, gravel) are often found in shallow waters.

3 Current action can dominate in certain areas, such as in narrow straits and along the shoreline. Then the 'Hjulström-curve' (see Fig. 1.5) gives the relationship between critical erosion and critical deposition of materials.

4 Slope-induced (gravity) turbidity currents appear on bottoms inclined more than about 4–5% (Håkanson 1977), and bioturbation generally prevails in oxic sediments (see Table 4.4), where the macro- and meiofauna cause a mixing of the sediments.

4.2.2.2 Determination of bottom dynamic conditions in lakes

The following processes influence internal loading and bottom dynamic (E, T and A) conditions

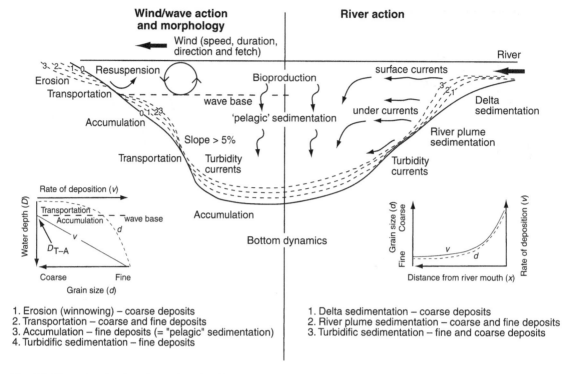

Fig. 4.8 Illustration of major sedimentological and bottom-dynamics processes in lakes. (Modified from Håkanson & Jansson 1983.)

in lakes (Håkanson & Jansson 1983): (i) an energy factor related to the effective fetch and the wave base (see the ETA diagram in Fig. 4.6); (ii) a form factor related to the percentage of the lake bed above the wave base (see Fig. 4.9); and (iii) a lake slope factor related to the fact that slope-induced transportation (turbidity currents) may appear on bottoms inclining more than 4–5%.

One approach to calculate the areas where resuspension occurs (the ET areas) is based on the wave base and the form of the lake (the form factor $Vd = 3 \cdot D_m/D_{max}$; D_{max} = the maximum depth, as illustrated in Fig. 4.9). The wave base (WB), which is set equal to the depth, D_{TA}, separating T areas and A areas is given by:

$$D_{TA} = (45.7 \cdot \sqrt{Area})/(21.4 + \sqrt{Area})$$

An evident boundary condition for this approach is that if $D_{TA} > D_{max}$ then $D_{TA} = D_{max}$.

The area above the wave base (A_{WB} = Area − Area$_A$) may be calculated from the hypsographic form of the lake, which in turn is calculated from the form factor (= volume development, Vd). An equation expressing A_{WB}, as a function of Vd, the area of the lake (Area in m^2) and the maximum depth of the lake (D_{max} in m), is also given in Fig. 4.9. The ET areas are generally larger than 15% of the lake area because there is always a shore zone dominated by wind/wave activities, at least in all lakes larger than approximately 1 ha. One can generally also in shallow lakes find sheltered areas and deep holes with more or less continuous sedimentation, i.e. areas that actually function as A areas, so the upper boundary limit for ET is often set at 99% of the lake area.

4.2.2.3 Post-depositional processes

4.2.2.3.1 Bioturbation Bioturbation is the mixing of the deposited materials from the movement (eating, digging and foraging activities) of the bottom fauna (zoobenthos). The sediment classification scheme in Table 4.4 focuses on the

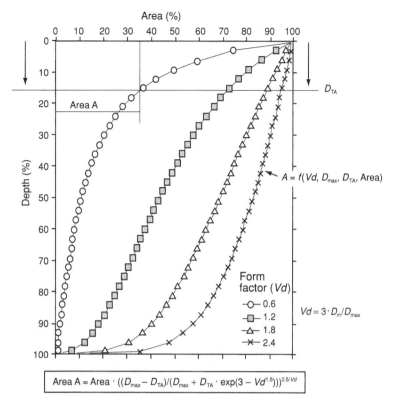

Fig. 4.9 An illustration of how the form factor (= the volume development), Vd, can be used to express the form, here given by the relative hypsographic curve (= the depth–area curve) of lakes. Shallow lakes with a small Vd have a relatively large area above the wave base (the surface-water areas), where processes of wind/wave-induced resuspension will influence the bottom dynamic conditions. Deep, U-formed lakes generally have smaller areas above the wave base (ET areas) and large deep-water areas. The equation describes how the accumulation areas (Area A) may be calculated from data on lake area (Area), maximum depth (D_{max}), the depth of the wave base (the critical depth, D_{TA}) and the form factor (Vd). (Modified from Håkanson 1999.)

$$\text{Area A} = \text{Area} \cdot ((D_{max} - D_{TA})/(D_{max} + D_{TA} \cdot \exp(3 - Vd^{1.5})))^{0.5/Vd}$$

oxygen status of the sediments. Zoobenthos will not generally survive if the oxygen concentration at the sediment–water interface becomes lower than about 2 mg L^{-1}. When the zoobenthos die, the biological mixing (bioturbation; see Fig. 4.10) of the sediments will stop. This has profound consequences for the lamination of the sediments and, hence, has implications for interpretations about the age distribution of sediments.

Generally, various types of zoobenthos live in sediments down to about 5–15 cm sediment depth (see Fig. 4.11). This upper part of the sediment column is biologically active in the sense that the bottom fauna can influence the physical, chemical and biological conditions of the sediments and cause bioturbation. Important groups of bioturbators in lakes are large zoobenthos such as worms and crustaceans. Benthic animals are known to display great areal, temporal, vertical and species-specific patchiness within lakes and great variability among lakes. It has been shown that bottom animals can eat sediment up to an average of seven times and this will,

evidently, strongly influence both the age and age distribution of the sediments (see Håkanson & Jansson 1983).

If the larger animals (macro- and meiofauna) die, bioturbation is halted and laminated (layered and unmixed) sediments may appear. Continuous sedimentation will cause the sediment layer to grow upward so that the bioturbation limit, i.e. the limit between the upper biological layer and the lower biopassive (or geological) layer, moves upward. As stressed previously, bioturbation will influence the age and the age distribution at any given sediment depth and this has evident and profound implications for sediment dating, i.e. when the sediments are used as a historical archive. When bioturbation is negligible, for example during anaerobic conditions, the age of the laminated sediments can be determined in a very straightforward manner by 'counting varves'. Section 4.4 will discuss some methods to determine the age of lake sediments and exemplify why this is important in lake studies.

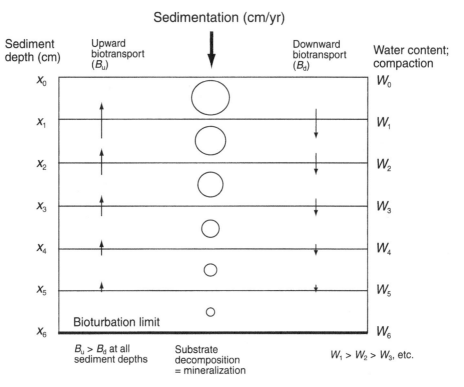

Fig. 4.10 Illustration of key processes related to bioturbated sediments (sedimentation, upward and downward biotransport, substrate decomposition and compaction). (Modified from Håkanson & Jansson 1983.)

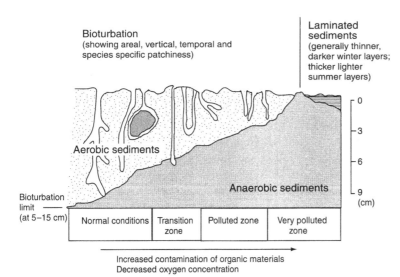

Fig. 4.11 Bioturbation and laminated sediments. Under aerobic (= oxic) conditions zoobenthos may create a biological mixing of sediments down to about 15 cm sediment depth (the bioturbation limit). If the deposition of organic materials increases and hence also the oxygen consumption from bacterial degradation of organic materials, the oxygen concentration may reach the critical limit of 2 mg L^{-1}, when zoobenthos die and bioturbation ceases and laminated sediments appear. (Modified from Pearson & Rosenberg 1976).

4.2.2.3.2 Compactional and diagenetic processes A number of important processes influence sediment accumulation following deposition. These include diffusion, mineralization, compaction and burial. Diffusion is the chemical transport of dissolved substances from sediment interstitial water back to lake water regulated by concentration gradients. For many substances (e.g. phosphorus and caesium), the diffusive transport is highly dependent on sediment

Table 4.6 Data from sediment cores from lakes illustrating the relationship between physical sediment character (the water content of surficial sediment, 0–1 cm, and the sediment constant illustrating the vertical gradient in sediment compaction) and lake type (as given by the trophic and humic status) (From Håkanson & Jansson 1983.)

Lake	Lake type	Water content	Sediment constant	Sediment character
Ingen	Polyhumic, oligotrophic	95.2	−0.41	Very loose, small vertical changes
Trosken	Polyhumic, oligotrophic	95.4	−0.83	Very loose, small vertical changes
Skal	Polyhumic, oligotrophic	97.4	−0.64	Very loose, small vertical changes
Hjalmaren	Mesohumic, eutrophic	90.4	−2.99	Loose, clear vertical gradient
Freden	Mesohumic, eutrophic	87.7	−3.80	Loose, clear vertical gradient
Vasman	Mesohumic, mesotrophic	95.9	−3.95	Very loose, clear vertical gradient
Aspen	Mesohumic, mesotrophic	86.9	−5.78	Loose, strong vertical gradient
Vanern	Mesohumic, oligotrophic	85.5	−5.97	Loose, strong vertical gradient
Vattern	Oligohumic, oligotrophic	92.5	−13.6	Very loose, very strong vertical gradient

redox-conditions – the lower the redox potential (and oxygen concentrations), the higher the diffusive fluxes (see Håkanson 1999). Diffusion may be a dominant flow of phosphorus in highly productive lakes. The function of bacteria in lakes is the decomposition of organic matter, and mineralization (which produces dissolved substances during early diagenesis) is the name for this process. Compaction in sediments concerns the vertical change in sediment water content and bulk density due to the accumulated weight from overlying sediments. The water content may change from about 85% in the uppermost sediment layer in a lake to about 70% at a sediment depth of 15 cm as a result of compaction (see section 4.2.3 below).

As a result, substances deposited on a lake bed may be returned from sediments back to water by diffusive and advective (= resuspension) processes. There is, however, a sediment depth beneath which substances will not return. Instead, they will be buried by the constant deposition of new matter.

4.2.3 Variations in lake sediment deposits

As lake sediments are influenced by materials supplied from the catchment area (allochthonous matter), from materials produced in the lake and from matter precipitated out of solution in the lake (autochthonous matter), there are great differences among lakes in sediment characteristics, such as elemental composition, physical properties and bottom fauna communities (see Håkanson

& Jansson 1983). This will be exemplified here by presenting information in two tables.

Physical sediment characteristics in lakes of different trophic and humic status are shown in Table 4.6. The vertical changes in sediment water content (or organic content or bulk density) may be expressed by the following relationship:

$$W(x) = W_{0-1} + K_s \cdot \ln(2 \cdot x)$$

where $W(x)$ is the water content (in percentage wet weight – % ww) of a 1-cm-thick layer at sediment depth x (i.e. $x \pm 0.5$ cm), W_{0-1} is the water content (in % ww) of surficial sediment (0–1 cm) and K_s is an empirical sediment constant illustrating the change in water content with sediment depth.

If K_s attains a high numerical value (such as −10), it means that there is a strong vertical gradient, i.e. the water content decreases very significantly with sediment depth as a result of compaction. If K_s is small (such as −0.5), it is characteristic of lakes with loose sediments deep down in sediment cores. One should note that the K_s value is site-specific as well as lake-specific. The value depends on prevailing bottom dynamic conditions (erosion, transportation and accumulation). It varies comparatively little within open water areas (areas of accumulation) in lakes, and much more in erosion and transportation areas. The value of K_s can vary widely between lakes depending on the chemical, physical and biological characteristics of the lake. From Table 4.6 it can be noted that humic lakes

Table 4.7 A chemical classification of elements in lake sediments. (Modified after Kemp et al. 1976.)

1. Major elements (Si, Al, K, Na and Mg) make up the largest group of the sediment matrix
2. Carbonate elements (Ca, Mg and CO_3–C) constitute the second largest group; about 15% of the materials
3. Nutrient elements (org.-C, N and P) account for about 10% in recent lake sediments
4. Mobile elements (Mn, Fe, P and S) make up about 5% of the total sediment weight
5. Trace elements (Hg, Cd, Pb, Zn, Cu, Cr, Ni, Ag, V, etc.), the smallest group accounting for less than 0.1% of the sediments

Table 4.8 Chemical characteristics of lake sediments from different regions of the world (data from Jones & Bowser 1978).

Region	Lake/group type	P	Organic C	Fe	Mn	Ca	K	Si	Al
Minnesota lakes	Low organic	0.13	7.6	5.0	0.6	0.9	1.2	–	–
	High organic	0.17	21	3.8	0.1	2.0	0.7	–	–
	High carbonate	0.14	9.9	2.2	0.15	15.2	0.5	–	–
	Low carbonate	0.08	6.1	2.5	0.4	6.8	0.4	–	–
African lakes	Kivu	–	–	5.3	0.09	9.5	–	19	5.0
	Tanganyika	–	–	5.0	0.03	1.2	–	26	10
	Edvard	–	–	2.4	0.03	3.0	–	29	4.3
	Albert	–	–	6.2	0.1	1.4	–	25	12
Great Lakes	Ontario	0.07	–	3.7	0.06	0.4	2.3	24	5.1
	Erie	0.06	–	2.8	0.06	0.35	2.2	26	4.8
	Michigan	0.08	–	1.5	0.08	11	1.3	25	2.8
	Superior	–	2.3	2.5	0.05	1.2	0.5	24	2.4

generally have loose sediments with high water contents and low K_s values, whereas eutrophic lakes often have sediments with a water content of surficial sediments in the range 83–93% ww and K_s values of about –2 to –5. Mesotrophic and oligotrophic lakes cannot be distinguished from eutrophic lakes by the water content of the sediments, but often by the sediment constant, which generally attains higher numerical values for low-productivity lakes. This depends, however, on many things, for example supply of minerogenic matter, which is relatively high in oligotrophic lakes, and on bioturbation, which is generally higher (causing higher K_s values) in sediments rich (but not too rich) in organic matter and food for the bottom fauna.

Chemical analyses of sediment cores can offer a good key to the history of lakes, and also to the present conditions. A chemical classification of elements in lake sediments is given in Table 4.7. Sediments affect and reflect the characteristics of lakes. Many studies have been presented related to the chemical composition of lake water and sediments and to interpretational codes of

what this means in terms of lake processes (see Håkanson & Jansson 1983; Pedley 1990). Table 4.8 is included here to stress this point. It provides data on sediment chemical properties (P, organic carbon, Fe, Mn, Ca, K, Si and Al) for lakes in Minnesota categorized into four groups (low and high organic, and low and high carbonate groups), for four large African and four large American lakes. From Table 4.8 it can be noted that some of these substances do not vary a great deal among these lakes (e.g. P and Si), whereas other substances (such as Ca and organic C) vary a great deal. This is an important point. Sediment variables, just like water variables, can vary among lakes owing to differences in catchment area characteristics (geology, soils, land-use, vegetation, etc.) and within lakes depending on climatological factors, season of the year, changing winds (such as frequency of storms) and bottom dynamic conditions. This means that some variables can reflect typical lake conditions much better than others, Sediment phosphorus, manganese and iron are well known (see Table 4.5) for poorly reflecting typical lake

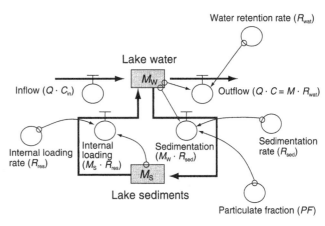

$$V \cdot dC/dt = Q \cdot C_{in} - Q \cdot C - M_W \cdot R_{sed} \cdot PF + M_S \cdot R_{res}$$
dC/dt = change in concentration in lake water
$C = M_W/V;\ V$ = lake volume
$T_w = V/Q;\ Q$ = water discharge
C_{in} = tributary concentration

Fig. 4.12 The basic mass-balance equation for a lake with internal loading.

conditions, because the sediment concentrations of these substances mainly reflect sediment redox-conditions and low redox values may appear in sediments of most lakes. On the other hand, sediment Ca concentrations more closely reflect typical lake properties on a scale from low to high calcareous conditions.

4.2.4 Mass-balance modelling for lakes

The basic aim of mass-balance calculations is to quantify fluxes so that large and important fluxes may be identified and differentiated from small fluxes. The aim of this section, however, is not to give a full mathematical account of all the processes shown in Fig. 4.2, only to give a brief introduction to mass-balance modelling and to illustrate some basic concepts with a focus on sedimentation in lakes.

A simple mass-balance model for a lake is depicted in Fig. 4.12. A typical mass-balance model envisions the lake as a 'tank reactor' in the sense that the lake mixes completely during an interval of time dt. The flow of suspended particulate matter (or a given contaminant) to and from such a lake, and net sedimentation may be described by the following equation (this is the steady-state solution to the equation given in Fig. 4.12):

$$0 = Q \cdot C_{in} - Q \cdot C - M_W \cdot R_{sed} \cdot PF + M_S \cdot R_{res}$$
$$\ \ \ \text{(in)}\quad \text{(out)}\text{(sedimentation)}\text{(resuspension)}$$

where V is the lake volume (usually m^3 or km^3), C is the concentration of the substance in the lake water (units usually g L^{-1} or kg m^{-3}; C is the M_W/V), C_{in} is the concentration of the substance in the tributary (C_{in} has the same dimension as C), Q is the tributary water discharge to the lake (usually expressed as m^3 yr^{-1} or m^3 month^{-1}), R_{sed} is the sedimentation rate of a given substance in the lake (like all rates, R_{sed} has the dimension 1/time and its unit is usually 1/day, 1/month or 1/year), M_W is the mass (= amount) of the substance in the lake water (units often in g or kg), PF is the particulate fraction (dimensionless), the only fraction that can settle out in lakes due to the influence of gravitation, M_S is the mass (= amount) of the substance on the lake bed (units in g or kg), R_{res} is the internal loading rate, or resuspension rate (units usually in 1/day, 1/month or 1/year). The steady-state assumption means that a change in lake concentration (dC) of the given substance per unit of time (dt; usually in g L^{-1} month^{-1} or kg m^{-3} yr^{-1}) is set to zero.

The lake water retention time (T, in days, months or years) is a fundamental concept in lake studies. It is defined as the ratio between the lake volume and the water discharge to or from the lake (Q):

$$T = V/Q$$

The value of Q can be derived from time-series of measurements, or be given as a mean monthly value, or as a mean yearly value. In the latter case, T is generally referred to as the theoretical lake-water retention time.

The concentration in the lake water may then be expressed as:

$$C = (Q \cdot C_{in} + M_S \cdot R_{res})/(Q + V \cdot R_{res} \cdot PF)$$

This expression for the lake concentration is fundamental in lake science and management. When Vollenweider (1968) presented his first loading model for lake eutrophication (phosphorus), it meant a breakthrough not 'just' for lake management but also for lake modelling. Vollenweider simplified the mass-balance model first by omitting seasonal variations and instead gave the annual budget. He also omitted different nutrients and different forms of the nutrients, and instead made the calculations for total phosphorus. In addition, he disregarded internal loading (the $M_S \cdot R_{res}$ term in the given equation) and simplified the sedimentation term ($M_S \cdot R_{sed} \cdot PF$, which he approximated to $\sqrt{T}$). This gave the famous Vollenweider model:

$$C = C_{in}/(1 + \sqrt{T})$$

It is evident that substances with large R_{sed} values settle rapidly, near the point of discharge, and that substances with small R_{sed} values may be distributed over much larger areas. For most substances, it is important to determine or predict the R_{sed} value, which is related to the settling velocity, v ($v = z \cdot R_{sed}$, where z is the distance through which the particle sinks in the given time interval). The settling velocity, v, is generally given in centimetres per second or metres per year. Given R_{sed} or v, one can model or predict where high and low concentrations are likely to appear in water and sediments. This is the key to predicting where high and low ecological effects

may appear. It is important to note the difference between the settling velocity (v in cm s^{-1}) and the sedimentation rate (R_{sed} in L s^{-1}).

Stokes' law expresses the settling velocity (v) as:

$$v = ((d_w - d_p \cdot g \cdot d^2)/(18 \cdot \mu \cdot \varnothing)$$

where v is the settling velocity (usually in cm s^{-1} or m month^{-1}), d_p is the particle density (usually in g dw cm^{-3}), d_w is the density of the lake water (often set to 1 g ww cm^{-3}), g is the acceleration due to gravity (980.6 cm s^{-1}), d is the particle diameter (in m, cm or mm), μ is the coefficient of absolute viscosity (obtained from standard tables; 0.01 poise at 20°C) and $\varnothing$ is the coefficient of form resistance (set to 1 for spheres; Hutchinson 1967).

Stokes' law (Stokes 1851) is depicted in Fig. 4.13. The behaviour of material that follows Stokes' law (i.e. particles with a diameter between about 0.01 and 0.0001 cm) differs from that of the coarser fraction material and from that of still finer material. The sedimentological behaviour of the material is closely linked to the grain size of the individual particles (Einstein 1950; Allen 1970). The sedimentological behaviour of the very fine materials is governed by Brownian motion. These latter particles are so small that they will not settle individually, but will do so if they form larger flocs or aggregates that are dense enough to settle according to Stokes' law (Kranck 1973, 1979; Lick et al. 1992). The cohesive materials that follow Stokes' law are very important, because they have a great affinity for pollutants (Fig. 4.7). This group includes many types of detritus, humic substances and plankton. All play significant roles in aquatic ecosystems (Salomons & Förstner 1984).

The settling velocity (v) of a given particle, aggregate or particulate pollutant, and its distribution in a lake, depends on the density, size and form of the particle (and on the hydrodynamics of the flow of water in the lake). If the particle density, d_p, is close to 1, if the form factor, $\varnothing$, is large and if the diameter, d, is small, the settling velocity, v, and the sedimentation rate, R_{sed}, may be very slow. If R_{sed} is close to 0, the particle or aggregate is conservative in

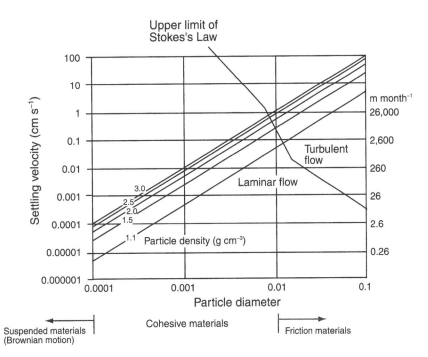

Fig. 4.13 The relationship between the settling velocity (v; of spherical particles) in water, particle diameter and particle density (at 20°C) as given by Stokes' law.

the sense that it may not be deposited in the lake. In many lakes (see Håkanson & Jansson 1983), Cl, Ca and alkalinity are typically conservative substances, and colour, organic matter, total P, Si, particulate P and suspended matter are more reactive.

4.2.4.1 The distribution coefficient

A very important part of most mass-balance models for chemical substances is the distribution coefficient. The lake distribution coefficient is also often called the partition or partitioning coefficient and gives the particulate fraction (*PF*) or the dissolved fraction (*DF* = 1 − *PF*). Related to sedimentological processes, there are three major categories of matter in lake water:
1 the particulate fraction (PF), which is the only fraction settling by gravitation;
2 the dissolved fraction, which is the most important one for direct biouptake by, for example, phyto- or zooplankton and often referred to as the bioavailable fraction;
3 the colloidal fraction, which is neither subject to direct uptake nor to sedimentation because the particles are too small.

The particulate fraction and the colloidal fraction may be subject to mineralization by bacterioplankton. Traditionally (see Santschi & Honeyman 1991; Gustafsson & Gschwend 1997), the K_d concept is used in these contexts; K_d is the ratio between the particulate (C'_{par} in g kg^{-1} dw) and the dissolved (C_{diss} in g L^{-1}) phases, i.e. $K_d = C'_{par}/C_{diss}$. The K_d ratio is often given in L kg^{-1}. This means that the dissolved fraction (D_{diss}) can be written as:

$$D_{diss} = 1/(1 + K_d \cdot SPM \cdot 10^{-6})$$

where SPM is the amount of suspended matter in the lake water (in mg L^{-1}). It is essential to distinguish between the dissolved and the particulate fractions for all substances. It is especially important to do so for the key nutrient in lake management, phosphorus, because phytoplankton takes up dissolved P and only particulate P can settle out. This means that there are different transport routes for the two fractions. The concentration of suspended particulate matter (SPM) influences the distribution of phosphorus into these two fractions. The settling velocity for particulate P in (m yr^{-1}) may be turned into a

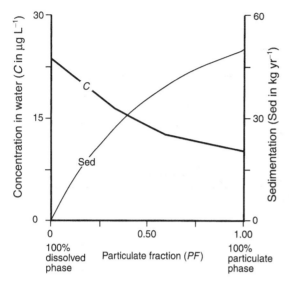

Fig. 4.14 Calculations illustrating the important role of the particulate fraction in mass-balance calculations (annual simulations for a lake with an area = 1 km²; mean depth = 10 m; catchment area = 10 km²; mean annual precipitation = 650 mm yr⁻¹; mean tributary concentration = 26 µg L⁻¹ of total phosphorus; and a settling velocity of 5 m yr⁻¹ for the particulate fraction).

sedimentation rate (dimension 1/time) by division with the mean depth of the lake. The sedimentation rate regulates sedimentation, and hence also internal loading in lakes. Figure 4.14 illustrates the very important role that the PF value plays for the concentration of phosphorus in water and for sedimentation of phosphorus. It is evident that the PF value influences the concentration in lake water: the higher the particulate fraction and the less in dissolved form, the higher the sedimentation and the lower the concentration in water, and vice versa. This is a general principle valid for all substances.

4.3 PROCESSES AND IMPACTS OF NATURAL AND ANTHROPOGENIC DISTURBANCE EVENTS

4.3.1 Storms and mass movements and earthquake records in lake sediments

Many studies have reported how storms, turbidity currents, earthquakes, landslides, etc. influence lakes and cause clearly distinguishable layers in the

sediment record (Sturm 1975; Sturm & Matter 1978; Ludlam 1981; Eden & Page 1998; Ambers 2001). These 'turbidites' can be deposited at any season of the year and may often be separated from 'ordinary' lamina by a greater thickness and often by a different colour or grain size distribution. It should be noted that turbidity currents triggered by subaquatic landslides can travel very fast and influence large parts of lakes (Inouchi et al. 1996; Evans & Slaymaker 2004). They generally appear more frequently in large and deep lakes; large and shallow lakes are dominated by wind-induced resuspension.

4.3.2 Lake-level fluctuations

Water-level fluctuations are common in many reservoirs, but also in natural lakes depending on changes in, for example, land-use practices or precipitation (Thompson & Baedke 1995; Kato et al. 2003). The situation in Lake Aral is a famous example of this, but there are many less well-known situations, for example Lake Kinneret, Israel (Håkanson et al. 2000). Evidently, there are close links between changes in water level and lake morphometry. If the water level goes down, the lake area will be smaller, the wave base lower and the sediments and substances continuously accumulated below the previous wave base will be exposed and influenced by wind-induced wave activities. So, resuspension is likely to increase for some time after the lowering of the wave base. Connected to this, the concentration of suspended particulate matter (SPM) is also likely to increase. This means that the water clarity will also be reduced, and this will influence the primary production of phytoplankton and benthic algae. If there are changes in primary production there are also probable alterations in secondary production of zooplankton and fish.

4.3.3 Lake sediment pollution

This section addresses the transport of pollutants to, within and from lakes. In soils, water is the main transport medium for pollutants and the migration of chemical pollutants is therefore tightly coupled to the hydrocycle. This section attempts to give a general introduction of the basic principles and processes regulating the migration of chemical pollutants. Possible anthropogenic sources of pollutants and types of deposition to the ground will be mentioned only briefly. The sources generally can be divided into activities related to: (i) combustion (e.g. for energy production or waste incineration), (ii) technical processes (e.g. metal and chemical industries) and (iii) other anthropogenic additions (e.g. distribution of pesticides and fertilizers) (e.g. Baudo et al. 1990; Butcher et al. 1992).

Different pollutants have different ranges of atmospheric distribution, owing to variable atmospheric residence times (e.g. mercury has a relatively long residence time in the atmosphere compared with many other metals). The spatial scales may be local, regional or global. After transport in the atmosphere pollutants are deposited either together with rain (wet deposition) or without (dry deposition). The rate of atmospheric deposition depends on the type of pollutant and soil cover. A coniferous forest, for example, has a large active surface area per square metre forest floor and generally catches pollutants effectively. Pollutants are transferred to the soil as direct deposition, throughfall or litterfall. After deposition to the ground, the pollutants mainly migrate through the environment with water. It is, therefore, of interest to study processes influencing the transport from the watershed divide to the lake.

Some elements (such as copper) are micronutrients and essential for life. Such elements are needed within a given concentration range and are toxic outside the range. The effect is, thus, a function of concentration or load. Different ecosystems do not, however, necessarily show the same response to the same given load. Sensitivity (or vulnerability) parameters include water pH, water hardness or lake colour, which will influence the effect of a given pollutant concentration (see section 4.6.2). Many pollutants such as cadmium, mercury and most synthetically manufactured organic pollutants do not have any known physiological functions. They are non-essential to life.

During migration through ecosystems, pollutants might be physically or chemically transformed and more or less toxic forms may appear. There are three main pathways of transformation.

1 Degradation including (i) photolysis, mediated by sunlight and (ii) biologically mediated mineralization of organic matter into its inorganic basic components (energy locked within more complex chemical structures is used as a source of energy by degrading microorganisms).

2 Radioactive decay of radionuclides at the rate of their physical half-lifes. The decay produces daughter nuclides, which in turn decay. This continues until a stable element is reached.

3 Chemical speciation (e.g. oxidation/reduction, metal methylation) due to changed environmental conditions.

For mass transport of chemical pollutants there are two basic physical processes, diffusion and advection. Nature has an inherent reluctance against structured orders, such as differences in concentrations, for example across different interfaces such as air–water and water–sediment. This is one interpretation of the second law of thermodynamics and the force of diffusion. The diffusive flux of pollutants is, thus, proportional to the difference (gradient) of concentrations. Volatilization/evaporation is one example of a diffusive flux. Advection, on the other hand, is the mixing by stirring. The force might be external, for example wind-driven waves in lakes, or internal, for example density differences in stratified waters, which causes mixing. Advective transport can be laminar or turbulent.

Retention is the opposite to migration, a term expressing the storage of a pollutant within the system in relation to the load. Pollutants with long retention times (a high retention) are potentially more hazardous because it means longer exposure times within the ecosystem. The retention is affected by several factors. Some of the most important are the chemical characteristics determining the distribution coefficient, K_d, and the physical factors influencing the flushing of pollutants through the system. Influences of the physical environment on the retention of pollutants mainly include the rate of flushing though the system. This is expressed by the theoretical

Chemical threat	Ecological effects
Acidification	Increase in filamentous algae
	Reduced reproduction of crustaceans, snails, bivalves
	and roach
Eutrophication	Decrease in Secchi depth
	Increase in chlorophyll a and hypolimnetic oxygen demand
Contamination:	
metals	Increased concentration in fish for human consumption
radionuclides	Decrease in reproduction of key organisms, e.g.
	zooplankton, benthos and fish
organic toxins	

Table 4.9 Different chemical threats to aquatic ecosystems and some examples of ecological effect variables. There are also many physical threats to aquatic ecosystems, like the building of dams, piers and marinas, and many biological threats, such as the introduction of new species.

water residence time T. The theoretical residence time of a substance, T_r, is defined by:

$$T_r = (V/Q)/(C/C_{in})$$

where C is the concentration in the system and C_{in} is the concentration in the inflow.

The residence times are related to time-dependent processes such as the settling of fine-grained particles trough the water column and the sorption of pollutants to carrier particles (and other chemical processes), which are examples of processes favoured by long residence times.

4.3.4 Toxicity of chemical water and lake sediment pollutants

In many contexts of lake management, there is a focus on the ecosystem-scale (i.e. on entire lakes), and on the following three major chemical threats to aquatic ecosystems (Table 4.9): (i) acidification, (ii) eutrophication and (iii) contamination (by metals, organic toxins and radionuclides). This section will examine fundamental principles and processes regulating the spread, biouptake and ecosystem effects of contaminants (see Munawar & Dave 1996).

The well-known environmental pollutant mercury belongs to a group of elements often referred to as heavy metals (i.e. metals with a density > 5 cm^{-3}). These metals generally form oxides and sulphides, which are often very hard to dissolve, and they tend to be bound in stable complexes with organic and inorganic particles,

the 'carrier particles', which is an important concept in sediment–water systems. The great interest in heavy metals in aquatic ecotoxicology derives from the fact that some of these elements are supplied to water systems in great excess by humans, and that some of them are hazardous to aquatic life (see Bowen 1966; Förstner & Müller 1974; Förstner & Wittmann 1979; Salomons & Förstner 1984).

A traditional way of determining toxicity of metals and other toxins is to establish the LC$_{50}$ or LD$_{50}$ value, where LC stands for lethal concentration and LD for lethal dose. The value is obtained for the concentration that exterminates 50% of the test sample relative to a control group of test organisms during a certain time span. More than 200 monographs on various toxicological test systems have been published (e.g. Cairns 1981; Burton 1992). A crude rule of thumb states that the least hazardous elements appear with the highest concentrations in water, sediments and biota, and vice versa, the 'abundance principle' (see Håkanson 1980). Elements appearing on the ppb-scale (parts per billion, 10^9), i.e. with extremely low natural concentrations, are, for example Hg, Ag and Cd (Table 4.10). Elements on the ppm-scale (10^6) are, for example As, Co, Cr, Cu, Mo, Ni, Pb, Sn, V and Zn. Elements on the mg-scale (10^3) are, for example Al, Ca, Fe, K, Mn and Na. Pollutants are also classified accordingly into water soluble (hydrophilic elements and compounds) and organic (soluble in organic solutes; liphophilic elements and compounds). Liphophilic compounds are generally 'bioavailable'.

Table 4.10 The abundance of various elements (in ppm) in igneous rocks, soils, fresh water, land plants and land animals (from Bowen 1966).

Element	Igneous rock	Soils	Fresh water	Land plants	Land animals
Ag	0.07	0.1	0.00015	0.06	0.006
Al	80,000	70,000	0.25	500	4–100
As	1.8	6	0.004	0.2	≤ 0.2
Cd	0.2	0.06	< 0.08	0.6	≤ 0.5
Co	25	8	0.0009	0.5	0.03
Cr	100	100	0.0002	0.25	0.075
Cu	50	20	0.01	15	2.5
Fe	55,000	40,000	0.65	140	160
Hg	0.08	0.03–0.8	0.00008	0.015	0.045
Mn	1000	900	0.01	650	0.2
Mo	1.5	2	0.00035	0.9	< 0.2
Ni	75	40	0.01	3	0.8
Pb	15	10	0.005	2.5	2
Sn	2	10	0.00004	< 0.3	< 0.15
V	150	100	0.001	1.5	0.15
Zn	70	50	0.01	100	150

Surfaces of fine-grained particles are chemically active. Surface sites are positively or negatively electrically charged or neutral. Adsorption reactions between charged surfaces and pollutants in ionic form include different types of physical/chemical bindings. Pollutants may occur as free uncomplexed species or as various complexes. Pollutants are attracted to different types of carrier particles depending on their inherent chemical properties. Fine-grained carrier particles have a large surface area per weight unit and are, thus, important for the migration of pollutants. Carrier particles exist in a wide size spectrum. From 1 nm size (10^{-9} m) up to approximately 0.45 μm (10^{-6} m) the particles are generally called colloids. From 0.45 μm to about 0.15 mm, particles suspended in water settle according to Stokes' law by laminar flow (cohesive material). It should also be noted that suspended particles are constantly moving and they aggregate (flocculate) and disaggregate as a result of collisions and reactions with other particles and fluid shear. The carrier particles may be classified into organic (e.g. humic matter) and inorganic (e.g. clays, (hydr-)oxides, Fe/Mn precipitates) or according to genetic origin (e.g. biogenic (mainly organic) or pedogenic (mainly inorganic matter)).

The chemical environment can be expressed in terms of, for example, pH (i.e. the availability of H^+ ions), oxygen (O_2) and redox potential (Eh). These factors influence the pollutants in a number of ways. The formation of important carrier particles, for example iron (Fe) and manganese (Mn) precipitates, are, for example, enhanced by high pH and Eh. The presence of O_2 determines the presence of oxides, such as Fe-oxides, or reduced species, which is important for pollutant cycling in lakes. The survival of higher life forms, such as zoobenthos, also depends on the supply of oxygen. The pH value has a major influence on many chemical/physical processes, some of which are important for the migration of pollutants. The H^+ ions compete for carrier particle binding sites with pollutants. A lower pH generally decreases the K_d value. Flocculation of carrier particles also depends on pH. With increased pH the usually negative charges of natural particles increase and the particles become less likely to flocculate.

The distribution form of the metal in aquatic environments is very important for the toxicity and the potential ecosystem effects (Gottofrey 1990; Wicklund 1990). Generally, the toxicity is highest for the ionic species and proportional to the oxidation number, for example CrO_4^- is more toxic than Cr^{3+}. The potential toxic effects of metals can often be significantly reduced because the metals are bound to different compounds, which may camouflage the toxic properties. Low pH and high redox

Table 4.11 Some well-known chlorinated organics.

Contaminant	Details
TOCl	Total organically bound chlorine
AOX	Adsorbed organically bound halogen
EOCl	Extractable organically bound chlorine
EPOCl	Extractable (acid-)persistent organically bound chlorine
Dioxins	Polychlorinated dibenzo dioxins (PCDD); and furans; there are many dioxins and furans, of which, 'the dirty dozen', are considered of special interest in ecotoxicology
PCB	Polychlorinated biphenyls; lipophilic substances used, e.g. in oils; certain forms, such as planar PCBs are considered to be responsible for the sterility of Baltic seals
DDT	Dichlorodiphenyltrichloroethane; this group includes, e.g. lindane, aldrine, dieldrine and dichlorodiphenyldichloroethylene (DDE), all well-known from R. Carson's book *Silent Spring*
HCB	Hexachlorobenzene
HCH	Hexachlorocyclohexane

potential (Eh) often increase metal toxicity. The solubility of most heavy metals increases at decreasing pH and metals that previously have been bound in rather harmless particulate forms in the sediments may be recirculated to the lake water and express their toxic properties if sediment pH or Eh changes. The roles of heavy metals are, as stressed, complicated by the fact that some of them are essential in small amounts for the organisms.

Table 4.11 gives one example of how to structure the very complex group of organic toxins. For example, AOX stands for adsorbed organically bound halogen, TOCl for total chlorinated organic material, EOCl for extractable organically bound chlorine, EPOCl for extractable persistent organically bound chlorine, etc. The focus of many studies has been on emissions of chlorinated substances from paper and pulp mills (see Södergren 1992). In sediments, only about 2% of a sum-parameter such as TOCl consists of EOCl, and only a small fraction (< 1%) of EOCl consists of chemically identified substances, specific parameters, such as dioxins, polychlorinated biphenyls (PCBs) or dichlorodiphenyltrichloroethane (DDT). The toxicity of organic pollutants, such as EOCl, DDT and PCB, may be manifested in many different ways, for example increased fin erosion in perch, increased frequencies of skin ulcers in herring and increased skeleton deformations, such as deformed jaws in pike or spinal column bending in fourhorn sculpin (Bengtsson 1991). Södergren

(1992) gives a thorough evaluation of biological effects of bleached pulp mill effluents in the Baltic.

In summary, the major threats from chemical pollutants today concern nutrients (phosphorus and nitrogen) causing different types of eutrophication effects (Ambio 1990; Wallin et al. 1992), toxic substances such as metals (Cd, Pb, Cu; Förstner & Müller 1974), chlorinated organics (e.g. PCBs, DDT, dioxins; Södergren et al. 1988) and acidification of land and water, its ecological damage and economic consequences (Ambio 1976; Likens et al. 1979; Merilehto et al. 1988). Case Study 4.1 gives an example of EOCl contamination.

4.3.5 Climatic change

Many lakes are highly sensitive to changes in temperature and precipitation (both magnitudes and frequencies; see Intergovernmental Panel on Climate Change (IPCC) website), because this may influence fundamental processes related to lake ecosystem structure and functioning (e.g. rates of evaporation, lake water level and production and biomass of key functional groups or species). Under extreme climatological conditions, very shallow lakes may even disappear. Responses to climate change will also vary between lakes at different latitudes, altitudes and depending on physical geographical conditions. Lakes, and especially lake sediments, are good sources of information about past climatic/environmental

Case study 4.1 Organic toxins in sediments: Baltic Sea

Discharges of chlorinated organic materials from paper and pulp mills (PPMs) have attracted a great deal of attention in Sweden and many parts of the world (Södergren 1992). This section will give a sedimentological angle to this important environmental problem. Case Fig. 4.1a illustrates the geographical distribution pattern of EOCl in surficial (0–1 cm) sediment samples from accumulation areas, and Case Fig. 4.1b shows how extractable organically bound chlorines (EOCl) in sediment samples co-vary with two very important and specific toxins – dioxins and furans. Other features are evident from Case Fig. 4.1a.

1 There is a large-scale spread of EOCl in this water system, the Baltic. The larger the emissions, the larger the impact areas and the greater the potential ecological problems. The distribution of EOCl in surficial sediments shows a very characteristic pattern, with high concentrations close to the PPMs along the coast.

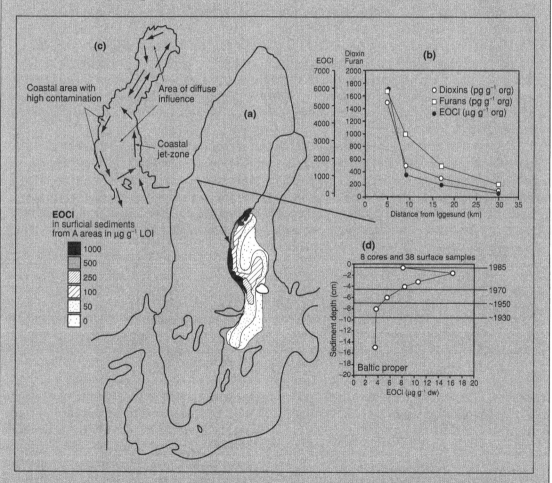

Case Fig. 4.1 (a) Distribution of EOCl in surficial A-area sediments in the Baltic. (b) The relationship between EOCl, dioxins and furanes in surficial A-area sediments taken at different distances from the Iggesund paper and pulp mill in the Bothnian Sea. (c) Illustration of the coastal jet-zone and the major hydrological flow pattern in the Bothnian Bay and the Bothnian Sea. (d) The historical development of EOCl in the Baltic (based on eight sediment cores and 38 surficial sediment samples). (From Jonsson 1992; Håkanson 1999.)

2 The dominating water circulation in each basin (the Bothnian Bay, the Bothnian Sea and the Baltic Proper) constitutes an anticlockwise cell, which distributes the settling particles, the suspended material and the pollutants in a typical pattern, reflecting the flow of the water (Case Fig. 4.1c). This anticlockwise cell is created by the rotation of the Earth (the Coriolis force), which deflects any plume of flowing water to the right in relation to the direction of the flow in the Northern Hemisphere (and to the left in the Southern Hemisphere). Thus, when a Swedish river enters the Baltic, the water turns to the right and follows the shore. The net hydrological flow is to the south on the west (Swedish) side of the Baltic. The currents are rather strong and stable close to land and weaker towards the centre of each basin. The figure illustrates only the net component of the flow – this means that the water would also flow in most other directions during the year.

3 A large number of sediment cores have been taken from Baltic accumulation areas and analysed for EOCl. Case Fig. 4.1d gives results reflecting average conditions. The increase in the EOCl concentration in the sediments, and hence the increase in sediment and water contamination, started in the late 1950s and coincides well with the general contamination and increase in eutrophication in the Baltic.

It should be noted that most of the emissions of EOCl to the Baltic have now been halted as a result of legislation and public awareness of the problem. The Baltic is recovering from this contamination. The time perspective of this recovery can be illustrated using the sediment data in Case Fig. 4.1d. It will probably take 20–40 years until the system has recovered so that the EOCl concentration in surficial sediments is back to the 'normal' values that characterized the system about 50 years ago.

This example illustrates the very close and important connection between pollution site and load, distribution of pollutants by water currents, which regulate where high and low contamination appears, and hence also where small and large potential ecosystem effects may be expected. This also shows that different contaminants from the same pollution site often (as in this case) are distributed in the aquatic environment together, because the distribution depends on the fact that both the dissolved and the particulate forms of the pollutants are distributed by the same hydrodynamic and bottom-dynamics processes.

Relevant reading

Södergren A. (Ed.) (1992) *Bleached Pulp Mill Effluents. Composition, Fate and Effects in the Baltic Sea.* Report 4047, Swedish Environmental Protection Agency, Stockholm, 150 pp.

conditions. Case Study 4.2 sets out an example of climate change impacts upon Lake Batorino, Belarus.

4.4 SEDIMENT DATING AND SEDIMENT RECORDS

4.4.1 Sediment dating using radionuclides

The sediments reflect what is happening in the water mass and on the bottom – they may be regarded as a tape-recorder of the lake's historical development and are often called 'the geological archive' (Zolischka 1998; Hammarlund et al. 2003). The sediments also affect the conditions in the water via, for example, resuspension processes and by the fact that the animals living in the sediments play a fundamental role in the ecosystem (Fig. 4.5). By extracting sediment cores and conducting a number of analyses, information is obtained on changes that have taken place in the ecosystem (Thomas et al. 1976;

Case study 4.2 Impact of global change on lake characteristics: Lake Batorino, Belarus

The presuppositions for this 'global change' scenario have been given by Håkanson et al. (2003). Wick (2000) has discussed how global temperature changes can influence vegetation and sediment records. The basic assumption is that global warming would increase the mean annual temperature (as given by the curve in Case Fig. 4.2a). Note that this is a simulation of a hypothetical sudden increase in temperature and that, realistically, global warming will lead to more gradual changes. In this scenario, the focus is on the final results rather than the path to the final results. Also note that global temperature changes may cause more extreme seasonal temperature variations. In this scenario, the mean annual temperature is raised by 2°C and an increased seasonal temperature variation has been assumed by applying an exponent > 1 for the weekly epilimnetic temperatures

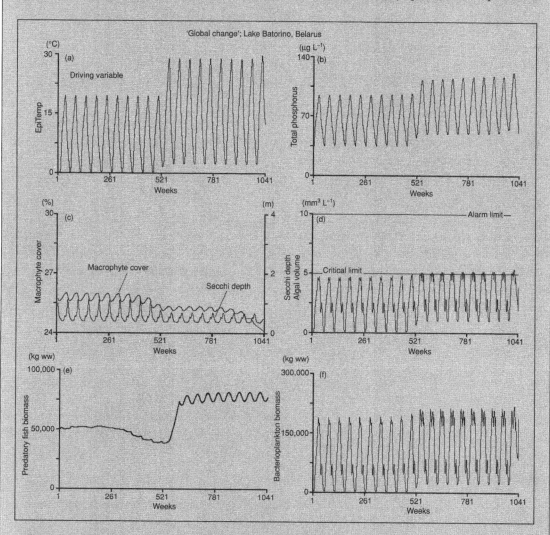

Case Fig. 4.2 Case study of the effects of global temperature changes on Lake Batorino, Belarus, assuming that there would be a hypothetical change in lake water temperatures (a). How would this influence fundamental lake characteristics, such as lake concentrations of phosphorus (b), macrophyte cover and water clarity (Secchi depth) (c), algal volume in relation to operational 'critical' and 'alarm' guideline values (d), biomass of predatory fish (e) and biomass of bacterioplankton (f)?

Case Table 4.2 Data on the case-study lakes. (From Håkanson & Boulion 2002.)

Lake	Catchment (km²)	Area (km²)	Mean depth (D_m, m)	Maximum depth (D_{max}, m)	pH	Colour (mg Pt L⁻¹)	C_{in} (µg P L⁻¹)	Latitude (°N)	Altitude (m a.s.l.)	Precipitation (mm yr⁻¹)
Batorino	92.5	6.3	3.0	5.5	8.0	54	120	54.5	165	650
Miastro	133	13.1	5.4	11.3	8.0	31	80	54.5	165	650

(i.e. EpiT^{exp}). In the following scenario, this exponent is set to 1.1. The temperature for the first 10-year period (weeks 1–521) has been kept normal and then for the second 10-year period a well-tested lake foodweb model (LakeWeb, see Håkanson & Boulion 2002) has been used to predict the response of the system, here Lake Batorino, Belarus (see Case Table 4.2 for lake data).

Under these presuppositions, Case Fig. 4.2a gives the driving variable, lake epilimnetic temperatures. The question is, how will this change in temperature regime influence important variables for lake management, such as lake total phosphorus (TP) concentrations (Case Fig. 4.2b), macrophyte cover and Secchi depth (Case Fig. 4.2c), algal volume (Case Fig. 4.2d), total fish biomass (Case Fig. 4.2e) and bacterioplankton biomass (Case Fig. 4.2f)? The following can be noted from Case Fig. 4.2.

• An increase in lake temperature will increase lake TP concentrations significantly, and for several reasons. More phosphorus, for example, will be bound in organisms with short turnover times (phytoplankton, bacterioplankton, benthic algae and zooplankton), which are included in the value given for the total lake TP concentrations. As a result an increase in mean annual temperatures will probably produce eutrophication (Case Fig. 4.2b).

• This also means that warmer climatological conditions will reduce the Secchi depth and hence also decrease the macrophyte cover (Case Fig. 4.2c). The clearer the water, the higher the macrophyte production.

• Warmer conditions will imply that algal volumes are likely to increase to values higher than the 'critical' limit (set by national environmental agencies) during the growing season (Case Fig. 4.2d).

• Warmer conditions will increase the fish production. This means a significant change in the structure of the lake foodweb (Case Fig. 4.2e).

• There are also clear changes for bacterioplankton biomass: the warmer the water, the more bacterioplankton will be produced (Case Fig. 4.2f).

The 'global change' scenario for Lake Batorino, therefore, indicates that major changes in the lake foodweb can be expected if the climate becomes warmer. A global warming in northern lakes would probably produce conditions similar to eutrophication. It would mean higher primary and secondary production, lower Secchi depths and hence also a reduced macrophyte cover. This may seem self-evident, but it is not evident how this is manifested in terms of quantitative changes in key functional groups of organisms and in target variables for lake management. Such changes, can, however, be calculated with the LakeWeb model.

Relevant reading

Håkanson, L. & Boulion, V. (2002) *The Lake Foodweb – Modelling Predation and Abiotic/Biotic Interactions.* Backhuys Publishers, Leiden, 344 pp.

Håkanson, L., Ostapenia, A., Parparov, A., et al. (2003) Management criteria for lake ecosystems applied to case studies of changes in nutrient loading and climate change. *Lakes and Reservoirs: Research and Management* 8, 141–55.

Wick, L. (2000) Vegetational response to climatic changes recorded in Swiss late glacial lake sediments. *Palaeogeography, Palaeoclimatology, Palaeoecology* 159, 231–50.

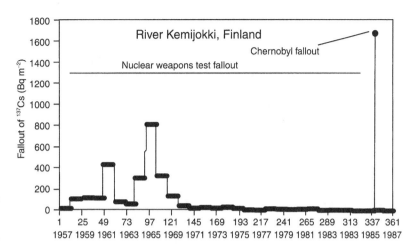

Fig. 4.15 Radiocaesium fallout from the nuclear weapons test and the Chernobyl accident on the catchment area of River Kemijokki, Finland (data from Ritva Saxen). These data can be used for dating lake sediments in this part of the world. Note that in bioturbated sediments, the age distribution at various sediment depths in a sediment core would be levelled out, whereas in laminated sediments the fallout peaks may be more easily identified and the dating more exact.

Golterman et al. 1983; Jonsson 1992). This section gives a brief discussion on methods to determine ongoing sedimentation and the age of recent sedimentary deposits and will not discuss palaeolimnological methods (such as pollen analysis and the radiocarbon method; see Reeves 1968; Goudie 1981).

There are two very useful and widely applied methods to determine the age of recent lake sediments, lead-210 (^{210}Pb) analysis (Robbins 1978; Appleby & Oldfleld 1979; Legesse et al. 2002) or the analysis of radiocaesium (^{137}Cs; see Pennington et al. 1973; Yan et al. 2002). Lead-210 has a physical half-life of 22.3 years and ^{137}Cs a half-life of 30.2 years, which makes these substances very useful for dating recent sedimentary deposits. To determine the sediment age using the ^{210}Pb method, one must also quantify all fluxes from the catchment to the lake, internal fluxes of ^{210}Pb (sedimentation and resuspension) and lake outflow. This means that mass-balance calculations are essential. Sediment dating using ^{137}Cs is simpler. For this, it is important to have access to a fallout curve, such as the one shown in Fig. 4.15 for a Finnish site. Radiocaesium in western Europe has two basic sources, fallout from the nuclear weapons testing (mainly between 1957 and 1975) and the Chernobyl accident (April–May 1986). By taking sediment cores and trying to identify these peaks in radiocaesium activity/concentration, a good estimate of the mean sedimentation rate and

the mean sediment age can be obtained. Evidently, this method is more likely to give more accurate estimates of sediment age in laminated than in bioturbated sediments.

4.4.2 Sediment dating using sediment traps

Sedimentation and sediment age can also be determined using sediment traps (see Håkanson & Jansson 1983). Generally, sediment traps are cylinders with a width:height ratio of 1:3 placed vertically in the water. Figure 4.16 gives an example where sediment traps have been used for sediment dating. From this figure, it is evident that the contamination of a river pollutant (here mercury entering Lake Ekoln, Sweden, from River Fyris) and the content of the contaminant in lake sediments offer an excellent key to the pollution history of the lake. In the same way, the areal distribution pattern of the pollutant in lake sediments can be used to evaluate transport patterns in the system and to identify the polluting site (or river).

4.5 MANAGEMENT AND REMEDIATION

4.5.1 Environmental consequence analysis and ecosystem indices

In environmental management it is important not to use personal viewpoints as criteria to rank

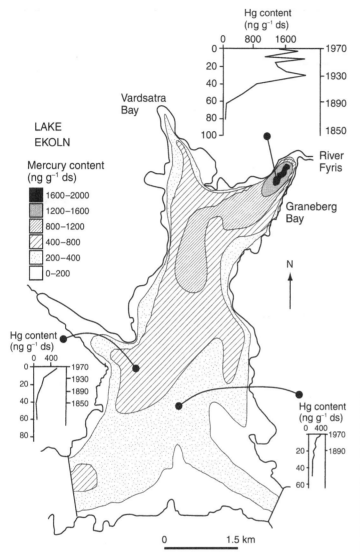

Fig. 4.16 Sediments as a historical archive illustrating both the areal and the temporal perspectives of ongoing changes, here exemplified by the mercury contamination in Lake Ekoln, Sweden. The sediment age has been determined from data on sedimentation in sediment traps. The main tributary to this lake is River Fyris and mercury is transported to the lake from different sources (hospitals, dentists, laboratories, etc.) mainly from the city of Uppsala. (Modified from Håkanson & Jansson 1983.)

threats as a basis for action, but to have a more objective approach. There is a growing awareness that much better individual 'indicators' and aggregated 'indices' of environmental health are necessary because they alone can provide a rational structure for decision-making in the environmental sciences (Bromberg 1990; OECD 1991). An index (an aggregated measure) is generally distinguished from an indicator (a single variable), and an ecosystem (a single instance, such as a lake or a field) from an ecosystem type (the summation of several to many ecosystems).

A frame of reference is required to assess the status of the environment. Since 1987, many countries have accepted 'sustainable development' as a goal for environmental and economic policy. The term was introduced in the final report of the Commission for Environment and Development (the Brundtland Commission). This phrase is empty, however, unless it is defined in terms of operationally measurable properties, desired goals and relevant data. There are alternatives to choosing ecosystems as the basis for environmental typology (Mackay & Paterson 1982; O'Neill et al. 1982; Cairns & Pratt 1987).

Instead, one might use different geographical areas or different media such as air, water and soil. There is, however, a clear international trend towards consideration of the 'health' of different ecosystems (Bailey et al. 1985).

The majority of threats involve chemicals. A set of ecological effect variables is expected to reflect such threats and the extent to which they affect the ecosystem. There is also a difference between biological effects for individual animals or organs and ecological effects for entire ecosystems. Practically useful, operational effect variables should be measurable, preferably simply and inexpensively, clearly interpretable and predictable by validated quantitative models, internationally applicable, relevant for the given environmental threat and representative for the given ecosystem.

Ecosystem indices would have the advantage of expressing the environmental status simply, but they simultaneously pose problems in that a great deal of valuable information is lost in aggregating the individual measures. This disadvantage is reduced if it is known exactly what an index represents, and if these individual components can be accessed as required. Ideally, the same basic framework would be used at both the national and regional scales. As problems and priorities cannot be completely congruent at different levels, however, the framework may be adapted to the different requirements of different levels. The national level may address large-scale threats, perhaps originating outside the country, such as acidification of soil and water, whereas the region can address more local problems, such as the eutrophication of lakes (Case Study 4.3).

Case study 4.3 Impact of eutrophication/oligotrophication: Lake Miastro, Belarus

The following case study concerns Lake Miastro, Belarus (see Case Table 4.2 for lake data; for further details about the scenario, see Håkanson & Boulion 2002) and uses the well tested lake foodweb model, LakeWeb, for the simulations. The aim here is to illustrate both the effect–load–sensitivity analysis and how changes in a critical load factor (the tributary concentration of total phosphorus, TP) will cause, first, changes in lake phosphorus concentrations, in concentrations of suspended particulate matter (SPM) and in sedimentation. It also shows how such changes will influence target operational effect variables in lake management, such as Secchi depth (a measure of water clarity) and algal volume (a measure of lake trophic level). In 1990 a drastic and sudden change in agricultural land-use practices occurred in the catchment area of this lake. The use of imported fertilizers was stopped as a result of political changes related to the fall of the former Soviet Union.

The presuppositions for this scenario are given in Case Fig. 4.3a, with the 'modelled values' curve illustrating the sudden change in tributary total phosphorus (TP) concentrations in week 261 (= January 1990; week 1 is the first week of 1985 and the simulation covers a period of 10 yr). There are good empirical data for this scenario (from Professor Alexander Ostapenia, Belarus State University, Minsk) giving mean characteristic monthly values for the period 1985 to 1989 for lake TP concentrations (see Case Fig. 4.3b) and TP concentrations in sediments (Case Fig. 4.3e). One can first note the good correspondence between modelled values of lake and sediment TP concentrations and empirical data. The following question is addressed in this scenario (from Case Fig. 4.3b), how will the changes in TP concentrations in the lake influence important sedimentological variables?

From Case Fig. 4.3c, one can note that the oligotrophication would likely imply a decrease in suspended particulate matter (SPM) concentrations because the algal volume would go down (Case Fig. 4.3g). This would also mean that sedimentation of matter decreases (Case Fig. 4.3d)

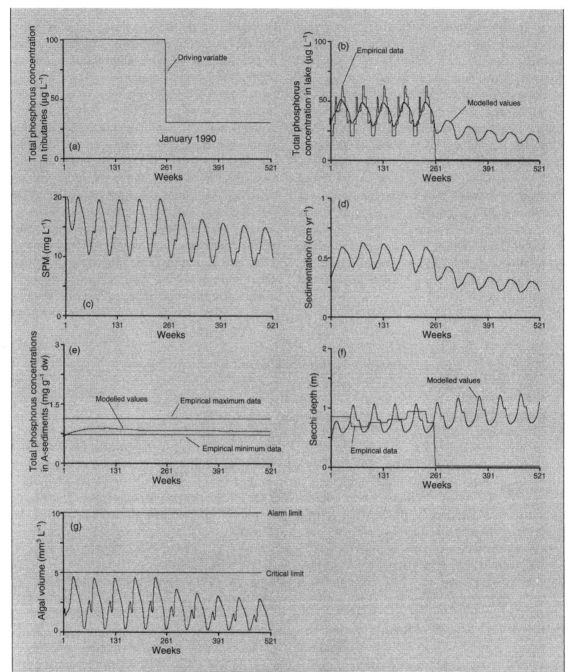

Case Fig. 4.3 Case-study on eutrophication/oligotrophication in Lake Miastro, Belarus. There was a drastic reduction in the use of fertilizers in argriculture in 1990. This has influenced the river concentration of phosphorus, as shown in (a), the driving variable in this scenario. It can then be asked how this will influence: lake concentrations of phosphorus (b), concentrations of suspended particulate matter (SPM; (c)), sedimentation of matter (d), phosphorus concentrations in A-sediment (e) and operational variables of ecosystem effect, such as the Secchi depth (f) and the algal volume (g).

and the Secchi depth increases (Case Fig. 4.3f). It is interesting to note there are no clear changes in TP concentrations in sediment (Case Fig. 4.3e) because phosphorus is a very mobile element in lake sediments and the phosphorus concentration in sediments depends more on the sediment chemical 'climate', i.e. on redox conditions, than on deposition (see Table 4.5).

The drastic reduction in phosphorus loading to this lake can have both positive and negative consequences for the lake ecosystem depending on management objectives and criteria. From a sedimentological perspective, one can conclude that the SPM concentrations probably will decrease, the water clarity will be higher and the deposition of matter also will decrease. These are signs of oligotrophication.

Relevant reading

Håkanson, L. & Boulion, V. (2002) *The Lake Foodweb – Modelling Predation and Abiotic/Biotic Interactions.* Backhuys Publishers, Leiden, 344 pp.

An environmental index must be based on the status of some crucial characteristics of chosen ecosystem types. These are the six basic ecosystem types.

1 Forests
2 Agricultural land
3 Natural land
4 Freshwater
5 Coastal areas
6 Urban areas

Generally, it is extremely difficult to distinguish cause and effect in natural ecosystems. One cannot base an environmental consequence analysis on a full understanding of the ecosystem. In complex ecosystems 'understanding' at one scale (e.g. the ecosystem scale) is generally related to processes and mechanisms at the next lower scale (e.g. the scale of individual animals and/or plants), and the explanation of phenomena at this scale is related to processes and mechanisms at the next lower scale (e.g. the scale of the organ), and so on down to the level of the atom and beyond. In environmental management, a balance must often be found between answering interesting, often important, questions of understanding, and delivering a practical tool to society. If an ecosystem index were based on a causal analysis of what takes place at the cellular level, then at levels involving organs, individuals, populations, and finally at the ecosystem level, it would be an eternity before the index could be developed. For the foreseeable future, ecosystem indices are more likely to be based on practical considerations of predictive power and sampling ease, rather than full causal priority.

What, then, should the strategy be for developing an ecosystem index? The first problem is that each ecosystem type, for example fresh waters, is not a single entity. It consists of many sub-ecosystems (Fig. 4.17). A general resolution about the basis of this approach is probably impossible, but questions about the appropriate hierarchical level of analysis are relevant to specific threats. Figure 4.17 lists the 12 general environmental threats. If one starts with the threat to fresh waters, it is clear that contamination from, for example, metals and radionuclides threatens fresh waters and that these threats might be manifested in, for example, reduced biological diversity and contaminated fish.

4.5.2 Effect–load–sensitivity analysis

Elevated concentrations of contaminants that cause no visible or measurable ecological effects would generally be of less interest for practical water management, and for remedial strategies, in the situation faced today in ecosystems with multiple threats. The aim of effect–load–sensitivity analyses (ELS) and ELS models is to provide a

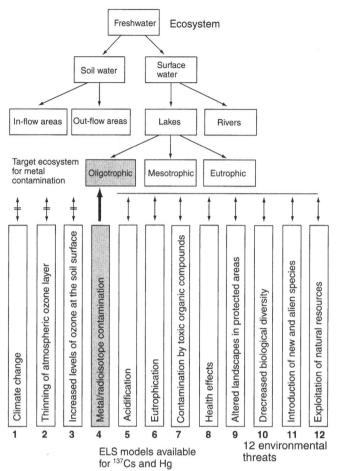

Fig. 4.17 The fresh water ecosystem may be divided into several subecosystems (such as oligotrophic, mesotrophic and eutrophic lakes) where different key organisms prevail. As part of this differentiation process, one must decide which subsystem should be used as the target ecosystem for a given chemical threat, which target organisms, functional groups or effect variables should be used for a given threat, and what load and sensitivity variables should be used relative to a given effect variable. The target ecosystem for metals (such as Hg) and radionuclides (such as ^{137}Cs) is low-productive (oligotrophic) lakes (shaded in this figure): ELS, effect–load–sensitivity. (From Håkanson & Peters 1995.)

tool for quantitative predictions which relate operationally defined ecological effects to compatible load and sensitivity variables (Håkanson 1999).

Differential equations and mass-balance models are often used to handle fluxes (e.g. g X yr^{-1}), amounts (g X) and concentrations (g X m^{-3}) of all types of materials (such as gases, carbon, toxins and nutrients), but not ecosystem effect variables (E). Statistical methods, such as regressions based on empirical data, are generally necessary to relate concentrations of chemicals to effect variables (E). In theory, both these model approaches (see Fig. 4.18) may be used for the ELS models, provided that at least one operationally defined ecological effect variable relevant for the load variable(s) in question is included in the model. Ideally, the E variable

should express the reproduction, abundance, mass or status of defined functional organisms (preferably fish at the top trophic level) that characterize the given ecosystem. Such ideal effect variables cannot normally be predicted with differential equations and mass-balance models. If the more ideal effect variables cannot be determined, then in practice one has to do the second best and define operational effect variables, such as toxic concentrations in fish eaten by humans, the oxygen concentration in the deep-water zone and the Secchi depth (see Table 4.10). The sensitivity variables express how different lake characteristics (such as pH, colour, total-P concentrations and lake mean depth) regulate the 'road between load and effect'. One and the same load will cause different ecosystem effects in lakes of different sensitivities. For example, in lakes

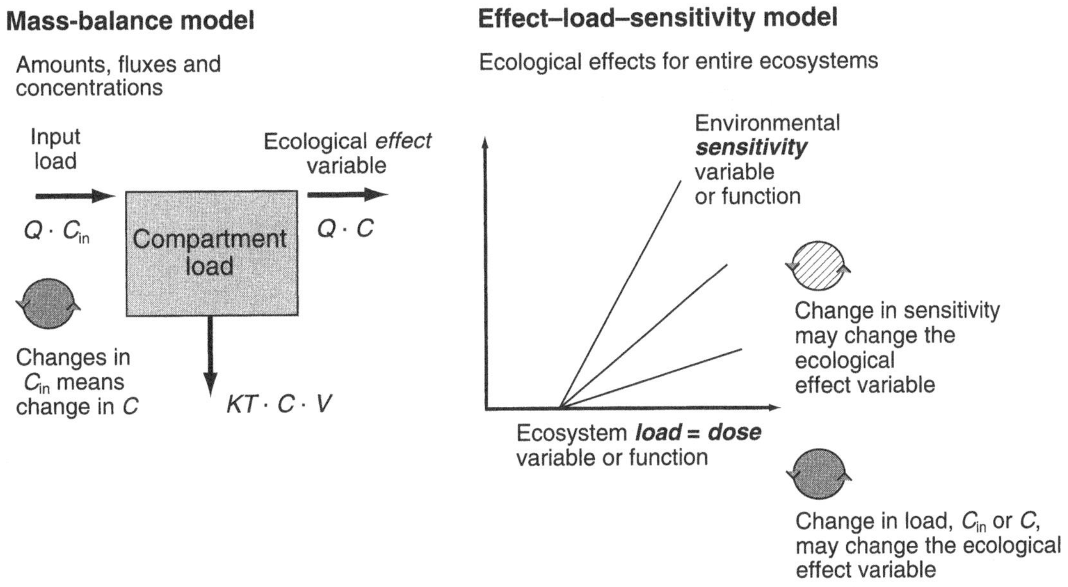

Mass-balance model

Amounts, fluxes and
concentrations

Effect–load–sensitivity model

Ecological effects for entire ecosystems

Fig. 4.18 Illustration of the fundamental difference between dynamic, mass-balance models and effect–load–sensitivity (ELS) models. The three wheels indicate that by means of remedial measures one may reduce the load variable in dynamic models and the load and the sensitivity variables in ELS models. KT is the sedimentation rate. (From Håkanson & Peters 1995.)

with low pH and low bioproduction, a given load of mercury will cause significantly higher biouptake of mercury and hence also higher mercury levels in fish, than in lakes with high pH and high bioproduction. There are many documented cases like this (see Håkanson 1999). One classic way to develop ELS models is to use dynamic mass-balance models to handle concentrations of pollutants and empirical models (such as regressions) to link these concentrations to the operational effect variables.

4.6 SUMMARY

Lake sediments are an important, integral part of the lake ecosystem. They reflect changes in land-use and lake characteristics and are often regarded as a historical archive. They also affect the present structure and function of lake ecosystems. Suspended particulate matter (SPM) in

lakes mainly comes from tributaries (allochthonous matter), from living and dead matter produced in the lake (autochthonous matter) or from resuspended materials. The SPM regulates the transport of all types of water pollutants in dissolved and particulate phases; it regulates water clarity and the depth of the photic zone, and hence also primary and secondary production; it regulates bacterioplankton production and biomass, and hence also mineralization, oxygen consumption and oxygen concentrations; and it regulates sedimentation, and hence also the use of sediments as a historical archive, for example of water pollutants. These matters are discussed in this chapter. The aim of the information given is to structure existing knowledge on the factors regulating variations among and within aquatic systems of suspended particulate matter in a rational manner. This knowledge is fundamental for an understanding of the function and structure of aquatic systems.

5

Arid environments

Anne Mather

5.1 INTRODUCTION

Within arid landscapes sediment transfer begins with the breakdown of bedrock to transportable material. This can be achieved by weathering processes, which in arid areas are dominated by moisture and salt or insolation. Owing to the role of gravity in the transportation processes, most source areas will be located within the upland areas of the arid landscape. These areas are typically dominated by zones where the erosive capability (erosivity) of water is high, such as steep slopes. Other sediment source areas are dominated by zones with sediment that is susceptible to erosion owing to its weak material characteristics (its 'erodibility'), such as exposed lake sediments (Fig. 5.1). These available materials will be transported by wind or water to (i) areas of temporary storage within transport zones, such as rivers, or (ii) a more permanent storage area of net sediment accumulation, such as an untrenched alluvial fan, a sand sea or a lake. These sediment stores will themselves be susceptible to recycling

within the sediment transfer system (Fig. 5.1). This chapter examines the processes involved in this transfer of sediment within the arid environment, and the environmental hazards that occur en route. It provides an insight into the recent sedimentary deposits of arid environments and how they are affected by changing external environmental controls.

5.1.1 Definition of arid environments

In this chapter arid environments will be considered to embrace regions described both as 'drylands' and 'deserts'. Deserts are more qualitatively defined by a range of physical criteria, whereas drylands are quantitatively defined and classified using meteorological data. As the definition of drylands is based on modern climate data records, however, it is difficult to rigorously apply this definition to older historic to geological time-scales. As a result both terms (drylands and deserts) have tended to be used interchangeably in the literature,

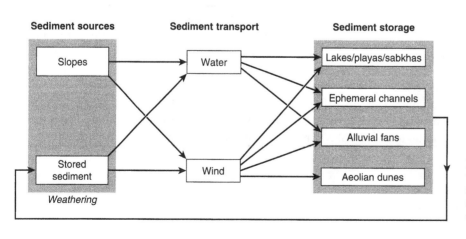

Fig. 5.1 Simplified flow diagram of sediment production and routing in arid environments.

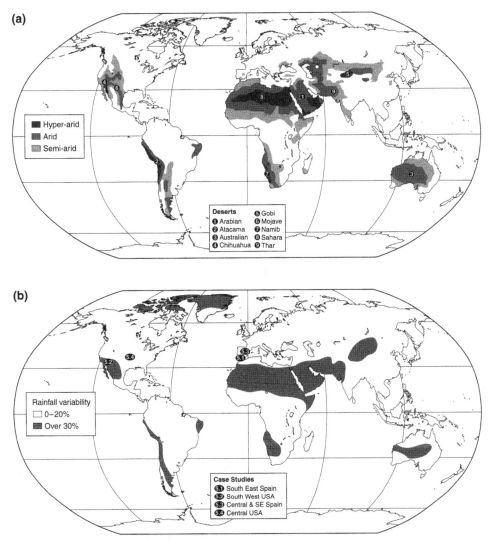

Fig. 5.2 (a) The global distribution of arid environments and the location of the major deserts referred to in the text. (Modified from Goudie & Wilkinson 1978.) (b) Rainfall variability and locations of Case Studies used in this chapter. (Modified from Dregne 1983.)

with deserts describing the hyper-arid to arid definitions of drylands. Thus, for the purpose of this chapter arid environments will be considered to represent the more extreme drylands (hyper-arid to arid; 4% and 15% of the Earth's land surface respectively), which can be recognized in the older landscape records, and those drier areas at greatest risk of degradation by human activity (semi-arid, representing 15% of the Earth's land surface; Fig. 5.2a). A fuller definition of the terms 'desert' and 'dryland' is given below.

Deserts are inhospitable, barren, poorly veget-ated areas devoid of water. There are numerous scientific definitions of deserts available based on a range of physical criteria from erosion to vegetation type (see Heathcote (1983) and Thomas (1997a) for a summary). Deserts there-fore cover a range of climate categories, ranging from cold arid (Gobi Desert) to hot arid (central Sahara), and some deserts are so large they may cover several climate types – the Sahara includes hot (central Sahara), mild (southern Sahara), and cool (northern Sahara) climates. Most people

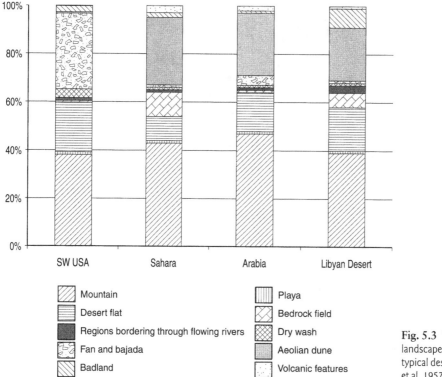

Fig. 5.3 The percentage of landscape units found within four typical deserts. (Data from Clements et al. 1957.)

associate deserts with areas dominated by extensive sand dune systems (sand seas). In reality, however, these landscape elements are relatively minor compared with other elements such as mountainous areas (Fig. 5.3).

Drylands are classified quantitatively into hyper-arid, arid, semi-arid and dry-subhumid using meteorological data. These classifications are defined using the balance of moisture inputs (precipitation) and losses (evapotranspiration) expressed as an aridity or moisture index. This index is typically expressed as the ratio of P (precipitation) to PET (potential evapotranspiration). Classifications vary, however, depending on the data used (Table 5.1). For example Meigs (1953) used aggregated monthly to annual moisture surplus and deficit data to represent P and the Thornthwaite approach to calculate PET. These data are not widely available from arid regions. More recent definitions therefore have tended to use annual precipitation for P and the Penman method of calculating PET, which utilizes an understanding of the

diffusion of water vapour and aerodynamic functions, data which are universally available. To further facilitate a comparable global coverage the data sets used in the most recent definitions also tend to be time-bounded to avoid data bias from regional data sets where the records are less continuous. The time period spans some 30 years to take account of inter-annual and interdecadal variations in climate. Variation in the data sets used means that figures on the modern extent of drylands range from 26.3 to 47.2% of global land area, depending on the classification used (Thomas 1997a).

5.1.2 Causes of aridity

5.1.2.1 Climate

Arid environments are typified by variable rainfall (characteristically more than 30% of the mean, Fig. 5.2b). This rainfall will range from a number of low-intensity events, which may

Table 5.1 Quantitative classifications of arid zones. (Based on Thomas 1997a.)

Classification and use	Data used	Classification	Comments
UNESCO 1979	Uses Meigs' (1953) classification together with Thornthwaite's (1948) indices of moisture availability (Im) $Im = (100S - 60D)/PET$ where S is the moisture surplus, M the moisture deficit and PET is potential evapotranspiration. S and M are aggregated on an annual basis from monthly data, taking stored soil moisture into account	Semi-arid Arid Hyper-arid	Does not include areas too cold for crops (e.g. polar deserts)
Grove 1997	Meigs 1953 classification, with rainfall figures	Semi-arid = 200–500 mm yr^{-1} Arid = 25–200 mm yr^{-1} Hyper-arid = < 25 mm yr^{-1}	Approximate rainfall figures. Figures available over widely varied time-scales spatially
UN 1977	P/PET index: where P is annual precipitation and PET is potential evapotranspiration	Semi-arid $0.20 \geq AI < 0.50$ Arid $0.03 \geq AI < 0.20$ Hyper-arid $AI < 0.03$	PET calculated by Penmans formula using data not globally available
UNEP 1992	Aridity index (AI) $AI = P/PET$ where P is annual precipitation and PET is potential evapotranspiration. PET is calculated using the Thornthwaite method. Data are taken from time-bounded study	Dry-subhumid $0.50 \geq AI < 0.65$ Semi-arid $0.20 \geq AI < 0.50$ Arid $0.05 \geq AI < 0.20$ Hyper-arid $AI < 0.05$	Includes dry subhumid areas

occur over several days but generate only limited runoff, to high-intensity rainstorm events, which generate flash flood events but may occur only two or three times in a 100-year period. Regions of such restricted rainfall are most commonly related to four main individual climate causes, which may be related in part to tectonics (see section 5.1.2.2). In addition the individual causes listed below may interact with each other, reinforcing conditions of aridity.

1 *Global atmospheric circulation*. Areas dominated by high-pressure cells and hot (such as the tropics of Cancer and Capricorn) or cold dry subsiding air (the North and South Poles) will have limited rainfall capability.

2 *Continentality*. In large continents (e.g. the interior of Asia) available moisture may have been precipitated out in the more coastal areas of the landmasses.

3 *Rain-shadow created by high mountain areas*. In coastal ranges such as the Rockies, USA, moist air derived from evaporation over the ocean

rises and is cooled adiabatically. Thus vapour saturation is reached quickly and precipitated as rain or snow. The desiccated air will then descend across the mountain range, becoming warmer and drier into the area of rain shadow.

4 *Cold upwelling ocean currents*. The most prominent examples are the Humbolt current of the Atacama Desert and the Benguela current of the Namib Desert. Both these ocean currents are derived from the colder, southern polar latitudes and flow northwards. Their associated cool moist ocean air reaches relatively warmer land, where its relative humidity decreases and its capacity to absorb moisture is increased. Thus the oceanic air mass tends to desiccate these coastlines. The main source of moisture in such areas is coastal fogs such as the 'camanchaca' of the Atacama.

Moisture in arid areas will not be solely sourced from rainfall events. Coastal fog may be important in arid regions that border oceans (e.g. the Atacama Desert of South America and the Namib Desert of southern Africa; Fig. 5.2). Gobabeb research station (Namibia) data records show that coastal fog is a significant contributor of moisture, even though the research station is some 60 km inland and 408 m a.s.l. Here fog moisture may well exceed rainfall in some years. Another significant source of moisture may be dew. Fewer records on dew are available but data from the Negev, Israel, suggest that dew occurs on an average of 195 nights per year at the ground surface, and provides 33 mm yr^{-1} of moisture, although only 0.35 mm for any one dew night (Evanari et al. 1982).

5.1.2.2 Tectonics

Arid environments can be found in a range of tectonic settings. These include currently stable intraplate settings such as cratons (e.g. the Australian Desert) and passive margins (e.g. the Namib). Arid regions can also be found in extensional tectonic settings such as the Mojave and Chihuahua deserts, and compressional continental margins such as the Atacama and Thar deserts. Some deserts, such as the Sahara, are so large that they may include several tectonic settings (the Sahara spans all of the above settings).

Despite the apparent lack of correlation between tectonic settings and arid region occurrence, tectonics are a major control on the location and persistence of arid environments. Tectonics control the regional and global topography of the continents, which in turn has significant effects on the local climate. The main impact of tectonics as a control on the development of arid areas is through the two processes listed below.

1 *Orogeny.* Tectonics can generate mountains, which create local rain shadows. On a larger scale, large orogenic belts such as the Himalayas can divert major air circulation, leading to aridity. The associated uplift will also lead to an increase in the continentality of an area.

2 *Continental drift.* Tectonics can affect latitude over many millennia through continental drift (consider for example Britain which was much closer to the Equator 250 Ma in the Permo-Triassic, and subject to desert conditions). Plate tectonics can also lead to the closure of oceanic gateways, which can affect oceanic currents and thus coastal climates. It has been postulated that the closure of one such gateway reinforced the hyper-aridity of the Atacama Desert of South America (Hartley 2003).

5.1.2.3 Anthropogenic agents

Humans have been credited with forming and expanding many modern deserts (Ehrlich & Erlich 1970). The expansion of desert areas through human intervention is known as 'desertification' (*sensu* Mabutt 1978, 1985). This process, and its definition, however, has become increasingly disputed (e.g. Williams 1994). There is no doubt that mismanagement of land through overgrazing and soil salinization has occurred since the Holocene increase in human population. The human population is estimated to have increased from 10^7 at the start of the Holocene Epoch (10 ka) to 10^8 by 5 ka and 10^9 by 2 ka (May 1978). Typically, however, the increase in desert areas is temporary and associated with intervals of drought, which in most cases return to their natural non-desert habit once the drought has ended and recovery set in. In 1992 the Rio Earth Summit thus defined desertification as 'land degradation in arid, semi-arid and dry subhumid areas ['susceptible drylands'] resulting from various factors, including climatic variations and human actions' (UNEP 1994).

5.1.3 Variability of arid environments through time

The geological record of aeolian dunes and evaporite sediments indicates that arid areas have been present since the Precambrian (Glennie 1987). The Atacama is probably the oldest extant desert in the world (Hartley et al. 2005), dating back to the Jurassic (some 150 Ma), but most deserts have retained their position over the last 2 Ma. Although the locations have remained relatively fixed over these geological time-scales, the actual surface area has changed. For example, some of the drier areas of Africa appear to have been expanding as a result of persistent droughts. Typically, however, these fluctuations are temporary. Quaternary records from tropical and

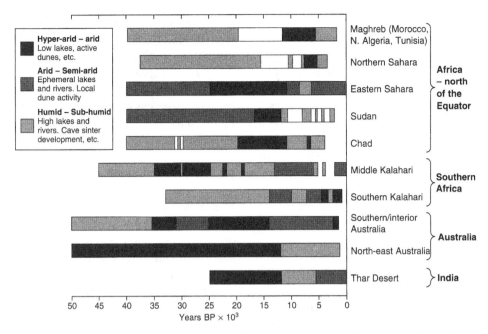

Fig. 5.4 A simplified chronology of late Quaternary expansion and contraction of tropical and subtropical arid zones derived from geomorphological and sedimentological data. Blank areas in the bars represent periods of uncertain dominant climate (due to climatic instability or a lack of data). (From Thomas 1989.)

subtropical regions, however, do indicate that the spatial distribution of arid regions has changed over this time frame (Fig. 5.4).

5.2 SEDIMENT SOURCES AND TRANSPORT

This section will examine how the natural arid landscape functions from source to storage, following the dominant transport processes and *in situ* modification (weathering) that can occur after deposition. Weathering is highly variable, depending on localized differences in climate (mainly temperature and humidity), geology (mainly lithology and structure) and inherited weathering artefacts (Smith & Warke 1997). Weathering is important as it weakens the bedrock ready for erosion and transport.

5.2.1 Significance of weathering in arid regions

Arid environments are typified by weathering-limited sediment supply. Exposed bedrock is common and the lithological and petrographic characteristics of the geology are more important than in humid environments in controlling local-

ized rates of weathering and sediment production. This also means that any inherited weaknesses, for example developed from weathering in past climate regimes, become more significant as these features are key to the *in situ* weathering development and thus, ultimately, erodibility of the sediments.

5.2.1.1 Insolation

Large temperature ranges (which can exceed 50°C, Goudie 1997) associated with the diurnal temperatures of deserts have been attributed with expansion and contraction of rocks *in situ*. Early workers suggested that this was the main factor behind rock breakdown. Experimental work, however, has failed to reproduce the same degree of weathering under laboratory conditions, leading to much debate about the possible role of insolation (see Goudie (1997) and discussions therein).

5.2.1.2 Moisture

Although by their definition arid areas have limited water availability from rainfall, moisture from other sources may be significant (section 5.1.2.1). Such sources of moisture tend to be alkali as they are sourced from salt-rich waters such as the sea or salt lakes. In the latter cases pH may be as high as 9.9 (Van Gulu, Turkey; Goudie 1997). This high pH greatly increases silica mobility during rock weathering. Where moisture is trapped within rocks the thermal expansivity of the rock (Goudie 1997) is increased. Heating of 10–50°C can develop 250 atmospheres of tensile strength (Winkler 1977), thus exacerbating breakdown of the rock.

5.2.1.3 Salt weathering

In arid environments salt can be sourced from a number of potential areas. These may include:

1 sea water or relict sea water which may be found in bodies of water (e.g. lakes) formerly connected to the sea;

2 'cyclic salts' derived from atmospheric inputs such as dust and rainfall or volcanic emissions;

3 the release of salts through rock weathering, especially where the bedrock is evaporitic in origin (e.g. halite).

These abundant salts in arid environments mean that salt weathering is not only key in weathering, but also creates problems in the built environment (section 5.6.3).

Weakening of sediment by salt weathering predisposes many types of sediment in arid areas to deflation. In some cases the hollows generated by deflation attract grazing animals, which in turn increase erosion (Goudie 1989). Erosion of the pans may be limited at depth by groundwater (Goudie & Wells 1995), which will also contribute to the salt weathering. Salt weathering can occur through a number of processes.

1 *Salt crystal growth.* This may result from changes in solubility with temperature, evaporational concentration of solutions, and mixing of salt solutions with the same ion (see Goudie 1997).

2 *Hydration.* Some salts will hydrate and rehydrate in response to changes in temperature and humidity. As salt changes to its hydrated form it takes up water, increasing its volume. Some of the largest hydration pressures occur

during anhydrite (dehydrated calcium sulphate) to gypsum (hydrated calcium sulphate) transformation (Winkler & Wilhelm 1970).

3 *Thermal expansion.* When halite (sodium chloride) is heated from 0°C to 60°C it expands by 0.5%, whereas granite minerals only expand by up to 0.2%. This differential expansion can contribute to rock disintegration (Goudie 1997).

5.2.2 Zones of net erosion

Within arid environments sediment production will be dominated by source areas susceptible to erosion. These most commonly include areas that generate conditions of high erosivity (slopes) or areas of high erodibility reflecting lithologies less resistant to erosion, such as lake-bed sediments.

5.2.2.1 Slopes

Many arid areas are associated with landforms such as pediments, cuestas and mesas that are protected by a caprock. This caprock may be part of the geological sequence, or it could be a crust developed *in situ* (section 5.3.5). Whatever the cause of the crust, caprock failure is not uncommon, producing slope talus and a potential sediment supply. Rates of scarp retreat, and thus sediment production, vary according to the local balance between the geological characteristics and the climate. Lower rates tend to be reported from hyper-arid regions, for example 100 mm kyr^{-1} from limestone has been reported from the Sinai, Israel (Yair & Gerson 1974). However, the overall control on rates of retreat is local lithology. This varies greatly between different lithologies, with the highest retreat rates reported from conglomerates (6700 mm kyr^{-1}, Lucchitta 1975) and sandstone (500–6700 mm kyr^{-1}, Schmidt 1980, 1989). Similarly retreat rates within the same overall lithology may vary dramatically as a function of microlithological variations in the caprock, and overall stratigraphy. For example, limestone scarp-retreat rates of 100–2000 mm kyr^{-1} have been reported in hyper-arid areas (Sinai, Israel; Yair & Gerson 1974) and 160–400 mm in semi-arid areas (Arizona, USA; Cole & Mayer

1982; Young 1985). Talus will accumulate as a function of mechanical failure of the caprock and fall impact. Typically coarse talus tends to be associated with lithologies that break into large chunks, i.e. lithologies composed of large component parts, such as conglomerates, or lithologies of greater mechanical strength, such as limestones. It is not uncommon for poorly cemented sandstones, such as aeolianites, to be associated with no talus at all, the sediment being carried away by contemporaneous aeolian processes (Schumm & Chorley 1966).

Once the talus slopes have been created they may generate their own runoff (Yair & Lavee 1976) in rainstorm events. This typically occurs close (within 10 cm) to the surface. Depending on the balance between talus supply and removal, the talus may be eroded to expose the area previously protected by them, revealing talus 'flat irons'. This is common in some parts of Utah, USA. Alternatively in hyper-arid settings the talus may form a 'fossil' accumulation in the modern landscape.

5.2.2.2 Exposed lake sediments

Lake-bed deposits in arid regions range in thickness from a thin veneer to hundreds of metres. As water bodies expand and contract in response to changes in the water balance, so former lake-bed sediments may become exposed to further erosion by surface processes such as mass wasting or gully erosion, being cannibalized into the lower lake levels.

Lake sediments contain a range of grain sizes, depending on the extent and depth of the lake system, and the characteristics of the surrounding catchments feeding the lake with water and sediment. Thus lake sediments may have been sourced originally from terrigenous, allochthonous sediments external to the lake water body. These clastic sediments are typically transported by runoff and river flow and thus can contain a range of grain sizes from clay to cobble. Other sources of lake sediment are autochthonous and originate within or beneath the water column or within the lake-bed sediments after deposition. These sediments are typically fine-grained, but often well-cemented carbonates. The finest of these lake sediments will be prone to dust erosion (section 5.2.4). Where this occurs sediment will be exported out of the basin system. One such relatively recent example of this is Owens Lake, California, which is recorded in some detail in Reisner (1986) and Knudson (1991). Owens Lake was a relatively small (110 km^2) water body that was located to the south of Owens Valley (Case Fig. 5.2A). Early European settlers were attracted to the area and it became important for agriculture, with irrigation farming introduced. In the 1900s, however, pressure on available water resources in the growing conurbation of Los Angeles was high. This led to the purchasing of the water rights of this area, by members of the Los Angeles Water Company posing as cattle ranchers. In 1913 many of the local streams that fed the lake were diverted into the Los Angeles Aqueduct. In 1927 the lake was nothing more than a small pond. The exposed dry, fine-grained playa sediments contain halite, trona, thenardite and mirabilite and are susceptible to erosion by strong summer and winter winds. Dust derived from this 110 km^2 area accounts for 1% of the total dust production in the USA every year. In February 1989 a concentration of 1861 µg m^{-3} was recorded, which was 37 times higher than the health standard in the State of California.

Where lake sediments become elevated in respect to the main lake level as a function of tectonics, then mechanical erosion of all grades of lake sediment material becomes possible from gully erosion. This is currently occurring in Lake Burdur in central Turkey, where tectonically elevated Pliocene lake sediments are being eroded by gullying, and redeposited in the smaller, and constantly shrinking, modern lake system.

5.2.3 Sediment transport by water

Channels within arid areas may be one of two main types: (i) perennial or (ii) ephemeral. Perennial channels have flow for most of the year and exist within arid areas where rivers within the arid basins are sourced by external inputs from mountainous regions. For example, in the Euphrates 90% of the total runoff comes from the mountains of north-eastern Turkey, with virtually zero runoff from areas such as Iraq (Beaumont 1989). Similarly the Nile flows from highlands in Ethiopia, which supply most of its water, but most of its course is through the arid lands of Egypt. Before the construction of the Aswan dam the Nile annual discharge ranged from a low of 1000 cumecs (cubic metres of water per second) in spring to 11,000 cumecs in late summer/early autumn (Beaumont 1989). Perennial rivers in general form the focus of Chapter 3, so will not be discussed further in this section. Ephemeral channels dominate in arid zones and thus form the focus of the rest of this section. These channels are only occupied by water following a rainstorm as they are mainly supplied by storm runoff. They tend to show less regular discharge patterns than perennial rivers, and suffer large transmission losses downstream.

5.2.3.1 Flow characteristics of ephemeral channels

Arid river systems cover a diversity of forms on a variety of scales. Cooper Creek, Australia, for example, is the largest internally draining catchment in the world (1.3×10^6 km²; Knighton & Nanson 1997) and lies entirely within an arid zone. Owing to the hydrological characteristics of the slopes feeding ephemeral channels (section 5.4.3), runoff is a typically more significant component in streamflow than in more humid areas, where throughflow and groundwater may dominate. For example, in New South Wales, Australia, only 16 mm of rainfall is required to produce runoff in the arid west, whereas 35 mm is required in the more humid east (Cordery et al. 1983). In Australia large, tropical monsoonal storms can penetrate large distances inland of the arid zones. In most arid systems, however, rainfall is derived from small, often convectional storm cells of less than 10–14 km width (Renard & Keppel 1966; Diskin & Lane 1972). Even though these may activate only a small proportion of a larger drainage basin, large floods can be generated rapidly. In the USA the top 12 largest floods recorded have all occurred within arid/semi-arid areas (Costa 1987). So, although

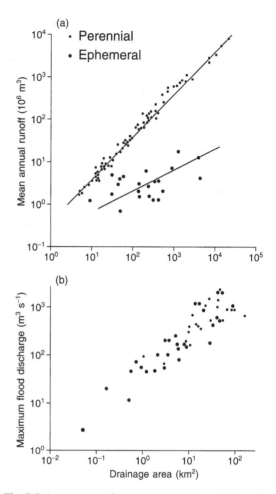

Fig. 5.5 A comparison of perennial (Maryland) and ephemeral (California) channels. (a) Differences in mean annual runoff, with largest values associated with perennial rivers. (b) A demonstration of the similarity in the size of peak flood discharges generated from perennial and ephemeral drainage areas. (Adapted from Wolman & Gerson 1978.)

mean annual runoff is lower for most ephemeral streams, compared with perennial streams, peak-flood discharges are similar (Fig. 5.5).

Owing to the transport-limited nature of the system, arid-zone rivers tend to have high peak values of suspended sediment concentrations and they are much more efficient transporters of bedload material than humid systems. In part this is reflected in the wide channels that typify ephemeral river systems (Fig. 5.6). Thus a unique aspect of ephemeral rivers is their multiple terrace and channel cross-section. Large events cut the widest channel and upper 'terraces' of

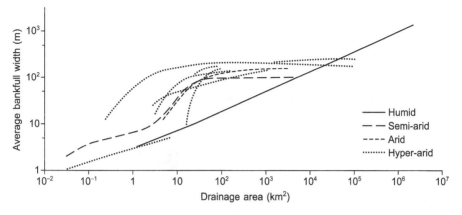

Fig. 5.6 The relationship between bankfull width and drainage area under different climatic environments. (After Wolman & Gerson 1978.)

the outer channel. Successively more frequent, lower magnitude floods occupy the topographically lower and narrower parts of the river channel (Fig. 5.7). Burkham (1972) recorded changes in river width over a series of floods in the Gila River, Arizona (Fig. 5.8). The largest of these floods had a recurrence interval of 300 years and increased channel width by 600%. Recovery here took some 50 years, but in more arid areas it can take much longer for vegetation to re-establish itself. Thus relationships based on bankfull discharge (considered the dominant channel forming discharge in perennial rivers by many authors) are of less relevance to ephemeral arid river systems.

One of the main differences between perennial and ephemeral river systems is in the propagation of the flood wave. For the reasons discussed above, the flood wave will be highly variable as each rainstorm event potentially activates a different area of the catchment. Also the small size of the rain cells (< 14 km), typically distributed 40–60 km apart (Sharon 1974), means that for increasingly larger basins a smaller percentage of the catchment will remain active, unless the region receives rainfall from larger weather systems (e.g. Australia already discussed). In addition to this the rain cells will migrate as they precipitate. Thus, tributaries within catchments may be activated at different times. Runoff generation typically takes only a few minutes in arid environments, with gully

systems activated in around 20 minutes but main tributaries may take hours, depending on the scale of the system. Schick (1988) suggests a piggyback contribution of individual tributaries generates the multipeak appearance of many arid flash floods. This spatial difference in rainfall may be increased where topographic variance is maximized; for example, in basin and range style topographies.

The flash-flood hydrograph is dominated by a steep rising limb. Normally the first observation is a flood bore. These vary in height. Hassan (1990) reported flood bores of 4–40 cm in height, and moving at 0.2–1 m s^{-1}, with velocity increasing with bore height. It is notable in gravel-bed rivers, however, that preceding the surface bore is a subsurface percolation of water through the river-bed coarse material. Flood bores can arrive suddenly and wreak devastation (Hjalmarson 1984) as they are rapidly followed by the flood peak within 10–23 minutes (Renard & Keppel 1966; Schick 1970; Reid & Frostick 1987). The flood peak often consists of several spikes reflecting the temporal and spatial activation of tributaries within the system (Fig. 5.9). Within a few minutes of the bore average stream velocities may have risen to 3 m s^{-1}, and water-stage height has been reported to rise by 0.25 m min^{-1} (Reid et al. 1994). The resulting flood may last only a few hours. This is a function of the lack of sustaining rainfall and thus surface runoff, combined with a lack of subsurface flow

Fig. 5.7 An ephemeral river bed (Rio Aguas, south-east Spain) displaying multiple channel sizes and levels associated with varying magnitudes of floods. Photographs (a) and (b) represent the same cross-section, occupied by a low magnitude, annual event in (b). The whole width of the channel visible here was last activated by a 1 in 500 year storm event in 1973, documented in Thornes (1974).

and high transmission losses through the bed of the river. Transmission losses will vary according to the characteristics of the channel-fill sediments, the length of the flood and the width of the flood. These will vary downstream. Overall water discharge will decrease downstream as a result of transmission losses. It is thus clear that ephemeral streamflow is highly complex and variable in nature.

5.2.4 Sediment transport by wind

Although winds in arid environments are no stronger than those in other regions, the sparse vegetation protection combined with large periods of transport inactivity by water mean that sediment supply is typically abundant and thus the winds carry more sediment than any other geomorphological agent (Cooke et al.

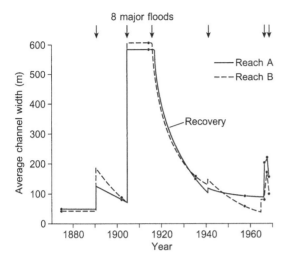

Fig. 5.8 Historical changes in channel width (1875–1970) for the Gila River, Arizona. (Adapted from Burkham 1972.)

1993). Aeolian dust (suspended particles typically less than 50 µm) is significant in global transport of sediment and it has been demonstrated that on a global scale the quantities of dust in motion are of the same order of magnitude as the quantities of sediment carried by rivers (Livingstone & Warren 1996). In areas of net sediment deposition, rates of accumulation of aeolian-derived sediment have been recorded at up to 70 mm kyr^{-1} (Dunkerley & Brown 1997) and dust can be transported over large distances,

for example, 10,000 km from the Gobi Desert to Alaska (Rahn et al. 1981). The major source of this dust is from the semi-arid areas of the world, particularly those desert environments containing floodplains, alluvial fans, wadis, salt-pans and former lake beds (Middleton 1997).

5.2.4.1 Transport zones

The key zones where sediment transport by wind dominates are in sand seas, sheets and stringers, with highest rates in sheets and stringers. These act as part of local and regional-scale aeolian transport zones from source areas to sinks (Lancaster 1995) and are found most commonly in areas with mean rainfalls of < 150 mm yr^{-1} (Wilson 1973). In such environments vegetation effects are limited. Transport rates in duneless sand sheets and stringers have been quoted as high as 62.5–162.5 m^3 per metre width per year in Mauritania (Sarnthein & Walger 1974) whereas mobile barchan dunes are reported at 3.49 m^3 per metre width per year (Thomas 1997b). These zones of transport reflect local topography, and may be sinuous and river-like in form (Zimbelman et al. 1995). Typically net accumulation of sands and bedforms (sand dunes and seas) will occur only where accumulation exceeds net transport. In cases where net transport exceeds accumulation then sand sheets and stringers will dominate.

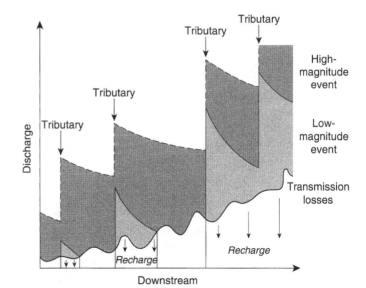

Fig. 5.9 Conceptual model of transmission losses and their relation to flow in an ephemeral channel for storms of two different magnitudes. The shaded area represents discharge. Note the intermittent channel flow for the low-magnitude event, and continuous but variable discharge for the high-magnitude event as a function of tributary input and transmission losses. (After Thornes 1977.)

5.2.4.2 Transport processes

As with water, sediment movement will occur only where the erosivity of the wind is greater than the erodibility of the surface materials. The erodibility will be determined by a number of factors. Vegetation will enhance surface roughness and modify local wind-velocity profiles. In dune systems this can reduce ground-surface velocities by as much as 200% (Wiggs et al. 1994) thus limiting entrainment and encouraging deposition. Surface slope will have an impact on both the threshold of sediment movement and the rate of sediment transport, and may be more important than initial theories suggest (Hardisty & Whitehouse 1988). Moisture increases surface tensions in sediments, reducing erodibility, although the impact is strongly affected by the grain-size characteristics of the sediment (McKenna-Neumann & Nickling 1989). In the case of moisture the most rapid changes for erodibility occur at about 8% moisture content. Thus near-surface groundwater tables may limit sand transport (Stokes 1968) as can periodic flooding, both of which are particularly common factors in playa, sabkha and coastal sand-sea situations (Fryberger et al. 1988). Moisture contents up to 14%, however, will have no real impact on sediment already in transport (Sarre 1989, 1990). Salt and algal crusts (section 5.3.5) may also protect surfaces from erosion.

Once entrained the sediments may be transported by a number of different modes. Suspension applies to smaller particles (< 0.6 mm). Very fine sediment may remain suspended for many days and travel great distances. Creep involves larger particles (0.5–2 mm), which roll as a result of wind drag and the impact of saltating grains. It has been estimated that creep accounts for 25% of aeolian bedload transport rate (Willets & Rice 1985). Reptation represents the transitional state between creep and saltation and is generated by the physical impact of high-velocity saltating grains on near-surface grains. Saltation is the most researched mode of transport. Saltating sediments (typically 0.06–0.5 mm) move in a number of steps using a ballistic trajectory.

Although individual saltating grains can reach 3 m into the air (Pye & Tsoar 1990), most (80%) of all such transport takes place within 2 cm of the ground surface (Butterfield 1991). The impact of saltating grains on the surface may lead to reptation, or, where sufficient momentum is transferred, may induce mass transport by saltation in a cascading system (Nickling 1988). Transport will be dependent on a number of factors including grain shape. For example, Willets (1983) showed that platy grains have a tendency for lower and longer trajectories than more spherical grains.

5.2.4.3 Landforms created by aeolian transport processes

Ventifacts are common in areas dominated by wind erosion, such as adjacent to slopes. They are generated mainly by saltating, rather than suspended grain movement. The latter tend to be swept around an object rather than impacting with it. Sand-grade material in saltation thus accounts for most wind erosion on the windward side. Smaller particles of dust, although having less impact energy, can, in sufficient load (Whitney & Dietrich 1973), erode on the lee side of objects, utilizing wind eddies (McCauley et al. 1977, 1979). There is, however, much controversy on the dominance of dust versus suspended load and its role in abrasion (see discussion in Breed et al. 1997). Larger scale versions of abrasion structures are known as 'yardangs'. These have the form of an inverted boat, with a high windward side and streamlined lee side. They range in size from a few metres to several kilometres. The largest yardangs on Earth are up to 30 km long and developed in the Tibesti Plateau of the central Sahara (Peel 1970).

5.3 SEDIMENT ACCUMULATION PROCESSES

Sediment deposition can occur temporarily within transport zones, or more permanently in areas of net accumulation (sediment storage; Fig. 5.1). This section will examine the sediment accumulation processes that affect these zones.

5.3.1 Desert lakes, playas and sabkhas

Lake basins act as sediment sinks, particularly in arid environments. Many arid zone areas either lack integrated drainage systems or are dominated by endoreic (internal) systems. Thus, topographic lows act as foci for runoff collection and associated sediment deposition. For example, erosion hollows generated by deflation may occur where crusts (section 5.3.5) are broken in arid regions. These hollows can later act as collection areas for runoff, forming playas. In most cases these water bodies tend to be supersaturated with salts (> 5000 mg L^{-1}) and are ephemeral. The associated landform is known as a 'playa' or 'pan'. The playas and pans vary in size from a few square metres to several thousand square kilometres. Their spatial significance varies regionally, but accounts for only 1% of drylands (Shaw & Thomas 1997). There are a wide range of terminologies used for the same feature, with the term 'pan' used for small areas originating from geomorphological (e.g. a depression between alluvial fans) rather than geological processes (Goudie & Wells 1995) and 'playa' for a depression with a saline surface (Rosen 1994).

Sediments trapped in pans and playas are typically fine-grained and represent the suspended sediments of distal flood inputs from adjacent catchments and fan environments. Fine materials are also supplied by aeolian transport. In addition evaporitic deposits may accumulate, sourced by evaporation of flood water or groundwater sources and the resulting concentration and accumulation of salts (Fig. 5.10). The flat morphology typically associated with playas is a function of infrequent inundation by water, which evens out the microtopography through a combination of deposition and dissolution. Playas with a highly irregular topography (e.g. from salt growth or sand-dune development in the centre) suggest extremely infrequent inundation by water, although an uneven topography from alluvial fans and drainage courses around the margins is not unusual (Fig. 5.10). To maintain a playa depression it is essential that accumulation does not outweigh erosion (typically dominated by deflation). Deflation is particularly effective, as the sediments are highly erodible when exposed due to a combination of dispersive agents (sodium), lack of protection from vegetation and availability of fine sediments (section 5.2.2.2).

5.3.2 Ephemeral streams

Ephemeral systems (both sand and gravel bed) tend to act as temporary stores for sediment

Fig. 5.10 Death Valley playa with well developed salt accumulation in the foreground. Note the irregular surface. The background consists of alluvial fans feeding into the edge of the playa. Polygons are approximately 1 m across.

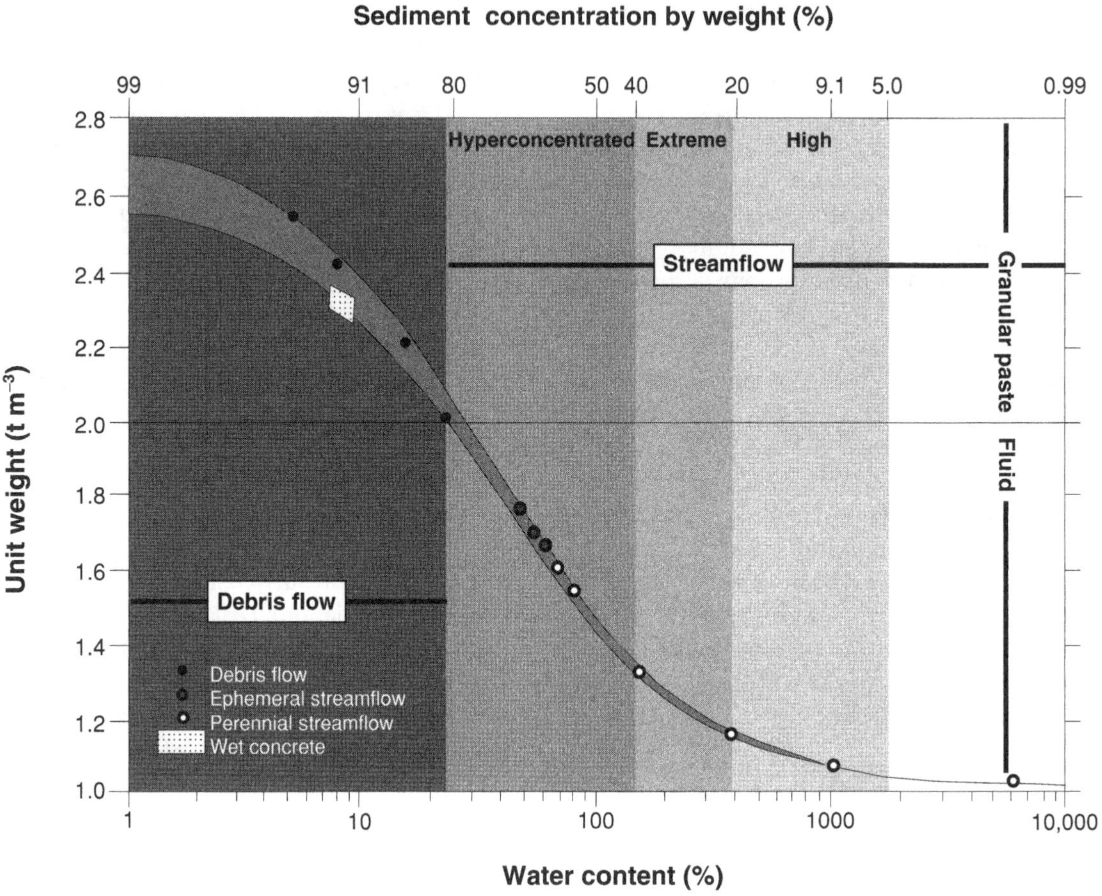

Sediment concentration by weight (%)

Fig. 5.11 The continuous spectrum of sediment concentrations from sediment-rich ephemeral rivers to debris flows. (Simplified from Hutchinson 1988.)

within the arid environment, unless they cross or terminate in an area of relative tectonic subsidence. In addition, during periods of no flow they may also act as a sediment source for aeolian processes. Ephemeral streams are dominated by large amounts of scour and fill within single flood events (Leopold et al. 1966). The period of subcritical flow during a flood event in ephemeral streams is short-lived (Reid & Frostick 1987). Thus the mechanism for the scour is also short lived and may relate to plane beds (Frostick & Reid 1987) and antidune migration (Foley 1978).

The bed scouring, together with abundant sediment supply, means that ephemeral rivers carry very high sediment loads (described as 'too thin to plough and too thick to drink'

by Colorado farmers; Beveridge & Culbertson 1964). These flows can be extremely hyperconcentrated in sediment (Fig. 5.11). As the flood discharge increases, so does the sediment load. Sediment supply exhaustion is unlikely in arid catchments, and ephemeral streams carrying 35 to 1700 times the sediment concentration of perennial rivers are not uncommon (Frostick et al. 1983). In addition, suspended sediment concentrations tend to remain more in-phase with flow characteristics than is observed in perennial rivers. Bedload transport in ephemeral rivers is also much greater than in perennial rivers. These differences may be 10^6 times larger at the threshold of entrainment and 10^1 at moderate shear-stress levels (Laronne & Reid 1993; Reid & Laronne 1995). In addition this transport is more predictable, probably reflecting the lack of armouring in arid-zone rivers, as the high supply of slope sediment ensures that fines are not preferably removed from the bed (Dietrich et al. 1989; Laronne et al. 1994). Large clasts (pebble to cobble) can be moved as much as 3 km in one flood event (Leopold et al. 1966).

General bedforms associated with rivers are discussed in Chapter 3. Here the characteristics more commonly associated with the deposits of ephemeral rivers are highlighted. In these rivers tractional bedload tends to be dominated by transverse and longitudinal mid-channel barforms. These comprise imbricated, sorted sediments associated with upstream dipping low-angle beds representing the bar top. If water depth is sufficient an avalanche face and associated cross-strata may develop. As a barform is typically associated with a coarser, upstream bar head and a finer downstream bar tail, migration of the bar head over the bar tail will commonly lead to a coarsening upwards within the bar deposits. The finer part of the bedload and coarser element of the suspended load tend to be deposited as horizontal lamination in sand beds. These may develop from upper flow regime plane beds or from pulses of sediment-rich water superimposed on the overall flood wave. The rapidly waning flood stage and abundant fines (typically muds) that make up the extremely concentrated water:sediment ratio tend to deposit thick (up to 10 cm) clay drapes. Once exposed and baked, these clay drapes have a tendency to desiccate and curl and be reworked in ensuing flows as mud clasts.

5.3.3 Alluvial fans

Alluvial fans are fan-shaped bodies dominated by coarse sediment that has been transported from steep upland catchments (Fig. 5.12). They require the juxtaposition of an upland source area and a lowland sediment accumulation area, where the fan will develop. Their other main requirement is a source area capable of producing course material, and high, flashy, peak discharges. These conditions are commonly met in arid environments and so fans form an integral

part of arid geomorphological systems (Harvey 1997). Alluvial fans, as long as they remain untrenched, tend to trap the deposits of each successive flood event originating from the associated fan catchment. Once alluvial fans become incised (trenched) throughout they will begin supplying coarse sediment to adjacent environments such as playa and river systems. Thus untrenched fans provide a useful record of processes operative within mountain areas bordering basins in arid regions.

In general (i.e. in the absence of any major external influences such as climate) alluvial fans tend to be dominated by debris flows in the younger, more proximal parts of the alluvial fan. As the fan system matures, supplies of fine material become exhausted, the drainage network expands and the conditions for the generation of fluvial processes are increasingly met, leading to the dominance of fluvial deposition in the fan dispersal area. As the fluvial processes require lower gradients than the debris flow processes to maintain transport, trenching of the proximal fan surface is often a common feature associated with mature fans (Harvey 1990).

5.3.3.1 Streamflow processes

Within alluvial fan accumulation areas the transport and deposition of sediment by streamflow may occur in (i) clearly defined channels such as the fan head trench (Fig. 5.12), (ii) in wide, ill-defined channels or (iii) as unconfined flows (sheetfloods). The last of these is a relatively rare phenomenon (section 5.4.3). In arid regions, where sediment movement is typically transport-limited, sediments are deposited from flows that are hyperconcentrated with sediment (40–70% by weight, Costa 1988) as well as more normal streamflow events (less than 40% sediment by weight, Costa 1988; Fig. 5.11).

'Normal' streamflow leads to the deposition of imbricated, sorted sediments (see e.g. Nemec & Steel 1984; Costa 1988). Where sediment is of a suitable calibre and water depths are sufficient (i.e. in confined areas of flow such as the fan-head trench), bars with an avalanche face may develop. Elsewhere, particularly in less confined

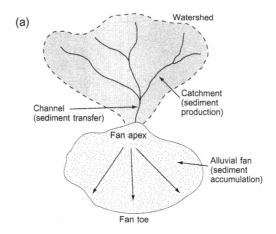

Fig. 5.12 (a) Schematic representation of an alluvial fan system showing the sediment production (catchment), sediment transfer (channel) and sediment accumulation (fan) elements and (b) a view from the toe to apex of a typical alluvial fan in northern Chile. Note the recent debris flow down the right-hand side of the fan.

areas of flow, sediment movement will be predominantly as diffuse gravel sheets or longitudinal bars (Smith 1974). Fluvial processes will be encouraged where the supply of fine material is limited, typically in larger catchments where stream-side supply of sediment dominates.

Hyperconcentrated streamflows are common on alluvial fans (Fig. 5.11). These flows may have some shear strength although there is some controversy as to their exact nature (Reid & Frostick 1994). Resulting sediments may contain weak internal stratification, normal and reverse grading, weak imbrication and clasts may be matrix-supported (Costa 1988).

Research has shown that fluvially dominated fans may be restricted to certain lithologies such as those associated with high-grade metamorphics (see for example the work by Harvey 1984, 1987), which source few clays and are dominated by coarser material. The topographic constraints are typically larger, lower relief catchments, which can produce higher water to sediment ratios from the slopes. These types of drainage basin become increasingly abundant with the progressive erosional development of the catchment (i.e. as weathered slope material is progressively exhausted and larger, more open, lower relief catchments develop). The spatial extent of fluvial alluvial fans is typically greater than for those deposited by other processes as the potential transportation distance for the fluvial flows is greater, but the gradient needed to maintain transport is less. Thus fluvial-dominated fans tend to be larger and possess a gentler apex-to-toe gradient than those dominated by other flow processes (Harvey 1987, 1990).

5.3.3.2 Debris-flow processes

The transport and deposition of sediment by debris-flow processes (Fig. 5.11) may be (i) confined within channels, or (ii) as an unconfined lobe. Debris flows behave as a plastic (Costa 1988) and may be channelized with levees in the upper parts of the fans, and more unconfined and lobate in the more distal parts of fans. The types of flow will be determined by the water to sediment ratio and the type of fine sediment available. Essentially the flows can be (i) cohesive or (ii) non-cohesive, depending on the shear strength of the flow resulting from the sediment concentration (typically 70–90% by volume, Costa 1988) and nature of fines. The resulting deposits of debris flows typically lack sorting, may possess inverse grading (towards the base) and normal grading (towards the top), are matrix-supported and lack much in the way of internal structure, but will vary in their internal organization depending on water content and cohesivity of the flow (see Nemec &

Steel (1984), Postma (1986) and Costa (1988) for a discussion of mass flow processes). Debris flows are associated with catchments that generate sufficient fine material to lend shear strength to the flow and are supplied predominantly by slope material, i.e. small, steep catchments (Wells & Harvey 1987). These conditions are typically met when the catchment is developed on sedimentary and low-grade metamorphic geologies (Harvey 1990), and before the drainage network has had time to reduce relief in the fan catchment. Debris flows require a steeper gradient to maintain transport and tend to have a smaller run out distance than comparable fluvial flows, so that debris-flow-dominated fans are typically smaller and steeper than corresponding fluvial-dominated fans.

5.3.3.3 Aeolian processes

Aeolian processes are not commonly associated with alluvial fans in the literature, yet observations of fans in arid areas contradict this. It is not uncommon to find impeded dunes (section 5.3.4.2) developed against fan apexes, or developed in topographic lows at the fan toe. The aeolian material may subsequently be reworked by debris and streamflow processes on the fan surface or buried by debris flows. Thus in some areas aeolian processes can be significant in sourcing external fine material, and locally affecting flow processes on the fan.

5.3.4 Aeolian bedforms

Aeolian sand deposits in arid environments cover approximately 5% of the Earth's land surface and aeolian silt (loess) covers a further 10%. Dunes can vary in size from < 1 m to > 200 m in height and can represent the deposition of one windstorm event or, for larger systems, the deposition of sediment over millennia. Dependent on source area, dunes are typically composed of quartz sand, but can include any material capable of being wind blown (e.g. gypsum, volcanic ash, shell fragments). Where dunes are composed of carbonate or gypsum, cementation of the dunes may occur early in their development.

Table 5.2 Free, simple dunes. Note that surveys of dune types indicate that linear dunes are the most common, followed by: transverse, parabolic (Table 5.3) and star dunes. Features such as dome dunes are comparatively rare (see data reported in Fryberger & Goudie (1981) and Thomas (1997b)). For an illustration of morphology see Figs 5.13 & 5.16.

Category	Type	Slip faces	Major control on form	Nature of movement	Internal structure
Transverse (consistent unidirectional winds)	Transverse ridge Barchanoid ridge Barchan	1	Some directional variability in the wind and abundant sand supply Less directional variability in the wind and abundant sand supply Limited sediment supply	Forward migratory	High angle cross-beds inclined downwind (more tabular sets in transverse dunes and a tendency to wedge shaped sets in barchanoid dunes). Narrow range of orientations of cross-beds. Cross-beds frequently truncated by low-angle erosion surfaces that dip gently downwind
Linear (bi-directional winds and more variable wind direction)	Linear ridge	1–2	A variable unidirectional or bi-directional winds, more limited sand supply than transverse dunes	Extending (sandpassing forms)	When viewed in cross-section perpendicular to the wind, sets of cross-beds are seen to dip in opposite directions from the crest
Other (various wind directions)	Seif	2	Bi-directional winds		As above but with more complex bedding pattern due to the sinuous crest
	Reversing	2	Opposing bi-directional winds	Will show a net migration if one wind direction is dominant	Mix of cross-bed orientations reflecting the bi-directional movement, although one direction may dominate
	Star	3+	Wind directions may vary seasonally, or where transverse dunes migrate into a region with multidirectional winds. Sand supply is abundant and wind direction unstable	Sand accumulating, vertical forms	Variable and complex cross-beds
	Zibar	0	Composed of course sand, which limits movement to highest velocities. Various wind directions	Limited	?Weak
	Dome	0	Various wind directions	Limited	?Weak

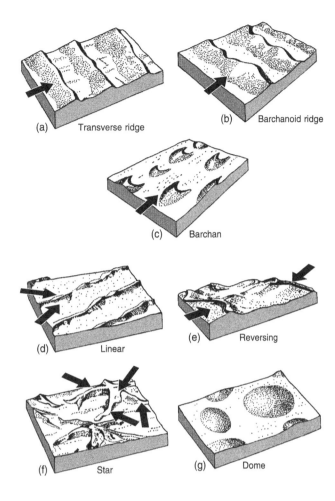

Fig. 5.13 Morphology of the main free, simple dunes described in Table 5.2. (Adapted from McKee 1979.)

Dunes may occur as 'simple' dunes (where the individual dunes create discrete forms), compound dunes (where simple dunes of the same type coalesce and merge) or complex dunes (where different simple dunes merge). Here the focus is on simple dunes, which, for the sake of simplicity, can be further subdivided into free dunes (Table 5.2 and Fig. 5.13), with a form dictated primarily by wind characteristics, and impeded dunes (Table 5.3 and Fig. 5.14), a form that is significantly affected by surface roughness or topographic barriers.

5.3.4.1 Free dunes

Free (or mobile) dunes are controlled primarily by wind regime and sediment supply. Without available sediment or without wind to move available sediment, dunes cannot form. The

Table 5.3 Impeded, simple dunes. For an illustration of morphology see Fig. 5.14.

Dune type	Major control on form
Blowout	Disrupted vegetation
Parabolic	Vegetation anchoring
Lunette	
Shrub-coppice (Nebkha)	
Lee/fore dune	Topographic barrier
Climbing/falling dune	
Echo dune	

nature of this relationship is illustrated in Fig. 5.15, which plots wind direction variability (wind regime) against equivalent sand thickness (sand availability). Wind direction variability is measured by calculating the frequency and strength of winds blowing from different directions. A high value indicates a single predominant

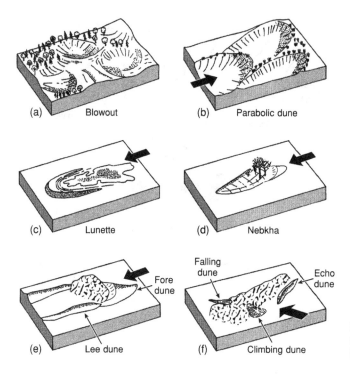

(a) Blowout

(b) Parabolic dune

(c) Lunette

(d) Nebkha

(e) Lee dune Fore dune

(f) Climbing dune Falling dune Echo dune

Fig. 5.14 Morphology of the main impeded, simple dunes described in Table 5.3. (Adapted from Summerfield 1991.)

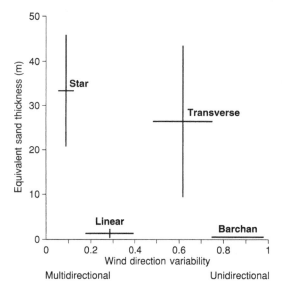

Fig. 5.15 Relationship between wind variability and equivalent sand thickness (thickness of sand if it were spread evenly over an area). For explanation see text. (After Wasson & Hyde 1983.)

wind. A low value indicates a highly variable wind regime. Equivalent sand thickness represents the thickness of dune sand in an area if it were spread evenly, so the greater the thickness, the greater the overall volume of the dune sand. Figure 5.15

shows that the largest dunes on Earth, star dunes, form in areas with greatest wind variability and greatest sediment supply. The calibre of available sediment is also important in the generation of dune types. For example, coarse sand is usually associated with dunes with low-angle morphology and limited height (typically up to 4 m). These dunes may be linear (zibar dunes) or mound-like (dome dunes) depending on wind regime. Strong wind velocities are required for transport, and movement of the dunes is limited.

Initial dune-sand accumulation may be initiated by a surface roughness or by temperature gradients in the atmosphere (sand accumulation tends to occur where airflows are ascending and relatively slower). The accumulation and movement of sand will eventually develop a stoss- and slip-face. Where there is abundant supply of suitable sediment and unidirectional wind, transverse ridge dunes will develop (Fig. 5.15). If two opposing prevailing winds of similar strength and duration exist, reversing dunes may develop as active slip faces change with wind direction. As topographically lower dunes can move more quickly than higher dunes, a sinuosity develops, creating a sinuous, barchanoid ridge

(Fig. 5.16b). Eventually this pattern will give rise to a dune network of nearly linked ridges trending at 10–20° to the prevailing wind direction (see Warren (1979) for more details on airflow over dunes).

In contrast, under similar unidirectional wind conditions to those that form transverse and barchanoid ridges, but where sand supply is less abundant, barchan dunes will form (Figs 5.15 & 5.16a). The lower outer ridges will move more quickly than the higher, central portions, leading to significant arms that extend downwind. The width and spacing of the dunes will reflect the effects of airflow patterns, which may be related to upwind dune forms.

In areas of high wind variability, but relatively limited sediment supply, linear dunes are created (Fig. 5.15). Linear dunes (also called longitudinal dunes, and including sinuous linear dunes or seif dunes) are by far the most common dunes on Earth. They are highly elongated in form and contain two opposing slip faces either side of a crest line. Linear dunes can be kilometres in length and coalesce downwind into Y-shaped junctions. They are typically associated with obliquely converging wind directions, although the details of their formation are still debated (see discussions in Thomas 1997b).

Star dunes can be 300–400 m in height and contain the largest sand volume (Figs 5.15 & 5.16c). This dune form requires abundant sediment supply and multiple dominant wind directions (Fig. 5.15). The latter wind directions are typically associated with seasonal variations, which cause other dunes to merge and modify. Thus star dunes are common in the centre of sand seas, although they are also associated with topographic barriers that influence regional windflow (Lancaster 1994). Once sufficiently large, star dunes can modify their own windflow patterns.

5.3.4.2 Impeded dunes

Impeded (or anchored) dunes are dominated by topographic barriers that inhibit sand movement and lead to sand build-up. The variety of forms is illustrated in Table 5.3 and Fig. 5.14.

The most common type of these dunes is the parabolic dune. Parabolic dunes tend to develop where blowouts occur. These are formed by the deflation of sand (often where vegetation has been disturbed), which leads to areas of higher wind speed over the associated lower roughness surfaces. The increased wind speed further dries the sand making it more prone to mobilization. With increased deflation the blowout will increase in size and the leeward rim will migrate downwind, leaving trailing limbs which are 'anchored' by vegetation. These tend to limit the main sand movement in the higher, central part.

5.3.4.3 Internal structures of dunes

Dunes form through the accretion of tractional deposits or a drop in wind velocities (and thus reduction in shear stress) leading to grain-fall deposition. Once the initial sand has become trapped, sediment will continue up the stoss (windward) slope to the crest of the dune. Here material will avalanche down a slip face. Thus most dune types are associated with particular sedimentological characteristics (Table 5.2). The internal structure of sand dunes is dealt with in more detail in McKee (1979). This and other research is summarized in Nickling (1994).

Free dunes (which dominate in arid to hyper-arid climates) are typically associated with well developed cross-strata which dip downwind at 30–34°. Dune deposits are also associated with often massive, tabular to planar cross-strata that thin from the base upward. Another common feature is the development of bounding surfaces which occur between sets of cross-strata, which result from the migration of dunes over inter-dune areas.

Vegetated dunes (which are typically associated with semi-arid climates) tend to show a bimodal distribution of steeper angle cross-strata and minor truncation surfaces. Most dips are low (about 12°). Parabolic dunes are dominated by steeper cross-strata (similar to transverse dunes), which often have concave slip-faces. They may also contain some concave-upward sets that have been deposited in hollows. Vegetated dunes are

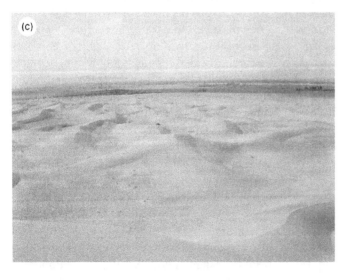

Fig. 5.16 Dune types sourced from wadi sand in the Draa river, southern Morocco: (a) barchan dunes, (b) barchanoid ridge (50 m from stoss to lee) and (c) star dune (300 m across – arm to arm).

Table 5.4 Main soil orders associated with arid environments in order of abundance. (Adapted from Dregne 1976; Dunkerley & Brown 1997.)

Soil order	Main characteristics	Total land occupied (10^6 km^2)
Entisol	Minimal horizon differentiation and modification of the bedrock (sedimentary material)	19.2
Aridisol	Dryness and/or salinity restricts plant growth throughout the year	16.6
Mollisol	Thick, dark base-rich and organically rich topsoil or surface horizon (epipedon)	5.5
Alfisol	Moderate base saturation, argillic horizon and seasonally available water for plant growth	3.1
Vertisol	Deep clay soils, characterized by cracking and shrink and swell characteristics	1.9

also associated with evidence of vegetation such as rhizoliths.

5.3.5 *In situ* landform modification

Once sediment has been deposited, depending on its residence time, it can undergo some form of *in situ* modification through pedogenic, biogenic or hydrological processes. These processes vary spatially in their intensity, creating a mosaic of surface runoff sources and sinks during a rainstorm. This is significant for sediment entrainment, erosion and deposition as is illustrated in section 5.4.3. The most significant *in situ* modifications of the sediment surface are discussed in sections 5.3.5.1 to 4 below.

5.3.5.1 *Arid region soils*

The USA soil taxonomy classification system recognizes five main soil orders (Table 5.4) for arid regions. It is evident from this table that arid areas are dominated by poor soils lacking in organic matter and associated with limited plant growth. Arid soils differ from soil developed in more humid regions in a number of ways. Typically accumulation is dominated by material from external sources (e.g. aeolian dust), rather than the *in situ* breakdown of parent material observed in more humid environments. In addition desert soils are typified by long time-scales of development. This facilitates the long-term build up of minerals such as carbonate, or gypsum in hyper-arid settings (Fig. 5.17a), developing indurated 'petrocalcic' horizons (section 5.3.5.5).

5.3.5.2 *Desert pavements*

'Desert' or 'stone' pavements are typified by a lag of stones, typically overlying a stone-free silt (Fig. 5.17b). Where developed the stones act to protect the surface from erosion, by breaking up runoff flow paths. It is widely believed now that the fines are largely aeolian in origin and concentrate around the stone lag, passing into the regolith, perhaps along desiccation cracks, displacing the stones by pedogenic and aeolian accretion (McFadden et al. 1984, 1987). The stones are too large to follow and remain at the surface. Some authors (e.g. McFadden et al. 1987) have argued that the enrichment with clay and carbonate restricts water penetration into the land surface in the initial stages of surface development, shedding more runoff and instigating gully incision, which will eventually isolate the surface, deactivating soil processes and facilitating further pavement development. Any associated coarse material at the surface can also affect the surface hydrology. Lavee & Poesen (1991) found through simulated experiments that in general increased stone cover will enhance runoff generation. The response, however, is complex, as the stone cover can also act as a mulch, trapping moisture and encouraging plant growth. In such situations vegetation may inhibit surface seal development and encourage infiltration.

5.3.5.3 *Microphytic crusts*

Although desert soils are typified by low vascular plant cover, they can have abundant

Fig. 5.17 *In situ* modification of sediments in arid environments: (a) gypcrete development and (b) desert pavement developed in the Atacama Desert of Northern Chile. In (b) note the uniformity of the main surface clast size and dominance of silt grade material beneath the surface in the linear surface scrape (light colour) to the left of the pencil. Note also the discoloured (red oxide, appears as light grey in the image) base of the upturned clast lying underneath the pencil. The darkish colouration of most of the surface clasts is due to the development of desert varnish (see Oberlander (1994) for a summary of how this forms). Hammer is 40 cm long and pencil is 15 cm long.

non-vascular cover (e.g. mosses, algae, lichens, fungi, bacteria – microphytes). Where microphytic crusts are developed they are typically irregular in topography and act to break up runoff on the surface, protecting it from rill erosion. In some environments these crusts act as permeability barriers, and increase runoff, but protect the underlying surface. However, the impacts of the crusts are widely variable (see the discussion in Dunkerley & Brown 1997).

5.3.5.4 Inorganic (soil or rainbeat) crusts

Inorganic crusts may develop on soil surfaces during or shortly after a rainstorm event. These relate to physico-chemical changes relating to water and clay content and rainfall characteristics. These can be grouped into (i) structural seals or (ii) depositional seals (Maulem et al. 1990). Structural seals develop as a result of the soil structure as a function of aggregate breakdown by rainsplash and slaking; washing of fines into adjacent, larger pores; particle segregation and compaction. Depositional seals are a function of the deposition of fine suspended sediments in a thin, comparatively impermeable film on the soil surface. Sealing intensity has been demonstrated to decrease with increasing slope angle as a function of higher erosion on higher slopes (Poesen 1986). Poesen also found that embedded rock fragments in sandy soils can generate seals, whereas those with rock fragments sitting on the top do not. This was attributed to the fact that non-embedded stone particles typically have 'overhangs' that protect the area immediately adjacent to the rock fragment from raindrop impact, thus a complete soil seal cannot be created. Once developed, surface seals may prevent the escape of air as a wetting front penetrates the soil, generating associated vesicular layers 1–30 mm below the surface (Chatres et al. 1985; Ringrose-Voase et al. 1989). These vesicles further reduce hydraulic conductivity of the soils.

5.3.5.5 Duricrusts

These are indurated surface/near-surface crusts. They may be exhumed relict deposits (e.g. ancient lacustrine deposits) or may relate to contemporaneous hydrological or pedological processes in the desert environment. Some crusts are ephemeral, for example salcretes (Table 5.5), others may persist for many millions of years (such as the gypcretes reported by Hartley & May 1998). Table 5.5 describes the duricrusts most commonly found in arid environments. Note that there are many more subtypes, such as the nitrate deposits in the Atacama Desert of northern Chile.

5.4 NATURAL IMPACTS ON PROCESSES

Arid environments, as described previously, are dynamic. Their spatial extent and degree of aridity can change over a range of temporal and spatial scales (Fig. 5.4). Although some of these changes can be accounted for by intrinsic controls, extrinsic controls can play an important role. These consist of external forcing mechanisms that cause the arid system to change. What is considered to be the dominant external control will change depending on the scale of the landscape element examined and the time-scale involved. For example, long term (million year) time-scales are typically dominated by tectonic controls due to the large spatial and temporal scale on which they cumulatively operate. Climate controls tend to dominate the medium scale (million to thousand year) and short term time-scales (annual) are dominated by factors such as storminess.

5.4.1 Long term time-scale controls: tectonics

Active tectonics can have an impact on arid environment sediment systems directly or indirectly. Direct effects may have an impact on localized areas of an arid landscape, for example they may generate fault scarps as a function of an earthquake event, which can then divert local drainages, or rejuvenate local erosion. These types of features, however, are typically ephemeral, and are removed in a matter of years, depending on the rates of erosion. Even within hyper-arid areas where water availability is low, *in situ* weathering, aeolian deposition and gravitational collapse can conceal the effects of faulting relatively quickly. Within alluvial fan systems tectonics can directly rejuvenate the catchment areas that produce and deliver the sediment to the alluvial fan, for example by increasing slope angles through uplift. Within the sediment accumulation area of the alluvial fan itself tectonics may locally directly modify the fan surface. The morphology and internal sedimentology of alluvial fans can be seen at its simplest as a balance between the discharge of sediment exiting the fan source area and the available accommodation space. Thus

Table 5.5 The major types of duricrust found in arid environments. Solubility increases from silcrete to salcrete. (Summarized from Watson & Nash 1997 and references therein.)

Crust type	Main mineral	Field appearance	Typical profile thickness	Modes of origin in arid environments	Distribution
Silcrete	Silica	Brittle, intensely indurated. Colour is variable with grey brown and green common.	0.5–3 m	Pedogenic and groundwater.	Found in areas not subjected to late Tertiary and Quaternary glaciation. Less globally significant than calcrete. Mantle low gradient slopes, but can also form beneath basaltic lava flows and in weathering profiles on a range of lithologies.
Calcrete	Calcium carbonate	Occurs as powdery through to nodular, laminated to highly indurated. Generally white, cream or grey with pink mottling and banding.	1–5 m	Highly alkaline evaporitic basins or hydrological or pedogenic origins.	Most widespread crust. Petrocalcic horizons cover 20×10^6 km^2 of the globe's landsurface. Typical of areas where moisture is deficient throughout the seasons. Pedogenic calcretes mantle undulating or gently sloping terrain. Can form on most lithologies.
Gypcrete	Gypsum	Occurs as 1) horizontally bedded crusts; 2) subsurface crusts with large (1 mm – 0.5 m diameter) crystals (desert roses); 3) mesocrystalline (crystals 50 μm –1 mm) or 4) surface crusts of alabastrine gypsum (cystallites < 50 μm). Occur as columnar crusts, powdery deposits or superficial cobbles. Colour ranges from white/grey to green or red.	0.1–5 m	Pedogenic, groundwater and lacustrine evaporites.	Typical of warm deserts with very low rainfall (< 250 mm a^{-1}). Pedogenic gypcretes mantle low gradient surfaces. They are also associated with hydrological basins. Can develop on a range of lithologies.
Salcrete	Halite	Typically white, but can be pink where micro-organisms are present.	0.1–5 m	Evaporation of waters in Sabkha settings, groundwater and pedogenic (the latter two being rarer).	Associated with areas of extreme aridity, particularly in association with saline basins and rainfalls less than 200 mm a^{-1}. Can form thin ephemeral crusts on dune sands.

spatially variable subsidence in the fan accumulation area can go some way to controlling fan thickness and morphology (Whipple & Trayler 1996). Viseras et al. (2003) found that in the Betic Cordillera of southern Spain in areas where the accommodation space for alluvial fan development is driven by relatively high tectonic subsidence (at > 1 mm yr^{-1}) the resultant alluvial fans tended to aggrade vertically and were not incised. Where the accommodation space was controlled by lower rates of subsidence (< 1 mm yr^{-1}), however, the resultant fans were narrower and prograded out into the basin with some incision.

Indirect effects of tectonics include situations where regional tectonics, such as epeirogenic uplift, create the necessary conditions for a secondary geomorphological process to operate such as river capture. This situation tends to occur where the development of regional gradients can increase the stream power, and thus erosivity, of one stream in relation to its neighbours. The headward erosion associated with these phenomena may be exacerbated where the geological structure generated by the tectonics also exposes sediments of higher erodibility. Combined with the runoff-dominated hydrology, high-intensity rainstorm events and minimal protection from vegetation cover associated with arid environments, and river capture becomes a prolific modifier of drainage networks. As such river capture can play a major role in sediment production, delivery and routing in tectonically active arid landscapes. Details of such an example are highlighted in Case Study 5.1.

Case study 5.1 Influence of tectonics on rates of sediment production, delivery and routing, the Sorbas Basin, south-east Spain

The Sorbas Basin is located within a basin and range topography within the Betic Cordillera of southern Spain. Southern Spain represents part of the plate boundary between the African and Iberian plates, which ceased subduction in the late Neogene. Since this time compression has dominated with tectonic movement being expressed through differential uplift within and between the sedimentary basins and the mountain ranges. Pliocene to recent average uplift rates are calculated to be in excess of 160 m Myr^{-1} for the Sierras Alhamilla and Cabrera, but are typically much lower for the basins (80 m Myr^{-1} for the Sorbas Basin; 11–21 m Myr^{-1} for the Vera Basin; Mather et al. 2002). This deformation has been significant in generating regional topographic gradients. Combined with the geological structure this has led to the development of abundant river captures as a direct function of related increases in stream power and accelerated headward erosion. These river captures are significant in affecting the sediment flux, routing and delivery within and between basins (Mather et al. 2002; Stokes et al. 2002). One such capture occurred in the upper Pleistocene and forms the focus of this case study.

The modern topography of the Sorbas Basin is dominated by an incising drainage network (the Rio Aguas), which is associated with many landscape instabilities such as extensive badland terrain and landslides (Case Fig. 5.1a). Landscape erosion is thus locally severe (Harvey 2001), with some areas of abandoned agricultural land undergoing gully headcut retreat of several metres in one rainstorm event. Erosion-pin experiments indicate localized surface lowering of several millimetres per year. In some lithologies piping is abundant and forms a pseudokarst system, which supplements and overprints surface runoff features in terms of sediment and water discharge and routing. Localized piping features may be as much as 15 m deep and 1–2 m in diameter. These badland areas are associated with material of low residual strength such as marls and silts. The main landslides, in contrast, tend to dominate in areas of stronger lithologies with higher unconfined compressive strengths such as limestones, or in pervious material, which in

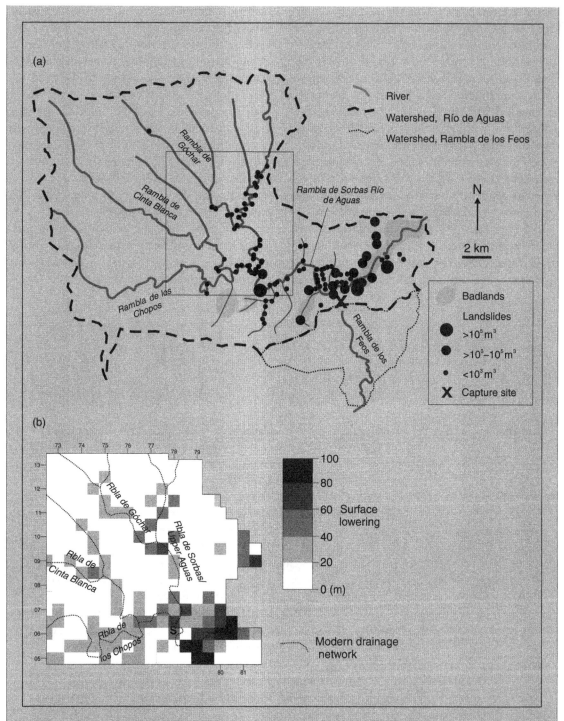

Case Fig. 5.1 (a) Distribution of erosional and landslide features in relation to the 100 ka river-capture site. Note the increase in frequency and magnitude of landslide occurrences in proximity to the river-capture site, and greater abundance of badland areas. The box outlines the area depicted in (b). (From Griffiths et al. 2002.) (b) Surface lowering above river-capture site. Note that the largest amounts of surface lowering are associated with (i) the valley networks, (ii) valley confluences and overall with (iii) the lower reaches of the drainage network proximal to the capture point. S. Sorbas. (Adapted from Mather et al. 2002.)

an arid environment has more resistance to erosion, such as gypsum. The landslide features vary in style from deep-seated rotational slips to block topples and may be up to 1 million cubic metres in volume (Griffiths et al. 2002).

The spatial and temporal distribution of the above features in the landscape is governed by their geographical location to the upper Pleistocene river capture site. The beheaded river (Rambla de los Feos in Case Fig. 5.1a) experienced a reduction in erosion rates. In contrast the captured drainage system (Upper Aguas; Ramblas Góchar, Cinta Blanca and Chopos in Case Fig. 5.1a & b), and the area below the capture site (Lower Aguas), incurred dramatic increases in net erosion. At the capture site base level was lowered by 90 m, starting the propagation of a rejuvenating wave of incision up the catchments and accelerating sediment production and delivery to the fluvial system (Mather et al. 2002; Stokes et al. 2002). The river capture event was thought to have occurred at around 100 ka. More recent U-series dating, however, suggests that it may be much younger, before 77.7 ± 4.4 ka (Candy et al. 2005). Over this time frame the impact of the river capture has reached 20 km upstream, at a decaying rate. Near the capture site a tenfold increase in incision was experienced. This changed the sediment delivery processes. Initially the generation of steep, rapidly unloaded slopes generated mass movement processes such as landslide failures (Mather et al. 2003). In weaker lithologies the dominant sediment delivery process later became dominated by more progressive slope erosion by surface runoff and subsurface piping processes (pseudokarst). Most of this accelerated erosion is still restricted to the main valley side-slopes, and has not yet reached the main drainage divides so that overall *surface* lowering is much less than that recorded by the localized valley incision (Case Fig. 5.1b).

The change in processes driven by the tectonically induced river capture has affected the Pleistocene to Holocene sedimentation within the study area in a number of ways. In the pre-capture state the then linked Upper Aguas and Feos fluvial terraces were dominated by large material (pebbles, cobbles and boulders) sourced from the Sorbas Basin and routed to the sedimentary basins to the south. Post-capture, however, the beheaded section of the river system (the Feos) underwent a reduction in sediment and water discharge and became dominated by localized and smaller scale slope erosion. This is reflected in a change in sedimentology of the fluvial terrace deposits in terms of sediment calibre, provenance and sedimentary style. Post-capture fluvial terrace deposits are dominated by more localized, typically smaller (granule and pebble) material than the pre-capture terraces and no longer receive sediment sourced from within the Sorbas Basin. In addition the pre-capture river terrace deposits are frequently buried by fine-grained colluvium in the more open parts of the valley and coarser, alluvial-fan material in the more valley constrained, basement sections of the river. In contrast the captured Upper Aguas and Lower Aguas have seen enhanced sediment production and delivery to the fluvial system, together with the re-routing of the high water and sediment discharges sourced from the Sorbas Basin to the sedimentary basins to the east. Adjacent to the capture site the sediment delivery from the valley sides has been dominated by a combination of landsliding and gullying.

Relevant reading

Candy, I., Black, S. & Sellwood, B.W. (2005) U-series isochron dating of immature and mature calcretes as a basis for constructing Quaternary landform chronologies for the Sorbas Basin, southeast Spain. *Quaternary Research* 64, 100–11.

Griffiths, J.S., Mather, A.E. & Hart, A.B. (2002) Landslide susceptibility in the Rio Aguas catchment, SE Spain. *Quarterly Journal of Engineering Geology and Hydrogeology* 35, 9–17.

Harvey, A.M. (2001) Uplift, dissection and landform evolution: the Quaternary. In: *A Fieldguide to the Neogene Sedimentary Basins of the Almería Province, SE Spain* (Eds A.E. Mather, J.M. Martin, A.M. Harvey & J.C. Braga), pp. 225–322. Blackwell Scientific, Oxford.

Harvey, A.M. & Wells, S.G. (1987) Response of Quaternary fluvial systems to differential epeirogenic uplift: Aguas and Feos River systems, south-east Spain. *Geology* 15, 689–93.

Mather, A.E., Stokes, M. & Griffiths, J.S. (2002) Quaternary landscape evolution: a framework for understanding contemporary erosion, SE Spain. *Land Degradation and Management* 13, 1–21.

Mather, A.E, Griffiths, J.S. & Stokes, M. (2003) Anatomy of a 'fossil' landslide from the Pleistocene of SE Spain. *Geomorphology* 50, 135–49.

Stokes, M., Mather, A.E. & Harvey, A.M. (2002) Quantification of river capture induced base-level changes and landscape development, Sorbas Basin, SE Spain. In: *Sediment Flux to Basins: Causes, Controls and Consequences* (Eds S.J. Jones & L.E. Frostick), pp. 23–35. Special Publication 191, Geological Society Publishing House, Bath.

5.4.2 Medium term time-scale controls: climate

Climate and humans can have very similar effects on landscapes vulnerable to change, such as those in the arid realm, as they both affect vegetation cover (style and abundance) and through this have an impact on the erosional, transportational and depositional elements of the arid system (Case Study 5.2). Typically the older the record then the less likely human impact is to be the main control. This reflects both the existence and abundance of people in sufficient numbers to drive change and also the technology available to implement change. Thus on Pleistocene time-scales climate is more likely to be the driving force.

Case study 5.2 Climate controls on sediment production and delivery, the pluvial lakes of the Basin and Range Country, American South-west

Many of the basins in the regions of Nevada and California were once the home of extensive lake systems. The high stands for these lake systems mainly correspond with Quaternary glacial periods in the high Sierra Nevada when the adjacent lowland areas (basins) were subjected to relatively wetter and cooler climate conditions. The lowest of these basins was (and is) Death Valley. There is some debate over the exact nature of connectivity between drainages, but it is suggested that possibly three rivers fed Lake Manly of Death Valley (Jannik et al. 1991). These were (i) the Amargosa (rising from the Spring Mountains), (ii) the Mojave River (San Bernadino Mountains) and (iii) Owens River (Sierra Nevada). This water passed through four pluvial lakes: Owens, China, Searles and Panamint (Case Fig. 5.2A). In some cases the overflow points between the lakes are clear, for example fossil falls (Case Fig. 5.2B) between Owens Lake and China Lake. In others, such as the link between Lakes Panamint and Manly, the evidence is less clear. Evidence of former lake levels comes from shorelines marked by benches, shoreline tufas and beach bar deposits (Case Fig. 5.2C).

Using dated shorelines as a framework it has been established that lakes fluctuated in level temporally (Jannik et al. 1991). It also can be demonstrated that for the same time period sediment supply changed spatially. Harvey et al. (1999) demonstrate that the Lakes Lahontan and Mojave (Case Fig. 5.2A) show quite different sedimentary responses to the same late

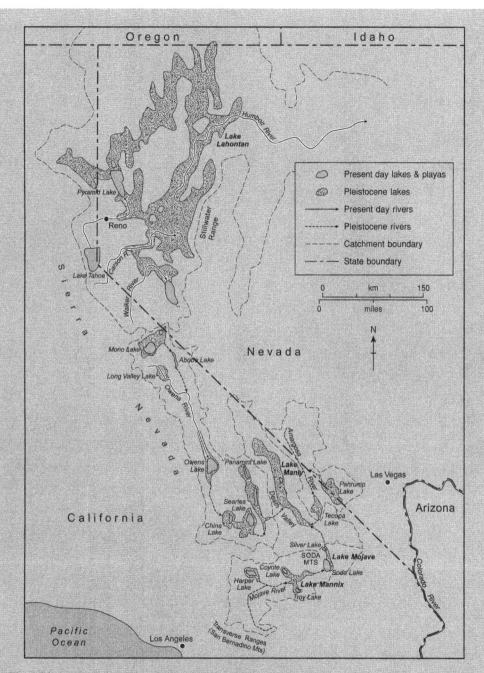

Case Fig. 5.2A Pleistocene lakes of the American South-west (Adapted from Jannik et al. 1991; Harvey et al. 1999.)

Pleistocene to Holocene climatic transition. The alluvial fan systems of Lahontan, sourced from the Stillwater Range underwent minimal slope erosion and fan sedimentation during that time frame. In contrast the fan systems of the Mojave, sourced from the Transverse Ranges of the San Bernadino Mountains, were far more active over that same period, first dominated by debris flows on the fan surfaces and then by ephemeral streamflow causing fanhead trenching and

Case Fig. 5.2B Fossil falls overflow between Owens and China lakes.

Case Fig. 5.2C (below) Former shorelines of Lake Manly, Death Valley. (a) Tufa lines (arrowed) at Badwater and (b) the Beach Ridge on the road to Beatty.

(a)

Case Fig. 5.2C (Continued)

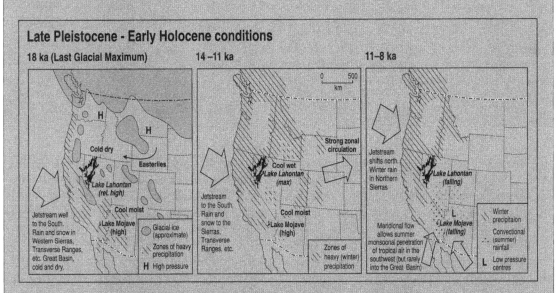

Case Fig. 5.2D Postulated Late Pleistocene to Holocene changes in airmass circulation. (From Harvey et al. 1999.)

distal alluvial fan sedimentation and progradation. Combining this information on vegetation cover with climate reconstructions (Case Fig. 5.2D) highlights the interaction between these two variables and their impact on sediment production and delivery. In the Late Pleistocene both Lahontan and Mojave were subjected to much wetter and cooler climates than today. Winter precipitation was dominant and heavy snowfall in the Sierras and the Transverse ranges maintained the pluvial lake levels. During the Late Pleistocene to Holocene transition the weakening of zonal circulation and the monsoonal incursion of warm moist tropical air from the Gulf of Mexico or eastern Pacific affected the Lake Mojave catchment. This meant that summer storms became more effective in this region. The Lahontan area was not affected by these storms. In terms of the vegetation cover, research using the distribution of remains from packrat middens demonstrates that catchment hillslopes in the Mojave supported a desert shrub vegetation. In contrast the Lahontan supported juniper woodland and grasses at low elevations and pine at higher elevations. Thus in Lahontan the more continuous vegetation cover reduced the generation of runoff and sediment production, leading to only rare debris-flow events associated with the winter storms. In contrast the sparse, discontinuous vegetation of Mojave allowed greater runoff generation from the winter rainfall, and associated enhanced overall sediment erosion and delivery in the form of debris flows from the winter storms. Incision and streamflow were probably generated from the high intensity summer storms, which generated more effective runoff.

With increasing aridification into the Holocene, climate and vegetation became similar between the two areas and sediment production and delivery differences became less pronounced. The lakes dried up and in most cases the reduced vegetation cover associated with the climate change enhanced effective runoff generation, leading to erosion of slopes and fluvial trenching of the upper parts of fan systems in both areas. This has led to the progradation of many of the fan systems, partly encouraged by associated drops in the lake (base) levels as the lakes dried up (Harvey et al. 1999).

Relevant reading

Harvey, A.M., Wigand, P.E. & Wells, S.G. (1999) Response of alluvial fan systems to the late Pleistocene to Holocene climatic transition: contrasts between the margins of pluvial Lakes Lahontan and Mojave, Nevada and California, USA. *Catena* 36, 255–81.
Jannik, N.O., Phillips, F.M, Smith, G.I, et al. (1991) A ^{36}Cl chronology of lacustrine sedimentation in the Pleistocene Owens River system. *Geological Society of America Bulletin* 103, 1146–59.

Alluvial fans may be affected by changes in climate, where changes in process are affected. For example, where climate either limits or enhances the weathering processes within the catchment the water to sediment ratio of the dominant fan process may change. Similarly, changes in the amount of effective runoff generated may again change the water to sediment ratio and thus modes of sediment transport and accumulation (Case Study 5.2). Streamflow processes require less of a gradient for transport than debris-flow fans, so a switch from debris flow to streamflow may lead to fan-head trenching. If the entire fan becomes trenched, the mountain supply catchment will then be able to supply sediment to the main basin area, altering basin sediment routing. Alternatively climate may control lake-levels in arid environments (Case Study 5.2). As with tectonic controls, lake-level fluctuations can control alluvial fan morphology and sedimentary architecture by controlling the accommodation space for alluvial fan development. Lake-level rise will typically lead to retrogradation of the alluvial

fan system, backfilling into mountain embayments. Viseras et al. (2003) demonstrated that this is associated with a reduction in the size of feeder channel. In these cases the fan morphology most strongly reflects the morphology of the mountain embayment in which the fan was developed, rather than the nature of the catchment area. This reflects the control that the embayment has on the distribution of debris and ephemeral streamflows over the alluvial fan surface (Viseras et al. 2003).

Arid aeolian systems are particularly sensitive to changes in precipitation and also to factors such as temperature, affecting evapotranspiration. The threshold of change for precipitation lies within the arid–semi-arid transition zone, where increased vegetation cover becomes more important. Thus if precipitation increases dune types may revert from free to vegetationally impeded (section 5.3.4) and existing dunes may become stabilized. In addition, aeolian environments will be susceptible to changes bought about by changes in wind strength and direction. Examination of Quaternary records in tropical latitudes suggests that the extent and location of arid and semi-arid regions have changed (Fig. 5.4). Generally greater aridity occurred in tropical latitudes during glacial periods in high latitudes. In mid-latitudes altitude played an important role, with greater aridity in non-glaciated lowland areas such as the Mediterranean during glacials, and relatively greater humidity during associated interglacial periods. Thus changes in relative aridity in these regions would have led to an expansion of areas susceptible to aeolian activity. It has also been suggested that increased wind speeds would have affected some areas during glacial periods, further exacerbating aeolian sediment movement. This in part is reflected in the extent of relict ergs to be found in areas such as Australia and the Sahara.

5.4.3 Short term time-scales: storminess

For shorter time-scales, such as the individual storm event, it becomes apparent that the simpler models of (net) water and sediment movement on slopes considered for longer time-scales are too simple. The problems are in monitoring the more complicated detail of storm events. Much work has been undertaken using rainfall simulation in the field, but there is much debate on what is a suitable scale for these plots to be of use. There is also much debate about the precise controls of sediment and water movement, and it would appear that the control is scale-dependent. In order to consider overall sediment transfer this section will concentrate on the hillslope scale (Case Study 5.3).

Case study 5.3 Runoff and sediment movement on a hillslope scale, results of the MEDALUS and IBERLIM European Union projects

Partly as a response to growing concerns over desertification in the drier (arid to semi-arid) regions of Europe, the European Union supported research into sediment and water transfer across the region, at the catchment and hillslope scale, in the 1990s. Examples of two of these projects are MEDALUS (Mediterranean Desertification and Land Use) and IBERLIM (Erosion Limitation in the Iberian Peninsula). The MEDALUS project was primarily involved in the examination of uncultivated sites, whereas IBERLIM was focused on cultivated (afforested) sites. The field results and field experimentation at the plot scale will form the case study into short term (individual storm) significance in sediment transfer and runoff in managed (IBERLIM, central Spain) and non-managed (MEDALUS, south-eastern Spain) semi-arid catchments.

The MEDALUS project developed one of its erosion monitoring sites in the Rambla Honda in the early 1990s. The experimental catchment is a first-order catchment developed on mica-schists and consists of bare hillslope components in the upper parts and alluvial fan material derived from the mica-schist in the lower part. In general the hillslope element was acting as a source of hydrological pathways (surface runoff) and sediment movement, and the alluvial material at the base as a sink. At the individual storm-event scale, however, it was found that widespread transfers of water and sediment were unusual and of short duration, even in extreme rainstorm events. In the latter events although local runoff generation may be high on slope lengths of 10 m, they decrease dramatically on longer, 50 m length hillslopes (Puigdefabregas et al. 1998). On the slope surface runoff occurred by infiltration excess in the early part of the individual events, but was dominated by local subsurface saturation of upper layers of the soil. Across the hillslope surface mosaics of plant clumps and bare earth became important. Runoff was generated from bare earth areas but lost to vegetation clumps, which acted as sinks. Runoff was thus laterally discontinuous, even in the larger observed areas in storm events. On the hillslope scale the only time connectivity between hillslope and channel elements can occur is when the saturation of the shallow surface layer occurs across the entire slope to enable the

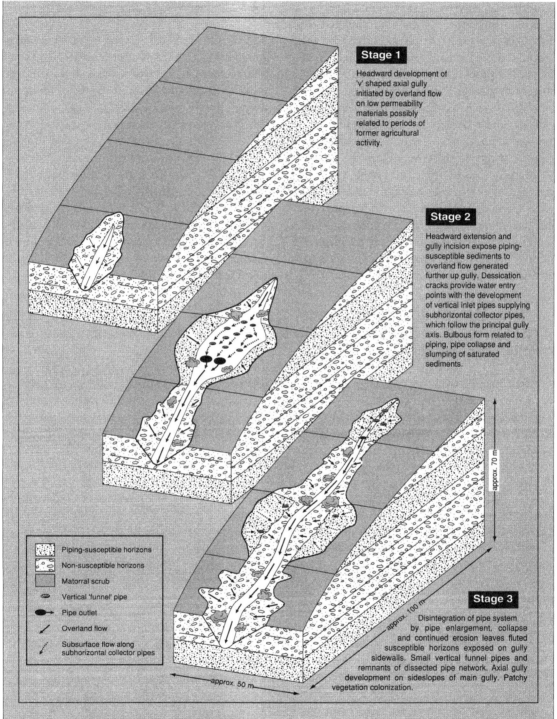

Stage 1

Headward development of 'v' shaped axial gully initiated by overland flow on low permeability materials possibly related to periods of former agricultural activity.

Stage 2

Headward extension and gully incision expose piping-susceptible sediments to overland flow generated further up gully. Dessication cracks provide water entry points with the development of vertical inlet pipes supplying subhorizontal collector pipes, which follow the principal gully axis. Bulbous form related to piping, pipe collapse and slumping of saturated sediments.

Stage 3

Disintegration of pipe system by pipe enlargement, collapse and continued erosion leaves fluted susceptible horizons exposed on gully sidewalls. Small vertical funnel pipes and remnants of dissected pipe network. Axial gully development on sideslopes of main gully. Patchy vegetation colonization.

approx. 70 m

approx. 100 m

approx. 50 m

Legend:
- Piping-susceptible horizons
- Non-susceptible horizons
- Matorral scrub
- Vertical 'funnel' pipe
- Pipe outlet
- Overland flow
- Subsurface flow along subhorizontal collector pipes

Case Fig. 5.3 Gully evolution and piping at the IBERLIM study site in central Spain. (From Ternan et al. 1998.)

connection of the hillslope elements to the first-order channel, and this occurs extremely rarely (Puigdefabregas et al. 1999), and was not witnessed during the project.

The IBERLIM managed site covers a 3.68 ha gully catchment in central Spain, located on the Raña Formation (Pliocene alluvial sediments). Prior to the 1940s the area was used as agricultural land, which then became abandoned and taken over by *Cistus mattoral* vegetation (Ternan et al. 1996a). In an attempt to combat erosion problems in the area an afforestation programme was undertaken and in the 1970s the catchment was bench terraced and planted with *Pinus*. The IBERLIM team began monitoring in 1992, on a variety of scales (Williams et al. 1995) from (i) gully (3.5 ha), (ii) erosion plot (10–21.5 m^2) to (iii) rainfall simulation plot (1 m^2). The research determined that the bulk soil had an overall low aggregate stability (Ternan et al. 1996a), which was exacerbated locally by high clay content (especially if expandable clays were present). Not all sediment and water erosion occurred at the surface, with a good deal of erosion occurring through subsurface pipes, which developed in layers within the sediment where dispersion, shrinkage and swelling were prolific (Case Fig. 5.3). This again coincided with clay horizons lacking in coarse material (typically less than 40% sand). The sediment dispersion was found to link closely with the chemistry of the porewater, soil chemistry and clay minerals. Horizons with sodium absorption ratios (SAR) of more than 0.4 and exchangeable sodium percentage (ESP) of more than 1.5 were found to be susceptible (Ternan et al. 1998). These horizons had been exposed with the excavation of the bench terraces within the woodland (Ternan et al. 1996b). These areas were thus sources of runoff and associated surface erosion early in a rainstorm event. Continuous runoff pathways were more prolific under wet weather conditions. It was demonstrated that to manage erosion in such areas, and also to assist in the minimization of flooding, both the overall wetness threshold at which runoff is generated *throughout* the catchment should be considered together with the spatial mosaic of runoff-generating source areas and sinks (Fitzjohn et al. 2002). If sinks could be used to disrupt continuous hydrological flow pathways both slope sediment and water transfer could be minimized.

The MEDALUS and IBERLIM studies serve to emphasize the mosaic nature of runoff generation on hillslopes in arid environments. The resulting data demonstrate how human intervention can exacerbate problems by increasing linkages of source areas on slopes and indicates that a knowledge of the linkages could be used to mitigate runoff problems by using sink areas to break up flow pathways.

Relevant reading

Fitzjohn, C., Ternan, J.L., Williams, A.G., et al. (2002) Dealing with soil variability: some insights from land degradation research in central Spain. *Land Degradation and Development* 13, 141–50.

Fitzjohn, C., Ternan, J.L. & Williams, A.G. (1998) Soil moisture variability in a semi-arid gully catchment: implications for runoff and erosion control. *Catena* 32, 55–70.

Puigdefabregas, J., Sole, A., Gutierrez, L., et al. (1999) Scales and processes of water and sediment redistribution in drylands: results from the Rambla Honda field site in Southeast Spain. *Earth Science Reviews* 48, 39–70.

Puigdefabregas, J., del Barrio, G., Boer, M.M., et al. (1998) Differential responses of hillslope and channel elements to rainfall events in a semi-arid area. *Geomorphology* 23, 337–51.

Ternan, J.L., Elmes, A., Fitzjohn, C., et al. (1998) Piping susceptibility and the role of hydro-geomorphic controls in pipe development in alluvial sediments, Central Spain. *Zeitschrift Für Geomorphologie* 42, 75–87.

Ternan, J.L., Williams, A.G., Elmes, A., et al. (1996a) Aggregate stability of soils in central Spain and the role of land management. *Earth Surface Processes and Landforms* 21, 181–93.

Ternan, J.L., Williams, A.G., Elmes, A., et al. (1996b) The effectiveness of bench-terracing and afforestation for erosion control on Raña sediments in central Spain. *Land Degradation & Development* 7, 337–51.

Williams, A.G., Ternan, J.L., Elmes, A., et al. (1995) A field study of the influence of land management and soil properties on runoff and soil loss in central Spain. *Environmental Monitoring and Assessment* 37, 333–45.

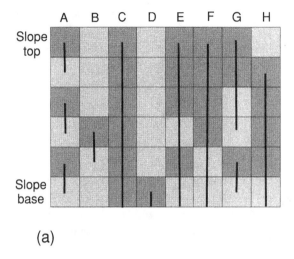

(a)

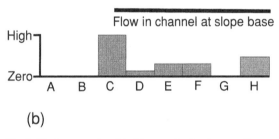

(b)

Fig. 5.18 Conceptual model of the mosaic of surface runoff pathways on a hypothetical slope (a). Each cell represents a source (dark) or sink (light). If it is assumed that (i) rainfall is even across the slope and (ii) that source and sink capabilities per cell (pixel) are equal, the contribution of runoff from the slope mosaic to the channel is displayed in (b).

Recent research has highlighted the significance of heterogeneities in slopes, which generate a 'mosaic' of hydrological characteristics. These mosaics can be considered as 'pixels' in a digital picture, where each pixel has a different individual response to a rainstorm event, even if the intensity and duration of the event were constant across the entire area (which is unlikely). The hydrological response of each pixel will be determined by its ability to store water. This will be determined through a range of factors such as porosity, permeability, organic matter content, slope angle, etc. As a function of these characteristics some pixels will act as 'sources' and some will act as 'sinks' (Fig. 5.18). The sources will respond to the rainstorm event and generate runoff. This runoff will continue downslope

until it hits a sink. Runoff will be continuous only where the source cells dominate the sinks (Fig. 5.18). As rainfall inputs vary then so will the number of cells acting as sources or sinks. The amount of sediment carried by this runoff will be dependent on the 'erodibility' of the sediment and the 'erosivity' of the rainfall. The latter will be determined mainly by the rainstorm characteristics (rainfall intensity and drop size), but may be modified by the intervention of vegetation. The erodibility of the sediments will be determined by the aggregate stability, which in turn is affected by the soil structure (amount of binding organic matter or presence of dispersing agents such as sodium salts). Thus, the nature of the sediment characteristics across and into the slope, combined with the slope morphology and cover characteristics, added to the spatial variability of an individual storm event will generate a spatially varied movement of sediment and water across a slope. This can be further complicated by the intervention of humans, modifying the local hydrological characteristics (Case Study 5.3). All factors considered it is not difficult to see why the geological concept of a 'sheetflood' rarely, if ever, occurs in the natural environment!

5.5 ANTHROPOGENIC IMPACTS

Humans tend to affect the landscape on the same time-scales as climate. Thus, the impact of humans is often difficult to unravel from the natural controls discussed above. This section will illustrate the significance that these impacts exert on controlling the modern landscape, which in turn has implications for how environments can be managed sustainably. The world's arid environments have long been occupied by humans. In more recent times, however, improvements in technologies of water management, improved healthcare and hygiene have meant an increase in population of these regions. In many cases this may be exacerbated by the availability of mineral deposits. The section examines the impact of humans on the erosion, transport and deposition of sediment in deserts.

5.5.1 Water resources and salinization

Salinization is the process of accumulation of soluble salts in the upper parts of soil horizons. Secondary salinization (where salinization is enhanced by human intervention) of soils in arid areas is a major problem. This is particularly so where groundwater levels are close to the surface and salts can be drawn into the upper parts of the soil where they naturally accumulate. A simple removal of vegetation cover may be all that is required to modify the hydrology of the soil (raise the water table) and place it in danger of salinization in an arid area. For example, in Australia native evergreen forest removed by settlers increased the salinity of over 200,000 ha to the point where they were no longer productive (Mackay 1990). Secondary salinization more commonly occurs through excessive irrigation. This is not surprising given the rapid increase in irrigation globally since 1800. In some countries (e.g. Egypt) nearly 100% of land in agricultural use is irrigated (Rhoades 1990). Approximately half of all irrigated areas suffer from problems of secondary salinization (Rhoades 1990).

Salinization can lead to cracking and puffing of salt-affected soils (Mabbutt 1986), affecting the hydrological characteristics of the soil. Where a salt crust or accumulation forms runoff can be exacerbated. The salts will also exacerbate weathering rates, producing abundant fine material which may then be deflated by wind action, leading to the formation of 'salt scalds' (Mabutt 1986). In hyper-arid areas where salts naturally occur in the soils, for example in the form of gypcretes (Table 5.5), irrigation may lead to rapid dissolution and collapse of the soil surface, affecting the urban area. As salt is a dominant weathering agent in arid landscapes (Doornkamp & Ibrahim 1990; section 5.2.1.3) it can also lead to significant impacts on engineering structures and urban areas (section 5.6.3).

5.5.2 Agricultural resources and soil erosion

Soil erosion by water has been recognized recently as a major contributor to erosion in arid lands. Natural vegetation clearance and increased populations have accelerated rates of soils erosion. Regions bordering the Mediterranean Sea have some of the highest erosion rates in the world (Woodward 1995), with some $1000 \text{ t km}^{-2} \text{ yr}^{-1}$ of suspended sediment generated. Of this, 70% can be attributed to anthropogenic agents (Dedkov & Mozzherin 1992). Within this region some countries have implemented the policy of reducing the number of people involved in agriculture and replacing them with mechanization. Such countries have been experiencing accelerated rates of erosion (Morgan 1994). In some cases this is related to the abandonment of agricultural terraces (e.g. Millington 1989). Mismanagement of the landscape can also lead to alterations in the hillslope hydrology, which can have an impact on runoff pathways and exacerbate soil erosion related to surface runoff (Case Study 5.3, the IBERLIM example).

Soil erosion by wind is another common phenomenon. It is particularly common in areas where water resources have been used for irrigation, leading to a lowering of the affected water bodies and exposure of lake sediments to deflation. Wind erosion is further exacerbated where secondary salinization inhibits vegetation growth, thus limiting the erosional protection benefits of vegetation cover. In other cases, soil erosion is related to overgrazing of land or overproduction in years of drought, which can lead to a loss of arable plant protection. Deflation can become a major issue where this coincides with fine-grained sediments such as glacial and river silts or loess.

5.5.3 Mineral resource extraction

Arid environments are often locations of mineral resources such as the diamonds of Namibia or the salts and nitrates of the Atacama of northern Chile. The impact of this mining on the environment is highly variable. In the hyper-arid climate of the Atacama of Chile extensive nitrate mining occurred across the Central Depression. The mining at its peak provided almost 50% of the income for the Chilean Government (Wisniak & Garces 2001). Most of the mines were abandoned in the 1930s and now litter the Central Depression in a fossilized landscape, but their impact upon sediment systems is considered to be minimal.

In contrast the landscape response to abandoned mining is more dynamic in semi-arid environments. The Rio Tinto fluvial system in the Province of Huelva, south-west Spain has a mean annual rainfall at Huelva of around 500 mm yr^{-1} (Instituto Nacional de Meterologia). The annual discharge ranges from close to zero in the summer to 11 m^3 s^{-1} in the winter. The lowest 20 km of the catchment is tidal, with a mean tidal range of 2.2 m. The waters of the river are acidic (typically pH 2.0–2.5), but become less acidic with the addition of Atlantic waters on flood tides. The Rio Tinto lies within the Iberian Pyrite Belt, one of the largest deposits of its type in the world. The main rich ore body is 5 km long, 750 m wide and 40 m deep. It contains abundant Fe, Cu, Zn, Pb, Ag and Au. Mining of the sulphide deposits has been occurring with increasing efficiency for some 5000 years, ceasing in the late 1990s. The impact of improved mining technologies and efficiencies can be demonstrated by comparing the amount mined in the Rio Tinto catchment in million metric tons (values taken from Davis et al. 2000). The Tartessans removed 3 million tonnes over 4200 years (5000–2800 yr BP); the Romans removed 24.5 in 200 years (2000–1800 yr BP); and the British (Rio Tinto Zinc) re-moved 1600 in the last 200 years. These mining operations have liberated abundant suspended sediment and solutes into the river system. Cores of 1 m depth taken throughout the catchment from the channel margins and dated with ^{210}Pb and ^{14}C show overall rates of sediment accumulation within the catchment of 0.3 cm yr^{-1}. High levels of pollutants (e.g. Cu at 970 ppm for 840 yr BP and 2466 ppm at 3640 yr BP) were found to pre-date the RTZ opencast mining (Davis et al. 2000).

5.6 LIVING WITH DESERTS: AN ENGINEERING PERSPECTIVE

Deserts provide engineering challenges where the arid environment may have to be traversed or inhabited. This section will consider the engineering management and remediation of sediment-based hazards in desert environments today.

5.6.1 Aeolian hazards

The physical detachment, entrainment and transport of solid particles has been enhanced in many regions where the erodibility of the soils has been exceeded by the erosive agents, which include humans. Soil erosion exacerbated by humans has been recorded from the time of Plato (fourth century BC) to more recent history (e.g. O'Hara et al. 1993). Aeolian removal of sediments is a major problem in arid environments. Wind erosion affects 39% of arid areas susceptible to human degradation (UNEP 1992). Agricultural mismanagement can lead to severe erosion and sources for available dust can also be exacerbated by human activity, especially where major water diversions take place. Mitigation of these processes ranges from physically removing or limiting the source area of the sands, building protective barriers and building to reduce the impact of structures (Table 5.6).

5.6.1.1 Sand dune encroachment

Arid environments are characterized by sparse or no vegetation. Thus free, mobile sand dunes (section 5.3.4) are common in these areas. Dune soils have proved popular for agriculture in some regions, as they are less prone to salinization than other areas. These resources can be easily overexploited so that vegetation is lost. Once the vegetation is removed, moisture loss is increased and restabilization unlikely. In some areas this is further exacerbated by quarrying, for example in Kuwait, for urban building projects. Thus, partially stabilized dunes may become reactivated. The drifting sand becomes a problem for road transport and can bury settlements.

Urban areas themselves pose a problem in arid environments as they offer an obstacle to sand movement. In addition, associated urban activities may exacerbate the magnitude of aeolian problems through the loss of vegetation, etc. Sand and dust problems are thus prolific in the literature on desertification. Areas dominated by sand dune encroachment experience the burial of structures and transport links, and features

Table 5.6 Mitigation of aeolian hazards. (Summarized from Cooke et al. 1982.)

Method		Comments
Avoidance of problem areas		Best approach
Removal of dunes		Expensive, but successful for small dunes. Need to be maintained.
Vegetational stabilization	Natural recovery	Protects the main area
	Artificial recovery	Successful but expensive, and needs to be maintained. Careful plant selection is required to survive the hydrological conditions
Surface stabilization	Water	Needs to be maintained. In some instances mineral precipitation associated with the evaporation of the water may enhance surface stability
	Lag	Material greater than 2mm diameter used as an artificial pavement. Can be costly to transport
	Oil	Successful but unsightly
	Chemical sprays	The chemical crust generated may be fragile and expensive
	Wood cellulose fibres	Sprayed as slurry with water fertilizer, grass seed and asphalt or emulsion. Fragile but cheap
Fences	Windbreaks	Can be living or artificial. Can be expensive
	Diversion fences	Set at an acute direction to the wind. Expensive and require maintenance (removal of lee sand)
	Impounding fences	A series of fence increments, which extend the life of the barrier from burial. Successful and relatively cheap
Architectural planning		Minimizing the impact of housing on wind flow and sand accumulation

such as pipelines, which are purposely exposed for maintenance, can become inaccessible.

5.6.1.2 Soil erosion

Loss of topsoil through deflation will inhibit fertility, and can undermine structures through scouring of footings. This can affect anything from telegraph poles to pipelines, railway sleepers and roads (Cooke et al. 1982). Typically once the sand source has been depleted, and an armoured surface has been reached, e.g. a gravel lag, or where desert pavement predominates, then deflation ceases to become a problem. It will only remain a problem in areas where the sand supply is continually renewed, e.g. in mobile dune fields or along unpaved roads. The Arizona Department of Transport (1975) reported an annual loss of silt and clay by deflation at 5–50 kg per vehicle per mile of unpaved track.

5.6.1.3 Sand abrasion

Transport of sand by aeolian processes leads to the problems of abrasion. Typically, abrasion is exacerbated over indurated surfaces such as tarmac and concrete (Cooke et al. 1982). Pitting

(surfaces more than 55° to the prevailing wind direction) and fluting (surfaces less than 55° to the prevailing wind direction) can occur on buildings, and telegraph poles may suffer sand blasting damage near their base. Glass becomes frosted and paintwork may be damaged on both buildings and vehicles. Air filters are also affected. In addition, visibility can be impaired by sand storms. Dust storms have been reported to create many problems including the transportation of pathogens, suffocation of cattle and interference with reception on transmitting and receiving devices.

5.6.2 Water hazards

Despite the arid nature of drylands, water is one of the main hazards in these regions. From an engineering perspective ephemeral arid streams provide a challenge. Peak sediment and water yields are commonly associated with the semi-arid environment, where drainage densities are best developed (wetter areas have channel development inhibited by plants and drier areas have insufficient frequency of events to maintain channels). Some authors (e.g. Farquarson et al. 1992) suggest strong similarities in flood frequency between ephemeral rivers in arid zones. Other data (e.g. McMahon 1979), however, suggest that specific peak annual discharges vary considerably.

A key problem in many arid areas is the slope erosion of natural and artificial cuts principally though gullying. This can lead to rapid erosion of cut surfaces. Mitigation of such factors typically takes one of two approaches:

1 reduction of the erosivity of the rainfall by reducing gradients of slopes, and where possible keeping slope lengths and hence catchment areas as small as possible;

2 reducing the erodibility of the sediments by protecting the surface with vegetation or artificial stabilizers.

In the main flood zone associated with larger gully systems, alluvial fan systems and ephemeral river channels, flash flood damage is the major hazard. Although canalization can be used to constrain flood events, the most successful approach is to avoid high risk areas such as ephemeral channels on alluvial fans and river systems. Arid rivers are associated with large amounts of scour and fill in individual events (Fig. 5.19). These processes have significant implications for the

Fig. 5.19 Deposition and scour impacts of individual flash flood events. (a) Coarse sediment deposition (railway line has been excavated from deposits at base of photograph) from the 1991 Antofagasta flood in Northern Chile. (Photograph courtesy of Guillermo Chong, University of Antofagasta.) (b) Bridge removal by scour in a bedrock tributary of the Draa river, southern Morocco.

loss of housing and transport routes through burial and erosion.

Ultimately, in many river systems, the suspended sediment will end up in a playa or lake system. Reservoirs in arid areas are thus subject to high rates of sedimentation, so that their half-lives are considerably reduced. For example, the Tarbella Reservoir on the Indus lost 6% of its capacity within 5 years of construction (Ackers & Thompson 1987). Thus, siltation is a primary concern for reservoirs in arid zones, with mitigation taking the form of slope stabilization by vegetation, check dams to retain sediments in gully systems and maintenance of the reservoir though flushing.

Another potential problem associated with water in arid environments is the development of karst and pseudokarst. The lack of water means that deposits with a low unconfined compressive strength such as gypsum can become scarp-forming caprocks. Arid environments also commonly supply the necessary geological and soil-chemistry conditions (e.g. unconsolidated sediments, abundant dispersive salts such as sodium) for the subsurface flow of water, leading to the development of subsurface piping or 'pseudokarst'. Where these features develop they offer problems for construction and, as they are liable to change though collapse and natural expansion (Fig. 5.20), can lead to the destabilization of existing structures.

5.6.3 Salt hazards

Salt weathering is a major feature of deserts (Doornkamp & Ibrahim 1990). Salt weathering accelerates the breakdown of source rocks to silt-grade material (Goudie 1984) thus making them more susceptible to other sediment erosion agents such as wind and water, and exacerbating dust problems. It also attacks the fabric of houses, roads (Cooke et al. 1982; Doornkamp & Ibrahim 1990) and important archaeological features such as the Sphinx in Egypt. In many cases accelerated salt weathering may be directly related to increased salt in the local environment as a function of irrigation (Goudie 1977) or building-site leakages. The main impacts of salt

Fig. 5.20 Megapipe developed in gypsiferous Quaternary deposits, south-east Spain.

weathering in urban areas are through (i) chemical alteration (e.g. corroding reinforcing bars in concrete and chemically altering concrete) and (ii) volumetric changes due to salt weathering (e.g. leaching of salts from foundations causing subsidence, or hydration leading to expansion and ground heave).

5.7 FUTURE ISSUES

The population of Earth has increased from some 2.5 billion people in the 1950s to 6.1 billion in the year 2000. This growth in population creates greater pressure to use resources in arid environments and increases the risk of desertification. In all it has been calculated that there are some 49 million square kilometres on Earth at some risk of desertification (Eswaran et al. 2001). Most of this degradation is associated with unsustainable land management practices, which have an impact on the soil and vegetation and are exacerbated during periods of drought, such as in the Sahel in the 1970s. However, there are problems in distinguishing natural dryland ecosystem variability from longer term natural fluctuations in climate. For example, Lamprey (1975) analysed data from 1958 to 1975 and concluded that as a function of overgrazing the Sahara was marching northwards at 5.5 km yr^{-1}. However, new data sources and approaches (e.g. Hellden 1988; Le Houérou et al. 1988; Tucker et al. 1991; Hulme 1992) indicate that the degradation (desertification) observed by Lamprey was in fact attributable to natural changes in weather conditions (i.e. drought), comparing wet years (1958) with dry years (1975). Quaternary records over the past 40 kyr show that the Sahara has naturally ranged in climate from humid/ subhumid to hyper-arid/arid (Fig. 5.4). Over the Holocene Epoch fluctuations in aridity (4.5– 3.5 ka) have been attributed to the socio-economic collapse of several cultures (Petit-Maire & Guo 1998). It is now accepted that dryland vegetation systems are dynamic and far from fragile, showing adaptation and resilience to environmental change. So although advancing sand dunes destroying agricultural land may be true locally, the issue of land degradation may be overestimated and confused with natural environmental variability (Thomas & Middleton 1994). These concepts are particularly pertinent in the context of future global climate change, which may alter the spatial pattern and extent of drylands and the severity of their associated problems for human habitation. Thus there is a need to better understand the dynamism of drylands and how they respond to environmental change.

To assess the extent to which current trends in dryland development are attributable to longer term controls such as climate, there is a need both to exploit the data stored in ancient records and to develop higher resolution Holocene records (Fig. 5.4). These provide valuable data stores on the past extent of arid realms. For example, aeolian dune sediments provide information on past wind regimes. The identification of aeolian dune sediments in palaeochannels may help us identify the presence of periodically drier conditions in the ancient record. Palaeolakes offer a wealth of information on former high and low stands reflecting changes in the hydrology of the lake basin. Rivers and alluvial fans may yield information on changes in climate relating to the generation of effective runoff (Case Study 5.2). Caves, karst, tufas and groundwater provide information on wetter versus drier periods within the climate record. Floral and faunal changes may indicate landscape and climate change, such as the pack-rat middens of the south-western USA (Case Study 5.2). Archaeological evidence of human behaviour may reflect changes in local and regional climate.

Where absolute dates can be applied to the above data then rates of change can be determined within the arid realm. The advent of new methods of dating enables higher resolution chronologies to be established, allowing us to better correlate and understand the larger scale and longer term impacts of aridity. The responses to climate change in arid lands can be complicated and, as demonstrated in this chapter, wetter climates do not necessarily mean more erosion. In contrast drier, more seasonal climates may tend to lead to greater erosion as a direct function of their effect on vegetation cover and, through that, erosion. It is also clear that an understanding of the long-term evolution of landscapes (Case Study 5.1) may improve understanding of both the spatial and temporal process dynamics within an area, and that if we are going to manage these landscapes to minimize the impact of human intervention these need consideration.

Prediction of future events relies on an understanding of both the palaeoenvironmental record and the contemporary record. As demonstrated in the case studies in this chapter, however, isolating cause is far from straightforward, even in the contemporary. In addition there are the added difficulties of the differing time-scales on which processes operate. For example the form of dryland valleys, which formerly has been explained by changes in surface erosional and depositional factors, may be more strongly related to other factors such as groundwater sapping and deep weathering (Nash et al. 1994) in some arid settings, which operate on different temporal scales.

Field experimentation on a range of scales has been attempted to try and deal with some of the above spatial and temporal issues. In the 1990s the EU-funded experimental field studies IBERLIM and MEDALUS (Case Study 5.3) generated field data that were used to generate computer simulation models (Kirkby et al. 1998) of surface runoff, vegetation and erosion and their interplay through space and time on both a hillslope (MEDALUS model) and large, up to 5000 km², catchment scale (MEDRUSH model). These were calibrated with the collected field data and are being used to explore future trends under different scenarios. All modelling, however, is fraught with difficulties of scale. Global climate models (GCMs) use cells of < 0.5° latitude and longitude, and yet previous work has shown that detailed knowledge is needed of a wide range of variables, much smaller than this cell size (such as rocky surface conditions and cover, Yair 1994), to have any hope of accurately predicting the magnitude and size of change.

Debates rage on prediction of future climate change and whether global warming or cooling will prevail. Predictions from the generated GCMs vary widely. Integration of the data from three of the main GCMs for doubled atmospheric CO_2 levels indicates that southern European drylands (bordering the Mediterranean) will have increases in summer temperature of + 4 to + 6°C in the summer, and + 2 to + 6°C in the winter, with associated overall significant decreases in precipitation and overall available soil moisture (Williams & Balling 1995). The Hadley Centre for Climate Prediction and Research (UK) provides a summary of some of the various types of models available through links on www.metoffice.com. These predictions (to the year 2100) point towards raised temperatures, lower precipitation and lower soil moisture in general, for the zone 30°N–30°S of the Equator. This zone contains many of the world's arid environments (Fig. 5.2). In areas such as the Atacama Desert of South America impacts of global climate change on the duration, magnitude and frequency of El Niño and La Niña need to be understood, as these phenomena are significant for rainfall generation in such regions.

Dryland landscapes and their associated vegetation systems, which help control rates of landscape change, are naturally dynamic and have a degree of resilience to environmental change. These systems thus have some natural resistance to human intrusion. With a burgeoning global population, however, our intervention with the arid environment will only increase as demand for resources increases, particularly if, as predicted by GCMs, the spatial extent of drylands expands/intensifies in response to factors such as global warming. Thus there is a need to address the issues of sustainability in such environments in order to avoid currently localized impacts becoming more global. This may require embracing remote sensing technology and integrating it with field studies to monitor dryland environments and the nature of changes within them. These data, together with longer term Quaternary data, can be used to try and better understand the complexities of dryland environments and their response to environmental change. If, however, this knowledge is to be successfully implemented in the sustainable management of dryland systems we need to ensure that scientists communicate with the decision makers more effectively (Thomas 1997c).

6

Urban environments

Kevin Taylor

6.1 INTRODUCTION AND DEFINITIONS

6.1.1 Introduction

The urban environment is one that is of increasing importance globally, with implications for both hydrological and sedimentological systems. It has been estimated that 50% of the Earth's population live or work in an urban environment (United Nations 2003) and this percentage is predicted to increase. Urbanization dates back several thousand years, with development taking place primarily along major rivers acting as waterways. It was not until the Industrial Revolution in Europe, and later throughout other parts of the world, however, that the growth of significant urban areas took place. The very nature of urban land surfaces and waterways results in unique hydrological, sedimentological and atmospheric environments, leading to a wide range of specific management and sustainability issues, many of which are not encountered in natural environments. The application of process models derived from observation and measurement of natural systems to urban environments is, therefore, of limited value. The issues of environmental pollution and sustainability have resulted in a plethora of research into the quality of urban air, groundwater and, to a lesser extent, surface water. Urban sediments, and the role of the urban environment as a sedimentary system, have been largely neglected. It is increasingly being recognized, however, that particulates in urban environments are a major factor in human health, through their impact upon air quality, waterways and biodiversity, through the role of sediments as vectors for contaminant transfer and

reservoirs for contaminant storage. As a result, our knowledge base on urban sedimentology, and the interactions with hydrology, air quality and biodiversity, is rapidly increasing. This chapter describes urban sedimentary environments, the sources of sediments to these environments, the physical and chemical characteristics of the resulting accumulated sediment, and discusses the impacts of intrinsic and extrinsic processes upon these sediments. Finally, the sustainable management options for urban sediments and their impacts will be discussed.

6.1.2 Definitions

The term *urban* is used widely throughout both social and scientific literature, and has come to mean different things to different people. A definition suggested here for urban areas is 'those areas where the ecosystem is significantly modified by human settlement and associated activities; they are characterized by a unique modification of the physical, chemical and biological environments, resulting from the construction of buildings on a large scale'. One key element of the urban environment that distinguishes it from the environments discussed in the other chapters of this book is its highly engineered nature. Such engineering results in land surfaces with distinct physical and chemical properties. Although natural environments have inevitably been altered by anthropogenic activity, these have not resulted in the significant, and wholesale, change in fundamental processes of sediment and water movement and accumulation seen in urban environments. These include increased storm peaks and shortened

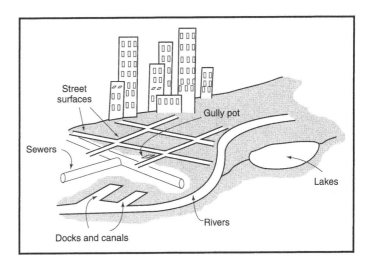

Fig. 6.1 Schematic diagram of the urban sedimentary environments covered in this chapter.

lag times, with greater opportunity for scouring, erosion and transport of sediment. Furthermore, the extensive presence of transport and, to a lesser degree, industry in urban environments leads to increased contaminant sources and enhanced pollution pressures.

The term *urban sediment* is commonly used and is used in a range of different contexts throughout the literature. In the past *urban sediments* has been a term commonly used to refer to sediment accumulation on street surfaces, although in this chapter the term is used in a broader manner to mean any sediments present within the urban environment. They, therefore, comprise sediments accumulating on street surfaces, in gully pots and sewer systems, and in receiving water bodies (rivers, canals, docks and lakes; Figs 6.1 & 6.2).

The term *road dust* has also been used extensively in the past, but the term *dust* implies fine, respirable material (i.e. grains less than 10 μm). Numerous studies have shown that street sediment is composed of a full range of particle sizes, commonly biased towards coarse material and, therefore, the use of the term *dust* is not appropriate. The terms *street sediment* or *road sediment* have also been used in the past, but with no formal definition. More recently the term *road-deposited sediment* has been used (e.g. Sutherland 2003). This term is favoured here, owing to its clear descriptive nature, and

will be adopted throughout this chapter. Road-deposited sediment (RDS) can be thought of as predominantly subaerial in nature and contrasts with the subaqueous (or aquatic) sediments that are present within urban rivers, canals and docks. Sediments accumulated within gully pots and sewer systems are neither wholly subaerial nor subaqueous, but experience both conditions episodically dependent on weather conditions. These are, therefore, described separately in this chapter from other sediment types.

6.1.3 Historical development of urban sedimentology

The study of urban sediments from the perspective of environmental sedimentology is a young one. Research into urban particulates originated out of concern for pollution and human health, and focused on road-deposited sediment (RDS). Early work focused on the levels of lead in RDS, as this is a major vector for lead consumption by children through hand-to-mouth activities. This research documented the high levels of lead in RDS, compared with crustal abundance (Farmer & Lyon 1977; Duggan & Williams 1977; Harrison 1979). The importance of RDS as a vector for Pb uptake was shown by the correlation between RDS-Pb and blood-Pb levels in children (Thornton et al. 1994). This was strengthened by epidemiological studies that showed a

Fig. 6.2 Photographs of the typical environments, and their characteristics, associated with urban sediments. (a) Street surfaces. (b) Accumulation of road-deposited sediment (RDS) around a traffic island on a street surface. Note the highly variable grain-size of this material. (c) An urban river, showing the classic engineered and steep-sided nature of such rivers. (d) An urban dock, displaying surface debris accumulation due to prevailing wind directions.

clear link between high lead intake and intellectual development in children (Needleman et al. 1979). Partially as a direct result of this pioneering research, the use of leaded petrol was phased out in Europe. Following on from this, the levels for a wide range of metal and organics contaminant have been studied in RDS from around the world and this aspect of urban sediment study is now well-established (section 6.3.1).

The pressures of water quality regulations, biodiversity issues and regeneration of waterside areas led to the second main area of urban sediment research – aquatic urban sediments. It has long been recognized that sediments act as the major vector for the transport of contaminants in most aquatic systems (see Chapter 1) as contaminants are preferentially associated

with the particulate phase. It has also been well-established that sediments act as a major store of contaminants (e.g. floodplains and salt marshes). Such storage, and potential remobilization, is a major issue in urban rivers, canals and docks, as a result of a legacy of unregulated domestic and industrial discharge into these water bodies (Taylor et al. 2003). The study of aquatic urban sediments has, therefore, become an increasingly important area, not just as a research priority, but also for urban pollution management, sustainable river basin management and remediation. These studies have focused on sediment and contaminant fluxes in urban rivers, sediment quality in canals and docks, and understanding the short- to long-term pathways of sediment-bound contaminants through the urban environment.

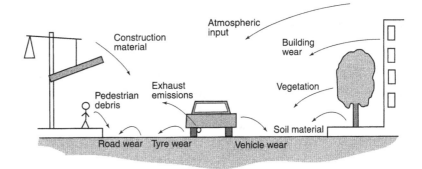

Fig. 6.3 Schematic diagram of the sources of sediment comprising road-deposited sediments.

6.2 SEDIMENT SOURCES AND SEDIMENT TRANSPORT PROCESSES

6.2.1 Sediment and contaminant sources

Sediments within urban environments originate from a wide range of sources, both natural and anthropogenic. In this chapter sediment sources to subaerial environments (road-deposited sediments) and subaqueous environments (rivers, canals/docks and lakes) will be considered separately. Although both environments receive a broadly similar range of source material, the differences in the physical and chemical characteristics between the environments lead to contrasting intrinsic and extrinsic sources dominating.

6.2.1.1 Road-deposited sediment

Compared with sediment in natural environments, road-deposited sediment (RDS) has a wide, and diverse, range of sources (Fig. 6.3). Sources are either intrinsic to the road surface, which are predominantly anthropogenic in nature, or extrinsic, which are predominantly naturally derived. Intrinsic sources include vehicle exhaust emissions, vehicle tyre and body wear, brake-lining material, building and construction material, road salt, road paint and pedestrian debris. Extrinsic sources are soil material, plant and leaf litter, and atmospheric deposition.

There are very few studies that have attempted to quantify the importance of these sources to RDS, partly as a consequence of the wide range of sources, and partly due to the heterogeneous nature of most RDS deposits. Volumetrically, the most important component of RDS comes from soils and building material. Soil material, which may be derived from a range of distances, contributes both minerogenic and organic material. For example, a study by Hopke et al. (1980) recognized that soil material formed approximately 75% of RDS. Building material contributes quartz sand, concrete and cement to RDS, and these are believed to be relatively inert. In urban areas undergoing extensive development, however, the volume of building material input to RDS can be large, and may have important consequences for air quality, sediment volumes and particulate-associated contaminants discharged into drainage systems and rivers.

Iron oxides are abundant in RDS and as a result mineral magnetic analysis has been applied to study source apportionment in these sediments. Mineral magnetic analysis can discriminate soil-derived iron oxide particulates from those derived from fossil-fuel combustion (Oldfield et al. 1985), and as such has great potential in urban sediment sourcing. Application of mineral magnetic analysis has shown the importance of non-soil derived material in RDS, including vehicle and industrial combustion, construction material and asphalt abrasion (e.g. Beckwith et al. 1986; Xie et al. 1999; Lecoanet et al. 2003). In a study of RDS in Manchester, UK, Robertson et al. (2003) concluded that vehicular sources were the primary contributor to iron oxide material (see Case Study 6.1). Although the application of mineral magnetic analysis has helped to identify

Case study 6.1   Road-deposited sediment, Manchester, UK

Road-deposited sediments (RDS) are a major component of the urban sediment cascade (see section 6.2.2) and are sourced from a wide range of material. An extensive data set on the physical, chemical and mineralogical composition of RDS exists for the city of Manchester, north-west England (see Robertson et al. 2003). The sediments contain high metal concentrations, in most cases significantly above average crustal abundances. Lead ranges from 120 to 645 $\mu g\ g^{-1}$, Cu from 39 to 283 $\mu g\ g^{-1}$ and Zn from 172 to 2183 $\mu g\ g^{-1}$. Temporal analysis has shown that Pb levels have decreased since the 1990s in response to the reduction in use of

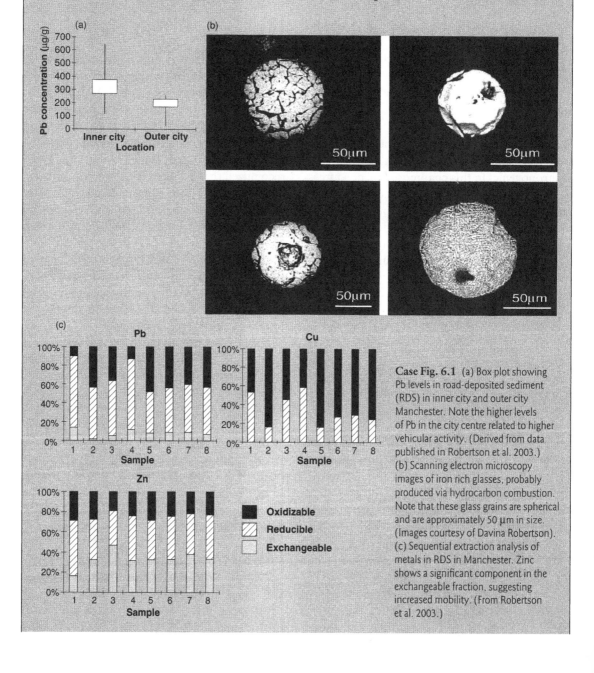

Case Fig. 6.1 (a) Box plot showing Pb levels in road-deposited sediment (RDS) in inner city and outer city Manchester. Note the higher levels of Pb in the city centre related to higher vehicular activity. (Derived from data published in Robertson et al. 2003.) (b) Scanning electron microscopy images of iron rich glasses, probably produced via hydrocarbon combustion. Note that these glass grains are spherical and are approximately 50 μm in size. (Images courtesy of Davina Robertson). (c) Sequential extraction analysis of metals in RDS in Manchester. Zinc shows a significant component in the exchangeable fraction, suggesting increased mobility. (From Robertson et al. 2003.)

leaded petrol (Nageotte & Day 1998; Massadeh & Snook 2002). Clear spatial distributions of metal concentrations are present. For example, lead concentrations are higher in sediments in inner city sites, than from those in outer city sites (Case Fig. 6.1a), reflecting the contrasting levels of vehicular activity. Analysis of the spatial distribution of RDS composition shows that metal levels are heterogeneous across the city, with different hotspot locations for different metals. This illustrates the localized nature of contaminant inputs into RDS. Mineral magnetic analysis of Manchester RDS (Robertson et al. 2003) reveals that the magnetic fraction is composed predominantly of ferromagnetic multidomain particles, indicating inputs of predominantly anthropogenic origin, derived primarily from automobiles.

Petrographic and mineralogical analysis shows that the sediments, in addition to quartz and clay grains, comprise a significant component of anthropogenic grains (Case Fig. 6.1b). These are either iron-rich combustion products (mostly iron oxides) or iron-rich glasses, probably also derived from vehicular combustion. These grains contain high levels of heavy metals (e.g. Pb and Cu up to 1500 ppm, and Zn up to 10,000 ppm) and are probably the major hosts for metals in the Manchester RDS. Sequential chemical extraction schemes (see Chapter 1) show that most metals are bound up with the reducible fraction (probably iron oxides and glasses), but in the case of Zn a significant fraction is bound in the much more environmentally available exchangeable fraction (Case Fig. 6.1c). This suggests that Zn is likely to be the most mobile metal in the RDS in Manchester and release into road runoff may take place; an observation consistent with storm runoff analysis. The RDS also contains significant levels of soluble nitrate, chloride, sulphate and phosphate. These anions are key components in river water, of which nitrate and phosphate are key nutrients. It is, therefore, likely that these sediments represent an input of dissolved species into urban rivers, with subsequent impacts upon water quality.

Relevant reading

Massadeh, A.M. & Snook, R.D. (2002) Determination of Pb and Cd in road dusts over the period in which Pb was removed from petrol in the UK. *Journal of Environmental Monitoring* 4, 567–72.

Nageotte, S.M. & Day, J.P. (1998) Lead concentrations and isotope ratios in street dust determined by electrothermal atomic absorption spectrometry and inductively coupled plasma mass spectrometry. *Analyst* 123, 59–62.

Robertson, D.J., Taylor, K.G. & Hoon, S.R. (2003) Geochemical and mineral characterisation of urban sediment particulates, Manchester. UK. *Applied Geochemistry* 18, 269–82.

sources of iron oxides in RDS it has also been recognized that the complex and heterogeneous nature of urban sediments limits the extent to which these measurements can be used as sediment source tracers. For example, Charlesworth & Lees (2001) consider that given the large number of sources of magnetic material in urban environments, and the processes that change magnetic signatures, unmixing individual components of the signal becomes very difficult.

The majority of contaminants to RDS are derived from intrinsic sources, and the major sources recognized are shown in Table 6.1. As stated previously, Pb is predominantly derived from leaded fuel (where tetraethyl-lead is used as an additive). Lead levels in sediment have declined, however, with the widespread reduction in use of leaded fuel (e.g. Nageotte & Day 1998). Copper and zinc have both been sourced to vehicle activity, with Cu coming from corroded car bodywork (Beckwith et al. 1986) and Zn and Cd being derived from tyre wear (Hopke et al. 1980). Chromium, bromine and manganese are also present in tyres and brake

Table 6.1 Contaminant sources to road-deposited sediment.

Contaminant	Source
Pb	Petrol combustion, paint, smelters, coal combustion
Zn, Cd	Tyre wear, galvanized roofs, abrasion of vehicles, lubricating oils, alloys
Cu	Brake linings, alloys, metal industry
Fe	Car exhaust particulates, corrosion of vehicle body work, background geology
Mn	Tyre wear, brake linings, background geology
Cr	Engine wear, vehicle plating and alloys, road surface wear
Ni	Engine wear, metal industry, background geology
Asbestos	Break clutch linings
Cl, Na	Road salt
PGEs (Pt, Pd, Os)	Catalytic convertors
Pesticides/herbicides	Garden application
PAHs	Biomas burning, petroleum combustion
PCBs	Petroleum combustion, industry
Bacteria	Sewage treatment works, animal faeces
Pharmaceutical compounds	Sewage treatment works

linings. Multi-element analysis of RDS, coupled with principal component analysis, has been used in some studies to aid in elemental source apportionment. De Miguel et al. (1997), in a study of Madrid, recognized distinct groups of elements. Those derived from vehicles and construction sources were Br, Cd, Co, Cu, Mg, Pb, Sb, Ti and Zn, whereas those deriving from natural soil material were Al, Ga, La, Mn, Na, Sr, Tl and Y.

The platinum group elements (PGEs) Pt, Pd and Rh are a relatively recent contribution to RDS, having been emitted into urban environments since the early 1990s. The PGEs act as catalysts in catalytic converters and, with the phasing-out of leaded fuel, are currently the metals of most concern emitted from vehicle exhausts. There is evidence from many studies for the widespread dispersion and accumulation of PGEs in RDS, as well as urban soils and airborne particulates. Concentrations of PGEs significantly above those of average upper crust values have been reported for RDS for cities in Europe and Australia (Wei & Morrison 1994a; Motelica-Heino et al. 2001; Whitely & Murray 2003). Although PGEs in metallic form are generally considered to be biologically inert, soluble PGE salts are indicated to be much more bioreactive (Farago et al. 1998) and so the presence of PGE in RDS is potentially of significant concern.

There is a whole suite of organic contaminants (so-called persistent organic pollutants) sourced to RDS. These include PAHs (polyaromatic hydrocarbons), PCBs (polychlorinated biphenols), hydrocarbons, dioxins, pesticides and herbicides. The sources of these are various and include both atmospheric and land-based sources. For example, PAHs were observed to be sourced from biomass burning and vehicular emissions in Vancouver, Canada (Yunker et al. 2002). Probably the largest source of organic pollutants are those derived from vehicular activity. Many of these are found in petrol or diesel (including benzene, toluene, naphthalene, PAHs), or associated with automobiles (including ethylene glycol, hydraulic fluids, styrene, oil lubricants). Pesticides and herbicides are applied directly to pavements or to urban soils in residential areas and gardens, where they can be removed from runoff or erosion and deposited in RDS.

6.2.1.2 River, canal, dock and lake sediments

The range of sediment sources for rivers and canals is greater than that for RDS, in that as well as the input of RDS into river sediments, upstream and downstream input of channel-associated material is a major contributor to these urban aquatic sediments. Although the origins of sediment in river basins have been studied extensively (see Chapter 3), very limited study has been made of river sediment sources in urban catchments. A recent exception is that of Carter et al. (2003) for the Aire–Calder river basin in eastern England (see Case Study 6.2). Through the application of well-developed statistical fingerprinting techniques (see Collins & Walling 2002) they showed that the sediment in the urban river sections was sourced from channel bank erosion (18–33%), uncultivated topsoil (4–7%), cultivated topsoil (20–45%), road-deposited sediment (19–22%) and sewage input (14–18%). The high contribution of urban sources (up to 40% sewage and RDS) illustrates the marked contrast of urban sediments to those in non-urbanized catchments. This is in general

---

Case study 6.2 Sources and dynamics of urban river sediments: Rivers Aire and Calder, West Yorkshire, UK

Limited information is available on the specific sources of sediment to urban rivers, in direct contrast to that of other river systems (Chapter 3). Exceptions are the Rivers Aire and Calder in West Yorkshire (eastern England; Case Fig. 6.2a), which has seen recent detailed analysis through research programmes directed at understanding fluvial and urban environments (The Land–Ocean Interaction Study (LOIS) and the Urban Regeneration in the Environment (URGENT) programmes of the UK Natural Environment Research Council). These rivers are part of a catchment that in its upper parts is predominantly rural in nature, but which in its lower parts runs through heavily urbanized catchments.

Sediment sources in the Aire and Calder have been shown to vary between the upstream rural-dominated and the downstream urban-dominated parts (Carter et al. 2003). Suspended sediment in the upper reaches is predominantly sourced from channel-bank material (43–84%) and from uncultivated topsoil (16–57%). This is in contrast to the lower, urbanized reaches, where road-deposited sediment (19–22%) and sewage treatment works (14–18%) acted as significant sources of suspended sediment (see Case Fig. 6.2a). Carter et al. (2003) also found that the proportions of sediment sources to the urbanized reaches varied over individual storm events. Road-deposited sediment sources were highest at the end of high-flow events (Case Fig. 6.2b) as a result of the time lag to flush sediment from road surfaces.

The quality of urban river water is commonly poor and the role of sediments, and sediment-borne contaminants, on water quality and contaminant fluxes is poorly understood. Goodwin et al. (2003) and Old et al. (2003) have quantified the nature of suspended sediment fluxes through the Bradford Beck, a tributary of the River Aire, which runs through the heavily urbanized city of Bradford. Goodwin et al. (2003) carried out continuous monitoring of the urban segments of the river, both for discharge and suspended sediment loads. The hydrographs show that the river responds rapidly to rainfall events and suspended sediment concentrations are also high during these events (up to a maximum of 1200 mg L^{-1}). They proposed that in the urban segments sediment was derived from road runoff and combined sewer outflows. As can be seen from Case Fig. 6.2c, suspended sediment loads were several orders of magnitude higher during storm events than during low flow times. Although Goodwin et al. (2003) do not publish compositional data for suspended sediments, as well as the physical impact of the suspended sediment, the flux of contaminants on these sediments is likely to be highly significant also.

For the same river, Old et al. (2003) studied the sediment dynamics for a single, large convectional summer rainfall event (Case Fig. 6.2c). They found that during this event, over a

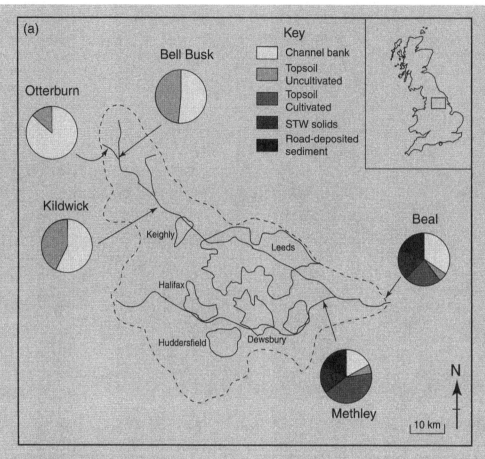

Case Fig. 6.2 (a) Location of the Aire–Calder catchment, eastern UK. Also shown are the relative source contributions to the suspended sediment in the catchment in both upstream and downstream sections. Note the significant component of urban-derived sediment (road-deposited sediment and sewage treatment works (STW) solids) in the sections downstream of the major urban conurbations. (Data from Carter et al. 2003.)

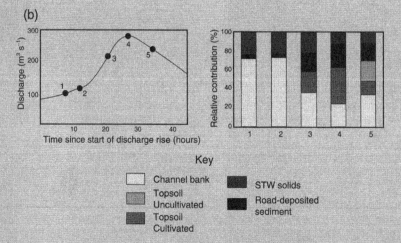

Case Fig. 6.2 (b) Variation in sediment sources to suspended sediment in the urbanized portion of the Aire–Calder over a single high-discharge event. (From Carter et al. 2003.)

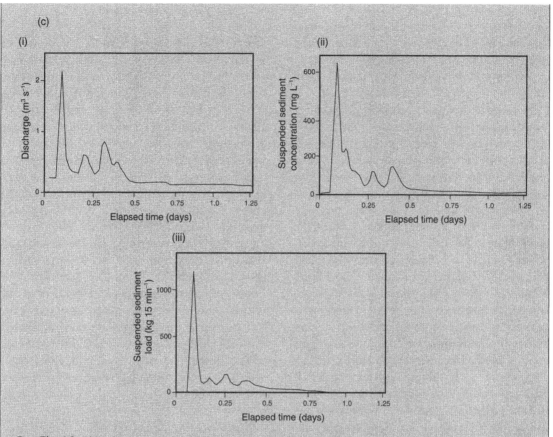

Case Fig. 6.2 (c) Discharge, suspended sediment concentrations (SSC) and suspended sediment load (SSL) for an urbanized portion of the Bradford Beck, Aire–Calder catchment, eastern England. Note the high levels of SSC and SSL associated with this single high-flow event. (Modified from Goodwin et al. 2003.)

period of 15 minutes, water discharge increased from 0.45 to 34.6 m^3 s^{-1}. During the same period suspended sediment concentrations reached a peak of 1360 mg L^{-1}. At the peak of this sediment discharge event the sediment flux reached 47 kg s^{-1}. Old et al. (2003) concluded that although the Bradford catchment represents only 3% of the catchment area of the River Aire, it can at times be a major contributor of fine sediment. Given the high contamination of such fine sediments, storm events are likely to have major impacts upon river quality downstream of urban areas.

Relevant reading

Carter, J., Owens, P.N., Walling, D.E., et al. (2003) Fingerprinting suspended sediment sources in a large urban river system. *The Science of the Total Environment* **314–316**, 513–34.

Goodwin, T.H., Young, A.R., Holmes, M.G.R., et al. (2003) The temporal and spatial variability of sediment transport and yields within the Bradford Beck catchment, West Yorkshire. *The Science of the Total Environment* **314–316**, 475–94.

Old, G.H., Leeks, G.J.L., Packman, J.C., et al. (2003) The impact of a convectional summer rainfall event on river flow and fine sediment transport in a highly urbanised catchment: Bradford, West Yorkshire. *The Science of the Total Environment* **314–316**, 495–512.

agreement with Nelson & Booth (2002), who found that as well as landslides and channel bank erosion, 15% of sediment in an urbanized catchment was from road surface erosion. Further such studies are required before a full appreciation of the relative contribution of sediment and contaminant sources to urban rivers can be gained. As well as contaminant input from road runoff, increased levels of nutrients (especially phosphorus) and micro-organic pollutants (e.g. pharmaceutical products) are sourced from sewage treatment works (Owens & Walling 2002; Warren et al. 2003). Industrial processes also source metal contaminants to urban rivers (Walling et al. 2003).

Compared with rivers, which receive sediment from a wide area, canal sediment is commonly dominated by material that is more locally derived, as a result of the limited transport of sediment in canals. This may be derived from industrial sources or sewage, as well as natural material eroded from nearby land and road surfaces. Major canal and dock systems that have significant water inputs from rivers, however, can have a significant sediment source from outside the system. For example, Qu & Kelderman (2001) showed that sediment, and associated contaminants, in the Delft canals, The Netherlands, have been derived predominantly from the River Rhine, with the remainder coming from urban and industrial sources. Urban docks and canals also commonly receive high levels of organic matter, discharged from combined sewer overflows, and contaminants derived from boat traffic, for example hydrocarbons and tributyl tin (e.g. Wetzel & Van Vleet 2003). Within urban lakes, sediment sources are generally a combination of both eroded soil material from the surrounding catchment and anthropogenic material from the urban environment. Atmospheric deposition may also be an important source of particulates and associated contaminants, especially for lakes with no direct river input (Charlesworth & Foster 1999). Sediments deposited within lake systems are probably highly catchment specific and observations made cannot readily be applied to other urban catchments (Charlesworth & Foster 1999).

6.2.2 Sediment transport processes

The transport of sediment in urban environments is complex. There is a relatively poor understanding of the pathways that sediments take from their source to receiving water bodies, the rate of sediment transport, the location of short-term and long-term sinks, and how these pathways have an impact upon the longer term fate and distribution of contaminants in the urban environment. The movement of sediment through the urban environment can be represented in what has been termed the *urban sediment cascade* (e.g. Charlesworth & Lees 1999; Fig. 6.4). This cascade recognizes the relationship between sediment sources, transport mechanisms and deposition (storage) of sediment. The urban sediment cascade is a highly dynamic system. As in many depositional systems there are stages of temporary deposition, or storage, prior to further transport. The top of the cascade is considered to be the sources of RDS. The deposition of sediment upon street surfaces is highly transient in nature, with remobilization taking place down the cascade.

The bulk of sediment transport in the urban environment takes place through the action of water, but local redistribution of sediment upon street surfaces may also take place by wind. Storm drains carry urban runoff from the street surfaces (as well as other impervious surfaces) to receiving water bodies (rivers, canals or docks; Fig. 6.5). In addition to this surface water runoff, the urban drainage system also has to deal with industrial and domestic wastewaters and sewage. In general, there are two main types of urban drainage: combined systems and separate systems. In combined systems road runoff is transferred to a common sewer, and from then onwards to a sewage treatment works. In this case, these combined sewers rarely have sufficient hydraulic capacity during storm events, and the water is discharged directly (including sewage) through a combined sewage overflow outlet (CSO) into receiving water bodies (Fig. 6.5). In more recently built drainage systems, a separate sewer system transports runoff directly to receiving water bodies, without involvement in the domestic

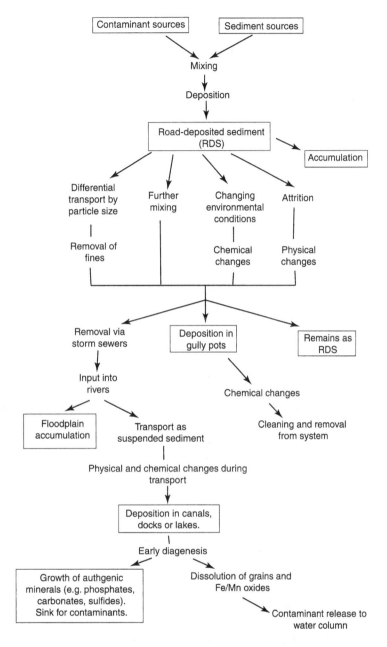

Fig. 6.4 The urban sediment cascade, showing the pathways of, and changes in, urban sediment particulates from sources to sinks. (Modified and extended from Charlesworth & Lees 1999.)

sewage system. Once within rivers, sediment transport takes place via channel transport processes typical for rivers in general (see Chapter 3). The temporal and volumetric scales of this process, however, contrast with rivers from non-urbanized environments (see section 6.3).

Street surfaces are subaerial and during dry periods resuspension of fine sediment particulates by wind can readily take place. The suspension of sediment particulates has major impacts on urban air quality. Miguel et al. (1999) attributed 5–10% of the allergenicity of atmospheric suspended particle matter in California to road dust emissions. Particles less than 10 μm (PM_{10}) and less than 2.5 μm ($PM_{2.5}$) have been measured extensively in urban air, and a major component of these particles, especially those above 2.5 μm in size, can be derived from road-deposited sediments (e.g. Hosiokangas et al. 1999; Harrison et al. 2001).

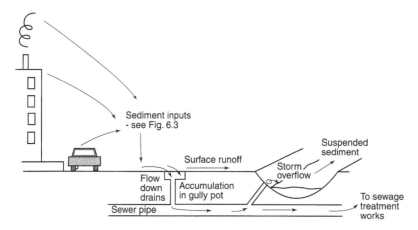

Fig. 6.5 Schematic diagram of the pathways of sediment transport in the urban environment.

In addition to street surfaces, the major sites of sediment deposition in the urban sediment cascade are gully pots and storm sewers, rivers, canals and docks, and lakes. Deposition (storage) on street surfaces, gully pots and storm sewers is short-term, as is the storage of sediment in urban rivers. In all of these systems sediment is moved during high-flow events associated with rainfall. Sediment storage in canals, docks and lakes is longer term and removal is generally by engineering activities (e.g. dredging). These are often termed receiving water bodies and sediment accumulation in these systems can lead to long-term physical and chemical impacts upon water and ecology, as well as providing a historical record of urbanization.

As well as physical transport down the urban cascade, chemical changes may take place in the sediment from source to final deposition. These take place as a result of chemical and biochemical interactions between water and sediment, including weathering, adsorption, desorption, mineralogical transformations and diagenesis. A study by Charlesworth & Lees (1999) concluded that metal levels in sediments changed during transport from street surfaces, through gully pots and sewers and into longer term sinks (rivers, lakes). Some elements (e.g. Cu) increased in concentration during transport through the cascade, probably as a result of their strong affinity for solids, and the dominantly fine-grained nature of transported sediments. In contrast, some elements (e.g. Cd) decreased in concentration in deposited sediments, probably reflecting their greater affinity for the aqueous phase.

In contrast to other sedimentary environments there has been little attempt to determine sediment budgets in urban sedimentary environments or model sediment movement. Exceptions are small-scale modelling of sediment transport in sewer systems from an engineering perspective (section 6.3.2), and temporal measures of suspended sediment transport in urban rivers (section 6.3.3). In particular, transport pathways of contaminants in urban sedimentary environments are complex and much work is needed to fully understand the longer term impacts of sediment-borne contaminants to atmospheric and aquatic systems. Furthermore, site specificity limits generalizations that can be made between different urban catchments.

6.3 SEDIMENT ACCUMULATION

6.3.1 Road-deposited sediment

Road-deposited sediments have probably received more attention than other urban sediments in recent years as a consequence of their potential impact upon urban air quality and urban runoff. They are also easy to sample and have the potential to act as a very good proxy for urban pollution levels. Road-deposited sediment is composed of a wide range of sediment grains, which are dominated by quartz, clay and carbonates.

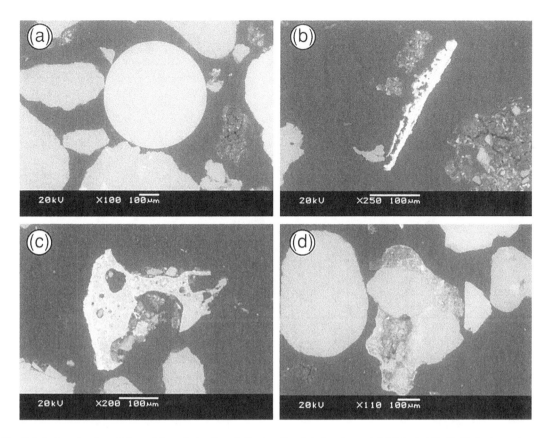

Fig. 6.6 Sediment grains within road-deposited sediments (backscattered electron images). (a) Spherical Ca–Si–Al–Fe glass grain. (b) Iron oxide fragment, probably derived through the oxidation of a metal Fe grain. (c) Metal-rich glass grain, probably derived from smelting processes. (d) Aggregated grain of quartz and clay particles; probably building material. All scale bars = 100 μm.

In addition, abundant anthropogenic grains are present, including glass particles from industrial processes and high temperature combustion, metal slags, cement grains, metallic fragments and iron oxide particles (Fig. 6.6).

Road-deposited sediments characteristically exhibit high concentrations of metals, an aspect of which most research has been directed towards (e.g. Beckwith et al. 1986; Kim et al. 1998; Charlesworth et al. 2003a; Robertson et al. 2003). Levels of metals in these sediments are commonly an order of magnitude higher than those of natural sediments. Table 6.2 shows typical values of metals in RDS for a number of urban environments. There is clearly a wide range of values, and Charlesworth et al. (2003a) analysed these data to show that a significant positive relationship exists between population size and metal levels in RDS. Particular study

has been made on the levels of Pb within urban street sediments, largely based on concern over human exposure to toxic levels of this contaminant. A significant enrichment in Pb levels in inner city centres has been documented (Duggan & Williams 1977; Thornton et al. 1994; Robertson et al. 2003), supporting other evidence, particularly Pb isotopic data (Zhu et al. 2001), that Pb is derived from petrol combustion in such sediments. More recent studies of RDS has shown a decrease in the levels of Pb, consistent with the reduced use of leaded petrol. In 1975 average lead levels were found to be 941 ppm in Manchester City, against a background level of 85 ppm (Nageotte & Day 1998). By 1997 this had fallen to 569 ppm (Nageotte & Day 1998). A more recent study (Robertson et al. 2003) has shown a further reduction to an average of 265 ppm in 2000 (see Case Study 6.1). These relatively

Table 6.2 Typical metal contents (μg g^{-1}) in road-deposited sediments in selected cities. (Data from Charlesworth et al. (2003a) and, for Manchester 2002, Robertson et al. (2003).)

City	Population	Cd	Cu	Ni	Pb	Zn
New York	16,972,000	8	355	–	2583	1811
Seoul	10,627,000	3	101	–	245	296
London	9,227,687	2.7–6250	61–512	32–74	413–3030	988–3358
Hong Kong	5,448,000	–	92–392	–	208–755	574–2397
Madrid	2,909,792	–	188	44	193	476
Manchester (1975)	2,578,900	–	–	–	970	–
Manchester (2002)	2,578,900	–	32–283	–	25–645	172–2183
Birmingham (1976)	2,329,600	–	–	–	950–1300	–
Birmingham (1987)	2,329,600	–	–	–	527–791	–
Taejon, Korea	2,000,000	–	47–57	–	52–60	172–214
Amman	1,272,000	2.5–3.4	69–117	27–33	219–373	–
Cincinnati	1,539,000	–	253–1219	–	650–662	–
Oslo	758,949	1.4	123	41	180	412
Bahrain	549,000	72	–	126	697	152
Hamilton	322,352	4.1	129	–	214	645
Christchurch	308,200	1	137	0	1091	548
Lancaster	136,700	3.7	75	–	1090	260

recent drops in sediment-Pb levels illustrate the transitory, short-term nature of these sediments within urban systems. Indeed, Allott et al. (1990), using radiocaesium from the dated Chernobyl fallout event, documented the residence time of sediment on street surfaces to be short, in the order of 150 to 250 days. Recently, with the introduction of catalytic converters, attention has been directed towards the levels of platinum (and associated elements) within urban street sediments (e.g. Wei & Morrsion 1994a). It has been documented that Pt levels are increasing in urban sediments, although health impacts of these increasing levels remain largely undetermined (Farago et al. 1998).

Data on the chemical speciation of contaminants within RDS have provided information on the mineralogical affinity and potential reactivity of contaminants (Fergusson & Kim 1991; Stone & Marsalek 1996; Charlesworth & Lees 1999; Robertson et al. 2003). Charlesworth & Lees (1999) found a low concentration of heavy metals associated with the exchangeable phase, results that have been reproduced by other studies. Hamilton et al. (1984), however, found Cd to be associated with the exchangeable phase, and Robertson et al. (2003) found Zn also to display a significant affinity to the exchangeable fraction. Therefore, RDS may be a significant source of Cd and Zn to urban runoff. Platinum in urban sediments has been shown to be in a form that may be soluble (Farago et al. 1998) and street sediments in gully pots have also been shown to be actively mobilizing Pt to the aquatic phase (Wei & Morrison 1994a). The majority of studies have found most metals to be associated primarily with the reducible (Fe and Mn oxide) fraction. Although this suggests that on street surfaces contaminant mobility is generally low, changes in pH and redox as a result of deposition in aquatic sediments or sediment water transport would possibly release metals back into aquatic environments. Copper has been shown to display a higher affinity to organic matter (Hamilton et al. 1984; Robertson et al. 2003). Charlesworth & Lees (1999) ascribed the preference of metals for the organic matter fraction in Coventry to high levels of organic matter in the sediments. Much less direct information exists on the role of individual minerals on contaminant behaviour in urban sediments. McAlister et al. (2000) documented the stabilization of weddellite (calcium oxalate dihydrate), derived from sewage, by interactions with metals in street sediments of Brazil. In this case, therefore, RDS acted as a sink for oxalate, exposure to which has significant potential impact on human health, including damage to the kidneys and nervous system. In contrast, Serrano-Belles & Leharne (1997) documented the enhanced release of Pb from RDS upon the addition of chloride, in the form of salt, probably as a result of the formation of chloro-lead complexes. There is significant scope for more detailed mineralogical analysis of urban sediments, and the role individual mineral phases play in contaminant uptake and mobility.

There have been only a few studies that have looked in detail at the grain-size distributions in RDS, and the distribution of contaminants between grain-size fractions. In general, RDS displays a wide range in grain sizes. Droppo et al. (1998) reported the mean grain size of RDS in Hamilton, Canada to be 227 µm. Sutherland (2003) found the < 63 µm fraction to dominate the mass fraction of RDS in Hawaii, accounting for 38% of the sample. Several studies have also been carried out to determine the contaminant loading on different grain-size fractions, with differing results. Biggins & Harrison (1986) showed that the mass-dominant fraction of Pb was in the 250–500 µm fraction, but that there was a range from 2 to 30% of the mass loading in the < 63 µm fraction. Stone & Marsalek (1996) found similar results for RDS in London, UK. Sutherland (2003) found a much higher Pb loading in the < 63 µm fraction for RDS in Hawaii, with this fraction accounting for an average of 51% of the Pb mass. In a similar manner to sediments from other environments, the increase in contaminant loading in finer grain sizes is generally believed to be a result of the increased surface area with decreasing grain size, providing greater surface area for metal sorption to clay minerals or organic matter. The recognition that contaminant loading is heterogeneously distributed is important when considering the management and pollution abatement of RDS (see section 6.6.1).

The spatial variability of RDS composition has been studied at a range of scales. Studies of RDS have shown that variability exists across the street environment, with different levels of contaminants being present in gutter samples from those in street centres and pavements (Linton

et al. 1980). Studies on city-scale variability have shown that Pb levels are lower in outer city locations compared with inner city sites, indicating the role that traffic plays in the distribution of this contaminant (Duggan & Williams 1977; Massadeh & Snook 2002; Robertson et al. 2003). Similar patterns have been observed for urban soil samples. For example, Madrid et al. (2002) showed that metal concentrations are higher in soils in the old quarter of Seville, rather than its outskirts. This was put down to vehicular-sourced metals. There has been a paucity of systematic spatial analysis of RDS composition over the scale of a city. A spatial survey of Manchester, UK in 2004 (Fig. 6.7; unpublished data) documented a large range in metal levels, with hotspots of Pb, Cu and Zn. These distribution patterns were related to both vehicle density and industry. Another spatial survey (in Birmingham, UK) by Charlesworth et al. (2003a) documented a large range in metal levels, with hotspots of Pb, Cu and Zn (Fig. 6.8). These distribution patterns were related to both vehicle density and industry.

There has been very limited study into the temporal variability of RDS. The limited studies that have been published indicate that there is temporal variability, especially in the input of anthropogenic material. Sodium and chlorine levels in RDS have been shown to fluctuate, with high levels being present in the winter months as a result of road salt application. Of particular importance is the weather, with contaminant levels (especially Pb) being highest following a number of dry days. Massadeh & Snook (2002), however, suggest that lower Pb levels in Manchester RDS in the summer is a result of lower traffic densities during the summer. Long-term data sets are limited to those on Pb that were highlighted in a previous section.

6.3.2 Gully pots and sewer systems

Gully pots and sewers are the key elements of subsurface urban drainage systems and the movement and storage of sediments in these has a marked impact upon both the physical and chemical aspects of urban drainage. The study of sediment build-up and pollutant loading in

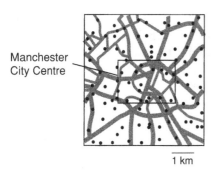

Manchester
City Centre

1 km

Fig. 6.7 Spatial distribution of metals in road-deposited sediment in Manchester, UK (Data courtesy of F. Carraz, unpublished).

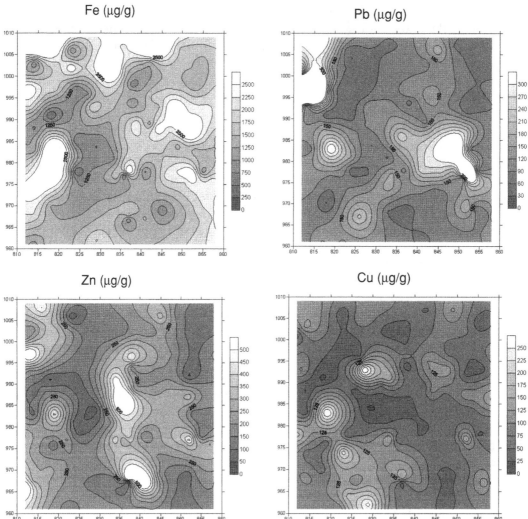

sewerage systems has been predominantly in the field of civil engineering and a large literature is available (for a good overview see Ashley et al. 2000). This chapter will focus only on the key aspects of this topic.

Gully pots are the first entry point of road runoff into the urban drainage network, and are designed to trap some of the sediment carried by the runoff. The design is usually one of a sump or a settling chamber, the entry of which is situated

(a) Low flow conditions

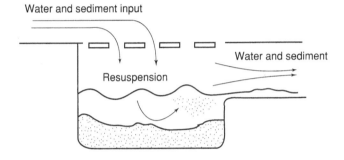

(b) Wet weather conditions (high flow)

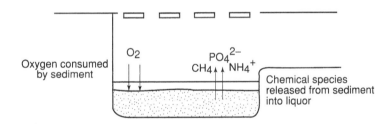

(c) Dry weather conditions (no flow)

Fig. 6.8 Schematic diagram of the sediment processes operating within a gully pot.

at the kerbside. They are a major feature of urban drainage networks, with more than 17 million present in the UK alone (Memon & Butler 2002a). The trapping of this sediment is desirable for two reasons. First, it minimizes the amount of sediment that enters into the sewerage system and, thereby, reduces the problems caused by sediment accumulation in sewers. Second, where gully pots are emptied frequently, it minimizes the amount of sediment that is potentially flushed out of the sewer system into rivers and receiving water bodies. The design and assessment of gully pots has been undertaken primarily in the field of civil engineering, where the term *sludge* is used

to describe the sediment in a pot and *liquor* to refer to the in-place water.

The processes acting within gully pots are complex. During runoff events (wet weather processes), denser particles in the water will settle out under gravity (Fig. 6.8). However, there is usually a high degree of turbulence within the gully pot, which not only limits the amount of sediment that will settle out, but may also lead to the erosion and re-suspension of existing sediment in the pot. As well as the physical processes taking place, major biochemical changes can take place within the gully pot (Fig. 6.8). Most biochemical changes take place during periods between runoff events, and are termed 'dry weather' processes (Fig. 6.8). Many of these are similar to those taking place during diagenesis in receiving water bodies (section 6.2.4), although at a smaller scale. Chemical studies of gully pot liquor have shown that the water reached negative Eh during dry period events, with major impacts upon redox-sensitive chemical species (e.g. Morrison et al. 1988, 1995). The major changes are the consumption of oxygen in liquor, changes in chemical oxygen demand (COD) and ammonium, and the release of contaminants and nutrients adsorbed onto redox-sensitive species into the pot liquor (Wei & Morrison 1994b; Memon & Butler 2002b; Fig. 6.8). Dry-weather processes overall increase the pollutant level of pot liquor, with subsequent negative impacts on runoff into water ways during wet-weather processes. The longer the period of dry weather the greater these changes (Memon & Butler 2002b).

Gully pots are the entry points into the urban drainage system, but sedimentation within the more extensive sewer system itself has marked physical and chemical impacts on urban drainage through the reduction of capacity, leading to sewer flooding, and the build-up of pollutants. Although many sewer systems date from the mid-1800s, there have been few advances in sewer design with respect to sediment movement. Attention upon sewer design is directed towards predicting the movement and accumulation of sediment. Using hydrological models of sediment movement, predictive models for sediment build-up in sewer networks have been developed, with some limited success (e.g. Gerard & Chocat 1998).

6.3.3 Urban rivers

Rivers are natural features of catchments, and are a major component of urban environments, many of which grew up around existing rivers. During the process of urban development rivers were modified to enable navigation, culverted to minimize flooding, and diverted to allow development. These changes led to the physical modification of rivers, to the extent that urban rivers commonly possess unique physical and hydrological properties. In addition to this physical modification, urban rivers became the major vectors of domestic and industrial waste removal. A direct result of this was high pollution levels, through both organic and inorganic contaminants. It is only with increased discharge legislation, and investment in sewer networks and treatment plants that urban rivers are now improving in water quality in developed nations. As highlighted in Chapter 3, however, the majority of the contaminant load in rivers is associated with the sediment fraction. The residence time of sediment in rivers is greater than that of water, and in spite of water quality improvement many urban rivers still possess low sediment quality. This legacy of pollution is one of the largest problems facing urban catchments.

Sediment within urban rivers is itself considered to be a major non-point source pollutant, as fine sediment causes a range of problems in rivers. An increase in fine sediment has impacts upon river turbidity, affecting biological processes and ecology. In particular, the impacts of high suspended sediment concentrations upon fish are marked. Impacts include effects on behaviour and health of fish through gill damage and damage to spawning grounds, through filling in spaces between sand grains, and reducing oxygen delivery to the fish eggs (Watts et al. 2003). In addition, fine suspended material contains the dominant load of metals, nutrients and other contaminants. An increase in coarse sediment has a physical impact causing channel aggradation, which may lead to channel volume reductions and flooding. Urbanization can also decrease the erosion rate of the land surface through its cover of concrete, thereby reducing sediment load in rivers. Furthermore, the changes in the flow properties of the river may lead to the downstream erosion of in-channel sediment due to increased flow events. This has major impacts upon the ecology and biodiversity of urban rivers.

Many of the processes that operate upon sediments, and their interaction with water, in urban rivers are similar to those for other river systems, and the reader is directed to Chapter 3 for full details. The specific study of heavily urbanized rivers, and the physical and chemical characteristics of the sediments therein, has been limited. It is only recently with the undertaking of integrated projects on urban catchments (e.g. the Natural Environment Research Council funded URGENT and LOIS programmes in the UK) that detailed information has become available. Two major differences can be recognized between urban rivers and those from other environments: the spatial and temporal scales of sediment movement, and the level of contamination of this sediment.

Sediment in urban rivers can be considered to be in two major forms. Channel-bed sediment is stored in the river channel and transported only rarely by traction and saltation, thus moving downstream slowly. Suspended sediment is carried downstream in suspension during regular flow, with high suspended sediment transport under higher energy flow conditions. The former acts as a major storage of sediments and contaminants in urban rivers, but the latter is the most important for sediment and contaminant flux through the river, especially on short time-scales.

Studies of urbanized rivers have shown the clear increased levels of contaminants associated with the suspended sediment fraction in urbanized river basins. Walling et al. (2003) showed that metals (Cr, Cu, Pb, Zn) and PCBs increased significantly in urbanized sections of the rivers Aire and Calder (north-east England). For example, Cr in suspended sediment increased from around $100 \ \mu g \ g^{-1}$ in non-urban sections to around $400 \ \mu g \ g^{-1}$ in urbanized sections, and PCBs showed a similar fourfold increase. Owens & Walling (2002) documented a clear increase in sediment-bound phosphorus (a major contributor to river eutrophication) as a result of urbanization in the same rivers. They documented changes in total phosphorus from $< 2000 \ \mu g \ g^{-1}$ in upstream sections to over $7000 \ \mu g \ g^{-1}$ in sections downstream of urbanization. They concluded that this increase represented point inputs of phosphorus from sewage treatment works and combined sewer overflows. The input of these point sources was further supported by a change in the inorganic phosphorus to organic phosphorus ratio from < 2 upstream to > 4 downstream of urban centres. This is significant in that inorganic phosphorus is more bioavailable than organic phosphorus.

Owing to both the impervious nature of urban land surfaces, and the engineered and culverted nature of urban rivers, the flow in such rivers responds rapidly to rainfall events. This results in a rapid rise and fall in the river level, and the river is said to display a flashy response to rainfall. This leads to high rates of fine-sediment transport in suspension during these flow events, often several magnitudes more than during low-flow periods. Suspended sediment concentrations in urban rivers can be very high. Gromaire-Mertz et al. (1999) compiled data showing mean suspended sediment concentration for high-flow events to be in the range $49–498 \ mg \ L^{-1}$. Gromaire et al. (2001) also reported sewer outlet suspended sediment concentrations in the range $152–670 \ mg \ L^{-1}$, illustrating the importance of sewer outfalls in urban river sediments. Sediment yields to rivers in urban catchments, calculated from suspended sediment measurement, are in the range of $93.6–479 \ t \ km^{-2} \ yr^{-1}$ (Goodwin et al. 2003). A study of the Bradford Beck, Yorkshire, UK (Old et al. 2003) documented the extreme levels of suspended sediment transport during a single storm event. Suspended sediment concentrations increased from 14 to $1360 \ mg \ L^{-1}$ over a period of 15 minutes. A peak sediment flux of $47.2 \ kg \ s^{-1}$ was recorded, illustrating the high levels of sediment that are transported by urban rivers during high flow, and that it is these short-lived events that dominate sediment movement (Case Study 6.2 and Case Fig. 6.2).

The role of suspended sediment in contaminant flux is further indicated in Fig. 6.9 for a small urban river. At low flow, suspended sediment concentrations are low, with suspended sediment loads of only $20 \ kg \ h^{-1}$. As a result these low-flow stages contribute only low levels of contaminant flux. At high-flow events suspended sediment loads of over $12,000 \ kg \ h^{-1}$ are observed, with resulting high levels of contaminant flux (e.g. over $3 \ kg \ h^{-1}$ of zinc). High-flow events, therefore, have a significant impact on contaminant input into receiving water bodies.

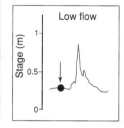
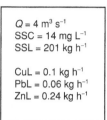

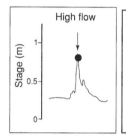
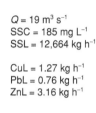

Fig. 6.9 An urban river in low flow and high flow, showing typical stage and suspended sediment relationships (River Medlock, Greater Manchester): Q, discharge; SSC, suspended sediment concentration; SSL, suspended sediment load; CuL, copper load; PbL, lead load; ZnL, zinc load. (Photographs courtesy of J. Coyle.)

6.3.4 Urban canals, docks and lakes

The major sites for sediment accumulation in urban environments are canals, docks and lakes, and these are often the terminal receiving water bodies in urban catchments. Canals and docks contrast with lakes in being artificial and heavily engineered so possess unique hydrological and sedimentological properties. Transport, especially for industry, was a major component in urban areas and as such navigable waterways and docks were built to accommodate this. These are most commonly freshwater in nature, but docks in coastal urban environments may be marine, and in the case of Venice, Italy, canals may also be marine in nature. For the majority of urban canals and docks, water and sediment are input via surface water flow or direct industrial or domestic discharge, and only occasionally are they fed directly by rivers and groundwater. Owing to the low-flow conditions in canals and docks, and the steep-sided nature of these water bodies, they are highly depositional in nature, resulting in the rapid accumulation of sediment. This leads to the requirement for regular dredging to preserve navigable status (see section 6.6.2).

Sediments deposited within urban canals and docks are predominantly derived from anthropogenic sources (section 6.2.1.2), and these anthropogenic sediments have mineralogical and geochemical compositions significantly contrasting to those of natural sediments. To date, however, very little research has been directed towards detailing the mineralogy and geochemistry of these sediments. Two aspects of urban canal and dock sediments are apparent: they commonly have a high organic matter loading as a result of historical sewage input (Boyd et al. 1999; Taylor et al. 2003), and have high metal loadings as a consequence of industrial waste and discharges (Bromhead & Beckwith 1994; Kelderman et al. 2000; Dodd et al. 2003). This high organic and contaminant loading may lead to significant impacts upon overlying water quality as a result of post-depositional chemical alteration (section 6.4; Case Study 6.3 and Case Fig. 6.3).

Very few published data are available on the mineralogical and geochemical association of these metal contaminants (Taylor et al. 2003). Canal and dock sediments undergo regular dredging to maintain water depths, and this material is now classed as controlled waste. The little published information on the contaminant geochemistry of these sediments is based on hazard assessment work with application to disposal or dredged material (e.g. Bromhead & Beckwith 1994). For example, Kelderman et al. (2000) showed that for the canals in Delft, The Netherlands, 95% of inner city canal sediments were classed as highly polluted, whereas only 33% of sediments in the outer city were highly polluted. There is a requirement for more detailed research into the *in situ* processes operating on such sediments, and their role in contaminant cycling in urban aquatic systems.

6.4 POST-DEPOSITIONAL CHANGES IN URBAN SEDIMENTS

As was stated in section 6.2.2, urban sediments undergo physical and chemical changes at many stages of the urban sediment cascade. The two environments in which sediment potentially undergoes major chemical changes (that are most likely to have an impact upon sediment reactivity and contaminant mobility), however, are gully pots and canals, docks and lakes. The former of these has been briefly dis-

cussed in section 6.3.2. This section will focus on the early diagenetic chemical and physical changes taking place in sediments within urban canals and docks.

Early diagenesis is the sum of the processes operating upon a sediment after deposition and includes physical, chemical and biological processes. The early diagenesis of aquatic sediments is dominated by a series of bacterially mediated redox reactions, which result in the oxidation of carbon species (organic matter) and the reduction of an oxidized species (Fig. 6.10). Within sedimentary environments that have an oxygenated water column, which includes virtually all urban water bodies, the primary reaction upon sediment deposition is aerobic oxidation, whereby O_2 dissolved in the water is utilized to oxidize organic matter. As O_2 is primarily sourced from the overlying water column, this oxygen is rapidly used up in the first few millimetres of sediment. The depth of O_2 penetration into the sediment depends on organic matter content, sedimentation rate and biological activity. In highly organic systems, such as sewage-contaminated water bodies and lakes (e.g. Boyd et al. 1999), the sediment oxygen demand through aerobic oxidation may be high enough to result in an anoxic water column, especially under low-flow conditions (Fig. 6.11). Such low-flow conditions are common in steep-sided canals and docks, and the high organic matter contents of the sediments can result in the rapid consumption of oxygen in the water column, leading to serious water quality problems.

Fig. 6.10 Early diagenetic reactions taking place within aquatic urban sediment in docks, canals and lakes. (From Taylor et al. 2003.)

	Reaction	Reactants	Products
Aerobic reaction	Aerobic oxidation	$CH_2O + O_2 \longrightarrow HCO_3^- + H^+$	
Anaerobic reactions	Manganese reduction	$CH_2O + 2MnO_2 + H_2O \longrightarrow 2Mn^{2+} + HCO_3^- + 3OH^-$	
	Iron reduction	$CH_2O + 2Fe_2O_3 + 3H_2O \longrightarrow 4Fe^{2+} + HCO_3^- + 7OH^-$	
	Sulphate reduction	$2CH_2O + SO_4^{2-} \longrightarrow HS^- + 2HCO_3^- + H^+$	
	Methanogenesis	$2CH_2O + H_2O \longrightarrow CH_4 + HCO_3^- + H^+$	

(a) Consumption of chemical species

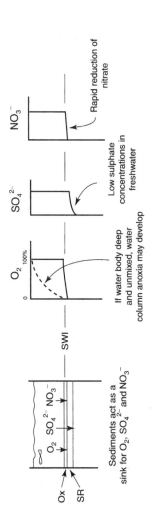

(b) Production of chemical species

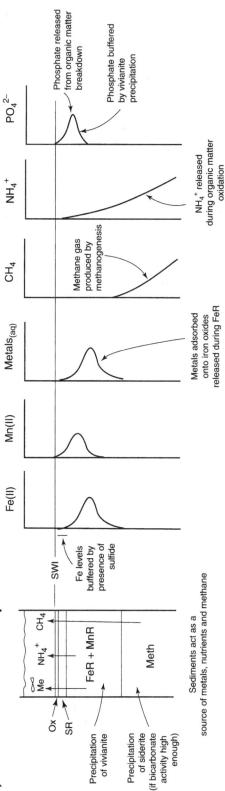

Fig. 6.11 Idealized summary of the results of early diagenetic reactions for sediment and porewaters in freshwater urban water bodies (docks, canals and lakes). Typical values and depths derived from data published in Taylor et al. (2003) and Dodd et al. (2003). Porewater profiles are idealized and will vary with variations in sediment composition and accumulation rate: Ox, aerobic oxidation; SR, sulphate reduction; FeR, iron reduction; MnR, manganese reduction; Meth, methanogensis; SWI, sediment–water interface.

Upon the consumption of O_2, a series of anerobic bacterial reactions are favoured, utlizing oxygen in species such as NO_3^{2-}, Fe_2O_3, MnO_2, SO_4^{2-} (Figs 6.10 & 6.11). Classically these were seen as taking place through a sequential set of reactions, giving rise to a series of diagenetic zones (Froelich et al. 1979; Coleman 1985). Such observations were based on thermodynamic considerations and observations from hemipelagic recent sediments. Such discrete zones are now recognized to be the case for slow sedimentation rate, low productivity environments. In urban aquatic environments the organic-rich nature of the sediment results in many of these reactions taking place simultaneously. These anaerobic early diagenetic reactions are many and complex, and a complete description of them is beyond the scope of this work. The major reactions are nitrate reduction, Mn(IV) reduction, Fe(III) reduction, sulphate reduction and methanogenesis (Fig. 6.11). All of these reactions break down organic matter and, therefore, lead to an overall decrease in organic matter content as sediments are buried. They also tend to result in the decrease in reactivity of organic matter with depth, having implications for dredging remediation.

These early diagenetic reactions have an impact upon the short- and long-term fate of contaminants in sediments through two principal mechanisms: release of contaminants into sediment porewater; and the uptake of contaminants into authigenic mineral precipitates. Within contaminated sediments, metal contaminants commonly co-precipitate with iron and manganese oxides. The chemical reduction of these oxides (FeR and MnR) results in the release of these adsorbed contaminants to sediment porewaters (Dodd et al. 2003; Taylor et al. 2003). These contaminants then become free to be moved into the overlying water column, through the process of molecular diffusion (Fig. 6.11). This process of contaminant movement from sediments into overlying water is commonly termed a *benthic flux*. This benthic flux has been recognized to be the most significant non-point source of pollution to water bodies. This release of chemical species into porewater during early diagenesis is not restricted to contaminants, it can also be a major pathway for nutrient release from sediments. One example is that of ammonium, which is released in the process of organic matter oxidation. In sewage-contaminated urban water bodies, large amounts of ammonium can be released into sediment porewaters, and then oxidized to nitrate in the water column. Gases may also be generated from sediment during early diagenesis. In freshwater organic-rich sediments, methane gas (CH_4) is released from sediments through the reaction of methanogenesis (Fig. 6.11). Methane is a flammable and noxious gas, and so can have major negative impacts upon water quality in urban canals and docks, both aesthetically and chemically.

The build up of chemical species in sediment porewater also leads to the precipitation of authigenic minerals in the sediment. Within marine and brackish sediments the predominant mineral formed in this way is pyrite (FeS_2). Pyrite has been observed in canal sediments (Large et al. 2002; Taylor et al. 2003) but the absence of sulphate in freshwater leads to this being a rare mineral in urban sediments. The limited studies of the diagenesis of urban sediments have shown the iron phosphate mineral vivianite ($Fe_3(PO_4)_2$) to be the most common mineral (Fig. 6.12; Dodd et al. 2003, Taylor et al. 2003). The importance of these minerals for contaminant mobility is that metals can be taken up by these minerals as they precipitate, thereby locking up contaminants in the sediment (Large et al. 2002; Taylor et al. 2003).

6.5 TEMPORAL CHANGES IN URBAN SEDIMENTS: NATURAL AND ANTHROPOGENIC IMPACTS

The recent nature of the data on urban environments is a limiting factor on the identification of temporal changes in response to internal and external factors. Indeed, in most urban environments there is an urgent need for the collection of baseline data to enable future changes to be determined. As is evident from other chapters in this book there has been extensive use of sediment records to gain insights into historical changes in sediment input, land use, climate and

Fig. 6.12 Minerals precipitated within canal sediment after deposition (backscattered electron (images). (a–c) Vivianite ($Fe_3(PO_4)_2.8H_2O$) taking the forms of radiating needles and laths. (d) Authigenic framboidal pyrite (FeS_2; p). Scale bars are shown for each micrograph.

pollution. Such sediments act as ideal archives for environmental change (e.g. salt marshes, floodplains). Many archives from fluvial and estuarine sediments have shown the increasing impact of urbanization on rivers and estuaries (e.g. Cundy et al. 1997; Walling et al. 2003). Many of these pollution changes are mixed in with inputs from industrial and mining activities, however, and thereby these studies do not give good, clear information on past trends of urban sediment quality and quantity. Although sediments do accumulate in urban environments, the engineered and disturbed nature of the urban environment has meant that continuous, undisturbed sediment records are uncommon. Many receiving water bodies have undergone dredging operations for shipping or remediation, or are too shallow to allow the accumulation of fine-grained sediments. There are, however, a limited number of studies that have provided information on short- to long-term changes in urban

sedimentary processes. Short-term records have come from road-deposited sediment monitoring programmes, whereas longer term records have been provided by sediment profiles in urban lakes and canals.

Although no long-term continuous monitoring data sets exist for RDS composition, individual studies over two decades on specific cities can be combined to provide useful data on temporal changes. One such example is Manchester, UK, where Pb levels have been documented to have fallen (see section 6.3.1). A more recent study (Robertson et al. 2003) has shown a further reduction to an average of 265 ppm in 2000. Similar results where found by Charlesworth et al. (2003a) for the city of Birmingham, UK, where Pb levels in RDS were documented to have dropped by one-third from 1987 to 2002. Recent analyses have also shown an increase in platinum group elements in RDS since the early 1990s (e.g. Wei & Morrison 1994a; Motelica-Heino et al. 2001;

Whiteley & Murray 2003). Future monitoring programmes on RDS composition are likely to further reinforce these temporal trends.

Urban lakes are a promising source of urban sedimentation records. Charlesworth & Foster (1999) showed that good records of urban sediment supply and composition were preserved in two small urban lakes in Coventry, UK. Sedimentation rates in these lakes had changed in response to very localized catchment processes, but both clearly showed an increase in metal accumulation after the mid-1950s. It was also observed that Pb inputs had decreased over time owing to the decrease in use of leaded petrol.

Sediments from urban canals and docks have also been shown to preserve a good record of localized changes in pollution and remediation. Generally, urban canals are too shallow (< 2 m),

and too frequently disturbed to allow a continuous sediment record to build up. In larger canals and docks, however, where water depths may exceed 5 m, in the absence of dredging such records can be preserved. One such example is from the Salford Quays in the UK where Taylor et al. (2003) documented the clear preservation of pre- and post-remediation sediments and associated contaminant levels (Case Study 6.3). In more general terms many studies have shown that surface sediments in urban canals contain lower levels of contaminants than deeper sediments, indicating a decrease in contaminant input into urban sediment in recent times (e.g. Kelderman et al. 2000; Bellucci et al. 2002; Taylor et al. 2003). This is generally put down to the environmental legislation reducing discharge and the cleaning up of the urban drainage and sewerage system.

Case study 6.3 Sedimentation in an urban water body: Salford Quays, UK

Docks and canals commonly form the terminal receiving water bodies for urban surface runoff, and as such act as significant sites of sediment accumulation and storage. These sediments can contain significant levels of past and present contamination and so any physical or chemical perturbations of the sediment can lead to the release of these contaminants back into the urban environment. The Manchester Ship Canal (MSC) was built in 1895 to allow for direct shipping access to the City of Manchester Docks at the eastern end of the Ship Canal (Case Fig. 6.3a). At its eastern end the MSC begins at the confluence of three rivers (Irwell, Medlock and Irk)

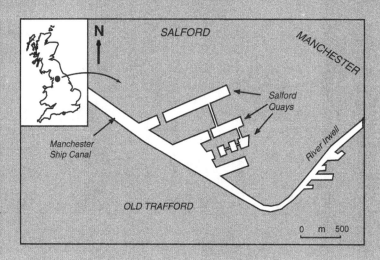

Case Fig. 6.3(a) Map showing the location of the Salford Quays, Greater Manchester, UK (from Taylor et al. 2003).

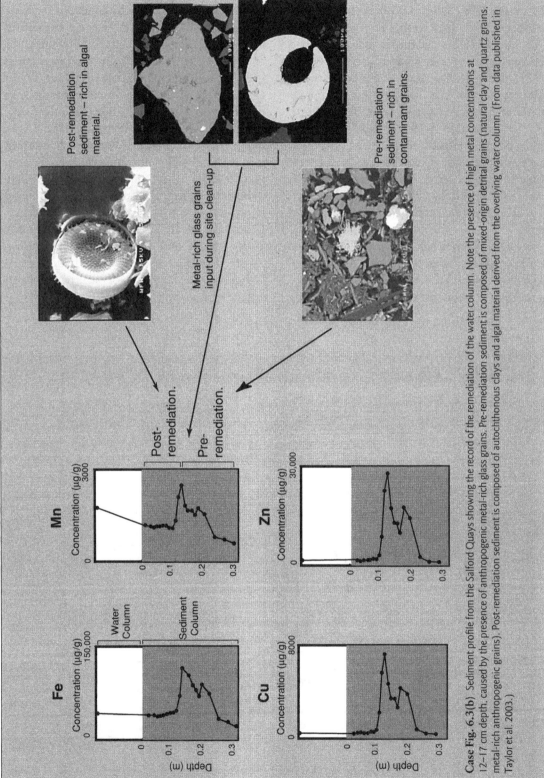

Case Fig. 6.3(b) Sediment profile from the Salford Quays showing the record of the remediation of the water column. Note the presence of high metal concentrations at 12–17 cm depth, caused by the presence of anthropogenic metal-rich glass grains. Pre-remediation sediment is composed of mixed-origin detrital grains (natural clay and quartz grains, metal-rich anthropogenic grains). Post-remediation sediment is composed of autochthonous clays and algal material derived from the overlying water column. (From data published in Taylor et al. 2003.)

that form the major drainage conduit for the Greater Manchester conurbation, and in the past combined sewer overflows have supplied domestic and industrial sewage, and road runoff to the canal. The Manchester Ship Canal is up to 8 m deep, steep sided and up to 50 m wide. Sediment is carried into the canal in shallow, fast moving rivers. As this river water meets the deep, slow flowing canal, sediment is deposited at the bottom of the canal, as flow strengths are not high enough to keep it in suspension. Owing to the urban nature of the catchment this sediment is organic-rich and highly polluted by sewage and metals. These contaminated sediments have caused a range of water quality problems (White et al. 1993). Aerobic bacteria oxidize organic matter using free oxygen in the water. This process, therefore, uses up oxygen from the water column, giving rise to what is termed anoxia – oxygen-depletion in the water column. Methanogenic bacteria also break down organic matter in anoxic conditions, and release methane gas (CH_4). This methane, a flammable and noxious gas, bubbles up to the water column, lifting up mats of sediment and sewage to the water surface (a process termed *sediment rafting*).

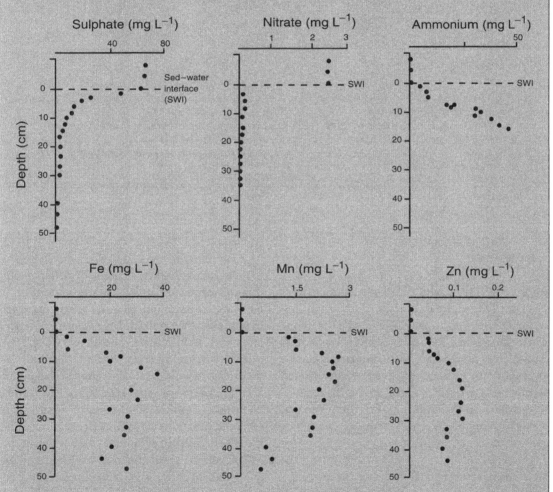

Case Fig. 6.3(c) Porewater profiles for the Salford Quays sediments showing the post-depositional consumption of chemical species, and the release of metals into sediment porewater. Note the release of Fe and Mn, probably via a combination of bacterial iron and manganese reduction and glass dissolution. (Modified from Taylor et al. 2003.)

As a consequence of this the water quality of the Salford Quays was improved through increased mixing of air into the water column (Boult & Rebbeck 1999).

The nature of sedimentation in the Quays has markedly changed in response to this water column clean-up (Case Fig. 6.3b). Pre-remediation sediments are composed of a range of natural detrital grains, predominantly quartz and clay, and anthropogenic detrital material dominated by industrial furnace-derived metal-rich slag grains. Post-remediation sediments are composed of predominantly autochthonous material, including siliceous algal remains and clays. At the top of the pre-remediation sediments and immediately beneath the post-remediation sediments is a layer significantly enriched in furnace-derived slag grains, input into the basin as a result of site clearance prior to water-column remediation. These grains contain a high level of metals, resulting in a significantly enhanced metal concentration in the sediments at this depth (Case Fig. 6.3b). Significant porewater peaks in Fe, Mn and Zn in the sediment (Case Fig. 6.3c) are most probably the result of dissolution of these furnace-derived grains in the sediments, possibly mediated by iron and manganese oxide-reducing bacteria. These species have subsequently diffused into porewater above and below the metal enriched layer, indicating the important control of diagenesis on the long-term fate of contaminants in urban sediments.

Relevant reading

Boult, S. & Rebbeck, J. (1999) The effects of eight years aeration and isolation from polluting discharges on sewage- and metal-contaminated sediments. *Hydrological Processes* 13, 531–47.

Taylor, K.G., Boyd, N.A. & Boult, S. (2003) Sediments, porewaters and diagenesis in an urban water body, Salford, UK: impacts of remediation. *Hydrological Processes* 17, 2049–61.

White, K., Bellinger, E.G., Saul, M., et al. (Eds) (1993) *Urban Waterside Regeneration: Problems and Prospects.* Ellis Horwood, Chichester.

6.6 MANAGEMENT AND REMEDIATION OF URBAN SEDIMENTS

As discussed throughout this chapter the impacts of urban sediment particulates on the environment are wide-ranging and include: impacts upon water quality of runoff and receiving water bodies, reduction of capacity in drainage systems, an increase in atmospheric particulate concentrations and a reduction in depth of navigable waterways. All this means that the active management of sediments is a key requirement in the sustainable urban environment.

6.6.1 Management of road-deposited sediment

As has been described previously, RDS can have a major impact upon waterways, both through the volume of sediment transported from these environments, and the potential for high levels of contaminants associated with the sediment. It is, therefore, a key element of urban pollution management that levels and composition of RDS are both monitored and actively managed to minimize their impact. Management practices attempt to address these issues in two ways. First, target levels for contaminant concentrations can be set, above which sediment removal is required. Second, routine physical removal of RDS can be undertaken. Although there have been studies on contaminant levels in RDS (see section 6.3.1) there have been no systematic studies on acceptable levels of contamination in RDS, and no guidelines or criteria exist for impacts of contaminants on urban runoff and water courses. Therefore, the first of these approaches is not currently utilized as a management strategy (Case Study 6.4).

Case study 6.4 Study of road-deposited sediment management: Tampa, Florida, USA

The management of road-deposited sediments (RDS) is an important component of urban pollution control. The most effective method of control is the removal of RDS through street sweeping, and disposal. There have been very few systematic studies of the effectiveness of RDS sweeping on improving urban runoff water quality. An exception is a study undertaken for the city of Tampa in Florida, reported in Brinkmann & Tobin (2003). Generally, street-sweeping frequency and policy for individual cities is not regulated by scientific information, and this study set out to address this.

Tampa is a mid-sized city, containing sectors of industry, commerce and medium-density residential housing. To assess the compositional variability of RDS within the city, street-sweeping samples were taken from residential, commercial and industrial areas of the city and analysed. The RDS from industrial areas exhibited the highest levels of zinc, copper and barium, and was linked to industrial activity (Case Table 6.4). Levels of strontium, nickel, chromium and vanadium were highest in commercial areas (Case Table 6.4) and this was linked to the increased vehicular activity in these areas. Overall, RDS element levels were mostly considered to be non-harmful, but concern over levels of copper, lead and zinc were raised. The fine-grained fraction of the RDS was greater in industrial and commercial areas than in residential areas, and this was linked to increased combustion of fossil fuels and industrial emissions. Organic matter was higher in residential areas, due to increased vegetation, and available phosphorus was also higher in these areas, probably due to increased garden fertilizer use.

Compositional analysis showed that the RDS sweepings removed from the city were composed predominantly of inert sands, cement and organic matter. Combined with a high nutrient content it was concluded that in many cases these street sweepings could be recycled as topsoil and in soil amendment. For those RDS containing high metal levels however, such a use was not appropriate. If the sweepings were treated by removing the fine fraction ($< 63\ \mu m$), however, then metal levels would be reduced enough to allow recycling.

A further aspect of this study was to look at the impacts of street-sweeping removal of RDS on sediment and water contamination levels. Sectors of the city were swept either every two weeks, weekly, twice weekly, or not swept at all. Three conclusions were reached. First, street sweeping was most effective at reducing RDS accumulation when carried out on a weekly basis. Lower levels of sediment and contamination were found on frequently swept surfaces. Second, street sweeping does result in a marked reduction in contaminant levels in runoff, with the most significant improvements seen with sweeping on a twice-weekly basis. Finally, maintaining a weekly street sweeping schedule is much more effective at reducing sediment and water pollution levels than other factors, such as weather conditions. The conclusion that can be drawn

Case Table 6.4 Compositional variation in road-deposited sediment in the city of Tampa. (From Brinkmann & Tobin 2003.)

Zone	Copper (ppm)	Zinc (ppm)	Lead (ppm)	Mass per cent $< 63\ \mu m$	Organic matter (%)	Available phosphorus
Industrial	125	96	65	27.3	5.57	353
Commercial	23	79	61	24.0	2.18	346
Residential	24	59	72	19.0	6.62	520

from this study is that a regular street-sweeping programme, which frequently removed RDS, should be an important component of a city's pollution management.

Relevant reading

Brinkmann, R. & Tobin, G.A. (2003) *Urban Sediment Removal: The Science, Policy, and Management of Street Sweeping.* Kluwer, Dordrecht, 168 pp.

The routine physical removal of RDS is carried out in many urban environments, both for aesthetic reasons and to limit the impact of RDS on watercourses and sewer systems. Removal is accomplished by mechanical street sweeping. Early street-sweeping procedures were carried out predominantly for aesthetic reasons, and removal efficiencies were low (Sartor et al. 1974). Subsequent studies have shown that regular removal of RDS by street sweeping may lead to a significant reduction in both sediment contamination levels and contamination of surface runoff (e.g. Sartor & Gaboury 1984; also see Case Study 6.4). It can, therefore, be a highly effective method of urban pollution management. It has been shown that street sweeping is most effective at removing pollutants in climates where long periods of dry weather lead to pollutant accumulation (Sartor & Gaboury 1984). Street sweepers can be based on vacuum systems or on a rotary brush system, and comparisons have been made on the effectiveness of each type for RDS removal. Generally, although the rotary type removes a greater proportion of RDS from street surfaces, vacuum-based models are better at removing the finer grain fractions (Brinkmann & Tobin 2003). This is an important consideration as it has been shown that the fine fractions contain the highest contaminant loading (see section 6.3.1). There is, therefore, a trade-off between an increased volume of sediment removal or more efficient contaminant removal. In general, street sweeping is much less efficient at removing the finer-grained fractions of RDS than the coarser-grained fractions. This has implications for pollution management as the finer-grained

fraction generally contains the highest loading of contaminants. In a study in Hawaii, Sutherland (2003) showed that street sweepers removed only 62% of Pb from RDS, primarily as a consequence of low efficiencies of fine-grain sediment removal (Table 6.3). The resulting waste produced from street-sweeping can be reused as ground cover or disposed of on land or to landfill, but this waste has not been widely assessed for its suitability for such. The limited studies that have been undertaken (e.g. Viklander 1998; Clark et al. 2000; German & Svensson 2002) have concluded that the sweepings material, owing to high contamination levels, should be treated prior to its reuse or disposal on land.

6.6.2 Management of sediment in gully pots

The management of gully pots, and their associated sediment, forms an important part of urban water quality management. Most authorities regularly empty and clean gully pots, thereby removing the sediment from the urban drainage system. This is commonly carried out to minimize flooding and drainage issues, rather than for pollution management reasons. Memon & Butler (2002a) modelled the efficiency of gully pots in urban drainage networks and showed that gully pots can reduce the suspended sediment content of water entering sewer systems (and ultimately receiving water bodies) by 40%, with even larger reductions being possible with improved gully pot design. The model also showed, however, that reduction in pollutants (such as ammonium and chemical oxygen demand) was minimal.

Table 6.3 Removal of lead load in road-deposited sediments by street sweepers in Palolo Valley, Oahu, Hawaii. (From Sutherland 2003.)

Grain size fraction (μm)	Mean Pb loading (%)	Mean street sweeper efficiency (%)	Pb load removal (%)
2000–1000	2.7 ± 1.6	83.7	2.2
1000–500	10.8 ± 6.2	81.7	8.8
500–250	13.4 ± 6.2	79.3	10.6
250–125	12.8 ± 6.3	75.0	9.6
125–63	9.7 ± 2.4	66.7	6.5
< 63	50.6 ± 14.9	48.7	24.7
Overall			62.4

6.6.3 Stormwater management ponds

Stormwater ponds are designed and engineered to remove pollutants from stormwater runoff, primarily through the settling out of sediments from the water column. As the majority of the pollutant loading is associated with the sediment, this leads to an improvement in the quality of the stormwater runoff, which can then be discharged to natural water bodies. Pollutant removal efficiencies of up to 90% have been reported for such ponds (e.g. Wu et al. 1996). Removal of the nutrient phosphorus has also been documented, but the release of nitrogen in the form of ammonia from the sediments to the water column may also take place (Hvitved-Jacobsen et al. 1984). The build-up of sediments in these ponds leads to a reduction in volume and, therefore, efficiency. Therefore, sediment needs to be removed periodically (see section 6.6.4), but these sediments are commonly contaminated. For example, Marsalek & Marsalek (1997) determined for a stormwater pond in Ontario, Canada, that sediment metal levels were such that the sediment could not be reused or placed in residential landfill without treatment.

6.6.4 Sediment dredging

Dredging of sediment is a management technique used on urban aquatic sediments for two purposes: to maintain draft in navigable canals and docks, and to remove contaminated sediment from waterways as part of pollution management. Commonly the two become interrelated, however, as once sediment is dredged for re-moval, sediment quality guidelines come into operation to determine its suitability for disposal. With the exception of inland urban docks, most sediment dredging from urban watercourses is undertaken to remove contaminated sediments, and to limit their impact upon water quality (for details of dredging issues at coastal ports see Chapter 1). Dredged material is either disposed of through land application (e.g. Chen et al. 2002) or, if contaminated, is disposed of to landfill.

An example of sediment dredging of urban canals for remediation is that of Birmingham in the UK (Bromhead & Beckwith 1994). Birmingham canals were built after 1770 and the banks of the canals were heavily industrialized, by for example metal manufacturing, chemical and engineering works. Inputs and waste discharge from these activities led to accumulation of highly contaminated sediments, with sediments having mean concentrations of 1.0% Cu, 0.7% Zn, 0.3% Cr and 0.15% Pb (Bromhead & Beckwith 1994). Dredging was undertaken to a depth of 1.5 m below water level and resulted in 24,000 m^3 of sediment removal. Once removed from the watercourses, such dredgings are generally treated as waste, and have to be disposed of within strict guidelines.

6.7 FUTURE ISSUES

The major issue of concern in urban sedimentary environments is that of increased urbanization and industrialization. In the developed world the input of contaminants into urban environments will at least stabilize, if not decrease, in response to ever stricter emission controls and environmental guidelines. As such, urban sediments are likely to become less important vectors for contaminant transfer through the sediment cascade. The volume of sediment is likely to remain high, however, unless controls and guidelines for sediment quantity are introduced. In contrast, in the developing world the ever increasing rate of urbanization, coupled with a lower level of environmental control, will mean that sediments in urban environments will continue to have a major impact on contaminant cycling and surface water quality.

In response to this, there needs to be a much more sustainable approach to urban development, and the consideration of sediments will need to play a key role in this. Sustainable drainage systems (SuDS) are increasingly being seen as the best way to manage surface water quality and quantity (Charlesworth et al. 2003b). These systems are designed to slow down the rate of surface runoff, through the use of permeable land surfaces (the 'porous' city). Currently, these focus principally on the hydrological aspects of urban environments, but potential exists for integration of sediments into SuDS. For example, the use of sediments as buffers for pollution (e.g. in artificial wetlands) is currently being pioneered in SuDS.

Climate change will also have a likely impact upon the hydrology, and hence sediment transport, of urban environments. Although numerous studies on the impacts of climate change on natural systems have been undertaken, little consideration has been given to engineered environments. Changes in climate may alter rainfall and snowmelt patterns, however, thereby having an impact on urban drainage. Semadeni-Davies (2004) modelled the impact of possible future climate change on a cold region and city and showed that frequency and volume of wastewater flows in an urban environment would be altered, with implications for drainage system design and management.

7

Deltaic and estuarine environments

Peter French

7.1 INTRODUCTION

The focus of this chapter is those coastal environments that are associated with the mouths of rivers: namely coastal deltas and estuaries. These dynamic sedimentary systems occur at the interface between terrestrial and marine systems and, consequently, their sediment supply, morphology and functioning are heavily influenced by both land-based processes, such as river flow and floodplain development, and marine processes, such as tides and waves. Also significant is the fact that with the exception of lakes, these environments receive all land drainage, inherent in which is consideration of water volumes, sediment loads and contaminants. This chapter examines these environments within the context of how they function as sedimentary systems, how they are influenced by changes in natural processes, and how they are impacted by anthropogenic activity. Useful additional literature relating to estuarine ecology (Adam 1990; McLusky 1989), the functioning and physical aspects of estuaries (Dyer 1997) and estuarine management (French 1997) are recommended. In terms of deltas, further details relating to the processes of delta formation can be found in many coastal geomorphology textbooks (e.g. Woodroffe 2003), and discussion of processes, because of the similarity between estuaries and delta channels, falls within the literature cited above. Tropical mangrove-colonized estuaries and deltas are referred to in Chapter 9.

7.1.1 Definition of estuaries and deltas

It is useful to discuss what is meant by the terms 'estuaries' and 'deltas', and what the difference is between them. Both environments represent areas of sediment accumulation at the coast, and both are linked to the mouths of river systems. The chief difference, however, is that estuaries are features that are formed of marine and terrestrial sediments within river mouths in response to their flooding by a rise in sea-level. Deltas, on the other hand, develop seawards of the coastlines where large volumes of fluvial sediment are carried seawards, at a rate that exceeds the sea's ability to erode it. Accepting this major distinction, however, there are many similarities, not least of which is that the processes of tides, waves and freshwater input that operate in estuaries also occur in the distributary channels of deltas. Therefore, a delta can be regarded as a coastal landform composed of a series of outlets to the sea that are, in effect, estuaries.

The term *estuary* originated from the Latin word 'aestus' meaning tide (Woodroffe 2003). The most commonly adopted definition was first used by Cameron & Pritchard (1963), when they referred to an estuary as 'a semi-enclosed coastal body of water which has a free connection to the open sea, and within which sea water is measurably diluted with fresh water derived from land drainage.' Although being a useful general overview of what an estuary is, this definition is generally regarded as an oversimplification as there is no reference to the tidal processes that are fundamental in shaping estuaries. E.C.F. Bird (2000) amends this and defines an estuary as: 'the seawards part of a drowned valley system, subject to tidal fluctuations and the meeting and mixing of fresh water and salt water from the sea, and receiving sediment from its catchment and from marine sources.' This is more useful in that it allows greater understanding of the processes that link to form the range of estuary types (see section 7.1.2).

The term *delta* was first used by Herodotus in 450 BC to describe the wedge-shaped landform, resembling the Greek letter delta (Δ), seen at the mouth of the River Nile. A definition based purely on morphology, however, is not useful when considering the range of delta shapes that can occur. Barrell (1912) and subsequently Bates (1953) altered this definition to include some understanding of riverine processes, before Wright (1982) combined the salient points of the two and defined deltas as 'subaerial and subaqueous accumulations of river-derived sediment deposited at the coast when a stream decelerates by entering and interacting with a larger receiving body of water.' This highlights the distinctiveness of deltas over estuaries, in that the importance of high sediment volumes and the accretion of sediment out onto the coast are emphasized. In terms of both shape and process, deltas, like estuaries, can be highly variable, dependent on volumes of river water and sediment, and the strength of marine currents (see section 7.1.2).

7.1.2 Nature and significance of deltas and estuaries

Deltas and estuaries, because of their location at the interface of rivers and seas, can be regarded as major sinks and stores of sediment, particularly terrestrial. In addition, the ability of waves and tides to rework and shape these sediments into recognizable landforms makes these environments very diverse in terms of morphology and ecology. The sediment that forms deltas and estuaries is also extremely variable, both in terms of mineralogy (reflecting sediment source) and grain size (clays to coarse grits). Considering that clays and silts are small sedimentary particles, their deposition in these environments of high tidal energy, waves and rapidly flowing river water, is a considerable achievement. A list of the depositional environments where clays are found would largely involve situations where sedimentation is possible from a still body of water (lakes and quiet backwaters). Estuaries and deltas do not readily fit into this category. Here we have a unique and significant characteristic of these environments. Estuaries and deltas are areas where fresh and salt water meet. Fresh water typically has a salinity of < 0.5 NaCl, whereas sea water is typically around 35. This means that when the two mix to form brackish water, the salinity will fall somewhere between these figures, depending on the relative proportions of salt and fresh water. The importance of this is that the presence of salt enables clay particles, through electrostatic attraction, to stick together to form large sediment grains. This is discussed in section 7.2.2, and is important because in addition to clay particle adhesion, pollutants can also be adsorbed, making estuarine sediments potentially rich stores for pollutants (see 7.2.3).

Sediment deposition occurs within estuaries and deltas as a result of the complexities of sediment delivery, current activity and marine reworking. Over time, sediment builds up and the surface becomes covered by tides for shorter periods of time, i.e. the flood tide will not flood the sediment surface until later in the tidal cycle, and will leave it sooner during the ebb tide. This means that ultimately, the sediment surface can start to be colonized by vegetation. This is fundamentally important in the future survival of the developing marsh because: first, root systems bind the sediment and help it resist erosion; and second, leaves baffle the sediment-laden waves and encourage more rapid sediment deposition, thus further facilitating vertical marsh growth.

Adam (1990) indicates that the first colonization by salt marsh plants generally starts to occur when inundation frequency falls below c. 500 times a year. However, the actual value varies between estuaries. In the Blackwater estuary, for example, colonization does not start until inundation frequency falls below 380 times a year (Burd 1995). This initial vegetation is salt tolerant (halophytic), because brackish water will still flood over it once or twice a day. As sediments build up further and tidal cover is gradually reduced, less salt-tolerant vegetation starts to colonize. Mid-marsh vegetation will colonize when inundation frequencies fall below *c.* 230 times a year, and mature marsh below 100 (Adam 1990). Ultimately, in the upper reaches of the higher marsh, where tidal inundation is rare, freshwater vegetation will appear.

A further significant aspect of deltas and estuaries is a more negative one. The formation of extensive areas of vegetation has, historically, seen this land being regarded as prime for development. This development has taken many forms. Initially large areas of salt marshes were drained and converted to farmland; such areas in the UK include the Fenland around The Wash (47,000 ha), and bordering major estuaries, such as the Severn (*c.* 8000 ha). Both of these areas have an almost continuous history of land claim from Romano-British times. In other examples, *c.* 2000 ha of marshes have been claimed since the nineteenth century in the Ribble estuary; and in the Dee estuary, around 6000 ha have been claimed since the eighteenth century (Davidson et al. 1991). More recently, with reduced demand for agricultural land, marshes around major estuaries and deltas have become prime sites on which to develop ports and marinas. Similarly, industrial growth has regarded such land as cheap and highly desirable. The Thaw estuary in South Wales, for example, was completely claimed in 1850 for construction of a power station, large areas of the Orwell estuary in Suffolk were lost in the nineteenth and twentieth centuries to construct Felixstow docks, and successive piecemeal land claim since the mid-sixteenth century for naval and commercial port facilities has significantly reduced the salt marsh resource of Portsmouth harbour (Davidson et al. 1991) (see also section 7.5, Fig. 7.14).

Land claim does not only serve to reduce the areas of vegetation in estuaries and deltas, but also has an impact on morphology and functioning. Fundamentally, land claim constrains an estuary, in that it reduces space for water to occupy. The volume that the flood tide occupies is known as the tidal prism. Clearly, as more marsh is claimed, the space available for the flood-tide water to inundate is reduced, and so the tidal prism becomes constrained into a smaller space. The amount of water entering the estuary remains the same, however, with the result that the only way of accommodating it is for the height of the water surface to increase. This has some potentially serious implications. First, it increases the risk of defence overtopping; and second, it means that the remaining marsh will be covered more frequently, for longer periods, and to greater depths, potentially leading to vegetation loss. This situation is, in fact, analogous to that which occurs during sea-level rise (see section 7.3). Significantly, however, some examples of historic land claim are now proving to be short sighted. When claimed, salt marsh soils dewater and contract, falling in elevation relative to any marsh remaining outside the line of defence. Hence, anything built on this land is soon below high water mark. In a world of increasing sea-level rise, this is now a major problem.

From the above, it can be seen that estuaries are areas of intense human impact and influence, yet they are also areas where natural processes can be particularly dynamic. This diverse range of interests and demands placed on estuaries and deltas represents a key dichotomy. On the one hand, these sedimentary environments are seen as highly dynamic and ecologically important, yet on the other, they are pollutant and sediment sinks and subject to a range of human pressures, such as land claim and port development. As a result, many of the large deltas and estuaries of the world have had to be protected through designation of major conservation status.

7.1.3 The classification of deltas and estuaries

Variations in grain size, sediment supply, freshwater discharge, tidal range and wave activity will lead to considerable variation between estuaries and deltas. In estuaries, differences between the amount of sea water entering during each tidal cycle are critical in classification. At the simplest level, estuaries may be classified purely on the basis of tidal range (the amount by which the water level rises between low and high tide). Estuaries with a tidal range less than 2 m are known as microtidal, those with a range

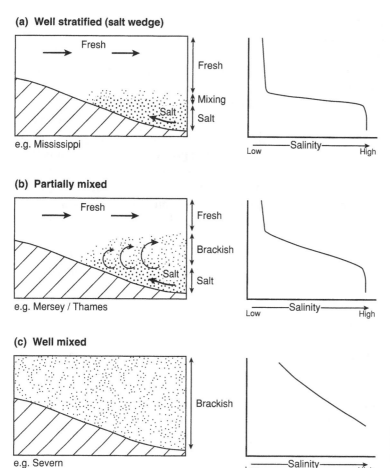

Fig. 7.1 The classification of estuaries based on the mixing of fresh- and saltwater bodies. (Modified from various sources.)

of between 2 and 4 m, mesotidal, and those above 4 m, macrotidal (Pethick 1984). Given that many estuaries have tidal ranges in excess of 4 m, the earlier classification of Davies (1964) classes macrotidal estuaries as being those with a tidal range between 4 and 6 m, whereas ranges in excess of this are known as hypertidal. Although both systems are in use, it is important when referring to a macrotidal system to make reference to which system is being used. A more complicated, but more useful system classifies estuaries on the basis of the amount of mixing that occurs between salt water and fresh water. Salt water is denser than fresh water, and so in a low-energy estuary there is a tendency for the fresh water to flow over the salt water, the two bodies of water maintaining their integrity, with the saltwater component forming a 'wedge' below

the fresh (Fig. 7.1a). This is a salt wedge, or well-stratified estuary. A typical profile will show that with increasing depth, the water will stay generally fresh, then rapidly increase in salinity. Bearing in mind the importance of salt and fresh-water mixing for the deposition of clays, the only place where this will occur is where the salt wedge meets the fresh water. Clearly, as the tide comes in and goes out, this point will move up and down the estuary, but regardless of where it is, it is marked by an area of higher turbidity caused by flocculated clays settling out (see section 7.2.2). This is called the turbidity maximum.

As tidal range and energy increase, the increased turbulence causes greater mixing of the fresh- and saltwater bodies. In a partially mixed estuary, there is still an identifiable salt- and freshwater layer but the boundary between

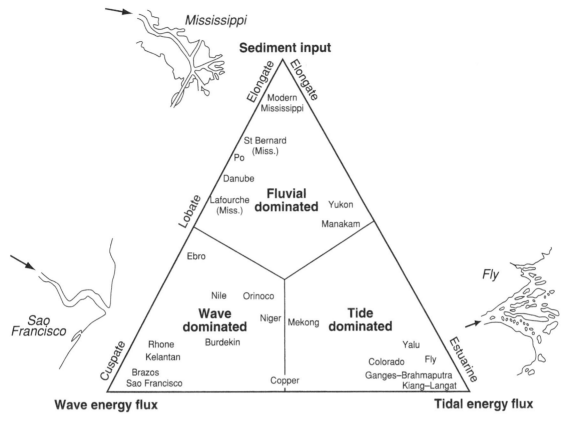

Fig. 7.2 The morphological classification of deltas based on the influence of river, tidal and wave activity. (Compiled and modified from various sources.)

them is more diffuse, and the salt wedge is less pronounced (Fig. 7.1b). With increasing depth, water will remain fresh, then show a more gradual salinity rise until it becomes truly marine. The final type is the well-mixed estuary produced by large volumes of sea water entering estuaries with significant river discharge. Because of the high energy conditions, water mixes completely and is generally brackish throughout its depth, although there will be some slight variation in actual salt concentrations with distance up the estuary (Fig. 7.1c).

The classification of deltas focuses on the degree of reworking of sediment by the sea. As previously outlined, deltas form when large volumes of river sediment enter the sea at a rate that exceeds the sea's ability to rework it. This does not, however, preclude that the sea (tides and waves) will rework the sediment to some extent. Hence, classification of deltas is generally done on the basis of the dominance of rivers, waves or tides (Fig. 7.2) (see also Carter (1988), Haslett (2000) or Woodroffe (2003) for examples). However, to complicate this, Wright & Coleman (1971, 1972) have suggested that there is a continuum of delta shapes related to the varying importance of tides, waves, and river power and that the tendency of some classifications to assign a delta to one 'type' is problematic because it fails to recognize this continuum. Hence, traditional terms such as 'lobate' (deltas with prominent sediment lobes), 'cuspate' (deltas with concave seaward margins caused by wave shaping) and 'birdsfoot' (deltas where the bifurcating river has extended seawards over the delta) are all well entrenched in the literature but tend to suggest that deltas are of one type or the other. Over time, however, these terms have not been used consistently, and have also been misinterpreted because of difficulties with field identification.

For example, lobate deltas may have distinct cusp-ate traits facing the prevailing wave fronts, yet not elsewhere (see Woodroffe (2003, pp. 324–5) for a discussion). The appearance of the term 'compound deltas' may have helped in this respect, but has tended to become a bucket term, where anything not conforming to the easily identifiable is placed. Further complicating the issue is the fact that many deltas change over time as they mature and as processes such as river input, continental shelf topography, or marine processes change. For example, the history of the Mississippi delta shows such a pattern (Coleman 1988) (see also section 7.3, Fig. 7.11). Initially, sediment lobes form (lobate trait) and undergo enlargement by seaward progradation. This lobate stage is followed by the development of a system of distributary channels (birdsfoot trait), which over time will switch and develop into a complex birdsfoot pattern. The final stage is for this active delta front to be abandoned and for a new lobate front to form elsewhere in the delta system. In the Mississippi, there are six major lobate to birdsfoot cycles, with a typical periodicity in the order of 1500 years (Coleman 1988).

Despite this complication, as with estuaries, different forms of delta classification may suit different needs. A shape-based approach, as in the Wright and Coleman model, remains convenient for morphological classification needs (Fig. 7.2). If studying how a delta functions in respect to sedimentation and process, however, other forms of categorization may be better suited. In this respect, the system suggested by Bates (1953) is useful as it is linked to processes, and in particular density differences, between the inflowing water and the relatively still receiving body. Where the inflowing river water is denser than the receiving water, the system is referred to as hyperpycnal, where they are the same, homopycnal, and where the river water is less dense, hypopycnal.

7.2 SEDIMENT SOURCES AND SEDIMENT PROCESSES IN DELTAS AND ESTUARIES

The source of delta sediments is predominantly terrestrial, whereas estuaries contain a combina-

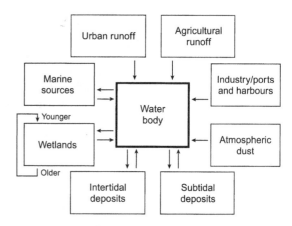

Fig. 7.3 The range of potential inputs to estuaries and deltas (water body). Note that some of these inputs are one way, whereas others represent a bi-directional exchange.

tion of terrestrial and marine sediments. From other chapters, it is evident that terrestrial- and marine-derived sediment may originate from different areas: for example, from upland areas via rivers (Chapters 2 & 3), from urban runoff (Chapter 6), or from the coast (Chapters 8 & 9) and offshore (Chapter 10) (Fig. 7.3).

7.2.1 Sources of deltaic and estuarine sediments

Figure 7.3 highlights the various potential inputs to estuaries and deltas. Actually quantifying these parameters is notoriously difficult owing to problems of measurement and accessing the more remote parts of the system, such as the incoming tide. In total, however, these components comprise the sediment budget. Basically, this is a summation exercise in which all of the inputs are totalled and offset against the losses. The resulting value, if positive, will indicate an estuary or delta with a net sediment gain, whereas if negative, it will indicate one with a net sediment loss.

In terms of source, the various inputs identified in Fig. 7.3 can be grouped as exogenic and endogenic, depending on whether they originate outside the system (from the catchment or sea) or within (through reworking of existing deposits or generated within the estuary through biogenic activity). Regardless of source, however, there will be considerable temporal variation in the relative contribution of each to the sediment budget. The reworking of material from within an estuary or delta (endogenic) is more likely during the more stormy winter months, or, in the longer term, during episodes of sea-level rise. In terms of biogenic material, both plants and animals can make a contribution. Plants are perhaps the greatest component of this, adding dead plant material to the marsh surface and representing an important element of marsh vertical growth (see section 7.6.1), although inputs are seasonally variable. Animal material is more variable. Although, in principal, skeletal material is potentially a large contributor (Frey & Basan 1978), it is argued that in reality actual amounts present in a sediment sequence are typically small because the durability of shell material is low, particularly when acidic conditions develop in the marsh sediment sequence (Wiedermann 1972). Faecal material is also an important contributor to organic sediment (Frey & Basan 1978).

Microscopic organisms such as diatoms and algae fill two roles. First, they contribute *post mortem* to the volume of organic remains in the sediment; and second, they play a key role in sediment stability. Diatoms and algae secrete carbohydrate-rich exopolymers, which serve to stick fine sediment grains together, thus increasing their resistance to erosion. Dyer (1998) demonstrates the importance of this process in the general salt marsh context, and Underwood & Smith (1998) in the Humber (UK), Kornmann & de Deckere (1998) in the Dollard (The Netherlands), and Riethmüller et al. (1998) in the Wadden Sea (The Netherlands–Danish Coasts); all demonstrate the importance in particular environments.

Input from outside the system (exogenic) is more likely to occur when soils are bare of vegetation or during periods of catchment land-use change. For example, increased sediment supply as a result of forest clearance in the catchment of the Mahakan delta, Kalimantan caused rapid delta growth (E.C.F. Bird 2000), and mining operations led to accelerated sediment supply and subsequent delta growth in the George River delta in Tasmania (J.F. Bird 2000),

the Pahang delta in Malaysia (E.C.F. Bird 2000) and the now disappeared Fal delta (UK) (Bird 1998). Conversely, reduced sediment supply can lead to reduced marsh or delta growth. The erosion of the Nile delta caused by the building of the Aswan High Dam is perhaps the classic example of this (see Case Study 7.2). Other examples where reduced river flow has caused a reduction in sediment supply include the Rhône delta, France, the Dneiper delta, Ukraine and the Barron delta, Australia (E.C.F. Bird 2000). Although the majority of the examples of reduced sediment supply are linked to dams and other water-control measures, other causes also exist. River dredging can remove large quantities of sediment from the local sediment budget, the cessation of mining activities can remove artificially high rates of sediment supply, and sediment increases following land disturbance can effectively 'run out'. As an example of the latter, the Argentina River delta, Italy, started to erode following reduced sediment supply after having initially experienced accelerated accretion as a result of increased soil-derived sediment inputs from land clearance. The subsequent erosion of this soil from the hinterland and the laying bare of the landscape to bedrock led to a cessation in sediment supply (Bird & Fabbri 1993).

In conclusion, the key factor for ensured stability is that whatever sediment the estuary or delta loses, whether via erosion to the sea or via decreased inputs, it is balanced by newly deposited sediment, i.e. a positive sediment budget. In the case of deltas, the continued growth of the delta, and seaward extension of the delta front, is dependent on the continued delivery of sediment from the hinterland. Similarly, the continued vertical accretion of intertidal flats and salt marshes is dependent on the accumulation of sediment on its surface to replace that removed by the tide and to compensate for rising sea-levels.

7.2.2 Controls on sediment accumulation and transport in deltas and estuaries

The sediment inputs described above need to be linked with mechanisms by which they can be

Plan view

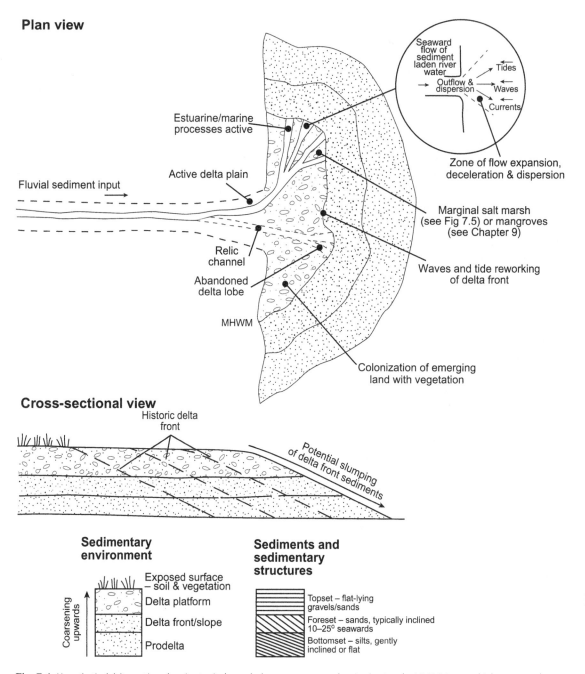

Fig. 7.4 Hypothetical delta section showing typical morphology, processes and grain size trends: MHWM, mean high-water mark.

stored and retained within the sedimentary system. A sediment grain will settle once its fall velocity exceeds the ability of the water to keep it in suspension. Hence, it is easier for heavier grains to settle than for lighter ones. Deltas, for example, often develop a characteristic stratigraphy with coarser sediments deposited close in shore, grading to finer with increasing distance from the coast (Fig. 7.4), such as is shown in the Mississippi delta (Scruton 1960). This land to sea sediment gradation is also marked by structural properties. The coarser delta platform tends to be typified by horizontal sediment layers (topsets), the finer sands of the delta slope by seaward dipping (10° to 25°) layers (foresets), and the finest sediments of the prodelta by horizontal to gently inclined layers (bottomsets) (see Fig. 7.4).

Although it is relatively simple to understand how a sand or gravel particle can settle, the settling of clay particles is more complex. Estuarine sediments typically comprise large volumes of intertidal or subtidal muds (mud flats). Clearly, the finer the grain size, the longer a sediment particle will take to settle. Stoke's Law, a method of estimating such settling times, states that for fine particles less than 100 μm (silts and clays) the settling velocity of a particle is proportional to the square of the grain diameter. For coarser particles of over 2 mm diameter (coarse sands and above), the settling velocity is proportional to the square root of the diameter. The implications of this are that fine clay particles in estuaries will settle very slowly. Pethick (1984) demonstrates that according to Stoke's Law, a 2 μm clay particle will take 56 days to settle through 0.5 m of still water. This is clearly not possible in a natural environment because water is only really still for very short periods. Theoretically, therefore, there is no way a clay particle could ever settle in estuarine or deltaic situations. To explain this apparent discrepancy a mechanism is required to enable fine clay particles to settle through significant depths of moving water.

The answer relates to the fact that estuaries are areas of brackish water. Clay particles behave rather differently from sand or silt grains in that they possess a surface attraction, which is amplified by the presence of only a few parts per thousand of salt in the water. Hence, when clays suspended in fresh water enter an estuary and mix with saline water, their attraction to each other increases. As clay particles adhere to one another they form agglomerations of particles called flocs. The name for this process is flocculation (Krank 1973, 1975). As more clay particles stick together and the flocs increase in size, their effective settling weight increases and so the flocs can settle much faster than the individual grains.

It is this process that explains why estuaries are such active sinks for muddy sediments. Importantly, this ability of clays to attach to other things in the water body is not only restricted to other clay particles. Clays may also attract to a range of contaminants, thus making muddy deposits potentially rich in a range of environmental contaminants (see section 7.4.2).

Although flocculation appears a good explanation for the occurrence of mudflats in estuaries, many researchers have identified a major discrepancy in that there appears to be too much mud deposited if flocculation was the only mechanism. McCave (1970) details studies from the German Bight that help explain this. He identified that the bottom of the water column is characterized by a viscous layer that moves just above the mudflat surface. This layer contains not just recently flocculated material that has sunk into it from above, but also material brought in by the tide, and reworked from elsewhere in the estuary. From this viscous sublayer, quasi-continuous sedimentation occurs throughout the tidal cycle (Fig. 7.5a). Thus, deposition of clays need not be restricted to the head of the salt wedge or to periods of slack water.

Flocculation and quasi-continuous sedimentation, therefore, make the formation of extensive deposits of mud within a relatively high-energy environment possible. There are, however, other controls over sediment deposition and retention. Even though clay flocculates and forms larger agglomerations, these will fall through the water column only when current velocities are low enough. On a tidal cycle, water moves at different rates depending on the stage of the tide (Fig. 7.5b). When the tide is fully out and on the turn, water velocity is at its lowest (theoretically, this has to be zero for a period of time for the water to stop moving out, and start moving back in). As the tide floods, it picks up speed before starting to slow again towards high tide. Similarly on the ebb tide, speeds increase as the tide ebbs, before slowing towards low tide. Current velocities, therefore, will be at their lowest at high and low water, and at their highest at some point between (note: owing to distortion of the tidal wave and the production of tidal

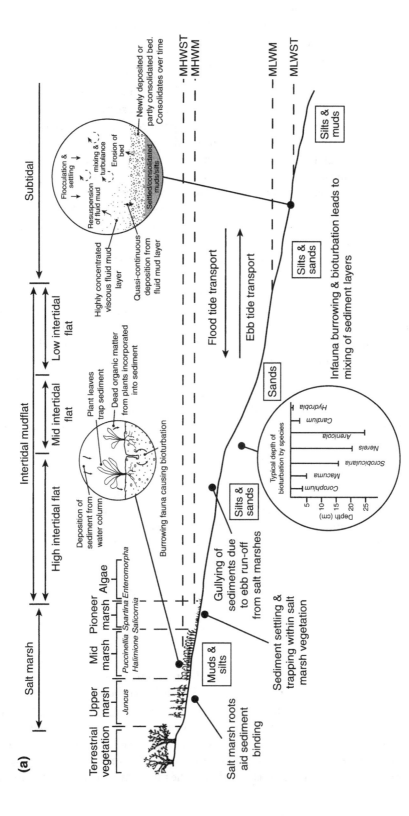

Fig. 7.5 Processes affecting the sediment deposition in estuarine and delta channels: (a) Cross-section of a hypothetical channel showing the main processes affecting sediment deposition and stability: MHWM, mean high-water mark; MHWST, mean high-water spring tide; MLWM, mean low-water mark; MLWST, mean low-water spring tide. (b) Diagrammatic representation of current velocity variation over a symmetrical tidal cycle (note: asymmetrical tides will skew this graph one way or another). (c) Resulting intertidal sediment deposits. (Parts b & c modified from French 1997, p. 45.)

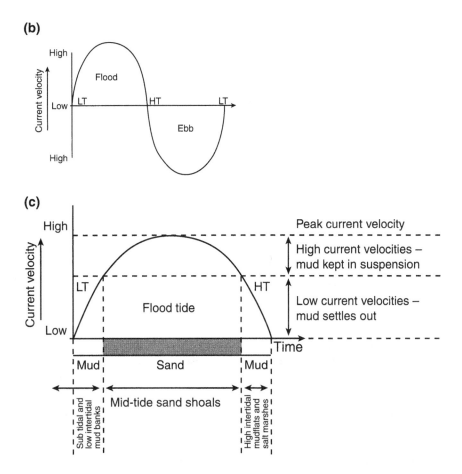

Fig. 7.5 (Continued)

asymmetry, this point of maximum velocity need not be at the half way mark). The patterns of sedimentation will reflect this (Fig. 7.5c). Where the current velocities are lowest, clays can settle out, but when they flow faster, clays are kept in suspension and only the coarser sediments can settle. Bearing in mind that this variation in current velocity is paralleled by a rising and falling of the tide, these patterns of deposition are shown spatially by areas of sediment accretion, such as subtidal muds (below low water), mid-tide sand flats where velocities keep clays in suspension, and high-tide mud flats where clays can again settle out.

A further aid to increasing sediment transfer from the water body to the sediment surface is the presence of vegetation (Fig. 7.5a). Sediment surfaces tend to be relatively smooth in respect of the ability to cause friction with the overlying water body. Hence, the energy lost by waves moving over a mud flat is relatively small. In contrast, vegetation is rough and provides much more of a barrier. Waves running over a vegetated surface will, therefore, use considerably more of their energy, which means that they will lose speed and drop more of their sediment load. This ability for vegetated surfaces (see also mangroves, Chapter 9) to dampen wave energy is important in estuaries and deltas because it not only facilitates vertical sediment accumulation, but also reduces the wave energy reaching the landward limit of the marsh (often a sea wall). Work on the North Norfolk (UK) marshes (Moeller et al. 1996; Shi et al. 2000) has shown that after crossing 180 m of a vegetated marsh surface, waves typically lose up to 80% of their energy. Not only does this represent an aide to coastal defence in the area, but it also means that

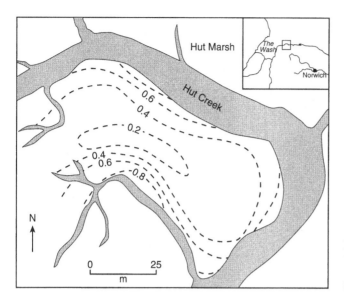

Fig. 7.6 Rates of sediment accretion (in mm yr⁻¹) in relation to creek proximity. Data based on Hut Marsh, North Norfolk (UK). (Based on data from Stoddart et al. 1989.)

the loss of such energy is coupled by a reduced ability to carry sediment, leading to increased sediment deposition.

Attenuation of wave energy due to baffling by vegetation will lead to the increased deposition of sediment directly onto the marsh surface. However, this need not be uniform across the marsh surface. Given that one of the prime mechanisms that increases deposition is the loss of wave energy, the areas of greatest deposition will occur in the parts of the marsh that create the greatest dampening effect to waves. Studies in North Norfolk (UK) have shown that the rate of sedimentation decreases with distance from both the seaward edge and creeks. At Hut Marsh on Scolt Head Island, North Norfolk, Stoddart et al. (1989) showed that accretion rates near major creeks reached 8 mm yr⁻¹, falling to 2 mm yr⁻¹ away from the creeks (Fig. 7.6). This is because as the water starts to flow onto the marsh (whether from the seaward edge or having overflowed onto the marsh surface following entry via a creek system), the greatest energy loss will occur over a relatively short period of time (see also Allen 2000). Other studies have also quantified this change in sedimentation rate. Richard (1978) has shown that for a marsh in Long Island, New York, vertical accretion rates are typically in the order of 9 to 37 mm yr⁻¹

along the seaward edge of the marsh, falling to between 2 and 4 mm yr⁻¹ at the landward edge. These figures, particularly for the low, pioneer marsh, show a wide discrepancy, and this relates back to the earlier point that the greatest rates of sedimentation will occur in those parts of the marsh that create the greatest energy loss. Even though the seaward edge of the marsh may be collectively referred to as low or pioneer marsh, there will still be considerable variation in vegetation height, density and exposure to waves and thus the loss of energy and corresponding increases in sedimentation rates will vary spatially. Similar depositional patterns and variation are shown for marshes along the New Brunswick coast, Bay of Fundy (Chmura et al. 2001), and along the Normandy coast of France, where mean vertical accretion rates fall from 5.5 to 4.1 mm yr⁻¹ with distance from the marsh edge (Haslett et al. 2003).

In deltas, the pattern of sedimentation along the seaward margins is controlled by the processes acting as the river discharges into the relatively still body of water (see Fig. 7.4 inset). The classification of deltas by Bates (1953) mentioned in section 7.1.3 is useful in this context. Where deltas form as rivers enter a freshwater lake, water densities are the same (homopycnal). Here, mud deposits are rare because there is no increase in salt to drive flocculation (an exception here would be a salt lake). In more typical situations, deltas are likely to involve the discharge of fresh water into saline (hyper or hypopycnal), and so flocculation of fine sediment occurs. Salt wedge conditions (see Fig. 7.1) would be analogous to a hypopycnal delta (e.g. Mississippi) and the sediment-laden river water will be carried further seawards across the top of the wedge (cf. Fig. 7.1a), being deposited further from the delta front. Tides and waves may then rework this sediment back onto the delta, or along the coast. If the delta is hyperpycnal (e.g. Yellow River delta), the incoming sediment load will sink at the front of the delta, to be redistributed by tides along its seaward edge. A detailed discussion of these relationships can be found in Woodroffe (2003).

The distribution of sediment type and the depositional morphology of deltas and estuaries are, therefore, controlled by a combination of river flow, vegetation coverage, tides and waves. It has been assumed so far that tides are simple events that involve the rising and lowering of sea-level. Although this is essentially the case, further complications vary the importance of tides as morphological agents. Tides can facilitate the reworking of sediment onto a delta front, or can control the amount of marine sediment entering or leaving an estuary. Hence, the incoming tide brings sediment inshore, and the outgoing tide takes it offshore. If the ability to move sediment is the same on both incoming and outgoing tides, the net impact will be no change in the net sediment budget. Frequently, however, this is not the case. Due to variation in, and shallowing of, local sea bed topography, the incoming tidal wave can become distorted by interaction with the long shore profile, outrushing river flow, and estuarine/delta channel shape, to produce asymmetrical tides. The implications of this are that the time taken by the flood- and ebb-tide can vary. In a flood-tide dominated setting, the flood-tide takes a lot less time to fill the estuary than the ebb-tide takes to empty it. The implications here are important because although the time taken to fill the estuary is less than to empty it, the volume of

water involved is the same. Thus, water has to move a lot faster on a shorter flood-tide. The result of this scenario is that a shorter, higher energy flood-tide will bring a lot of sediment into the estuary, and the slower, lower energy ebb-tide will allow a lot of it to settle and be retained in the estuary. The converse of this, i.e. a longer flood-tide and shorter ebb, can also occur, thus resuspending much of the sediment brought in by the flood tide and moving it out of the estuary. The former situation will lead to the net gain of sediment in the estuary (positive sediment budget), whereas the latter will result in a net loss (negative sediment budget). Dyer (1997) indicates that the degree of asymmetry is determined by the relationship between an estuary's volumetric and tidal characteristics. In general, he shows that ebb dominance tends to occur in estuaries that are microtidal and hyposynchronous, whereas those that are macrotidal and hypersynchronous tend to favour flood dominance. In this context, synchronicity relates to the convergence and frictional resistance of an estuary. Hypersynchronicity occurs when convergence exceeds friction, and hyposynchronicity when friction exceeds convergence.

7.2.3 Sources of anthropogenic inputs into deltaic and estuarine sediments

Anthropogenic inputs refer to any substance delivered to deltas and estuaries that is not of natural derivation. This includes contamination and ultimately pollution (see Chapter 1 for definitions) in the widest sense, i.e. waste products from industrial and urban areas, increased sediment input as a result of dredging or mining operations, and agricultural products. Figure 7.3 shows the range of sediment sources that can enter estuaries and deltas. All of these sediment sources, however, can also be linked with contaminants. What is immediately noticeable is that the range of potential contaminants is very wide, reflecting the fact that estuaries and deltas are sourced from environments that receive the majority of land-based drainage as well as tide-derived marine inputs.

Fig. 7.7 Deposition of coal dust in the intertidal zone of the Severn estuary (Ogmore Beach, South Wales). This material represents the reworking of contaminants from historic mining activity linked to the South Wales coalfield (French 1990).

Contaminant problems in estuaries and deltas can, therefore, be particularly acute because they act as key sediment sinks (see section 7.2.2). In a similar way, any contaminant that enters the system in solid (particulate) form will behave as a sediment, and become deposited in the same way. Such contaminants may derive from dredging, where subtidal sediment is resuspended; from agricultural activity, where exposed soils may get washed into streams and eventually estuaries; from industry, where tailings may enter through discharge pipelines; from ports, where the loading and offloading of ships leads to spillage; and from mining operations, where erosion of spoil tips leads to inputs of waste materials.

Particulate contaminants may pose aesthetic problems (Fig. 7.7), but of greater significance from a water and environmental quality aspect are the contaminants that enter an estuary or delta as part of the water body (e.g. metals, nutrients). Such contaminants may pass directly out of the system in solution, or they may become adsorbed onto clay particles or organic debris (Fig. 7.8). Similarly, organic contamination such as oil or sewage will break down using oxygen in the water body, thus increasing the biological oxygen demand. The longer a contaminant is present within the estuary, the greater the chance of it being broken down or adsorbed onto a clay particle and incorporated within the sediments. In terms of the latter, this means that as clay flocs settle out to form mudflats, they can carry with them significant amounts of contaminants adsorbed to their surface. The chemistry of this process can be very complex (see Stum (1992) or Andrews et al. (1996) for further details), but in essence metals (positive charge) are transferred from the water body to sediment (negative charge) (Fig. 7.8), and thus contaminants become associated with particle surfaces. Other modes of contaminant storage

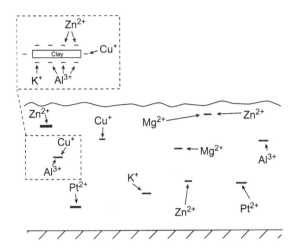

Fig. 7.8 Hypothetical representation of the adsorption of metals to clay particles. The positively charged metals are attracted to the negatively charged clays. As the clays flocculate, so the metals are laid down with the sediments.

involve chemical combination of metals with sulphides and sulphates, or within mineral lattices (see Chapter 1).

The main issue with metal contaminants is that they cannot be readily observed, and therefore attract little attention. The fact remains though, that many muddy estuaries and deltas contain significant quantities of metals stored within their

sediments. Figure 7.9 shows a typical example from a salt marsh in the Severn estuary. What is noticeable here is that the concentration of individual metals varies with depth. This trend is a historical one, with the levels of metals deposited in each layer of sediment reflecting that in the estuary at the time of deposition. Hence, sediment deposited before 1850 contains relatively little contamination, whereas that deposited in the 1950s has significantly higher levels. Such a trend is typical of many industrial estuaries and can be divided into three zones. The lower zone (I) represents background contaminant levels, when the estuary or delta sediments were being deposited in times of no industrialization, and reflects levels from natural erosion. The middle zone (II) marks a period of rapidly increasing contaminant levels, and represents a time of rapid industrialization and declining environmental quality. In the UK, the start of this zone can be linked to the onset of the industrial revolution in the mid-nineteenth century. The uppermost zone (III) shows declining levels of contamination, and links to the cleaning up of the environment, increased environmental awareness and legislation, and a general decline in heavy industries.

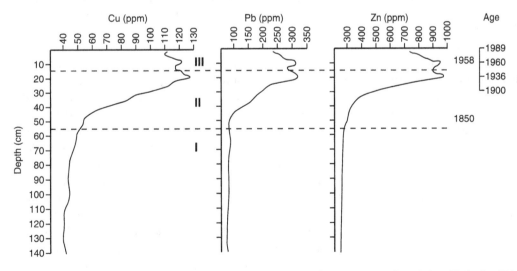

Fig. 7.9 Metal depth profiles from a salt marsh deposit in the Severn estuary, south-west UK. Note the variation with depth, which reflects the changing level of contaminants within the estuary over time. I, II and III represent 'chemozones'. These are distinct zones that can be identified in industrialized estuarine sediments. Chemozone I represents static, background levels of the uncontaminated, pre-industrialized estuary; II represents increasing contaminant levels during industrial growth; III represents declining contaminant levels representing industrial decline and increased emission legislation (see text). (Modified from French 1996.)

When still resident in the water body, or even after deposition, these contaminants can have a significant ecological impact, and thus be regarded as pollutants (see Chapter 1 for definitions). For example, toxicity may cause fatalities and endanger parts of the food chain or, if at sub-lethal levels, may have an impact on the ability to reproduce. Such issues may occur whether a pollutant is in solution within the water body, or within the sediment where it remains bio-available. Furthermore, some species can toler-ate pollutants and store them within their body tissues, a process known as bioaccumulation or biomagnification (Clark 1992).

Although the environmental levels of contam-inants/pollutants have declined since the mid-twentieth century (see Fig. 7.9), developments in chemical engineering have resulted in the environ-mental presence of a new suite of compounds known as organohalogens. These are generally used as pesticides or in electrical equipment but, importantly, are unable to be broken down and, unlike metals, can be toxic at very low concen-trations. The organohalogens include such groups of compounds as the PCBs (polychlorinated biphenyls), HCB (hexachlorobenzine), HCH (hexachlorocyclohexane, lindane), and DDT (dichloro-diphenyl-trichloroethane) (McLusky 1989). These compounds show the classic trends of bioaccumulation and biomagnification, with the greatest impact on the higher predators. More significant for estuarine systems is that much of the world's waste and surplus applica-tion of these substances has drained into them and, because of the accumulatory nature and storage potential associated with clays, have become stored.

Another significant pollutant in deltas and estuaries, particularly in terms of its visibility, is oil. By their very nature, estuaries and deltas lend themselves to the establishment of ports, harbours and oil refineries. Indeed, some of the larger deltas (e.g. Mississippi) also contain oil fields of their own. Oil spills are a very emotive issue and whereas the media may not appear too concerned to discuss metal pollution in salt marsh sediments, they will readily report on a large oil tanker spill. Although an oil tanker accident may appear more newsworthy, a factory releasing small quantities of oil on a daily basis is likely to be more significant in the long term than one major release. Nelson-Smith (1972) cites one example. A typical refinery effluent may contain small traces of oil (c. 10–20 ppm) that are not readily detectable with the naked eye. If, however, this is linked to a discharge of 455,000 L min^{-1}, then on a daily basis this equates to 6825 L of oil. Ironically, the major accident which receives the greatest publicity is somewhat of a rarity. Farmer (1997) reports data from the International Tanker Owners Pollution Federation (ITOPF) which shows that only 12% of marine (i.e. not neces-sarily delta or estuarine) oil is derived from tanker accidents. The greatest source (37%) is from industrial sources and urban runoff, such as that reported by Nelson-Smith (1972), and directly affects estuaries and deltas.

Despite the continued inputs of oil, perhaps the most remarkable aspect of estuarine and deltaic environments is their resilience. Although the oiled parts of plants may die, they gener-ally grow again once the oil has broken down. DeLaune et al. (1994) studied the impacts of oil on salt marshes by artificially oiling areas of *Spartina* marsh and trying different methods of cleaning, notably leaving the oiled marsh to be cleaned by flushing with sea water, applying a dispersant, and the cutting and removal of the oiled growth. They found that after 95 days of monitoring, there was no major difference in the plots and hence concluded that the best course of action is to leave marshes to recover and regrow naturally. Such conclusions have also been supported by Gilfillan et al. (1995), who showed that 15 years after the 1978 Amoco Cadiz spill on the coast of Brittany, France, the areas of marsh where recovery was most suc-cessful were those that had been left alone. Teal et al. (1992) also showed that 20 years after a large spill in Buzzard's Bay, Massachusetts, marsh growth was as good as areas unaffected by the spill. However, in both cases, regeneration took up to 15–20 years to achieve, suggesting that regrowth may initially be at a slower rate than pre-spill.

Table 7.1 Types of change and the associated impacts in deltas and estuaries.

Nature of change	Impact
Reduced sediment supply	Inability of vegetation surfaces to keep pace with sea-level rise. Loss of sediment supply to mudflats. Loss of supply to delta front, increasing net marine erosion
Increased sediment supply	Infilling of channels, levee breaching in deltas. Burial and mortality of infauna and vegetation
Changes in tidal range	Emergence or submergence of vegetated surfaces. Variation in water table. Changes to status/position of salt wedge. Changes in mixing. Increased saline penetration upstream
Changes in storminess	Increased erosion and delta/marsh recession (see Case Study 7.1). Increased defence overtopping
Increasing sea-level	Coastal squeeze, marsh loss, increased water table. Landward movement of salt wedge. Increased tidal penetration up-river
Decreasing sea-level	Delta/salt marsh advance. Falling water table. Seaward movement of salt wedge. Down river advance of freshwater habitats
Land claim	Increased tidal constriction producing net increased sea-level (see above). Loss of flood areas. Coastal squeeze
Dredging	Increased tidal prism with potential net drop in sea-level (see above). Modified currents and tidal flood/ebb patterns
Increased wave activity	Cutting back of delta front or marsh edge. Increased sediment input

7.3 PROCESSES AND IMPACTS OF NATURAL CHANGE IN DELTAS AND ESTUARIES

7.3.1 Erosion of intertidal sediment substrates

Deltas and estuaries are natural systems and, therefore, undergo change as the processes controlling them change (Table 7.1). Such changes can be simple, such as increased wave activity, which may erode more sediment from the delta front or marsh edge, or changes in the volume of sediment brought down stream to supply the delta. More subtly, changes in storm frequency may alter the erosion and accretion patterns of a salt marsh, or increasing sea-level may result in changes to water table conditions, and thus the stability of stored pollutants, or the stability of marsh flora. The increase in wave activity caused through increased storminess will result in higher energy conditions at the delta front or marsh edge. This could lead to an increase in net erosion and the loss of delta or marsh sediments over a period of many years. In the Severn estuary (UK), for example, the salt marshes show a series of accretion and erosion events that can be correlated with episodes of increased storminess (see Case Study 7.1).

Case study 7.1 Accretion and erosion cyclicity in the Severn estuary, UK

The Severn estuary is a large, macrotidal estuary system in south-west England. Its maximum spring tidal range is 14.8 m and extensive freshwater drainage from its catchment means that it is a well-mixed and extremely high-energy estuary. The intertidal morphology is complex in that it reflects a series of individual salt marsh units overlying or banked up against other marshes. The visual effect is that in many locations marsh surfaces descend, step-like towards the present channel. Allen & Rae (1987) first studied these marshes and categorized them stratigraphically. The oldest marsh unit, termed the Wentlooge Formation (Case Fig. 7.1a(i)), comprises a complex series of clays and silts, interspersed with peat layers of varying thickness.

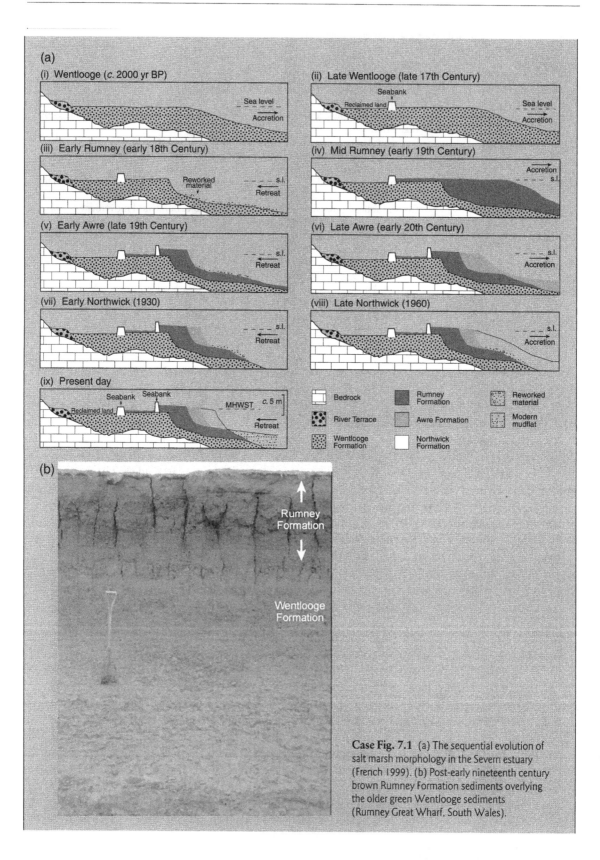

Case Fig. 7.1 (a) The sequential evolution of salt marsh morphology in the Severn estuary (French 1999). (b) Post-early nineteenth century brown Rumney Formation sediments overlying the older green Wentlooge sediments (Rumney Great Wharf, South Wales).

Case Fig. 7.1 (c) Stair-like descent of marsh surfaces towards the present channel. This image shows the vegetated surfaces of (from right to left) the Rumney, Awre and Northwick sediments (marsh edges arrowed) (Littleton Warth, Severn estuary).

These sediments started accreting around 5000 years ago and represent a complex series of different marsh units overlying each other, although the detail of this is difficult to determine at the present time owing to limited exposure and almost uniform sediment type between different marsh units (Allen & Rae 1987).

Much clearer, however, is the more recent morphology. The Wentlooge Formation continued accreting until the end of the seventeenth century (Case Fig. 7.1a(ii)). By the early eighteenth century, these marshes were eroding (Case Fig. 7.1a(iii)), remobilizing large volumes of sediment. At this time, the quantities of pollutants reintroduced through this reworking were minimal because this period occurred before the industrial revolution and, hence, the polluting of the estuary. Erosion continued through much of the eighteenth century but by the early nineteenth century rapid sediment accretion had resumed, leading to the deposition of the Rumney Formation (Allen & Rae 1987) (Case Fig. 7.1a(iv)). These sediments were slightly coarser (silty clays rather than clays) than those of the Wentlooge, suggesting a new input of silts to the system. Great thicknesses of Rumney sediments were deposited, leading to the burial of the old Wentlooge surface. This relationship can be seen in many marsh sections, such as those in South Wales (Case Fig. 7.1b). Importantly with regard to the pollution history of the estuary, the accretion of the Rumney Formation spanned the industrial revolution, meaning that the levels of pollutants both in the estuary and stored in sediments were rapidly increasing (see Fig. 7.9). In addition, the need to develop industry, agriculture and harbours was also increasing, leading to considerable land claim in the estuary at this time. Hence the construction of sea walls (Case Fig. 7.1a(iv & v)).

Despite the great thicknesses of sediment deposited, the period over which the Rumney Formation accreted was relatively short. By the end of the nineteenth century, the Rumney marshes were being cut back by renewed erosion (Case Fig. 7.1a(v)), releasing further quantities of sediment into the estuary but, importantly, sediments which were now contaminated with pollutants. This period of reworking, however, lasted just a few decades as by the early twentieth century the estuary had switched to being accretional again, with the deposition of the Awre Formation. These sediments again show an increase in coarseness over their predecessors, with sandy clays to clayey sands dominating these deposits (Case Fig. 7.1a(vi)). Importantly, however, the deposition of these sediments incorporated pollution levels present in the water body from continued industrialization as well as those reworked from the Rumney Formation. This marsh unit, however has only minimal development in the estuary because by the 1930s there was further erosion (Case Fig. 7.1a(vii)) before accretion began again in the 1960s. At this time, the last recognizable marsh unit present in the estuary, the Northwick Formation, began to form (Case Fig. 7.1a(viii)) with the deposition of silty clays and silty sandy clays. At the present time, the estuary is eroding again, cutting back into the Northwick deposits.

The causes of this clear erosion and accretion cyclicity are not fully understood, but they do correspond to periods of increased storminess in the estuary, when larger waves from the prevailing westerly wind direction enter the estuary and cause erosion to dominate. Interestingly, the period of time between erosion and accretion cycles appears to be shortening, possibly a response to deepening water caused by a combination of sea-level rise and channel restriction due to land claim. The appearance of the estuary today is of a series of marshes stepping down towards the channel, but often terminated in a marsh cliff, several metres in height (Case Fig. 7.1c).

Although this apparent storm-driven erosion and accretion cyclicity is particularly evident in the Severn, the last decade of the nineteenth century also led to considerable erosion and defence breaching in many other UK estuaries, such as those of the Essex coast, and the Medway (French 1999). This storm-driven erosion and accretion has important repercussions for human use of estuaries and deltas. Once human activity has claimed part of an estuary or delta for some use, it is considered sacrosanct. Hence, the erosion of these areas as a result of natural cyclicity inevitably leads to demands for protection and intervention in these natural processes.

Relevant reading

Allen, J.R.L. & Rae, J.E. (1987) Late Flandrian shoreline oscillations in the Severn Estuary: a geomorphological and stratigraphical reconnaissance. *Philosophical Transactions of the Royal Society of London, Series B* **315**, 185–30.

French, P.W. (1999) Managed retreat: a natural analogue from the Medway estuary, UK. *Ocean and Coastal Management* **42**, 49–62.

Erosion of vegetated surfaces does not just include uniform retreat following wave attack. Stripping of the vegetation surface and the exposure of underlying sediment to erosion is another way in which marsh loss can occur (Fig. 7.10). It is well accepted that vegetation roots are effective in increasing sediment resistance to erosion. These roots, however, penetrate only to a certain depth, for example, up to 1 m for *Spartina* and up to *c.* 20 cm for *Salicornia*, and if the vegetation is largely the same across the marsh surface, then root depth will be uniform. Figure 7.10 shows the effect of this at Silverdale marsh in Morecambe Bay, north-west England. Storm

Fig. 7.10 Erosion of salt marshes caused by the stripping of protective vegetation. Note how the salt marsh vegetation has been peeled back from the sediment surface, exposing the underlying sediments to wave attack (Silverdale salt marsh, Morecambe Bay, UK).

waves have stripped the vegetation surface from the marsh, leading to increased lateral marsh retreat (Pringle 1995).

7.3.2 Effects of changes in sediment supply

Although marsh erosion is significant, perhaps the greatest threat to deltas and estuaries is to reduce the volume of sediment available for deposition. The relationship between a prograding delta front and an eroding one, or an advancing salt marsh as opposed to a retreating one, is closely linked to sediment supply. Figure 7.3 shows the complexity of sources and inputs to an estuary or delta region, each varying in importance in terms of what they contribute to the sediment budget. Some sources are more easily varied than others. For example, hinterland activity can change the importance of urban or agricultural runoff, which will lead to increases or decreases in sediment delivery to the estuary, whereas port development or coastal defences can alter the importance of marine inputs.

A reduced sediment supply can also be caused by variations in the current activity which previously delivered that sediment to its site of deposition. For example, in a delta, a main channel may deliver its load of sediment to one part of the delta front. Were the bulk of that water supply to switch, and start to flow to another part of the delta, the original section of delta front would lose its sediment supply and could cease to function as an active part of the delta, potentially becoming erosional. The diverted sediment supply could lead to the formation of new active areas of delta accretion. Over time, therefore, a delta may reveal a series of accretional phases, each being marked by a delta lobe. A good example of this is the Mississippi, a large delta system with a series of lobes dating from *c.* 7500 yr BP (Fig. 7.11). In intertidal parts of estuaries, a similar process of current change as a result of

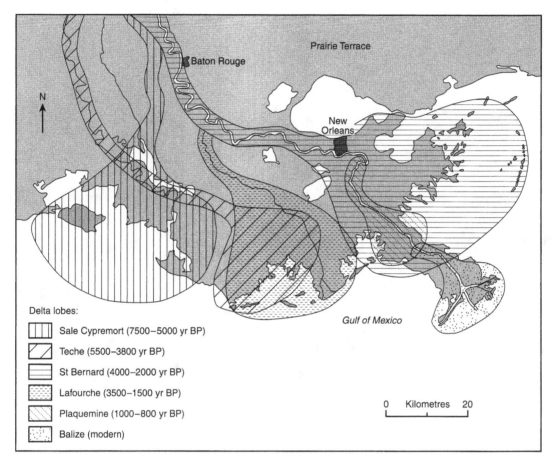

Fig. 7.11 Representation of delta lobe formation in the Mississippi delta. (Compiled and modified from E.C.F. Bird (2000) and Woodroffe (2003).)

channel switching can occur. At the mouth of the Kent estuary in Morecambe Bay (UK), the main flood-tide channel used to focus sediment accretion towards the eastern side of the mouth, where marshes developed rapidly and grew seawards at the rate of several metres a year (Pringle 1995). On the opposite side, little sediment was deposited and marshes were absent. Around the early 1970s, this situation changed following the blocking of old channels and the opening of new ones, to produce a situation of preferential sediment delivery on the western side of the estuary mouth. This led to the growth of salt marshes here, and the cutting off of sediment delivery to the eastern side, which is now experiencing erosion in the order of several metres a year (Pringle 1995) (see also Fig. 7.10).

7.3.3 Response of intertidal sediments to sea-level fluctuations

Although loss of sediment input and changes in currents can be caused by a variety of individual factors over a range of time-scales, sea-level rise has the potential to alter the entire functioning of estuaries and deltas because the whole basis for vegetation establishment, tides and wave energy will change. For example, the channel network in the Rhine–Meuse delta system (Germany) has shown several fundamental changes over the past 9000 years as a result of changing sea-levels (Törnqvist 1993, 1994; Beets & van der Spek 2000; Berendsen & Stouthamer 2000; Woodroffe 2003) and these have been linked to a fundamental change in how the system functions. Around 7000 years ago, the channel pattern changed from meandering to anastomosing. This led to the formation of sediment bars within channels which, as a result, frequently became blocked and changed course (Törnqvist 1993). Consequently, the method of delivery of sediment to the delta front became less reliable. Larger, more permanent meandering channels deposited sediment at fixed points around the delta front, but a frequently changing system of delivery in an anastomosing channel network meant that the point of sediment delivery became less certain and dependable. Around 4000 years ago, the system reverted to meandering.

The extent to which sea-level rise will cause an impact will depend on the ability of the system to respond. Bruun (1962) produced a model by which to predict shoreline behaviour under sea-level rise (see Chapter 8 for a discussion of its validity). In essence, a shore profile, such as a mudflat to salt marsh sequence, will erode and reform so that it remains at the same position relative to the tidal frame. Such movement is seen on the ground as an upward and landward relocation of vegetation zones. In many situations, however, this landward movement is restricted by coastal defence structures, meaning that although sea-level rise is forcing the seaward marsh edge landwards, the landward edge cannot move into the hinterland (Fig. 7.12). This process is known as coastal squeeze.

As salt marsh vegetation communities develop at an elevation controlled by the depth, period and frequency of tidal inundation (Adam 1990), different plant species can tolerate different periods and frequencies of inundation. Those that can tolerate the greatest inundation occur on the lowest parts of the marsh, whereas those

Fig. 7.12 Restricted salt marsh development along the southern banks of the Humber estuary, UK. The man-made embankment will prevent landward migration of salt marshes as sea-levels rise leading to loss of marsh area. (Photograph by Chris Perry.)

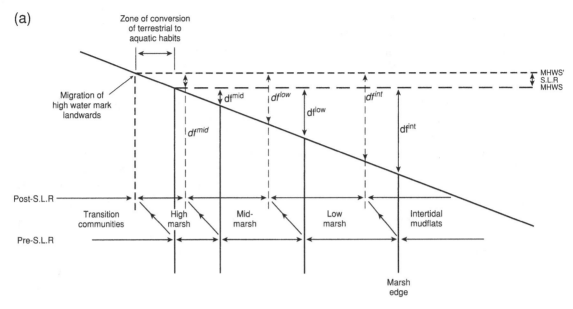

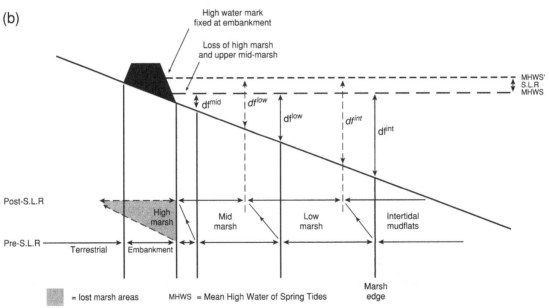

Fig. 7.13 Impacts of sea-level rise on estuarine intertidal zonation. (a) No landward constraints allow a gradual shift in intertidal zones and maintenance of vegetation community succession. (b) The sea wall to landwards results in coastal squeeze and loss of habitats. (Modified from French 2001, p. 278.)

that can tolerate only short or periodic inundation occur on the highest marsh. Figure 7.13a shows an idealized section across a marsh with low, mid- and high marsh communities. In a simplistic sense, low marsh vegetation is stable between the depth (d) and inundation frequencies (f) depicted by df^{Int}, df^{Low}, where df^{Int} represents the upper limit of the intertidal mudflat and df^{Low} represents the upper limit of the low marsh. Similarly, the mid-marsh communities are stable between df^{Low} and df^{Mid}, where df^{Mid} represents the upper limit of the mid-marsh. Landward of this point, higher marsh and transitional communities dominate. Under conditions of sea-level rise (MHWS to MHWS'), water depths in the estuary increase meaning that all zones of the marsh are covered more frequently and for longer periods (Fig. 7.13a). Clearly this will lead to plants becoming unstable in their new conditions causing die-back and regrowth further inland where water depths and periods of inundation match those of pre-rise conditions. Hence, the junction between low and mid-marsh communities moves inland to the depth it was at prior to the onset of sea-level rise. In the case of the scenario in Fig. 7.13a, this position moves from df^{Int} to df^{Int}, from df^{Low} to df^{Low} and from df^{Mid} to df^{Mid}. With continued sea-level rise, this relocation is ongoing and gradual, only ceasing when sea-level reaches a new stable level. Evidence for such landward migration can be seen in Maryland, USA, where freshwater forest communities are being replaced by salt marsh (Darmody & Foss 1979); and in Chesapeake Bay, where terrestrial meadows are being invaded by halophytic vegetation (Bird 1993). In both of these examples, the process is represented in Fig. 7.13a by the landward movement of the high-marsh–terrestrial communities limit.

When an estuary is in its natural, undefended state, any long- to medium-term rise in sea-level can be compensated for by these landward 'shifts' in vegetation zonation provided that suitable back-marsh areas for colonization exist (Fig. 7.13a). As soon as sea defences are built, however, any landward shift in these zones is restricted by a physical barrier (Fig. 7.13b), which effectively fixes the high water mark. Therefore, with time, although low and mid-marsh communities can shift landwards, there is no space for the high marsh, which becomes squeezed against the sea wall and will eventually disappear (Fig. 7.13b). As the depth, frequency and period of inundation increase further, this process will continue, with the possibility that a marsh will actually revert to lower marsh species, or even mudflat. Although such a loss has major implications for the ecology of an estuarine system, it can also have implications elsewhere in the system. It has already been seen (section 7.1.2) that vegetation plays an important role in the trapping of sediment and promotes vertical marsh accretion. With the loss of such vegetated surfaces, less sediment will be trapped, and so more will be retained in the water body. This surplus sediment has, in some situations, caused additional problems. In Tampa Bay, Florida, for example, increased turbidity caused by increased suspended sediment has reduced plant growth elsewhere in the estuary.

Rising sea-levels are not the only cause of marsh loss as vegetation may die back for other reasons. One way, which remains largely unexplained, is the natural die back in *Spartina* marshes. These are low marshes, in that they are the first real vegetation to form as the marsh develops from the mudflat. In the 1930s, *Spartina* marshes along the south coast of England started to die, leaving large areas of bare mud. In many areas this has been replaced by other low marsh vegetation, such as *Zostera* (Haynes & Coulson 1982; Adam 1990). Although one species replacing another may not be seen as a problem, especially when ecologists favour *Zostera* because of its increased biodiversity and better nutrient dynamics, *Spartina* is by far the more efficient species in terms of ability to trap and retain sediment (French 2001).

Although sea-level rise is very much a global issue, some coastlines are experiencing a relative sea-level fall. Whereas a sea-level rise will reduce the space for estuaries and deltas to occupy unless they can relocate inland, a sea-level fall will make more space available and, where sediment input remains high, allow expansion of the sedimentary and vegetation zones. The delta front will move further onto the continental shelf, and intertidal zones in estuaries will start to encroach on the original intertidal area and shift seawards.

7.4 PROCESSES AND IMPACTS OF ANTHROPOGENIC ACTIVITIES IN DELTAS AND ESTUARIES

Although estuaries and deltas are subject to considerable natural sediment dynamics, they also represent some of the most intensively exploited environments. The main reason for this is their proximity to, yet shelter from, the open sea, and

Table 7.2 Human activities and related impacts in estuary and delta environments.

Activity	Impact
Land claim	Loss of intertidal habitat, coastal squeeze, increased need for flood defences
Coastal defence	Loss of intertidal habitat, increased erosion, changes to natural sediment cycling, coastal squeeze, alteration to wave and tidal processes
Tourist development	Increased visitor pressure, habitat loss, increased contamination/pollution, increased need for coastal defences
Industry	Contamination/pollution, habitat loss, increased need for defences, increased need for shipping access (dredging)
Barrages	Loss of intertidal habitat, changes to tidal regime, contaminant retention, greater brackish water penetration up river, sediment retention, changes to water table
Sediment extraction	Loss of submarine habitat, changes to tidal and wave currents, loss of wave protection, re-suspension of sediment and remobilization of pollutants
Waste disposal	Contamination/pollution, increased turbidity
Dams	Reduction in freshwater supply/increase in net salinity, reduction of sediment supply from catchment
Catchment land use changes	Changes in sediment supply, changes in pollution levels, changes in freshwater input

their suitability for industrial and agricultural development. Human activity has, therefore, produced significant modification to many systems (Table 7.2).

7.4.1 Anthropogenic impacts on rates and styles of sedimentation

Reduced freshwater flow will produce significant changes in the functioning of deltas and estuaries. In the former, reduction in sediment supply can lead to increased marine influence, and thus increased erosion, whereas in the latter, an estuary may experience changes in flow structure, and ebb/flood dominance. One of the main ways of affecting the amount of riverine flow is by the construction of dams. Completion of the Aswan High Dam in 1964, for example, reduced the supply of sediment to the Nile delta from 124 to 50 million tonnes a year (Carter 1988). The net result was a dramatic increase in coastal erosion along parts of the delta edge (see Case Study 7.2). In a similar way, closure of the Akosombo Dam in 1961 completely blocked sediment supply to the Volta delta in Ghana (Ly 1980). Apart from initiating large amounts of coastal recession, the salt wedges in the delta channels moved further landwards, altering local ecology. Furthermore, pollutant flushing was reduced leading to increased resi-

dence times for pollutants in estuarine channels, and sediment deposition patterns were modified by weakened ebb currents (Collins & Evans 1986). The only sediment that now reaches this delta comes from erosion of sediments down stream of the dam. Dams on the Ebro (Spain) and the Rhône (France) have also reduced sediment supply to coastal deltas by 96% and 90% respectively (Viles & Spencer 1995).

As well as affecting the amount of sediment entering the estuary or delta, dams also retain fresh water, thus reducing the quantity that enters the estuarine area. Stratified and partly stratified estuaries (see Fig. 7.1) rely on fresh water to drive their circulation and create the stratification. This freshwater-induced circulation also controls the movement of nutrients on which primary and secondary productivity of the system depends (Mann 2000). More fundamentally, many brackish water species have, through long-term adaptation, adjusted to cope with seasonal fluctuations of freshwater input, caused by seasonality of flow from river basins. One impact of dams is that they often reduce or eliminate this seasonality, in that water is stored at times of high flow, and released at times of low flow.

Whereas dams are built to constrain fresh water in rivers, barrages are built across estuaries to withhold water on the ebb-tide for the purposes

Case study 7.2 The impact of dams on deltas and estuaries: the Nile delta and the Aswan High Dam

BACKGROUND

The Aswan Dam was built in order to provide irrigation and drinking water, via the construction of a large reservoir (Lake Nassar), to the Nile catchment, and also to facilitate the development of hydroelectric power to the growing population of Egypt (Rashad & Ismail 2000). Such a supply was considered vital for a country whose population (*c.* 48 million) had complete dependence on the Nile. Although its completion in 1964 facilitated this, environmental impacts were also common, not least in the delta area downstream of the dam. This was not the first structure to be used to regulate water along the Nile, however, with various schemes, such as construction of the Aswan Low Dam in 1902, having gradual accumulatory impacts on the delta.

As with any coastal landform, deltas undergo alternate constructive and destructive episodes (see Fig. 7.11 for the example of the Mississippi). Cores taken in the area have shown that there have been a series of deltas at the mouth of the Nile, which have accreted and eroded throughout Pliocene and Pleistocene times (i.e. over the past 5 million years) (Stanley & Warne 1998), with the present delta starting to form *c.* 8000 to 6500 years ago. In the early stages of the formation of the present delta, seven major channels drained across the delta surface, each with its own lobe at the coast (Badr & Lofty 1999). Significantly, by 2000 years ago, two of these had disappeared, and by 1000 years ago, a further three had gone (McManus 2002), leaving the two we now recognize as the Rosetta and Damietta. This loss of channels was partly the result of natural evolution of the delta, but also early attempts at irrigation and management.

Over the past 150 years, the delta has been in an erosional phase (Stanley & Warne 1998) with the majority of the 260 km long delta face undergoing erosion and landward retreat (Frihy 1996). Only small areas of localized accretion now occur, such as in association with jetty construction at Damietta Harbour (El-Asmar & White 2002) (Case Fig. 7.2). The reason for the widespread erosion has, to a great extent, been more organized water regulation, which has affected the balance between sediment input, coastal sediment drift and subsidence in the delta area. Despite earlier management efforts, the Aswan High Dam has, arguably, had the greatest impact on delta decline. Pre-closure in 1964, the Nile supplied *c.* 124 million tonnes of sediment a year to the delta area (Carter 1988). Such a flow of both water and sediment was sufficient to maintain the two major channels, but also a series of minor distributary channels, which provided an annual deposit of fertile silt on farmland. Following closure, the amount of sediment delivered per year dropped to *c.* 50 million tonnes (Carter 1988). This sediment was notably finer grained than pre-closure (Stanley & Wingerath 1996), and largely derived from the river downstream of the dam, and also from the delta hinterland itself. More significantly, however, the volume proved insufficient to compensate for that lost to erosion. As a result, the delta front became net erosional.

As well as reduced sediment load, the reduced river flow led to the silting up of many of the distributary channels, leading to major changes in delta morphology and ecology (Case Fig. 7.2). For example, reduced discharge and the easterly drift of sediment at the coast blocked many of the smaller estuaries (Frihy 1996). Chesworth (1990) shows that of the *c.* 55.5 billion m^3 of water released from the dam, only 17.5 billion m^3 reaches the delta coast, the rest being either extracted or lost to evaporation. Of the two main channels, the Damietta and Rosetta, the former is generally dry for much of the year, and the latter carries the bulk of what flow remains (Bird 1985).

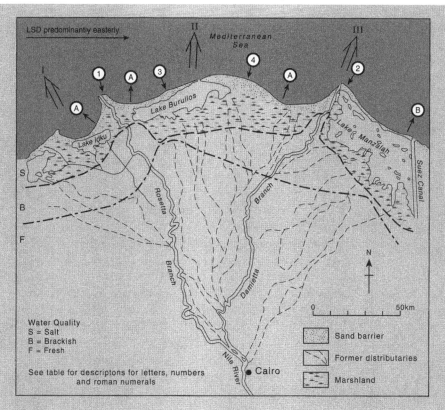

Erosional sectors

Site	Erosion rate (m yr⁻¹)	Measurement period	Reference
1	>50	1909 to late twentieth century	Stanley & Warne (1998)
	>100	1971 to mid-1990s	Frihy (1996)
	113	1985–1991	White & El Asmar (1999)
	52–88*	1971–1990	Frihy et al. (2002)
	3–13†	1990–2000	Frihy et al. (2002)
	118.6	1984–2000	Ahmed (2002)
	18–230	Last century	Ahmed (2002)
2	Unspecified	1909 to late twentieth century	Stanley & Warne (1998)
	10	1909 to mid-1990s	Frihy (1996)
	10	1971–1990	Frihy et al. (2002)
	15	1984–1991	White & El Asmar (1999)
3	6.5	1971 to mid-1990s	Frihy (1996)
	1.3	1909–1989	El-Fishawi (1994)
	2.1	1984–1991	White & El Asmar (1999)
4	6.5	1971 to mid-1990s	Frihy (1996)
	3.8	1909–1989	El-Fishawi (1994)
	5.0	1984–1991	White & El Asmar (1999)

*Pre-defence construction in 1990. First figure to west of promontory, second to east.
†Post-defence construction. First figure to west of promontory, second to east.

Accretional sectors

Site	Accretion rate (m yr⁻¹)	Measurement period	Reference
A	Up to 13	1971 to mid-1990s	Frihy (1996)
	14	1971–1990	Frihy et al. (2002)
B	Unspecified	1909 to late twentieth century	Stanley & Warne (1998)

Offshore sediment transport

Site	To offshore (m³ yr⁻¹)	Measurement period	Reference
I	3.2×10^6	Current	McManus (2002)
II	1.48×10^6	Current	McManus (2002)
III	1.8×10^6	current	McManus (2002)

Case Fig. 7.2 The Nile delta showing areas of erosion and accretion: LSD, longshore drift. (Modified from Carter 1988, p. 483.)

Over the past 150 years, the Nile delta has thus changed from a constructive, wave-dominated delta that had formed over the past 7000 years, to an erosive coastal plain (Stanley & Warne 1998). This has been chiefly caused by a negative sediment budget that is insufficient to maintain the relative height of the delta surface following subsidence. With increasing development and population growth, including the construction of a new road along the seaward margins of the delta to open up less accessible areas for development (Frihy 1996), delta degradation is likely to continue, with increased subsidence and greater risk of marine inundation and flooding. Furthermore, contamination of groundwater is likely to increase and to move further inland. The future is, therefore, bleak. The only way that the Nile can regain its role as a functioning delta is to reinstate water flow and sediment supply. In reality, this is not likely and will result in continued shoreline erosion and land subsidence.

Relevant reading

Ahmed, M.H. (2002) Multi-temporal conflict of the Nile delta coastal changes, Egypt. *Proceedings of the Conference Littoral 2002: the Changing Coast*, Porto. Eurocoast/EUCC Portugal 2002.
Badr, A.A. & Lofty, M.F. (1999) Tracing beach sand movement using flourescent quartz along the Nile delta promontories, Egypt. *Journal of Coastal Research* 15, 261–5.
Bird, E.C.F. (1985) *Coastline Changes: a Global Review*. Wiley, Chichester.
Carter, R.W.G. (1988) *Coastal Environments*. Academic Press, London.
Chesworth, P.M. (1990) The history of water use in Sudan and Egypt. In: *The Nile* (Eds P.P. Howell & J.A. Allan), pp. 65–79. Cambridge University Press, Cambridge.
El-Asmar, H.M. & White, K. (2002) Changes in coastal sediment transport processes due to construction of New Damietta Harbour, Nile Delta, Egypt. *Coastal Engineering* 46, 127–38.
El-Fishawi, N.M. (1994) Relative changes in sea level from tide gauge records at Burullus, central part of the Nile Delta coast. *INQUA MBSS Newsletter* 16, 53–61.
Fahim, H.M. (1981) *Dams, People and Development: the Aswan High Dam Case*. Pergammon, Oxford.
Frihy, O.E. (1996) Some proposals for coastal management of the Nile delta coast. *Ocean and Coastal Management* 30(1), 43–59.
Frihy, O.E., Debes, A.D. & El-Sayed, W.R. (2002) Processes reshaping the Nile delta promontories of Egypt: pre- and post-protection. *Geomorphology* 53, 263–79.
McManus, J. (2002) Deltaic responses to changes in river regimes. *Marine Chemistry* 79, 155–70.
Rashad, S.M. & Ismail, M.A. (2000) Environmental impact assessment of hydro-power in Egypt. *Applied Energy* 65, 285–302.
Stanley, D.J. & Warne, A.G. (1998) Nile delta in its destructive phase. *Journal of Coastal Research* 14(3), 794–825.
Stanley, D.J. & Wingerath, J.G. (1996) Nile sediment dispersal altered by the Aswan High Dam: the kaolinite trace. *Marine Geology* 133, 1–9.
White, K. & El-Asmar, H.M. (1999) Monitoring changing position of coastlines using thematic mapper imagery, an example from the Nile Delta. *Geomorphology* 29, 93–105.

of power generation, or for amenity uses. Their impact, however, is just as severe. By reducing the mixing of fresh and salt water, barrages can affect the salinity structure of the estuary, such as occurred at the Rance tidal power station in France (Shaw 1995). Sediment retention may also increase upstream of the barrage, as it would in a dam reservoir, such that the sediment budget may be affected downstream. The exact nature of this impact will depend on whether the estuary is ebb or flood dominant and thus whether the prime sediment source was land-based or marine (see section 7.2.2). Changes in water movement and reduction of fresh-water flow may also cause contaminant levels to change in response to water storage upstream of the barrage. On the one hand, riverine inputs may achieve a greater dilution owing to the increased volumes of water stored in the tidal reservoir, but on the other, changes in water depth, current activity, flushing and oxygen levels can all affect contaminant breakdown and flushing. In the Tawe estuary, South Wales, flushing has been reduced significantly since barrage construction. Although incoming water may overtop the barrier, being denser than fresh water, a captive salt wedge has formed inside the barrage, which has led to stagnation behind the barrage during neap-tide periods (Dyrynda 1996). Also in the Tawe, contaminants have continued to enter upstream of the barrage, leading to organic enrichment, algal blooms and metal enrichment. Overall, the water quality has deteriorated significantly and has led to the need for artificial remediation, such as periodic draining of the lake through sluices.

Other issues relating to tidal barrages include modification to currents and local increases in sea-level. Estuarine ecosystems owe their existence and spatial variation to the frequency and period of tidal inundation. Barrages will change all of these parameters and, hence, will affect the flora and fauna of the system. In many situations there will be an increased segregation between marine and riverine communities. In the case of La Rance in France, the construction phase (1963–66) saw the complete destruction of the estuarine ecosystem due to the sealing of the estuary. However, post-commissioning restocking has allowed a new, if not 'natural', ecosystem to develop (Rodier 1992).

Although dams and barrages provide the greatest impact on the overall functioning of deltas and estuaries, other human activities can also lead to major changes. Tides, waves, river flow and currents all combine in the functioning of an estuary or delta, and thus any activity that changes these has the potential to produce some form of modification to process and form. Such activities include shore-normal structures (ports, jetties), bridges, coastal defence structures, wrecks, dredging, land claim, wetland creation (managed realignment) and changes in land use.

7.4.2 Reworking of contaminants

Section 7.2.3 and Fig. 7.8 demonstrate how contaminants can become attached to sediment grains, and thus become incorporated into the sediment record (see e.g. Fig. 7.9). One key issue for water quality is how stable these contaminants remain, and whether they may re-enter the water body from the host sediment at a later date. Oil, for example, will break down in the aqueous environment. When treated with detergents to break up a slick, however, the oil residues often sink through the water column to the sea floor and become incorporated into sediments via pore spaces, burrows or through burial. Similarly, ships that sink will generally take their contents with them, again transferring oil to the sea floor. During the Gulf War of the early 1990s, around 14 million barrels of oil were released into the Gulf environment (Alam 1993). Some of this oil, particularly that from wrecks, was released at the sea floor, and much of this filtered into sediments through pore spaces and burrows, effectively putting it into storage. Over the following years, natural reworking of these sediments caused some of this oil to leak out, recontaminating some areas. In section 7.2.3 it was reported that oiled salt marshes can regenerate after an oil spill, and an example was cited from Buzzard's Bay, Florida (Teal et al. 1992). Although this marsh had fully recovered after 20 years, a major concern relates to the fact that large quantities of degraded oil remain within marsh and mud-flat sediments. Even though at the current time this is not a problem, any process that disturbs these sediments, such as erosion, could well cause the release of this oil store, estimated to be large enough to damage fauna (see also Chapter 9).

The storage and future release from sediment storage can occur for any of the contaminants mentioned in this chapter. Contaminants will stabilize to the environmental conditions present at the time of deposition, and their chemical form may adjust accordingly through reaction with other substances, or with organic/mineral sediment grains. If conditions change, however, these contaminants may become unstable, and re-enter the system. Such occasions could include: erosion, where material is physically removed from a store back to the water body; sea-level rise, where a rise in water table could destabilize oxic sediments or increased salinity up river may destabilize freshwater sediments; and vegetation death, where pollutants locked up in vegetation re-enter the system as material decays. Some of these processes may be gradual, such as sea-level rise, but others, notably vegetation die-back or a storm causing a large retreat of a marsh, can produce contaminant flushes, whereby large quantities of stored pollutants enter the water body at the same time (see Case Study 7.3).

Such changes in the stability of marshes and their constituent pollutants can also be caused by human activities. Increasingly, coastal management is turning to soft engineering to protect the coastline from flooding. One such technique used in estuaries is managed realignment, a process where previously claimed land is allowed to flood and become intertidal again (see French 2001). The implications here are that the sediments forming the claimed land were deposited under estuarine conditions, and contain levels of pollutants reflecting the estuary at the time of deposition. When these sediments were initially claimed, they were drained and became freshwater

---

Case study 7.3 The recycling of contaminants from eroding salt marshes

Sediments laid down in estuaries and deltas incorporate a range of contaminants which reflect those present in the water body at the time of deposition (see section 7.2.3). In time, these sediments and contaminants accumulate to produce an environmental record of contaminant status of the system (see Fig. 7.9 and Case Fig. 7.3). However, when these sediments start to erode, the contaminants contained within them will be reintroduced to the water body. This case study represents an assessment of the contaminant input from one such eroding marsh.

The Severn estuary (UK) is a large, macrotidal estuary that has undergone a series of erosion and accretion phases (see Case Study 7.1). At the current time, these sediments are eroding, re-introducing sediments and contaminants to the estuary. The marsh (Case Fig. 7.3) represents a 1.12 m sequence of sediments of the Northwick Formation, located on the southern shore of the estuary at Northwick Warth. The marsh was sampled via 1-cm slices, each of which was analysed to determine levels of copper, lead and zinc (full analytical method is given in French 1996).

The resulting data set provides a record of the levels of contaminants present in each successive 1-cm slice, and thus it is possible to estimate the quantities of these materials that will re-enter the estuary as this marsh retreats landwards. It is known that the average retreat rate of this marsh is 0.17 m yr^{-1}, and that the length of the eroding frontage is 3.2 km. It has to be assumed that the marsh retreats uniformly landwards, and that the height is constant throughout the length. These assumptions having been made, the volume of eroded sediment (in cubic metres) from each 1-cm (0.01 m) slice is 5.44 m^3 yr^{-1}. To determine the levels of metal input, it is necessary to consider the concentrations within each of the 112×1-cm-thick slices. For the topmost layer (0–1 cm), analysis has shown that there is 43 ppm Cu, 96 ppm Pb and 300 ppm Zn (French 1996). Using these values, the volumes of sediment eroded from this topmost slice are adding 248 g of copper, 554 g of lead and 1730 g of zinc per year.

Using this approach to calculate the amount of copper, lead and zinc in each layer, then adding the total for each together, the salt marsh at Northwick Warth is contributing 33,953 g of copper, 80,707 g of lead and 239,294 g of zinc, per year, back to the waters of the Severn estuary. This analysis will not show in what form these metals occur, in terms of chemical stability or bio-availability, but it does present an amount that can be used in determining the Environmental Quality Standard (see section 7.5.2). The key factor, however, is that the salt

Case Fig. 7.3 Eroding sediment of the Northwick Formation, Northwick Warth, Severn estuary. The erosion of this marsh, accreting since the 1930s, is now contributing stored contaminants back in to the water body. (Data from French 1996.)

marsh investigated here is only a small part of the salt marsh resource of the estuary. The two banks of the Severn estuary combined cover a distance of over 400 km, much of which is salt marsh. Hence, the release of material from these sources is potentially large.

Relevant reading

French, P.W. (1996) Long-term variability of copper, lead and zinc in salt marsh sediments of the Severn Estuary. *Mangroves and Saltmarshes* 1(1), 5968.

dominated. This would have represented the first major stability change. Managed realignment reintroduces brackish water to these areas, increasing the saturation of the soils and causing further remobilization of many pollutants. The processes by which remobilization occurs are diverse and include metal transformation by processes such as methylation, change of chemical phase due to the changing oxygen availability, salinity, redox, bioturbation and acidity/alkalinity. Bryan & Langstone (1992) provide a full review

of these processes, and Emmerson et al. (2000) demonstrate the role of changing acidity and chlorinity in the release of sediment-bound metals in a field-based situation at Orplands, Essex, UK. Blackwell et al. (2004) look at the immediate post-breaching change in the Torridge estuary, Devon, UK.

Problems with mobilization can become severe if levels of a particular contaminant are present in sediments in high concentrations. This may be the case where sediments are associated with industrial activity. Fraser (1993) reports on dredging operations in New Bedford Harbour (USA), where sediments had been contaminated with waste from the adjacent electronics industry. Particular concerns focused on PCBs and metals, PCB levels being the highest recorded in any estuary in the USA. When the sediments remained *in situ*, these contaminants remained stable and locked up in the sediments. As such, they were not considered an environmental threat. Natural reworking, however, led to the reintroduction of some material and, as a result, fishing and shell fish collection were stopped. Dredging, however, threatened large-scale destabilization of the stored contaminants and the large-scale reintroduction to the water body. As a result, dredged material was treated as high grade industrial waste and had to be containerized and isolated from the environment. Such situations are rare, but the New Bedford Harbour example highlights some interesting issues relating to estuarine and delta sediments. Most notable are the concerns of reintroduction and reactivation in the contemporary environment (see Case Study 7.3). Section 7.5 looks at management and the idea of emission quotas. Although this works well for industrial point source emissions, it often fails to adequately account for the release of contaminants through sediment reworking.

7.5 MANAGEMENT AND REMEDIATION

Humans have continually strived to alter estuaries and deltas to facilitate their own needs. Ever since industrialization began, intertidal sediments have been receiving increased quantities of contaminants, and agricultural, industrial and urban growth have been claiming large areas of the intertidal zone, and development in these regions has necessitated greater amounts of coastal defence and flood protection. The Tees estuary is a classic example of an area that has undergone progressive land claim for industrial development (Fig. 7.14). In the mid-nineteenth century, the estuary contained large areas of salt marsh and intertidal flats yet by the mid-1970s, *c.* 3300 ha, or 83% of this area had been

claimed for industrial and port activity (Davidson et al. 1991). Through this process, the intertidal areas of many estuaries and deltas have become highly altered in terms of morphology, functioning and sediment/water quality.

The range of impacts associated with this activity is large and diverse, and poses potentially serious problems for managers (see Table 7.2). In addition, many of these impacts are accumulatory. For example, land claim may be carried out in a piecemeal way over several centuries, but overall represents a large collective loss of salt marsh and mud flat. Although pressures for industrial and port development may be increasing, other threats are diminishing. The change in agricultural economics in the 1980s and 1990s has reduced the threats posed by agricultural land claim and, indeed, this legacy is actually seeing a reversal as agricultural land is returned to the intertidal zone through managed realignment schemes. In addition, increased environmental protection measures aimed at safeguarding intertidal areas for their wildlife interest have grown in stature, and, significantly, people have increased understanding of how these systems work and function. This means that it is now possible to gain a greater understanding of how systems change over time in relation to external forces, such as sea-level rise, and how they may react to human-induced changes, such as dam construction and land clearance. Once understanding of action–impact relationships develops, management becomes easier as it can start to be proactive, rather than reactive.

These human-induced changes mean that estuaries and deltas are now afforded a higher degree of protection than has hitherto been the case. Although this may be fine for future management, it does not help overcome some of the legacies left to modern managers by previous land-use policies. In considering this historical legacy, it becomes necessary to think of management in a variety of ways. Previous changes to deltas and estuaries, such as dam construction or the claiming of salt marshes, have resulted in considerable adjustments to the natural systems. Therefore, it is critical at an early stage in the management cycle to decide whether

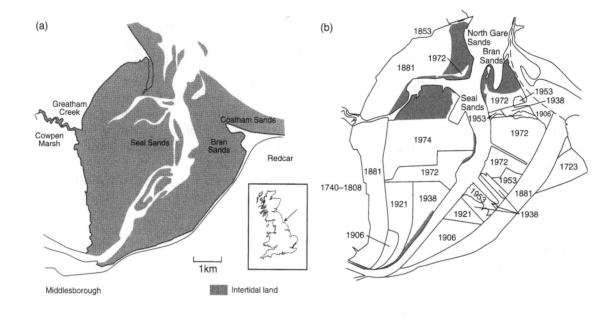

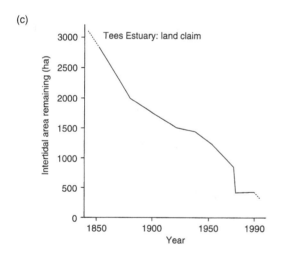

Fig. 7.14 The sequential claiming of the Tees estuary, UK: (a) Tees estuary in the mid-nineteenth century; (b) phases of subsequent land claim; (c) loss of intertidal area between 1850 and 1900. (Modified from Davidson et al. 1991.)

management should be a means of maintaining a form of *status quo* (i.e. protect what these systems have become and prevent further human-induced change), or whether it should be a means of managing what they once were, i.e. to regain a natural estuary or delta. The latter is impractical in many situations owing to the extent to which humans have developed and interfered with natural functioning. The former is perhaps the most workable, but the issue then becomes what to take as the *status quo*. This debate becomes complicated. For example, the Nile delta is still

changing in response to the construction of the Aswan High Dam, so do managers wait until it has stabilized, or begin to manage the area as it is at the present time, given that the current situation is only a snap-shot representing one short section of the conveyor belt of gradual adjustment and change?

The key debate here is the idea of conserving or preserving. Natural systems need to change and adapt to changing environmental conditions. Whether such changes are natural or anthropogenic is, to some extent, unimportant.

The key thing for management today is to protect environments from further avoidable change (i.e. caused by human actions), and to allow them to adjust to ongoing forces of change, i.e. sea-level rise. Hence, management is about stopping issues such as land loss and contaminant input, and providing for other needs, such as increased sediment to compensate for negative budgets and the freeing up of land to allow it to respond to sea-level rise. To do this, it is necessary to determine what the anthropogenic forces of change are. Viles & Spencer (1995) present a case study relating to wetland loss on the Mississippi delta, in which they cite a range of management issues. These are split into three categories: 'origin and evolutionary development'; 'degradation and loss'; and 'human activities'. The former includes issues such as storm activity and channel switching, which it could be argued are natural processes that form part of the ongoing evolution of the delta. The second contains management issues that arise from the first, such as loss of sediment due to channel switching. The final category contains the impacts of human activities, such as dam building, land claim and drainage. In terms of management approaches, the final and to some extent the second categories are management problems, but the first is not. It is common, however, and not just in the Mississippi, that the first category becomes a management issue because these processes affect human development. In other words, it is necessary for processes of natural system evolution to be managed purely because if such evolution were permitted, it could have an impact on human activity.

This then puts the earlier point of whether to conserve or preserve into another context. Frequently the overriding strategy for management is that although it is considered a good thing, it is only good if it does not inconvenience human activity. For estuaries and deltas, and to a great extent coastlines, this is a critical issue because often management and remediation cannot progress according to the best environmental reasons because of the human interest in an area. It is relatively easy to manage some human activities based on our knowledge of their impacts. For contaminants and pollutants (see section 7.5.1), abatement and input reduction can be achieved easily providing that we are able to identify what these inputs are, and where they originate from (see previous discussion on contaminant reworking, and Case Study 7.3). Other aspects of human interference are more difficult. The reversal of land claim in order to facilitate sea-level rise is a case in point because freeing up land, often in areas where land use is intense, cannot be facilitated easily. Hence, in many cases, management will fail purely because managers cannot undo the legacy of the past and it may ultimately be that this legacy facilitates the demise of many estuarine and delta environments. One important criterion, however, is that whatever management we use, it should be aimed at facilitating a natural system, and not at managing natural processes that will further disrupt the system.

7.5.1 Managing for the prevention of environmental change

One of the main difficulties with management relates to where the management responsibility for such actions lies. One argument is that it should lie with interest groups and various government-based environmental authorities. With such an approach, however, there is significant potential for conflicting approaches to management. For example, an electricity generating company may manage a dam and its lake, but the water which emerges from that lake becomes the responsibility of a river authority. Hence, is the release of water based on the demand for electricity generation, or is it on the basis of what the estuary or delta downstream needs to support extraction, fish populations, or sufficient quantities of water to dilute permitted waste discharges? Such issues demand an overall management strategy covering rivers, estuaries and coasts. However, such systems are generally thought of as unwieldy and unworkable. Hence, smaller units of management have to be used.

The Estuary Management Plan (EMP) can provide for development and general land-use planning in the estuary (or delta); it cannot manage what goes on beyond the estuary limits. For deltas, this is critical because of the issue of sediment supply: issues that are linked to the freshwater system, and come under the Catchment Management Plan (CMP). For estuaries, what goes on to seaward is also important and again these fall within another management plan region, the Shoreline Management Plan (SMP). The risk of having three plans is that the lack of communication between different management groups could lead to one plan recommending a course of action that could be of detriment to the other. An obvious example here would be freshwater management and dam construction as cited above. Managers of these respective plans, however, are encouraged to meet and consult on the outcomes of their recommended activities, which is generally done through a liaison committee that facilitates dialogue at the planning stage, and also oversees the running and implementation of the plans (Fig. 7.15).

The formulation of management plans should be a dynamic process that develops and implements a co-ordinated strategy to achieve the conservation and sustainable multiple use of estuaries and deltas. Inherent in this is the idea of co-ordination, which should bring catchment, estuarine and shoreline management together. Such co-ordination is facilitated by a liaison committee with the role of promoting dialogue between different managers. Eventually, this should then fall within the realm of coastal management, of which estuaries and deltas are integral parts.

7.5.2 Managing for the prevention of anthropogenic inputs

The natural ability of estuaries to assimilate contaminants means that they can be used to treat some of the waste material generated by society. It is important to ensure, however, that the permitted inputs do not exceed the

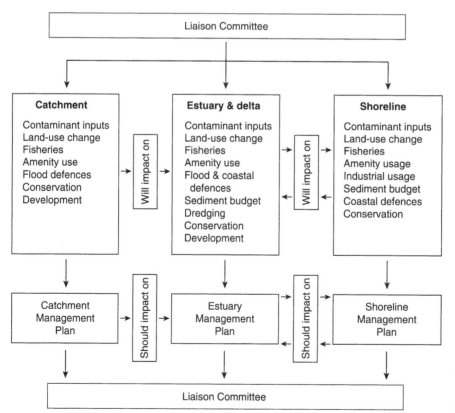

Fig. 7.15 Management structure and issues having an impact upon estuaries and deltas.

estuary's capacity to assimilate them. Hence, one method of ensuring safe inputs is to base this on a determination of how much of a contaminant a particular estuary can assimilate before functioning is impaired. Such a determination is a function of size, tidal range, freshwater input, floral/faunal/sediment assimilation and flushing ability.

McClusky (1989) makes the general point that '. . . the solution to pollution is dilution', whereby contamination in a water body can have less impact, and avoid becoming a pollutant, if it is diluted by a large enough volume of water. Thus, large estuaries with large tidal ranges and where large volumes of water are exchanged on each tidal cycle will facilitate greater contaminant assimilation. However, other factors also play an important part in the calculation of safe limits. The nature of the contaminant, its environmental residence time (is it conservative or does it break down), and the ecology of an area are all important. Hence, an estuary's ability to assimilate contamination is partly a function of its size and flushing ability, but importantly it is also a function of its other component parts, such as ecology, the contaminant itself, and also the spatial position along the estuary axis at which the discharge is made.

To determine how much of a particular contaminant an estuary can assimilate without impairing its functioning, and therefore the desired state of the estuary post-discharge, is known as the Environmental Quality Objective (EQO). Having determined this, it is necessary to determine the levels of discharge which can be safely discharged without threatening this objective. This is known as the Environmental Quality Standard (EQS). Examples would include the amount of a metal contaminant that could be introduced before toxic effects were felt by fauna and flora; how much sewage could be introduced before the biological oxygen demand increased to the detriment of fish; or how much sediment could be introduced before the deposition rate threatened to bury infauna. Having determined safe limits, it is then necessary to proportion this permitted amount amongst those groups that wish to discharge.

The EQO/EQS approach is increasingly used to determine safe levels of inputs into an estuary. There are key issues, however, which although appearing simplistic in the procedures of determining the EQO/EQS, are actually problematic. Determining safe levels of discharge to satisfy the EQS is difficult because it relies on measurement of the range of estuary-specific factors outlined above, such as natural dispersal and flushing. Here the problem lies with determining what is typical for the system concerned. At what point, for example, do you take measurements of tidal range, and what is assumed as a normal freshwater input? The use of mean tidal range and mean river flow is an obvious answer, but a further problem is that a mean is a combination of high and low values, which occur at different times of the year. Tidal range, for example, varies on a fortnightly cycle with neaps (lower) and spring (higher) tides, and also varies on a seasonal scale (equinox tides). Furthermore, tides also follow longer periodicity cycles. In terms of determining safe discharge limits, using a mean value will mean that there are times when discharge could be higher (i.e. when tidal range is above the mean), but others when it should be lower. The key issue here is that in the case of the latter, when discharge becomes more concentrated, will this still be 'safe' in terms of estuary functioning? Another key factor, and one that is regularly overlooked, is the amount of material that enters an estuary through routes which are not accounted for in the determination of the EQS. One mechanism is through the reworking of stored contaminants (see Case Study 7.3). This could, in certain situations, lead to significant inputs above those considered safe both by the management authority and by the EQO.

7.6 FUTURE ISSUES

The introduction to this chapter highlighted the fact that estuaries and deltas are formed by different processes. However, once channels form on a delta surface, processes tend to converge, leading to many of the issues pertaining to sediments, human impacts and pollutants being largely common. Susceptibility to changes in these issues makes both systems highly vulnerable, both in the long and short term. The last section discussed management and the prevention of some elements of change. Remaining problems include: contaminant management and the legacy of stored contaminants in many historically industrialized estuaries; loss of habitat, largely now the result of erosion and land claim for development; freshwater management due to the growing conflict between the need for increased water resources for a growing population, and the need to maintain freshwater flow into brackish systems; and sea-level rise.

Some elements of change, such as erosional and depositional cycles in marshes and in delta channels, may show a cyclicity (see e.g. Case Study 7.1). There is an increasing element of directional change, however, which could see these systems under increasing threat in the future. The major decision for managers is to recognize the point at which management has to give up trying to maintain a *status quo*, and allow change to happen. It is a great mistake for contemporary scientists and morphologists to regard all of our current environments as being at the end of their evolutionary cycle. It is much more likely that much of the current change is part of a process of evolution and must, therefore, be allowed to continue for the sake of the future stability of the environment. A key factor to determine, however, is to what extent will human activity accelerate or alter the rate and direction of change. This final section will discuss some of these key issues.

7.6.1 Impacts of climate change and sea-level rise

There is some irony in the fact that estuaries represent the drowned mouths of river valleys, as this immediately ties their origin into sea-level rise, yet today there is concern about the threat of this very same process as an agent of change. Sea-level rise, however, represents one of the greatest issues in relation to contemporary coastal systems. Any area that experiences tides will experience changes caused by deepening sea water and increased tidal penetration inland (see

Chapter 1). This is exacerbated for estuaries and delta channels, which are intensively used for development and, because of historic trends of sea-level rise and of coastal erosion and flooding, have seen increased use of coastal defence measures to protect human interests. Ironically, this policy now threatens to increase the vulnerability of habitats to loss through coastal squeeze. In the UK there is not one estuary that will not be vulnerable to coastal squeeze under sea-level rise. This is because they either have extensive defences or they are rock-bound (ria) type systems. This has major implications for intertidal mudflats and salt marshes. Similarly, the coastlines of many deltas, and the margins of their major distributary channels, are often heavily defended and are thus also susceptible to squeeze.

Impacts of sea-level rise can be seen over a range of scales. On a large scale, the increased loss of fringing salt marshes and greater penetration of the salt wedge further upstream are all ways of affecting the ecological and morphological structure of the system as estuaries widen and deepen. It is also feasible that the increasing volume of sea water entering on the flood tide may alter flood/ebb dynamics and affect sediment depositional patterns. Furthermore, increased volumes of sea water within estuaries may impede freshwater flow for longer periods, leading to the possibility that in some large systems, where the volume of sea water is great, there could be an increased risk of riverine flooding further into the catchment (Bird 1993). Other, more subtle changes can be equally important. Sea-level rise has caused salinity increases in some estuaries, which have led to faunal and floral changes. In Louisiana, for example, salt marshes have replaced marginal reed swamps as water salinity and frequency of inundation have increased (de Sylva 1986), and in the coastal lagoons of the Nile delta, such as Lake Burullus (see Case Study 7.2), increasing salinity of surface and groundwater has affected fish populations (Bird 1993). Similar salinity increases also occur in coastal groundwater aquifers, thus posing an increased threat to drinking water supplies.

With respect to deltas, continued sea-level rise may have an impact on the seaward edge, in that the sea will start to dominate over the processes of river discharge and fluvial sediment input. It is even argued that in many situations, the progradation of deltas will cease (Bird 1993) and that the low-lying lands of the delta will be more prone to flooding as a result of storm-wave overtopping and increased brackish water penetration up channels. Bird (1993) argues that a rise in sea-level of 1 m would cause the submergence of most of the seaward parts of the Nile delta, with the coastline moving several kilometres inland. Similarly, Milliman et al. (1989) suggest that much of the delta region of Bangladesh would be lost. Clearly, such change is dramatic and difficult to prevent. Defence provision is one response, but to cope with a sea-level rise of 1 m, and given that storm-wave height would also increase correspondingly with respect to current levels, the potential cost of the hard engineering necessary would be extreme.

On a smaller, but no less significant level, the survival of vegetation communities faced with sea-level rise is linked to two factors. First, the issue of coastal squeeze can cause marshes to experience lateral retreat and loss of higher marsh communities (see Fig. 7.13). Second is the threat of drowning. For a marsh surface to survive in the face of sea-level rise, it needs to grow vertically and keep pace with the deepening sea-level. Allen (1997, 2000) developed a simple model for quantifying this process. Given that the vegetation surface receives sediment in two forms, minerogenic and organic (see Allen (2000) for detailed review), then the total amount of material received has to exceed the amount of sea-level rise. Hence

$$\Delta E = \Delta S^{min} + \Delta S^{org} - \Delta M - \Delta P$$

where ΔE is the net change in marsh surface elevation, ΔS^{min} is the thickness of minerogenic sediment added to the marsh surface, ΔS^{org} is the thickness of organic sediment added to the marsh surface, ΔM is the change in relative sea-level and ΔP is the height change resulting from compaction. If ΔE is negative, then the rate of sea-level rise is greater than vertical marsh growth, and the marsh is in danger of drowning. Conversely, if ΔE is positive, then the vertical growth is out-pacing sea-level rise. Although such rates can be measured for a contemporary system, prediction of the likelihood of future marsh survival is difficult. This is because there are many unknowns with respect to estuarine response to progressive sea-level rise, particularly in respect of the sediment budget and hydrodynamics of the system.

Another significant aspect of sea-level rise is that the associated increase in marsh erosion and increased height and salinity of the water table will accelerate the reworking of a range of environmental pollutants. Case Study 7.3 detailed this idea from an erosion point of view, but chemical stability also becomes a factor when ground saturation and/or salinity increase. The key question to ask is where will this material go? Increased sediment loads due to marsh and mudflat erosion may be washed out to sea, or may remain in the estuary, depending on tidal symmetry. Either way, this may cause future management problems.

7.6.2 Impacts of increased anthropogenic disturbance

Coupled with sea-level rise, humans still see estuaries as areas of cheap land for development. Although with the advent of estuary management plans, development has become much more restricted, there are still future issues that pertain to this area of estuarine and delta impact. The growing leisure industry is imposing greater demands for marinas and sailing centres. As a 1993 English Nature report stated, there was not one major estuarine system in England that did not either have, or have plans for, a major marina development (Pye & French 1993). Similarly, the growth in ship size means that capital dredging needs to be carried out in many ports and harbours in order to develop berthing capacity. Again, this represents further capacity to change tidal and current activity. Furthermore, as population grows, so the demand for fresh water will see more dams being constructed in catchments.

In Cyprus, much of the coastline relies on fluvial sediment supply for its main source of sediment. Owing to increased population and tourist numbers, freshwater storage capacity increased from 6 million cubic metres to 297 million cubic metres between 1960 and 1988 (Smith & Marchand 2001). This has been achieved through increased reservoir construction and the building of dams. The result is that little sediment now reaches the island's estuaries, resulting in beach erosion along some coasts of up to 0.5 m yr^{-1}, ironically destroying the resource necessary for sustaining the tourist industry. The significance of this increased demand for fresh water cannot be underestimated. Aquifers and rivers are the prime sources of such water and increased extraction in both, but particularly rivers, will have key impacts in estuaries and deltas. The Nile delta (Case Study 7.2) is a good example, where important lessons should be learnt.

One further future issue to threaten estuaries in particular lies in the method being increasingly used to mediate against intertidal loss as a result of sea-level rise. Managed realignment is increasingly being seen as a way to increase the intertidal volume of an estuary, yet there are many unknowns about this technique that need to be investigated. As a comparatively new approach to coastal management, its increased usage has occurred against a background of little firm knowledge of the longer-term impacts and a range of issues have arisen from early uses of the method. Most notable appears to be the role of pre-realignment vegetation on a site. Because vegetation is effective in enhancing the trapping of sediment on developing marsh surfaces, its presence can be seen as beneficial (Stoddart et al. 1989; Fig. 7.6). In some examples of realignment, however, vegetation has not served this function but has decayed following rapid burial (Macleod et al. 1999). This decay has led to the generation of anoxic conditions, which have actually delayed marsh initiation. Other factors include: larger scale changes in soil properties, often governed by what has happened in terms of improvement during its agricultural history; the change in groundwater dynamics, leading

to the release of stored contaminants; and general changes in ground water levels and salinity (Boorman & Hazeldon 1995; Crooks et al. 2002; Blackwell et al. 2004). Overall, the message here is that although realignment presents a logical and sound approach to increasing estuarine marsh loss, there are important lessons still to be learnt. Some of these can be informed from existing schemes, and some from looking at historic storm-breached sites. Overall, however, there are still major questions to be answered, with much of the required knowledge having to come from existing realignment sites.

Much of what has been written has been predicated by a presumption that scientists understand deltas and estuaries. In some areas of knowledge, this is the case, but in others, the ability to predict what may happen in the future is based on a poor understanding of the principles and functioning of these highly dynamic systems. The bottom line is that although we can measure and attempt to understand how estuaries work now, our ability to predict change is still restricted. Perhaps the key issue here is that estuaries and deltas are complex systems that contain a range of linking processes combining aspects of geomorphology, sedimentology and ecology, as well as human influences. One essential issue for the future, therefore, is to manage the whole system as an entity with a range of component parts, and not to manage the component parts individually. The ecosystem approach to management is a philosophy that is gaining popularity across a range of systems. By combining population dynamics with nutrient and sediment cycling, ecological production, sediment and water movements, and anthropogenic usage, management can be made much more effective. Mann (2000), however, reported that at the time of his writing efforts to produce predictive models of coastal ecosystems, such as estuarine and delta systems, have had limited success. One of the key issues here is the lack of adequate data sets to allow models of the complexity of estuarine interactions to be developed, and a second is to develop such a model that can cope with the spatial and temporal dynamics necessary.

8

Temperate coastal environments

Andrew Cooper

8.1 INTRODUCTION

This chapter addresses the sedimentology of the clastic depositional environments of open ocean temperate coasts. The environments that will be considered are associated with beaches, barriers and barrier islands composed largely of reworked terrigenous material. These environments are commonly sandy, but reference will also be made to coarser grained gravel-, cobble- and boulder-dominated systems. These types of environment are typical of the temperate latitudes. In the tropics they are often replaced by systems in which organic materials dominate (Chapter 9). Estuarine and deltaic environments are considered in Chapter 7.

The common characteristic of the environments considered is the non-cohesive nature of the constituent clasts. The consequent ability for clasts to readily be reorganized into different geomorphological forms in response to changes in environmental conditions is a key characteristic. It is commonly asserted that beaches, as loose accumulations of small grains of sand or pebbles, are among the most unlikely landforms to exist in the often-harsh conditions of open coastal areas (Pethick 1984). The very ability of these accumulations to change shape and absorb excess energy in response to storms is central to their ability to survive when more solid structures, such as seawalls, can be destroyed under the same conditions.

Thus open coast sedimentary environments, composed of a range of non-cohesive clast sizes, under a variety of energy conditions, take many geomorphological forms. The controls on their spatial distribution, the forms that they take and the controls on their development are outlined in this chapter. Morphological change over time is discussed in order to provide an appreciation of the control of changing environmental conditions on coastal morphology. The interaction of humans with these dynamic coastal systems adds an additional layer of complexity to the dynamics and future evolution of temperate coasts, and thus past human interventions and future management options are assessed.

8.2 NATURE AND SIGNIFICANCE OF TEMPERATE COASTS

Temperate coasts (Fig. 8.1) are widely distributed in the Northern Hemisphere on Atlantic and Pacific coasts of North America, Europe and North Africa, and the Far East. In the Southern Hemisphere they are limited to the Argentinean and Chilean coasts of South America, Southern Africa, and the southern coasts of Australia and New Zealand. They are bounded toward the poles by coasts with sea ice, and toward the equator by tropical coasts. Temperate coasts can be crudely categorized into open ocean coasts (swell- or storm-wave dominated – see section 8.2.3), sheltered seas and areas subject to tropical storms (Fig. 8.1).

Sedimentary environments considered here are beaches, barriers, barrier islands and coastal sand dunes (Fig. 8.2). A beach is an accumulation of wave-deposited, non-cohesive sediment that typically spans the subtidal–supratidal interface. Sustained beach sedimentation can give rise to a barrier (Woodroffe 2002). Barriers are

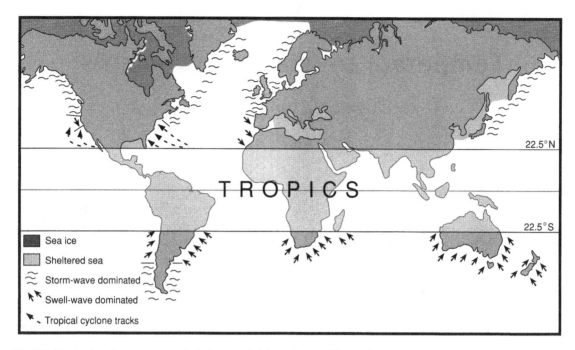

Fig. 8.1 Distribution of temperate coasts including wave heights and types. (After Davies 1980; Short 1999.)

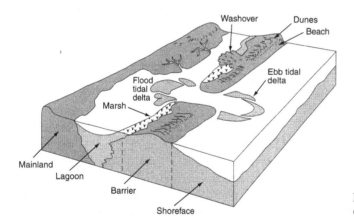

Fig. 8.2 Diagram of main sedimentary environments considered in this chapter.

large, shore-parallel accumulations of sediment (of which a beach forms the most seaward part), deposited under wave action and which separate a low-lying area (marsh, lagoon or estuary) from the ocean. Barrier islands are barriers that are surrounded on all sides by water. Coastal dunes develop in the supratidal zone by aeolian transport and deposition of sand from adjacent beaches.

The critical conditions for development of these environments are as follows:

1 adequate sediment supply;
2 suitable accommodation space (shaped by antecedent coastal morphology);
3 sufficient wave energy levels to move available sediment.

Sediment is derived from various sources and has varying texture. There is a broad latitudinal control in that gravel beaches are preferentially developed in glaciated or formerly glaciated (paraglacial) regions (Ballantyne 2002). Sediment texture is an important constraint on beach morphology and behaviour. Differing clast sizes influence packing, resistance to motion and porosity, all of which affect beach morphology. The accommodation space in which beaches form is controlled by the antecedent coastal geomorphology. Commonly beaches form in embayments, but the geomorphology of the shoreline is an important factor in their distribution. If too steeply dipping, sediment accumulation may not break the water surface and no beach will form. Sufficient wave energy to transport available sediment sources is an obvious requirement – insufficient energy will result in a variety of unmodified terrestrial deposits rather than beaches. The most important attribute of the environments described here is their ability to change morphology as energy levels change and to seek equilibrium with environmental conditions. This ability to accommodate change renders them important buffers against high-energy storm waves.

At a global scale, there is much variability in wave energy levels (Fig. 8.1). Swell waves on oceanic coasts are often fully refracted and arrive parallel to the shoreline, whereas locally generated sea waves in sheltered environments may arrive at an angle to the coast. Paradoxically, these lower energy waves may be more efficient in effecting net sediment transport through longshore drift. Tidal range exerts a secondary influence on beach morphology as it mediates the vertical and horizontal area in which wave energy is expended. Higher tidal range tends to reduce the relative influence of waves on beach geomorphology.

The defensive significance of such malleable natural environments to human developments at the coast has not been uniformly appreciated. There has been much historical degradation of beaches and dunes and in some localities this is ongoing. The recreational and landscape attributes of temperate coasts combine to draw increasing numbers of humans to the coast. The consequent infrastructural development is often poorly located and impedes the ability of natural beach systems to respond to changing conditions.

The environmental significance of temperate coastal sedimentary environments can be expressed or considered from several, often competing, human perspectives (physical, ecological and economic). In a *physical* sense, the environments provide a dynamic natural system capable of responding to and absorbing high levels of marine energy. These environments are often regarded as hazardous to navigation. From an *ecological* perspective, beaches have been recognized, over the past 20 years in particular, as important ecosystems (Brown & McLachlan 2002). The interstitial fauna and flora play a role in the ecology and sedimentology of adjacent inshore and dune ecosystems as do their distinctive vegetation and related fauna.

The *economic* significance of beaches is evident in the extent of seafront development and the vast numbers of visitors to resort beaches. As a coastal defence against wave action, beaches are also economic assets. Beaches too provide access to the coast and they have also been seen as ready sources of aggregate (Carter et al. 1991). The human values attached to beaches are clearly not always compatible with each other or with their ecological and physical value. The nature of human pressure on beach and dune environments changes with time (Nordstrom 2000). Much of this is technology-dependent. For example, widespread kelp collection in the north-west British Isles during the Napoleonic wars for iodine extraction undoubtedly affected the development of drift lines, so necessary for dune development. Similarly, changes in farming practice have controlled the patterns of small-scale sand removal from beaches in Ireland (Carter et al. 1991).

Human pressure on the coast takes many forms and the continuing migration to the coast is set to increase those pressures. Cohen et al. (1997) estimated that over 2 billion people (37% of the global population) live within 100 km of a coastline. Dramatic increases have been noted in the temperate regions of the Mediterranean and the USA ocean coasts. In the Mediterranean, the coastal population was estimated at 146 million in 1990 and the urban coastal population alone is projected to rise to 176 million by 2025, with an additional 350 million tourists (Hinrichsen 1998). In the USA, 55–60% of the population live in the coastal counties of the Atlantic and Pacific coastlines and coastal population density rose from 275 to 400 people per square kilometre between 1960 and 1990 (Hinrichsen 1998).

8.2.1 Environmental sedimentology of temperate coasts

The world's temperate coasts are among the most densely populated and developed in the world. As a consequence, large stretches of temperate coastlines have been subject to intensive anthropogenic modification with attendant sedimentological impacts. In many cases these involve direct impacts on the sediment volume (e.g. dune removal, sand mining) or sedimentary processes (e.g. jetty construction interfering with longshore drift). In other instances, indirect influences result from, for example, the armouring of coastal cliffs, which formerly supplied sediment to adjacent beaches. In all cases, human development of the coastline has the potential to foreclose future management options. Intensive infrastructural development at the coast almost inevitably leads to the desire to armour the shoreline in the face of coastal erosion. Armouring fixes the shoreline position, thus setting a human limit to the acceptable extent of shoreline processes. This constrains the ability of the beach system to operate naturally and passes the fate of the beach into the human decision-making sphere. This has given rise to the study of 'developed coasts' as a distinctive field of investigation (Nordstrom 2000). The distinction is somewhat blurred as even remote, rural beaches often exhibit significant human alteration and impact (Power et al. 2000).

8.2.2 Overview of issues, processes and problems

There are many contemporary management issues in the temperate coastal zone, most of which involve conflicts between human utilization and natural processes of change. Even in an environment with predictable dynamics and a fully understood coastal system, the issues would still be difficult to resolve because of the human value system, politics and economics that influence human responses to pressure (Pilkey & Dixon 1996). In reality, the dynamics of the coastline are unpredictable (storms occur chaotically) and the coastal system is not well understood (the complexities and feedbacks in a coastal system are largely unresolved). Furthermore, sea-level, the plane at which coastal dynamics operate, is changing on a global scale, and global climate change may be altering the frequency and intensity of storms. This is manifest in many parts of the world as a propensity toward coastal recession (Bird 1985). Added to this is ongoing deliberate and accidental modification of sedimentary processes through human intervention.

8.3 SEDIMENT SOURCES AND SEDIMENT ACCUMULATION PROCESSES

8.3.1 Sources and characteristics of sediments

Temperate coastal environments receive sediment from various sources and inevitably beach morphology is strongly influenced by the nature and volume of available sediment (Fig. 8.3). The main sources are rivers, the continental shelf, coastal erosion and redistribution (see Chapter 1) and there are important regional differences in the type of sediment supply. In the western USA fluvial sediment supply is dominant because of the steep hinterland gradient, whereas on the eastern coast, continental shelf sources predominate.

Fluvial sources are a key component of temperate sediment supply, especially adjacent to large rivers or rivers that drain steep hinterlands. Fluvial sediments may be reworked within a delta complex, often to form barrier islands, for example the Nile delta (Pilkey 2003). In other instances, sediment is transported from the river mouth to adjacent coastal depocentres (Cooper et al. 1999). Fluvial sediment supply is often episodic, particularly where a seasonal discharge regime exists (see Chapter 3). Thus coastal sediment supply may fluctuate between periods of high fluvial supply followed by intervals

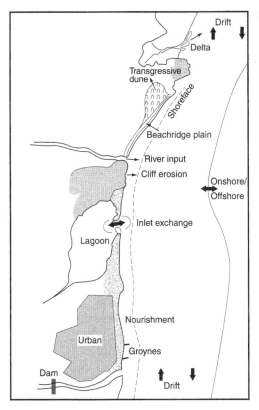

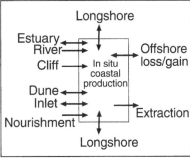

Fig. 8.3 Diagrammatic representation of coastal sediment budget showing main sources and exchanges both natural and anthropogenic.

during which sediment is reworked by coastal processes (Cooper 1993).

The continental shelf is a vital sediment source for many temperate coastlines. Here relict terrestrial, marine or coastal sediments from lower sea-levels or past glaciations may be reworked landward by contemporary wave action. Offshore sediment abundance was accompanied by millennial-scale coastal progradation in southeast Australia (Thom 1984) and decadal-scale beach ridge accumulation in south-east Spain (Goy et al. 2003). In conditions of sediment scarcity, wave action may, however, erode the sea-floor to transport older deposits to the shoreline. In North Carolina, grains eroded from underlying Tertiary lithologies are a component of contemporary beach sediment (Pilkey et al. 1998). Sediment transport on the shelf varies according to wave conditions and during storm-wave action extends to deeper water. Thus the shelf sediment supply may also be temporally variable (see Chapter 10).

The coastline itself is a source of sediment to adjacent beaches. Often, longshore drift of sand from adjacent beaches and cliffs is a major element in the sediment supply, although this is diminished in strongly embayed coasts. Erosion of relict glacial deposits yields abundant gravel and boulder beaches in the higher latitudes (Davies 1980). Elsewhere, the lithology and texture of coastal outcrops strongly influence the nature of eroded clasts that supply beaches. In southern England, for example, the preferential preservation of chert (flint) clasts eroded from less resistant limestone (chalk) is marked (Carr 1969). The supply of sediment is not constant and is controlled to a large extent by patterns of slope failure on coastal cliffs, and the capacity of waves to transport sediment. Slope failure depends on a range of factors including the porosity and permeability of the rock, stratification, rainfall intensity and the extent of wave undercutting (Emery & Kuhn 1982). Often sediment input is in the form of large-scale landslides

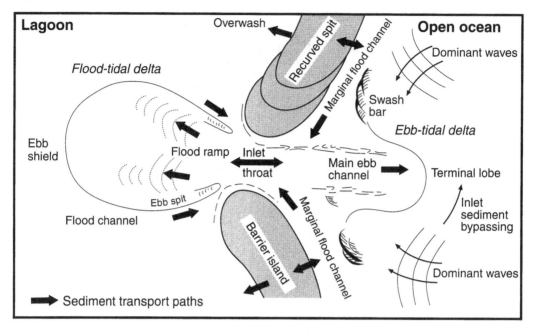

Fig. 8.4 Main geomorphological features and sediment transfer pathways at a tidal inlet. (After Hayes 1991.)

that occur episodically. Brunsden & Moore (1999) documented a landslide on the Dorset Coast (England) that blocked alongshore sediment supply to adjacent beaches.

Estuaries, depending on their hydrodynamics, may be either sediment sources or sinks in the coastal zone (Cooper 2001a) (see Chapter 7). Flood currents may produce landward transfer of bedload sediments (sand and gravel), whereas ebb-currents, often augmented by river discharge, induce their seaward transport (see Chapter 1). Reworking of sediments among different elements of the coastal system produces local sources and sinks of sediment in response to changing dynamics. Cycling of sediment between tidal deltas, inlets and beach-dune systems exemplifies the process (Fig. 8.4). Temporal lags in sediment transfer rates between the various parts of the inlet system may lead to periodic erosion and accretion on beaches, and episodic events such as storms strongly influence rates of sediment transfer (Morton et al. 1995).

The contribution of organic materials in temperate latitude beach systems is typically much less important than in the tropics. At sites of low terrigenous sediment supply, however, carbonate materials may dominate the coastal sediment. Barnhardt & Kelley (1995) documented a beach in Maine where sediment was derived from offshore mollusc fragments. In western Ireland, carbonate-rich beaches composed of foraminifera and coralline algae are locally abundant (Bosence 1980). Less persistent, but volumetrically substantial organic components of temperate beaches include woody debris and marine algae. Although these components eventually decay they provide important nutrient sources, and foci for subsequent aeolian accumulation. In areas where beaches are underlain by back-barrier sediments, eroded peat or mud may form clasts on contemporary beaches.

8.3.2 Temperate coastal depositional landforms

A suite of coastal landforms develops in temperate latitudes in response to sediment supply and a suitable sediment accumulation zone or trap. These landforms (Fig. 8.5) at the macroscale include barriers (Fig. 8.6a), barrier islands (Fig. 8.6c), and mainland-attached beaches in

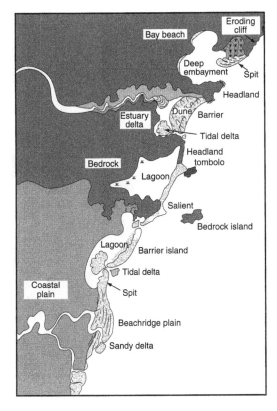

Fig. 8.5 Temperate depositional landforms (headland–embayment beach, spit, tombolo, salient, cuspate foreland, barrier, barrier island).

the form of cuspate forelands (Fig. 8.6b) and spits (Fig. 8.6d). Barrier islands are surrounded on all sides by environments that are regularly submerged. Thus a barrier island is fronted by the open ocean and backed by a sheltered lagoon (which may be marsh-filled). They are bounded at either end by inlets that permit exchange of water and sediment between the back-barrier environment and the open sea.

Barriers enclose a marsh, lagoon or estuary and, unlike barrier islands, are mainland-attached. An inlet may or may not be present, through which water flows between the back-barrier system and the ocean. In some instances these inlets are ephemeral or seasonal features dependent on freshwater flow (Cooper 2001b). Inlets on both barriers and barrier islands may or may not contain flood- and ebb-tidal deltas. The presence and morphology of tidal deltas are controlled by the relative strengths of waves and

tidal currents, coupled with the requirement for a suitable accommodation space in which to form (Oertel 1985).

Mainland-attached beaches are backed by dry land rather than water. These beaches (and some barriers) may be strongly influenced by the onshore and nearshore topography. A number of distinctive beach planform shapes develop in response. These include headland–embayment systems and cuspate forms (salients, forelands and tombolos) (Sanderson & Elliot 1996).

Coastal dunes constitute the aeolian component of many temperate beach systems and exhibit a wide range of morphologies (Fig. 8.7). In temperate regions many coastal dunes are vegetated. The dunes begin as accumulations around debris on the beach, typically drift lines of seaweed and other material. Initial colonization by strand-line vegetation partly stabilizes these dunes and may permit continued accretion as wind-blown sand is trapped by the vegetation. Dunes comprise varying areas of vegetated and unvegetated sand and are intimately linked to their source beach areas (Sherman & Bauer 1993).

On prograding systems with abundant sediment supply and strong vegetation growth, shore-parallel lines of foredune ridges may form. Less vigorous vegetation growth leads to a less organized, hummocky appearance. Very dense and vigorous vegetation can lead to effective sediment trapping and vertical dune growth. A steep nearshore slope probably accentuates this form of dune growth by precluding seaward progradation. In the absence of vegetation a transgressive sand sheet may form, which migrates landward across pre-existing features. Instability and vegetation damage may lead to blowout formation on vegetated dunes and, if persistent, will cause the development of parabolic dunes that migrate landward through the dune system. As a result, dunes must be regarded not only in terms of interactions between the beach and dune but as dynamic systems in their own right in which morphological change and sediment transfer may proceed intermittently.

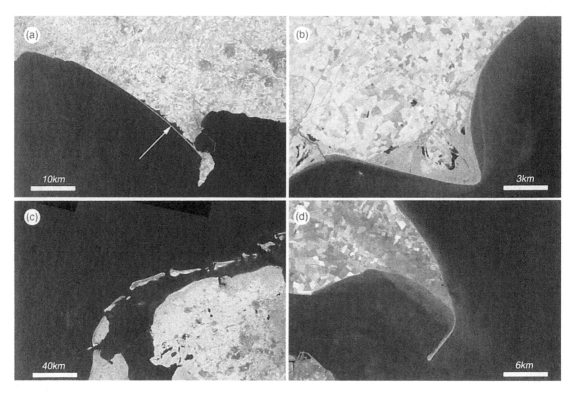

Fig. 8.6 Depositional landforms: (a) Chesil Beach, a barrier (arrowed); (b) Dungeness, a cuspate foreland; (c) The Frisian Islands, a barrier island chain; (d) Spurn Head Spit at the Humber Estuary mouth. (Images from NASA Earth Science Applications Directorate (https://zulu.ssc.nasa.gov/mrsid/).)

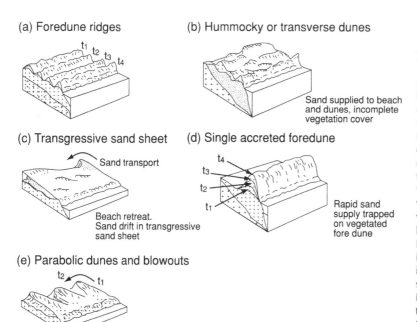

(a) Foredune ridges

(b) Hummocky or transverse dunes

Sand supplied to beach and dunes, incomplete vegetation cover

(c) Transgressive sand sheet

Sand transport

Beach retreat. Sand drift in transgressive sand sheet

(d) Single accreted foredune

Rapid sand supply trapped on vegetated fore dune

(e) Parabolic dunes and blowouts

Beach retreat. Blowouts result in parabolic dune migration

Fig. 8.7 Typical dune morphologies. On gentle gradient basement slopes (a), shore-parallel beach ridges form. Foredune ridges may develop on each beachridge to form a sequence of dune-topped ridges. If vegetation is patchy, the dune topography is more irregular (b) and if sparse or absent, a transgressive sand sheet (c) may form. On steep basement slopes with good vegetation growth, vertical accretion may give rise to a high, single dune ridge (d). Blowouts are associated with breaks in vegetation cover and which involve cannibalization of the dune sand by ongoing wind action (e). In time parabolic dunes may evolve that transport sediment landward through the dune system Temporal evolution is denoted by t_1–t_4. (After Woodroffe 2002.)

8.3.3 Controls on sediment accumulation and transport

Sediment transport in the nearshore zone is a highly complex phenomenon that occurs through a variety of interlinked processes operating at spatially and temporally variable intensity. A synopsis is presented here; for a more thorough review of sediment processes in the nearshore the reader is referred to Komar (1998) or Woodroffe (2002). The dominant force in sediment transport and geomorphological change in the coastal zone is wave action. Waves are formed by winds blowing over a water surface. They are typically described in terms of wave length, height and period. Once formed, waves radiate out from the generation area and the wave form propagates across the water surface. Typically, a wide range of wave sizes (defined by wavelength and period) are produced by winds at the source area. The waves formed directly in the area of generation tend to be of various sizes and the water is very 'choppy'. Larger waves are produced by winds of greater speed and duration, and greater fetch distances. As waves travel further from the generation area they become sorted and amalgamated. Longer waves travel faster and arrive at distant coasts first as a regularly spaced set of large waves (termed swell). Oceans have large fetches and hence large waves are produced. Coasts facing open oceans are thus typically dominated by such conditions. In semi-enclosed seas, most waves are generated by more proximal winds blowing across restricted fetches and hence the waves are smaller and more variable in their dimensions.

As waves move from deep to shallow water, friction with the sea-floor retards the wave, which maintains its energy by steepening. Waves with longer wavelengths 'feel' the sea-floor first and are slowed comparatively far from the shore. The point at which significant wave–sediment interaction begins is known as wave base (see Chapter 1). It occurs at variable depth depending on the length of the waves. Waves approaching a shoreline obliquely thus change the orientation of their crests in response to bottom bathymetry. This process by which the wave crests are reorientated is known as refraction and it is important in distributing wave energy along a shoreline. A related process of diffraction, by which wave energy is transferred laterally along the crest, also occurs as wave crests bend. Because swell waves have longer wavelengths they interact with the sea-bed further offshore and hence arrive more fully refracted than short wavelength waves. Full refraction tends to arrange wave crests parallel to the shoreline and to equally distribute energy alongshore. Short period, locally generated waves exhibit greater energy differences alongshore due to incomplete refraction.

Waves undergo further modifications as they approach the shore and a series of zones are defined according to wave behaviour near the coast (Fig. 8.8). As waves interact with the sea-bed and slow, they become higher. The zone in which this takes place is known as the shoaling zone. Wave steepening may lead to instability as the wave becomes too high relative to its length. It then 'breaks' as gravity collapses the waveform. A range of breaking criteria has been determined empirically (Komar 1998), and several breaker types are recognized in a continuum including surging, collapsing, plunging and spilling breakers related to the mode of energy release.

Landward of the breaker zone, wave energy continues to propagate shoreward in what is known as the surf zone (Fig. 8.8). This is a zone of often intense turbulence as waves reform, break again and generate secondary wave and current motions. Much wave energy is dissipated in the surf zone through breaking, turbulence and sediment transport. The surf zone is variable in width. Wide surf zones are associated with large waves, which dissipate their energy through a series of spilling breakers. Lower energy beaches have narrow surf zones in which wave energy is dissipated through breaking close to the shore. Reorganization of wave energy can lead to generation of secondary wave forms (edge waves) that have longer wave periods than incident waves. These have crests that are oblique or perpendicular to the shoreline. They can be stationary or may propagate along the shore and give rise to widely spaced inequalities

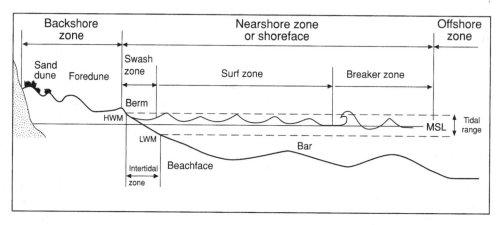

Fig. 8.8 Nearshore wave zones and profile morphology: HWM, high water mark; LWM, low water mark; MSL, mean sea level. In the offshore zone waves begin to interact with the sea-bed, and increase in height as the water depth diminishes. Waves break, generating bars as they do so, in the breaker zone and propagate across the surf zone toward the shoreline. The final wave energy dissipation occurs in the swash zone. Overwashing occurs when waves overtop the berm and flow landward across the backshore. Aeolian transport of sand from the beach may produce dunes at the rear of the beach.

in water level, often observed as regularly spaced, alongshore peaks in swash. In addition, secondary currents may be generated due to inequalities in the water surface elevation and energy density. These currents may be arranged into circulatory cells that redistribute water through shore-normal (rip) and/or shore-parallel (longshore) currents (Fig. 8.9).

Energy that propagates to the shore through the surf zone to the beach may run up the beach for some distance as swash. The elevation to which swash rises is dependent on wave height, the porosity of the beach material and the extent of saturation at any given time. Excess water that does not infiltrate the beach runs back down the beach as backwash. The high porosity of gravel beaches reduces or eliminates backwash. Swash is normally the zone in which final dissipation of wave energy takes place. On steep beaches, however, excess energy may be reflected seaward, or, if the swash reaches the beach berm, it may flow in a landward direction as overwash.

Waves are thus responsible for a variety of fluid motions and potential sediment-transporting mechanisms on temperate coasts. Commonly, for ease of understanding these are divided into cross-shore and longshore transport mechanisms, although in reality the system is fully

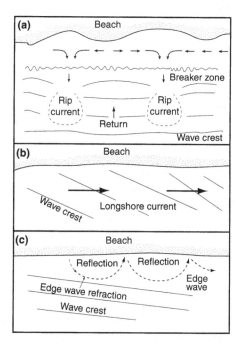

Fig. 8.9 Nearshore circulations in the surf zone take the form of circulation systems (a) comprising rip currents and alongshore currents, (b) shore currents generated by oblique incident waves and (c) edge waves produced by reflection of incident wave energy from steep beaches and subsequent refraction back towards the shoreline.

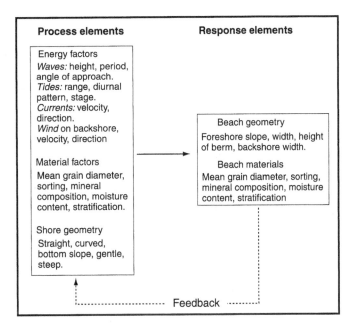

Fig. 8.10 Conceptual model of the coastal system illustrating process elements (energy, materials and geometry) and response elements (geomorphology and texture) and feedback between the two. Variability in all factors and the presence of feedback, render quantification of nearshore processes difficult. (After Krumbein 1963.)

three-dimensional. Longshore sediment transport takes place largely via waves that approach the shoreline obliquely. These move sediment in the direction of the dominant wave approach. The process is, however, highly variable. Larger waves interact with the sea-bed further offshore and extend the active transport zone. Cross-shore sediment transport involves the onshore or offshore movement of sediment in response to changing wave character and bed slope.

The main role of tidal water level variation on sedimentation on beaches is to mediate the plane at which wave processes operate and hence coasts are commonly categorized on the basis of tidal range (see Chapter 7). In areas of higher tidal range, wave energy is spread over a greater vertical range (and hence lateral extent) and is less effective than in areas of low tidal range. Low tidal range has often been associated with wave dominance on coasts (Davis & Hayes 1984), however, Anthony & Orford (2002) have recently drawn attention to the features of mixed-energy coasts with high levels of both wave and tidal control. Each of the wave-generated energy fluxes outlined above has the potential to transport and reorganize the non-cohesive sediments of beaches. The nature

and relative strength of these water motions will influence the direction and magnitude of sediment transport, as will the dimensions and texture of the beach sediment. It is also important to recognize the feedback relationships that exist between morphology, fluid dynamics and sediment transport (Fig. 8.10).

Although actual sediment transport takes the form of individual grains moving in suspension, saltation or bedload (see Chapter 1), the integrated transport volumes are of interest in understanding beach behaviour. The link between coastal landform and the formative processes has given rise to the field of study known as morphodynamics in which the relationship between dynamic forcing and morphological response is considered (Carter & Woodroffe 1994). The typical approach to sediment transport at the coast is to consider longshore and cross-shore components separately, although both operate together. Movement of beach sediment in a shore-normal direction is largely driven by inequalities in the landward and seaward velocity and duration of wave-induced currents. The dominant mechanism in longshore transport is movement of sediment by longshore-directed currents generated by oblique waves (USACE 2003). As most wave energy dissipation takes place in the surf zone, this is a zone of high sediment transport potential.

The processes that affect coastal sand dunes are quite distinctive as they rely not upon wave action, but wind. The presence of coastal dunes requires three criteria to be met. First, there must be a supply of sand. This is usually the adjacent beach and beach supply is enhanced if the sand is dry and transport is not impeded by coarse fractions. Second, the wind velocity must be sufficiently strong to entrain grains and must have a significant onshore component. The third requirement is an interruption in the airflow such that velocities drop and sand is deposited. This is commonly provided by a physical obstacle on the beach, but may also be related to topographic features or vegetation, which cause airflow to diverge. Potential constraints on aeolian transport on beaches include fetch distance, moisture, diagenesis (e.g. salcretes), surface roughness and armouring (Sherman & Bauer 1993).

8.3.4 Coastal morphology

Just as sediment transport is commonly considered in longshore and cross-shore contexts, the same is true of beach morphology. Below, the morphology of beaches is considered first in a cross-shore (profile) context and second from an alongshore (planform) perspective. In reality the two are closely related in the fully three-dimensional beach morphology.

8.3.4.1 Beach profile morphology

The cross-shore geometry of a beach is strongly influenced by its constituent grains. Boulder beaches (clast diameter > 0.25 m) have low gradients (typically 6–14°). This is lower than the natural angle of repose and results from the fact that they are flattened by extreme storms and there is no mechanism to rebuild them. Gravel beaches (clast diameter > 2 mm) in contrast tend to be steep (> 15°) because the large pore space promotes infiltration of water rather than surface backwash. In contrast sand

beaches are gentler in gradient. In general, sand beaches assume lower gradients (usually < 10°) with higher wave energy; the reduced gradient is associated with dissipation of wave energy. Paradoxically, this may result in lower residual energy at the coast than on steeper sand beaches associated with lower wave energy. In the latter case, residual energy is reflected from the beach. The terms dissipative and reflective have thus been used to describe sand beaches in a number of morphodynamic classifications. Beaches with mixed grain sizes tend to show some degree of spatial segregation of the grains. Typically this takes the form of a high tide, steep beach facet composed of coarse material, fronted by a low-gradient sand 'apron'. Such beaches comprise both dissipative and reflective elements.

Beaches show a range of features (Fig. 8.8) that are developed to greater or lesser extents depending on several factors. The beachface is the zone in which waves are altered as they approach the shoreline. The berm is the limit to which swash action typically extends, and is an aggradational feature. The back-beach typically slopes gently landward and is activated by overwash processes. It also acts as the source zone for dune sands. The beach may merge into a dune system in which ephemeral dunes give way to a foredune ridge that in turn may be backed by more stable dunes.

8.3.4.2 Coastal planforms

The processes of sedimentation outlined above have high potential to sort and arrange individual clasts into distinctive landforms. In most circumstances, there is a limited sediment supply, and in all cases, a topographic control on the distribution of sediments along the coast. Thus coastal sedimentary systems are discontinuous and form discrete systems. Although some sedimentary systems are physiographically distinct from a planform perspective (e.g. headland–embayment cell), others on linear clastic coasts are more difficult to compartmentalize. The concept of coastal cells enables identification of semi-closed zones within which material fluxes may be quantified (even if in

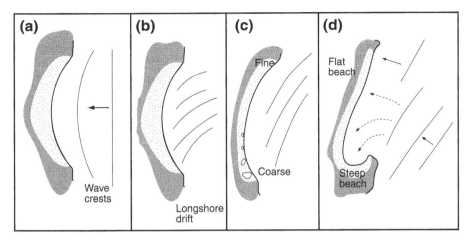

Fig. 8.11 Cell equilibrium types: (a) swash-aligned equilibrium where wave crests arrive at the shoreline with equal distribution of energy alongshore; (b) drift-aligned equilibrium where sediment flux equals the potential transport ability of waves; (c) graded equilibrium where grain size at any point is too great for wave transport; (d) log spiral or zeta bay.

broad terms). Such cells are commonly defined in terms of the two-dimensional horizontal plane arrangement of sediments, known as the coastal planform (Fig. 8.11). The simplified approach of May & Tanner (1973) in which longshore gradients in wave energy at the breakpoint are quantified, provides one approach to the delimitation of cells for any given wave condition. A number of equilibrium cell forms have been identified (Fig. 8.11) (Carter 1988).

In swash-aligned equilibrium, wave energy is distributed evenly along a shoreline such that there is no longshore wave energy gradient to produce sediment transport. In a graded equilibrium, the beach sediments become sorted in response to longshore gradients in wave energy. The phenomenon is best seen on gravel or cobble beaches and is manifest by alongshore gradients in clast size such that the clasts at any given point are too large to be transported by wave energy at that point. On coasts with dominantly oblique wave approach, drift-aligned equilibrium may be attained at any given point if inputs of sediment from up-drift sources are matched by losses to downstream sinks. Particular types of equilibrium forms (zeta and log-spiral bays) comprise both swash-aligned and drift-aligned sections and occur where a headland obstructs longshore drift incompletely (Woodroffe 2002).

8.4 PROCESSES AND IMPACTS OF NATURAL DISTURBANCE AND ENVIRONMENTAL CHANGE

8.4.1 Temporal change in temperate beach and dune morphology

Beaches and dunes are characterized by a wide range of types, and grain size provides a first-order discrimination among boulder, gravel and sand beaches. Although these types form a continuum there are distinctive modes of behaviour and response to environmental forcing for each. Boulder beaches (Oak 1984) are inactive under most coastal conditions because of the large clast sizes involved. Extreme (high magnitude, low frequency) events (e.g. storms and tsunami) thus create the only circumstances under which the beach morphology changes. The boulder beach gradient at any time reflects a combination of the intensity of the last dynamically effective storm, as well as the cumulative work of a series of such storms. An interesting feedback relationship thus exists in that the morphodynamic effectiveness of successive storms of the same magnitude is reduced by the beach having moved closer to equilibrium with that magnitude of storm.

Gravel beaches have been shown, largely by virtue of their porosity, which inhibits backwash,

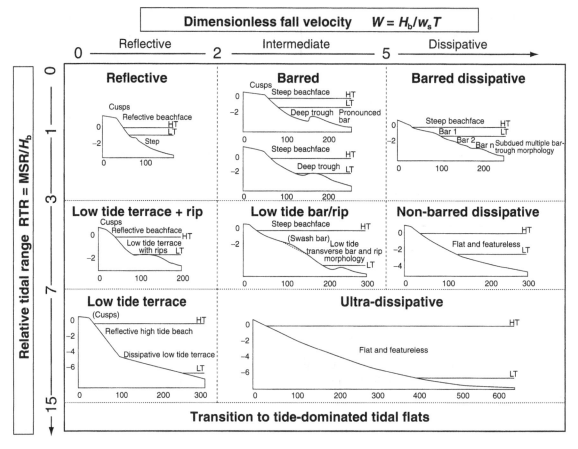

Fig. 8.12 Conceptual beach state models for micro- to macrotidal beaches: MSR, mean spring tidal range; H_b, breaking wave height; W_s, sediment fall velocity; T, wave period. With increasing breaking wave height and/or decreasing grain size, beaches become gentler in gradient and more 'dissipative' of incident wave energy. The relationship of tidal range and breaking wave height (relative tidal range) mediates the vertical range over which wave energy is dissipated. Thus beaches with larger tidal range become progressively more dissipative. (After Masselink & Short 1993.)

but also their comparative resistance to movement under low wave energy, to exhibit landward migration via barrier overwashing. High porosity precludes a seaward return mechanism for beach clasts and thus gravel beaches tend to be steep because of the higher angle of repose of coarse clasts. Sand beaches are the most widely distributed and morphologically variable of beach types. For this reason they have been studied more intensively than other types. Attempts to synthesize studies of beach morphology have been made, perhaps the best known resulting in the concept of beach state models (Wright & Short 1984; Masselink & Short 1993). These studies recognize spectra of beach types that

are arrayed on a dissipative (low-angle beach gradient) to reflective (steep gradient) continuum (Fig. 8.12).

Dissipative beaches are associated with long period, large waves with high incident energy. Energy is dissipated by wave breaking across a wide, low-angle shoreface such that waves at the shoreline have little remaining energy. Because such beaches are associated with long-period waves, these are often fully refracted by the time they reach the shoreline and hence temporal morphological variability on such beaches is comparatively low. Reflective beaches at the opposite end of the continuum are associated with coarser sediment and/or smaller waves of shorter period. These beaches tend to permit propagation of incident wave energy closer to the shore and thus waves break closer to the shoreline with more residual energy than on dissipative beaches. Excess energy is reflected from the beach face, which tends to be steep. Waves are often not fully refracted at the shoreline and are thus more likely to generate secondary wave motions and currents. The presence of beach cusps, and high longshore sediment transport rates are physical manifestations of such conditions. Intermediate beaches between these two extremes display both dissipative and reflective elements. On such beaches, a low-angle dissipative facet is backed by a steep, reflective facet, and wave energy is accommodated by a combination of dissipation and reflection.

Empirical data have been used to relate beach state to various morphodynamic indices including Dean's parameter (combining grain size, wave height and period) and surf scaling parameter (combining nearshore slope, wave height and period) (Wright & Short 1984; Masselink & Short 1993). These beach state models provide a set of 'expectation criteria' for beach morphology in a given dynamic setting. A number of authors (e.g. Hegge et al. 1996) have questioned the universal applicability of beach state models and it is likely that factors such as sediment supply, underlying topographic control and variations in wave climate may well create a wider variety of beach states than is accommodated within existing models.

Within the beach state approach, temporal changes in factors such as wave height are matched by corresponding changes in beach morphology that take place in a predictable sequence. These changes are largely described by variations in profile shape, driven by cross-shore sediment transport. With decreasing wave energy, changes take the form of bar migration onshore and eventual welding onto the beachface. Increasing wave energy leads to a lowering of beach gradient and bar formation on the shoreface. Beaches with strongly seasonal wave climate variation may traverse the full range of beach states (Shaw 1985), whereas others remain within the same state perennially. There

are many beaches that do not appear to fit the beach state models and the role of sediment transport thresholds remains a constraint on such models, whereby beach morphology might reflect only higher energy events.

8.4.2 Planform variability

The planform morphology of beaches and barriers in relation to sediment supply and accommodation space has been considered in the previous section. Some aspects of the gross morphology of beaches and barriers relate to spatial and temporal variations in dynamics. The most commonly recognized control is the relative importance of wave and tidal processes. This was identified in a regional study of the eastern USA (Hayes 1979; Davis & Hayes 1984). In terms of barriers and beaches these studies identified wave-dominated and mixed wave- and tide-dominated systems. Wave-dominated barrier islands tended to have relatively few tidal inlets (a result of small tidal prisms), small ebb deltas (a result of wave reworking) and well-developed flood deltas. Mixed-energy coasts in contrast tended to have more frequent inlets (larger tidal prisms requiring more inlets for tidal exchange), larger ebb deltas (stronger tidal currents) and more extensive marshes (a consequence of flood-tidal current deposition of fine sediments). Such a division has not been recognized on gravel barriers, perhaps because of their latitudinally restricted distribution.

Beaches and barriers are subject to many additional changes in their overall planform. Typically these changes are more gradual than the types of change envisaged in beach state models. Changes related to sea-level rise are discussed below. Here we are concerned with changes that may be effected independently of sea-level change. Beach planform adjusts, as do profiles, to minimize variations in wave energy. The planform adjustment is markedly constrained by the topographic setting of the beach, and the sediment supply. Abundant sediment supply leads to progradation (Fig. 8.13). Such progradation involves successive welding of bars, which provide sediment supply for

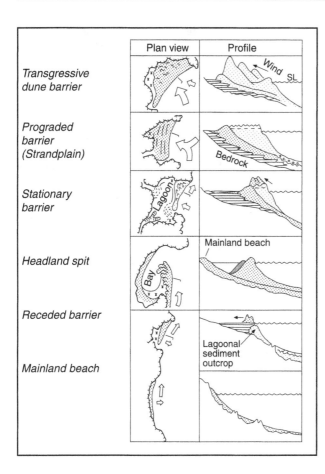

Fig. 8.13 Features of temperate coasts associated with abundant and scarce sediment. Progradation is associated with sediment retention in the beach and shoreface. With strong wind dispersal, a transgressive dune may form, dispersing sediment landward. Vertical dune growth may lead to a stationary barrier with onshore sediment transport. Longshore sediment supply may enable spit elongation and formation of recurves as the spit terminus advances. With sediment scarcity, barriers migrate landward over underlying lagoonal deposits. These may be exposed intertidally seaward of the barrier. Mainland-attached beaches may undergo narrowing and seaward dispersal of sediment. (After Roy et al. 1994.)

embryo dune and then foredune growth. With continuing sedimentation the foredune is increasingly isolated from sediment supply, and from maritime influences, and thus vegetation becomes increasingly terrestrial in nature and the dune increasingly stable.

Progradation at the end of a spit may take place through longshore transport of sediment and accumulation at the terminus of a sediment cell and thus spits may extend and eventually enclose embayments. Successive recurves may be preserved that mark previous positions of the spit terminus. Sediment abundance may lead to closure of tidal inlets if a period exists in which wave energy is able to seal an inlet against tidal currents. Such conditions are normally associated with storms (see below). If an inlet closes, a residual body of sediment may be left in the tidal deltas, which, in the exposed ebb delta, is likely to be redistributed by wave action and retained

in the beach sediment budget. Tidal delta sand in the sheltered, back-barrier environment may remain intact or be redistributed by locally generated lagoonal waves (Cleary et al. 1976).

In areas where the modern coastal sediment was originally derived from a relict source (such as the continental shelf, or reworked glacial sediments), the supply may eventually be exhausted. Under such conditions coastal behaviour involves reworking of a finite sediment volume, although it is inevitable that losses will occur through leakage either offshore (to the shelf), alongshore (to adjacent systems) and/or onshore (to dunes). In the absence of sediment supply, a beach or barrier may experience recession and landward dunes then become new sediment sources as they are eroded and their sediment is reintroduced into the beach sediment budget.

As a finite sediment volume adjusts to variations in wave energy along a beach, reorganization leads to localized erosion and accretion that is dependent on the ambient wave conditions and the antecedent morphology. Such a coast may exhibit spatially variable patterns of erosion and accretion that are to some extent envisaged in the coastal cell model of May & Tanner (1973). Long-term sediment scarcity may be manifest in barriers retreating, such that back-barrier sediments (e.g. lagoonal muds) are exposed on the beachface (Fig. 8.13).

Migration of tidal inlets may take place under sediment abundance, which promotes longshore extension of up-drift barriers, or under sediment scarcity, as sediment is reworked to accommodate changing wave energy. The migration of inlets causes reworking of the associated tidal deltas (Reddering 1983). A suite of models of barrier planform change has been presented (Carter et al. 1987; Carter & Orford 1993). Although based on gravel barriers, a number of the changes envisaged are evident on sand barriers, where the higher degree of dynamism makes changes less readily attributable to a particular forcing factor. Cooper & Navas (2004) measured planform changes in a headland–embayment system over more than 100 years, and attributed them to natural variations in bathymetry caused by sediment accumulation on the sea-bed. These, in turn, altered the pattern of waves approaching the shore, and changed the longshore drift directions at the coast.

8.4.3 Response to storms and tsunami

Storms have been cited as the most important sedimentary forces on temperate coasts. This is intuitively believable because of the high-magnitude processes that operate during storms and the potential for dynamic thresholds to be exceeded that may dominate coastal behaviour at the historical time-scale. Their location outside the tropics does expose temperate coasts to the impact of cyclonic weather systems, as well as to hurricanes that originate in the tropics but which make occasional landfall in temperate zones (Fig. 8.1). Storms produce large waves and elevated water levels (surge). Thus wave energy is not only increased but it operates at higher levels than normal. Thus, sediments in dunes and back-beach areas are exposed to the effects of waves and currents.

The importance of storms on boulder and gravel beaches has already been inferred, as they produce conditions necessary for sediment transport. In comparison, the importance of storms on sand beaches is more difficult to isolate from the effects of other processes. One important consequence of storms on barriers is the occurrence of overwash, which moves sediment landward of the active beach profile, effectively removing it from the system until the barrier migrates landward and it then re-enters the system. Storms, which have been difficult to monitor because of the high energy levels involved, have recently been identified as producing orders of magnitude differences in longshore transport rates and even change in the nature of processes operating compared with lower energy conditions (Miller 1999). Storms, too, lead to enhanced energy levels in the surf zone. This has been associated with the enhanced likelihood of infragravity waves and nearshore circulations, which give rise to rhythmic shoreline topography (Komar 1998).

Storms have also been important in raising wave energy levels such that they can overcome tidal currents and close inlets. The Kosi Lagoon in South Africa closed as a result of a tropical cyclone that temporarily enabled wave energy and wave-sediment transport processes to overcome the tidal currents generated by a massive tidal prism (Cooper et al. 1999). The ebb-delta was thus mobilized and reworked landward to seal the former inlet. Without human intervention the system was unlikely to have ever reopened. Storms have also been responsible for formation of new inlets in barrier island chains through erosive overwashing, which has lowered barriers to the extent that tidal flow is initiated through low points (Fig. 8.14).

Large-volume, near instantaneous sand transport during storms has been responsible for major changes in beach erosion and accretion rates at tidal inlets, with storms interpreted as the major determinants of tidal inlet behaviour at decadal time-scales (Case Study 8.1). Similarly, Orford

Fig. 8.14 Inlet formed during Hurricane Isabelle, September 2003, Outer Banks North Carolina. The inlet was closed artificially within a few weeks in order to facilitate road access. (Photograph courtesy of the Program for the Study of Developed Shorelines, Duke University, NC.)

Case study 8.1 Impact of storms on sediment movement: Texas Coast, USA

Measurements of sediment movement along coasts can be made at various time-scales. Short-term studies use instruments to provide information on the relationship between sediment transport and coastal dynamics. Longer-term studies involve comparison of beach profiles, maps, charts and air photographs in order to quantify change, which then may be related to records of wave activity, sea-level and climatic variables. In either approach, identifying the role of storms in the total sediment budget can be difficult because of their infrequency (which makes them difficult to capture on record) and their high magnitude (which destroys instruments).

Working on part of the barrier island coast of Texas (Case Fig. 8.1a), Morton et al. (1995) monitored beach sand volumes on two barrier islands separated by an inlet (Case Fig. 8.1b). The barrier islands are experiencing long-term sea-level rise and limited sediment supply and hence have been migrating landward over the past 100 years (Case Fig. 8.1b). The inner shelf is muddy and all sand movements in the nearshore can be clearly identified and volumes of sand calculated. Measurements over a 10 year period showed that volumes of sand lost from one barrier island (Galveston Island) did not match the volume gained by the adjacent, down-drift barrier island (Follets Island). Instead, sand eroded from the up-drift island was transported across the tidal inlet on the shoreface and deposited there. Subsequently it was transported by waves onto the down-drift island.

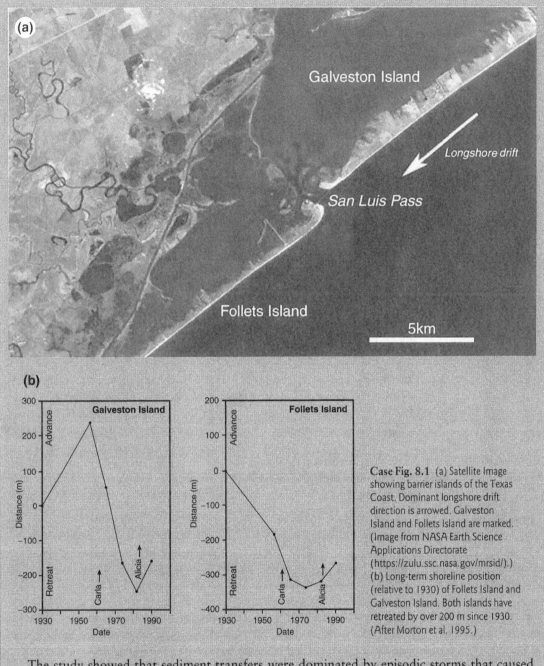

Case Fig. 8.1 (a) Satellite Image showing barrier islands of the Texas Coast. Dominant longshore drift direction is arrowed. Galveston Island and Follets Island are marked. (Image from NASA Earth Science Applications Directorate (https://zulu.ssc.nasa.gov/mrsid/).) (b) Long-term shoreline position (relative to 1930) of Follets Island and Galveston Island. Both islands have retreated by over 200 m since 1930. (After Morton et al. 1995.)

The study showed that sediment transfers were dominated by episodic storms that caused near-instantaneous erosion of barrier islands. A hurricane in 1983 (Hurricane Alicia) eroded between 50 and 70 m³ of sediment per metre of beach from Galveston Island and lesser amounts from Follets Island. Most of the eroded sand was transported south-west by strong currents associated with the hurricane and was deposited on the shoreface (only 12% of the volume lost was carried over the islands and deposited as overwash).

Post-storm recovery (Case Fig. 8.1c) took place over several years and involved phases of sand migration onshore, bar welding to the beach, and finally dune build-up as the beach became dry

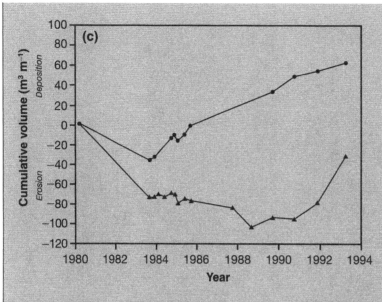

Case Fig. 8.1 (c) Cumulative volume changes of profiles measured on Follets Island (closed circles) and Galveston island (closed triangles) relative to 1980. Note how Follets Island continued to accrete for several years after Hurricane Alicia in 1983 whereas Galveston lost volume for many years before gaining sand. (After Morton et al. 1995.)

and wide. The beaches on Galveston Island never recovered their pre-storm volume whereas those on Follets Island continued to accrete as sand (carried south from Galveston during the storm) was transferred onshore under fair-weather wave conditions. The sand eroded from Galveston Island during the storm acted as a sediment supply for Follet Island and thus whereas Galveston experienced net retreat over the period 1980–1994, Follet experienced net progradation.

A subsequent hurricane in 1988 (Hurricane Gilbert) caused erosion and a temporary reversal in the recovery of Galveston Island, whereas on Follets Island, the post-Alicia sand still on the shoreface was pushed onshore, causing enhanced accretion. This study demonstrates the importance of antecedent conditions in determining coastal response to a storm. The accretion on Follets Island after Hurricane Alicia temporarily reversed the long-term erosional trend, however, when the sand supplied from Galveston Island is transferred onshore, it will be dispersed further south in longshore drift and the long-term migration of the island is predicted to continue.

Relevant reading

Morton, R.A., Gibeaut, J.C. & Paine, J.G. (1995) Meso-scale transfer of sand during and after storms: implications for prediction of shoreline movement. *Marine Geology* **126**, 161–79.

et al. (1999) identified three periods of rapid foredune recession followed by long intervening periods of accretion, which they attributed to the impact of high-magnitude storms on a modally high-energy beach. The observations pointed to storms as the major determinants of historical scale coastal behaviour, as storms produced large-scale, near instantaneous erosional events. Interstorm periods were characterized by decadal-scale phases of post-storm beach accretion.

Tsunami, as fast-moving solitary waves or groups of waves, have been recorded on many temperate coasts, but particularly on tectonically active coasts, from sedimentary evidence (Dawson 1994) and direct observation. These waves may reach heights of up to 30 m as they approach the shore and cause overwashing of barriers and mainland shorelines. Typically tsunami deposits contain a range of grain sizes and may include microfossils that indicate a continental shelf origin (Hindson & Andrade 1999). They are less well sorted than deposits associated with storms and may contain sequences deposited during both landward and seaward return flows (Nanayama et al. 2000).

It should be clear from the above discussion that the three-dimensional changes that occur in beach systems are multifarious in nature. The processes that drive them (dynamics, sediment supply and antecedent morphology) are diverse in type and in the time-scales at which they operate. The very dynamism of these systems hinders attribution of cause and effect and, as such, the identification of process–response relationships on beaches and barriers is difficult. Such relationships are perhaps most likely to be identified at the short time-scale through instrumentation. The difficulty lies in relating these observations to longer time-scales because of the importance of additional factors such as antecedent conditions and sediment supply and periodicity of episodic high-energy events. Relating site observations to other locations is equally fraught with uncertainty because the particular constraints on sediment movement and geomorphological response are not conducive to full quantification. The up-scaling of short term observations to longer time-scales suffers similar constraints for the same reasons, coupled with the non-linearity of process–response mechanisms, unpredictability (or impossibility of complete measurement) of forcing mechanisms and morphodynamic feedback. By its very nature, investigation of the role of sea-level change and climate change requires a long-term approach.

8.5 PROCESSES AND IMPACTS OF ANTHROPOGENIC ACTIVITIES

A range of human activities influence the natural functioning of temperate coastal systems. Perhaps the most pervasive impact results from the legacy of development at the shoreline. This impedes the ability of the coastline to respond morphologically to natural forcing because such responses may damage infrastructure. This situation also influences contemporary management approaches to the shoreline (see section 8.6).

In the pre-industrial era, coastal infrastructure threatened by the sea was abandoned. Submerged former landscapes and archaeological monuments in the Middle East (Sivan et al. 2001) provide evidence of this practice. This still happens in some areas but the intensity of development and its human value (both social and economic) has prompted a range of engineering interventions to defend infrastructure.

The problems posed by human intervention arise partly from construction inside the active profile of the coastal zone (*passive intervention*) – for example, failure to recognize that dunes and/or the back-beach form an active part of the system during storms. Long-term stability during interstorm periods may give a false impression of the coastal morphodynamic regime. Consequent removal of foredunes and their replacement with solid infrastructure renders that infrastructure vulnerable to erosion during storm conditions. Human alteration of sediment movement (*active intervention*) occurs through structures intended to alter wave and current patterns and intercept sediment in transport (e.g. groynes and jetties), and prevent sediment from being eroded (e.g. seawalls). Additional, less obvious impacts on sediment movement include denudation or stabilization of dunes. Such actions alter the natural sediment budget.

Human activities may also impede the ability of the shoreline to adjust to rising sea-levels. Developments built landward of the active coastal sedimentary system eventually find themselves within it, as a result of sea-level rise. Development at the shoreline imposes restrictions on the options available in response to rising sea-levels. This type of anthropogenic constraint on the boundaries of natural coastal systems poses a major problem in coastal zone management and is often stated as the 'erosion problem'. As noted above, however, erosion is an entirely natural part of the cycling of sediment along coasts. Only the presence of human infrastructure renders it a 'problem'.

There have been historical changes in human use of the coastline (Nordstrom 2000). The first impacts were probably vegetation destabilization in coastal dunes. Prehistoric occupation of sand dunes was widespread and alternating periods of occupation and abandonment have been documented in many dune systems, which probably relate to alternating periods of stability and instability in which humans themselves may have played a role (Gilbertson et al. 1996). Wilson & Braley (1997) described the nineteenth century engulfment of houses in Donegal, Ireland by blown sand as a result of overgrazing in adjacent dunes.

Navigation posed the next major human threat to natural coastal morphodynamics. Tidal inlets provide strategic links between inland systems and the open sea. Although inlets may be sufficiently deep, adjacent tidal deltas pose hazards to navigation. Dredging of channels and ebb deltas was the initial response to this hazard. Such alteration of the sediment budget causes readjustments, and, inevitably, the deltas reform and require further dredging. The next phase of control was the construction of jetties to fix the location of tidal inlets. In most cases, however, maintenance dredging is still required.

Increased levels of human utilization and occupation of the coast have been accompanied by developments in engineering that have promoted ever increasing interference with natural coastal functioning. Human interventions fall into three main categories.

1 *Planned modification* – where there is deliberate planned construction of harbours, reclamation of coastal lands, sea wall construction, etc. This is almost always an engineered approach and is often planned around a relatively short-term time-scale (decades). Generally, little attention is given to long-term morphodynamics at the site.

2 *Accidental modification* – often a knock-on effect from (1), where further along the coast there is a direct impact on the general wave patterns and sediment transport pathways from, for example, an engineered structure. In many cases, simple ignorance of coastal processes leads to a direct modification of the coast. For example, the removal of sediment from beaches can have a beach-lowering effect and therefore an increase in beach vulnerability from storm-wave attack where less energy is dissipated and backshore erosion takes place.

3 *Reactive modification* – in response to planned or accidental modification. A process of progressive coastal modification can take place when attempts to solve sedimentary problems using engineering solutions produce further sedimentary problems. These are then addressed by further engineering. Examples occur when longshore supply of sediment is halted by stabilization of eroding cliffs that threaten human infrastructure. Beaches, deprived of sediment are then 'stabilized' using groynes, which in turn starve other areas of the coast of sediment. Such situations produce progressive down-drift intervention.

The above types of intervention take place through four main groups of human activities, comprising coastal engineering, agricultural activity, the extractive industry and recreational activities. Each one of these will be addressed in turn but in many cases a combination of two of more of these activities may be present at the coastal site and there may be feedback between them as levels of activity change over time.

8.5.1 Engineering works

Off-site engineering has the potential to affect sediment supply to the coast. In cases where fluvial sediment supply is important, construction of impoundments, flow reduction and sediment abstraction from rivers can reduce the supply and have an impact on the coastal sediment system. At the Nile delta the reduction in sediment supply (through impoundments) has led to enhanced subsidence and relative sea-level rise (Stanley & Warne 1998; see Chapter 7). The sandy barrier islands at the leading edge of the delta are subject to both impacts. In California, impoundment of numerous small rivers has led to severe erosion on many beaches as the sandy sediment supply has been reduced (Sherman et al. 2002). Another form of off-site impact may occur through modification of back-barrier areas. Reclamation of the coastal fringe, for example, can reduce the tidal prism and in turn may induce changes in the tidal current regime and the adjacent coast (Case Study 8.2). Coastal engineers typically attempt to modify coastal processes in order to protect property/structures

Case study 8.2 Shoreline changes near Wexford Harbour, Ireland

Wexford harbour in south-east Ireland (Case Fig. 8.2a) is a large estuary partly separated from the Irish Sea by a sand and gravel barrier. Associated with its wide inlet are extensive flood and ebb-tidal deltas. Locally generated south-easterly storm waves and diffracted ocean waves produce a northerly drift. The beach in the south of the bay is backed by cliffs of glacial till. A resistant bedrock headland (Greenore Point) in the south provides a hinge for development of a crenulate bay that extends northwards and forms Rosslare Bay. At its maximum extent, the spit that forms the northern part of this shoreline was 8.3 km long but lost about 3 km of its length between 1925 and 1983 (Case Fig. 8.2b) as the spit was breached and eroded by storm-generated waves. Erosion of the spit was viewed with concern by the local authority because of the tourism associated with the beach, which was considered the only safe bathing area in the region. A variety of small-scale coastal protection structures were emplaced (Case Fig. 8.2c). Groynes were emplaced to reduce longshore drift and rock armour was constructed adjacent to the spit terminus in 1965–66. The apparent interruption of longshore drift by a pier at Rosslare Harbour (constructed in 1882 and extended by 1902) was perceived as the reason for the erosion. A sandy beach built up to the south of the harbour. In 1978 the harbour was replaced by an impermeable breakwater and seawall.

Subsequent research (Orford 1988) suggests, however, that the shoreline erosion was related to the realignment and failure of the spit rather than interruption of the longshore drift. Realignment of the coastal planform that included the mainland beach and the spit, by a

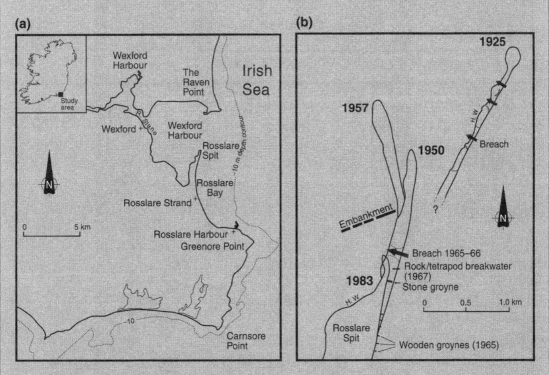

Case Fig. 8.2 (a) Coastal setting of south-east Ireland showing Wexford Harbour, Rosslare Bay. (b) History of spit disintegration: H.W., high water. (After Orford 1988.)

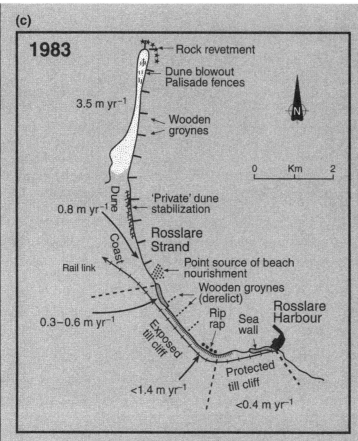

(c)

1983

Rock revetment

Dune blowout
Palisade fences

3.5 m yr⁻¹

Wooden
groynes

N

0 — Km — 2

'Private' dune
stabilization

0.8 m yr⁻¹

Dune
Coast

Rosslare
Strand

Rail link

Point source of beach
nourishment

Wooden groynes
(derelict)

Rip
rap

Sea
wall

Rosslare
Harbour

0.3–0.6 m yr⁻¹

Exposed
till cliff

Protected
till cliff

<1.4 m yr⁻¹

<0.4 m yr⁻¹

Case Fig. 8.2 (c) History and
distribution of coastal protection works
constructed in response to coastal
erosion. (After Orford 1988.)
(d) Reclamation in Wexford harbour
showing initial intertidal area and post-
reclamation area. The marked reduction
in tidal prism was associated with major
changes at the inlet and adjacent coast.
(After Orford 1988.)

(d)

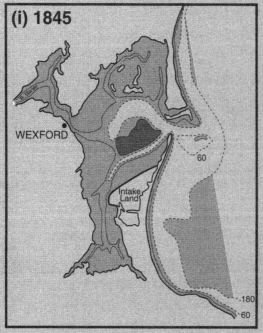

(i) 1845

WEXFORD

Intake
Land

60

180
60

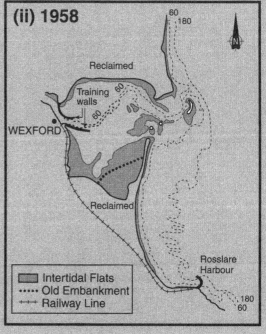

(ii) 1958

60
180

N

Reclaimed

Training
walls

60

60

WEXFORD

Reclaimed

Rosslare
Harbour

180
60

Intertidal Flats
Old Embankment
Railway Line

few degrees, caused large-scale coastline retreat. Three sets of sedimentological information supported this conclusion. First, the spit failed by reduction in supply to its distal end that was not matched by reductions in supply at the proximal end. This was inconsistent with drift interruption at Rosslare Harbour. Second, the relative immaturity of the bay planform is evidenced by observations of erosion and breaching before construction at Rosslare. The bay had not reached equilibrium with the ambient wave field. Third, a gross estimate of longshore transport supply from the eroding cliffs at Rosslare is approximately equal to the accumulation rates recorded south of the harbour and therefore there is not a major reduction in longshore sediment supply.

Orford (1988) also studied changes in nearshore bathymetry recorded on historic charts, and changes in the estuary of Wexford Harbour. In the harbour, a large area of intertidal land had been reclaimed in the mid-nineteenth century (Case Fig. 8.2d). The effect was to reduce the tidal prism of the estuary. This, in turn altered the balance of tidal and wave power at the inlet and caused large-scale sediment reorganization as the positions and strengths of the main flood- and ebb-directed currents shifted. These ultimately caused the demise of the spit and the changes in planform at Rosslare.

Relevant reading

Orford, J.D. (1988) Alternative interpretations of man-induced shoreline changes in Rosslare Bay, southeast Ireland. *Transactions of the Institute of British Geographers* 13, 65–78.

or to improve navigation. There is often an associated undesirable sedimentary consequence.

8.5.1.1 Hard engineering

Groynes are shore-normal or oblique structures that intercept the longshore drift (Fig. 8.15). They are frequently constructed in areas deprived of their sediment supply by other human interventions. They can be successful in trapping sediment where longshore sediment transport is naturally present (Komar 1998). At Cape May, New Jersey, construction of groynes to intercept longshore drift led to recession of the down-drift coast by over 800 m (Pilkey & Dixon 1996). There have been attempts to lessen the impact of groynes on down-drift coasts. For example, the vertical profiles of some groynes have been adjusted as well as their porosity in order to enable sediment bypassing when the desired beach profile has been achieved. A variety of different designs have been used, including changing the angle to the shore, and varying planforms (e.g. hammer-head

designs), all of which seek (with variable success) to lessen their undesirable effects.

Jetties also extend normal or oblique to the shore. Their purpose is to stabilize inlets and to prevent inlet migration. In so doing, they interrupt the longshore supply of sediment and cause the ebb-tidal delta to be destroyed and/or shift position. A range of ancillary activities usually accompany jetty construction. These include channel dredging and artificial bypassing of sediment around the inlet.

Sea walls are shore-parallel structures placed on a coast to prevent landward movement of the shoreline or to act as flood defences. Many seawalls were built with the purpose of providing promenades next to the sea. Various materials, slope profiles and slope angles are used to absorb wave energy. The sedimentary effects of seawalls (Fig. 8.16) include:

1 beach lowering – where wave energy is reflected from the wall and sediment is stripped away from the beach surface in front of the structure;

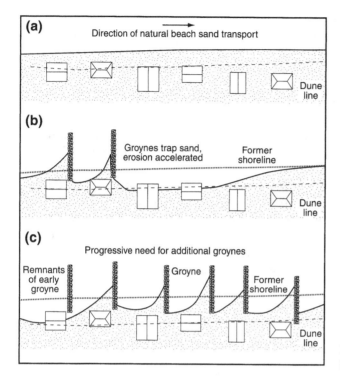

Fig. 8.15 Typical impacts and temporal pattern of groyne construction. (After Pilkey et al. 1998.)

2 edge erosion – where localized turbulence at the end of the wall causes severe erosion;

3 potential grain-size changes – natural sediment sorting can be significantly altered by the emplacement of a sea wall;

4 separation of dune and beach systems – where a seawall removes any dune or back-beach material from the active profile and precludes incorporation of such sediment into the profile during storms – this also causes loss of sediment supply from the beach to adjacent dunes;

5 new littoral currents and sediment transport may be generated by the presence of sea walls, which in turn create additional problems for sites further along the coast, and may threaten the stability of the seawall itself.

Seawalls are commonly used to protect infrastructure from long-term coastal retreat. In many cases they are also used inappropriately to combat seasonal erosion. More often than not this induces irreversible changes on the coast. Perhaps the greatest impact of seawalls is in fixing the landward boundary of the coastal sediment system. Under a rising sea-level (see 8.7) landward migration of beaches is precluded. Thus beaches

backed by seawalls become narrower and steeper, often disappearing altogether.

Offshore breakwaters are designed principally to modify wave patterns and to produce sedimentary effects on the shoreline. They absorb wave energy before waves reach the shore. This effect reduces onshore–offshore sediment transport (especially during storms) but can, in certain circumstances, reduce littoral drift and create cuspate accumulations or tombolos as a result of wave refraction effects. Attempts to simulate natural log-spiral forms of headland–embayment cells using artificial headlands have been made using both attached and offshore headlands (Silvester 1976). Tombolos and salients have also been produced through installation of offshore structures (French 2001). As with all engineering activities in the coastal zone, unexpected coastal events can cause undesired morphodynamic effects. Under storm conditions, for example, the water layer above the crest will no longer intercept energy from incoming waves. The height of the breakwater poses serious problems because if too low, inadequate protection will ensue, and if too high,

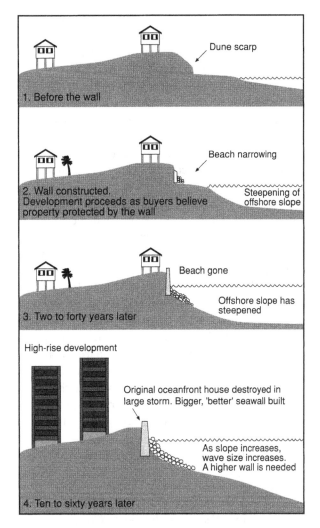

Fig. 8.16 Impacts of seawalls. Seawalls fix the landward margin of the beach, thereby inhibiting its ability to respond to storms or sea-level rise. They also may be constructed on the beach, thus causing effective loss of the active beach. Active loss occurs as waves reflect off the seawall and erode the beach, particularly during storms. Over the medium term, seawalls cause beaches to narrow as waves are reflected and sediment is lost. The beach may disappear and the seawall itself may be undermined as the shoreface steepens. (After Pilkey et al. 1998.)

interference with the shore processes will result in new problems.

8.5.1.2 Soft engineering

Beach nourishment is a relatively recent technique of shore protection and following some early twentieth century emplacements (Bird 1996) widespread application began after the 'Ash Wednesday Storm' of 1962 on the eastern USA. Beach nourishment tackles erosion by replacing lost sediment and relies in part on natural wave action dispersing introduced material along the shoreline and achieving equilibrium with ambient dynamics. Ideally therefore, the introduced sand should replicate the native sand texture

(Dean 1974). Beach nourishment often goes hand in hand with navigational channel maintenance where dredge spoil provides suitable sediment. A variety of emplacement modes have been employed to nourish beaches. They include: single direct emplacement from offshore dredgers, followed by mechanical profiling; trickle feeding, where sediment is introduced from a single source at small volumes; recycling of sediment from down-drift sinks to up-drift source areas; and emplacement in the nearshore for natural onshore transport (French 2001).

Nourishment has been seen as a panacea for erosion control. Once emplaced, nourished beaches adjust to natural dynamics and morphology by dispersing sediment and producing geomorphological forms (planform and profile) in response to wave energy. If sediment is removed from nearshore areas, the net effect is to steepen the nearshore profile, which in turn alters the relationship between beach morphology and wave dynamics. In certain cases, offshore breakwaters are constructed in front of the newly nourished beach to help break up incoming wave energy in order to reduce erosional effects. Scarcity and cost of available sediment is a major constraint on beach nourishment. There is an inherent lack of accuracy in predicting the longevity of nourished beaches and thus their cost–benefit relationship can be difficult to assess (Pilkey & Dixon 1996). Beach nourishment has the effect of widening the dry part of the beach, which in turn may lead to enhanced aeolian activity. Engineering history has shown that there is a strong regional difference in the durability of nourished beaches along the eastern coast of the USA. Leonard et al. (1989) noted that beach durability was probably related to many factors, most of which could not be isolated, but that storm frequency and overall wave energy were probably particularly important.

Other forms of soft engineering involve the use of ecological elements to deliberately induce morphological change. Examples include vegetation planting on foredune systems to induce additional sediment build-up (Woodhouse 1978) and to provide additional buffering against storm attack. Similar effects can be achieved by construction of various forms of wind trap that encourage aeolian sediment to accumulate. In Ireland, non-indigenous species (sea buckthorn, laurel and sycamore) have been planted in sand dune systems in order to stabilize blowouts and block access. In certain cases a scenario of 'overmanagement' may develop in coastal sand dunes. Attempts to stabilize the dune morphology have sealed the main landward corridors of sediment transport. The US National Parks Service artificially built up many foredune systems through trapping and marram planting on barrier islands during the 1930s. This resulted in a deficit in natural sediment and nutrient throughput within the dunes. In turn, this induced

a massive reduction in washover and blow-over processes and increased beach erosion on the seaward margin. By the 1970s there was a major reappraisal and a decision was made to allow the barriers to readjust naturally.

8.5.2 Agricultural activities

Domestic animals (sheep, cattle, horses, etc.) are frequently grazed on coastal dunes. The impact on the dune surface depends on stocking densities. Trampling and grazing can result in vegetation damage. The resulting instability may cause large-scale wind blow in dune systems. Low stocking levels can promote vegetation stability by stimulating vigorous growth via grazing and the addition of manure. Rabbits were imported for food into coastal dunes in Ireland from the twelfth century onward. Rabbit burrowing subsequently led to the destabilization of sand dune systems and the dynamics of the rabbit population became a major factor in dune system dynamics. The myxomatosis-induced population crash in the 1950s led to enhanced dune stability. The presence of bare sand in dunes has often been seen as an opportunity for large-scale afforestation. In Spain and Portugal significant stretches of coast have been planted with pine trees, which have resulted in almost total blockage of inland sediment movement.

8.5.3 Anthropogenic influence on sediment supply

Human influences on sediment supply can be substantial. The manner in which humans alter sediment supply and transport at the coast includes the following:
1 reduction or increase in sediment supply through river regulation and land-use change;
2 loss of sand by dune destruction or stabilization;
3 loss of sediment through dredging (navigation or aggregate extraction);
4 blocking of littoral transport by structures;
5 loss of sediment by stabilization of cliff sources.
In many cases, agricultural practices alter sediment supply. In central Italy widespread coastal erosion post-1950 was ascribed to gravel

extraction and land surface stabilization under agriculture (Coltorti 1997). Similarly, dams on rivers have been responsible for reduction of sediment supply through direct entrapment and through reduction of transport capacity by reducing water flows (Case Study 8.3), and large-scale urbanization may reduce sediment discharge from paved areas. Mining and quarrying may also locally alter sediment supply. The disposal of mine waste in Corsica introduced vast quantities of sediment to the beach systems and caused

shoreline accretion of over 400 m (Bernier et al. 1996). More widespread sediment inputs to the coastal zone arise from waste disposal – few beaches do not contain occasional bricks, glass fragments or other human debris.

Small-scale sediment removal from beaches and dunes has taken place for a variety of reasons, including aggregate, construction, fertilizer and animal bedding. Carter et al. (1991) reported 5000–6000 t of sand removal over a 10-year period from eleven Northern Ireland beaches and

---

Case Study 8.3 Impacts of reduced fluvial sediment supply to California beaches, USA

The sandy beaches of southern California are an important economic resource. Over the past few decades many beaches have experienced severe erosion that has reduced their amenity value and placed human infrastructure at risk. Griggs & Savoy (1985) estimated that > 80% of the California coast was eroding and c. 30% was in the high risk category. The reasons for the rapid coastal erosion relate largely to reduction in fluvial sediment supply, which was estimated to yield 70–85% of all beach sand (Sherman et al. 2002) (Case Fig. 8.3a). Additional reductions in sediment supply relate to interruptions of alongshore sediment movement by engineering structures (mainly jetties) and protection of eroding bluffs by seawalls (Liedersdorf et al. 1994). The fluvial sediment supply reduction was related to the construction of dams on sediment-yielding rivers in the steep, arid California hinterland (Case Fig. 8.3b). More than 500 dams impound more than 42,000 km^2 (38% of the surface area) of California (Willis & Griggs 2003). Sherman et al. (2002) calculated the sediment retention in 28 dams and 150 debris basins in southern California and concluded that these structures impounded > 4 million m^3 of sediment per year, equivalent to 3 m^3 of sediment per metre of shoreline in the five southern coastal counties in the State. Willis & Griggs (2003) determined that 70 dams (13%) were responsible for 90% of sediment reductions to the coast (Case Fig. 8.3c). Half the dams are old and have lost significant water retention capacity as a result of sedimentation and are also in need of maintenance. In view of their reduced functionality, a number of dams have been identified for removal.

In the face of rapid beach erosion, a study on the management options was undertaken (Coyne & Sterrett 2002). This study identified an annual tax revenue to the State of $4.6 billion based on beach tourism and recreation. Potential losses in revenue through beach erosion were calculated at $1 billion in taxes. There was thus a strong economic argument for beach restoration. The study recommended a twofold approach to beach restoration involving opportunistic beach nourishment and dam removal. Opportunistic beach nourishment involves the emplacement of 'sand of opportunity' that becomes available from construction or excavation. Previous experience had shown such emplacements to be successful in improving beach longevity. The more far-reaching recommendation is the removal of dams that are no longer serving a useful function, in order to increase the natural sediment supply to beaches. The economics of beaches versus the reduced value of dams was probably an important factor in enabling this management strategy to be adopted. A number of studies are presently underway in advance of dam removal on several sediment yielding rivers.

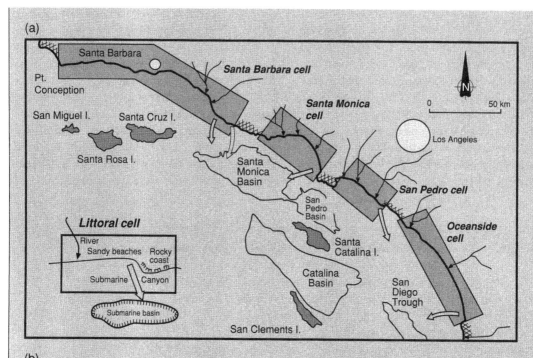

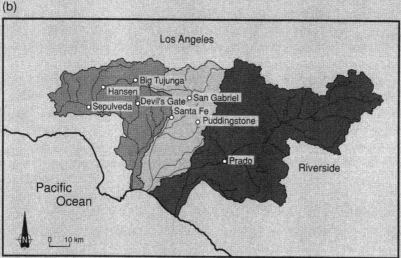

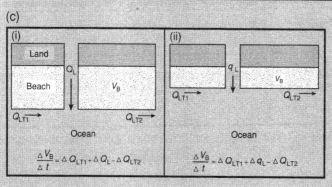

Case Fig. 8.3 (a) Sediment cells on the southern California coast. Each cell has identifiable sediment sources, sinks and transport pathways. (Adapted from Flick 1993.) (b) River catchments and dams in the Los Angeles sector of the San Pedro cell (see (a)). The large number of dams in the small, steep coastal catchments has significantly reduced sediment supply to the coast and contributed to coastal erosion. (c) Littoral sediment budget for Californian beaches. If sediment flux Q_L in (i) (natural conditions) diminishes to q_L in (ii) (post-impoundment) and littoral transport Q_{LT1} remains constant the initial beach volume must reduce to balance the sediment budget. (After Willis & Griggs 2003.)

Relevant reading

Coyne, M.A. & Sterrett, E.H. (2002) *California Beach Restoration Study.* California Department of Boating and
Waterways and State Coastal Conservancy, Sacramento, CA. (Available at: http://dbw.ca.gov/beachreport.htm)
Griggs, G.B. & Savoy, L. (1985) *Living with the California Coast.* Duke University Press, Durham, NC.
Flick, R.E. (1993) The myth and reality of southern California beaches. *Shore and Beach* 61, 3–13.
Liedersdorf, C.B., Hollar, R.C. & Woodell, G. (1994) Human intervention with the beaches of Santa Monica
Bay, California. *Shore and Beach* 62, 29–38.
Sherman, D.J., Barron, K.M. & Ellis, J.T. (2002) Retention of beach sands by dams and debris basins in
southern California. *Journal of Coastal Research* (Special Issue) 36, 662–74.
Willis, C.M. & Griggs, G.B. (2003) Reductions in fluvial sediment discharge by coastal dams in California and
implications for beach sustainability. *Journal of Geology* 111, 167–82.

Brunsden & Moore (1999) noted large-scale gravel removal from Chesil Beach. In areas of finite sediment volume, this may produce a serious erosion problem. Large scale, commercial beach and sand dune mining is also practised. In Australia and South Africa, natural concentrations of heavy minerals in dune sands are exploited. Offshore aggregate removal is a commercial operation that has led to a reduction in the nearshore sediment supply and also to changes in bathymetry (see Chapter 10). These in turn alter wave refraction patterns and may cause readjustments in shoreline morphology. The English village of Hallsands was eroded as a result of offshore aggregate extraction for harbour construction (Pearce 1996). Nearshore sediment extraction also takes place in the search for precious minerals (e.g. diamonds off the South African and Namibian coasts).

8.5.4 Recreational activity (human trampling, leisure vehicle activity)

The coastal zone has traditionally attracted human visitors who have utilized it as a food source, living area and more recently as a leisure area. Recreational activities can be a major factor in coastal sedimentology. In coastal dunes, human trampling and use of recreational vehicles can result in large-scale reactivation of fixed sediment (Gilbertson 1981). Car parking on beaches can induce compaction of the sediment and interfere with natural sediment transport in the intertidal and adjacent supratidal zones.

'Cleaning' beaches of litter including seaweed on drift lines can cause loss of potential dune nucleation sites and nutrients needed to promote plant growth.

8.6 MANAGEMENT AND REMEDIATION

8.6.1 Management approaches

Management of sedimentary coastlines involves consideration of (i) utilization of the coast such as to minimize human impact on the sedimentary system, and (ii) balancing human utilization with a naturally dynamic system. Management of sedimentary coastlines must therefore begin with an understanding of the processes that shape and maintain those coasts at time-scales relevant to management. The discussion above should immediately make clear that this is a difficult task. The range of processes and the physical constraints that operate at varying spatial and temporal scales are unlikely to be well understood at any coastal location, and hence a wide margin of error is inherent in sedimentological assessments related to coastal management. Existing constraints on a coastline imposed by human interventions (Fig. 8.17a), and the potential impact of further intervention on adjacent stretches of coastline, impose additional considerations for management of sedimentary coasts.

Applied sedimentological investigations in the coastal zone are therefore typically undertaken for two purposes. One is the design and

Fig. 8.17 (a) High-rise tourist development on a barrier spit at Mar Menor, south-east Spain fronted by a narrow, eroding beach. Such intensive development inhibits the ability of the shoreline to respond to changes in sediment volume. (b) Coastal defence structures on a shingle beach in Kent, England. A series of stone and wooden groynes has been used to trap and retain shingle and a revetment has been constructed on the back-beach, backed by an armoured slope. This 'hold the line' option seeks to counteract natural shoreline behaviour in the face of reduced sediment supply and relative sea-level rise. (c) Rapid coastal retreat on the Outer Banks of North Carolina (USA) near Nags Head has eroded the sand dunes on which these houses were constructed. The concrete septic tank was originally buried in dunes. The construction of sea walls is prohibited in this State and thus the coastline can migrate at the expense of poorly located developments. The houses are now abandoned and must be removed by the owners. This approach is consistent with the 'managed retreat' or 'do nothing' coastal management option.

environmental assessment of specific engineering projects, and the other is the development of management policies in response to environmental change. These are outlined briefly below and the techniques in use are described in the following section. The design and impact assessment of coastal infrastructure involves prediction of coastal processes and sedimentary response in the face of anthropogenic activities. Typical considerations include the design of groynes to interrupt longshore drift and the likely sedimentary consequences of their installation at relevant time-scales. In the face of actual or potential coastal change four policy options are commonly identified. These are as follows:

1 do nothing;
2 hold the existing line;
3 advance the line;
4 retreat.

The decision-making process related to coastal management policy of this type varies in its degree of formality and, depending on the local resources and infrastructure at risk, the coastline will be defended or otherwise. The approach has been formalized in the concept of shoreline management plans (SMPs) in England and Wales, whereby each section of the coast is subject to a range of analyses designed to inform policy choice. The 'do nothing' policy option is one in which natural processes are left to operate free of further human intervention. This option is typically adopted where no coastal infrastructure is at risk and permits the coast to fluctuate freely within current constraints.

In the 'hold the line' option, the loss of infrastructure is considered to be unacceptable and the coast must be defended. The Netherlands decision to maintain the 1991 shoreline position is an example of such policy at a national level (De Ruig & Hillen 1997). This option is usually associated with solid structures, although the beach nourishment approach is becoming more widespread. It is an unfortunate but unsurprising fact that most landowners prefer the 'defend' option, particularly if the expense is borne by others (Fig. 8.17b).

The 'advance' option is one in which a deliberate decision is made to claim intertidal or subtidal land and to defend this advanced position. In practice this option is rarely taken on the open coast and is confined mainly to areas that are actively prograding, and where infrastructural development follows the advancing shoreline. An important consideration in this policy is whether the progradation is likely to be sustained.

The retreat option is one in which the inevitability of shoreline retreat is identified and accepted. Structures that are undermined or collapse are not replaced, and might even be deliberately removed, and human infrastructure is relocated landward (Fig. 8.17c). Statutes that prevent seawall construction are usually indicative of the adoption of a 'retreat' strategy as in Maine or North Carolina (Pilkey et al. 1998). Selection of the retreat option is typically taken on economic grounds although it is increasingly popular owing to the conservation benefits of natural coastal systems. It is politically the most difficult option to pursue when infrastructure is at risk.

8.6.2 Sedimentology in coastal zone management

Understanding the sedimentary dynamics of the coastline is central to the adoption of management policy and design of human infrastructure. For the purposes of coastal management, the time-scale of interest is typically in the range of years to decades. A range of approaches are available to the applied sedimentologist to develop an understanding of the morphodynamics of temperate coastlines, and more than one approach may be necessary. The inability to upscale short-term measurements to longer time-scales is a key constraint on modern applied sedimentology. This in part lies in the feedback relationships and complexity of processes in the nearshore zone. It should therefore be acknowledged at the outset that it is impossible to quantify coastal behaviour in a generic sense and that coastal morphodynamics at meaningful time-scales will always be expressed in qualitative terms. With this caveat, three types of applied sedimentological study typically inform coastal management: field studies, historical geomorphological change/geoindicators and modelling. A key consideration of such studies are the relationships between sedimentary processes and geomorphological change at different time-scales.

8.6.2.1 Field studies

Field studies in applied coastal sedimentology utilize a wide range of techniques including, at the short time-scale, tracer studies of grain movements, sediment traps that capture some or all of the sediment flux at a given point and acoustic and optical backscattering to measure sediment concentrations. Each of these measures is usually accompanied by quantification of some element of the dynamic environment (e.g. wave parameters, currents) or geomorphological response (e.g. bed-level changes), and relationships are sought between dynamics and sediment transport. Over longer time-scales, repeat field measurement of beach profiles, nearshore topography or other elements of the coastal morphology are used to assess sedimentary behaviour over time. These too are often accompanied by measures of dynamic data or proxy dynamic data, usually at longer time-scales than those utilized in short-term studies. Recognition of the lack of a direct relationship between short and medium-term data hampers the integration of data gathered using these approaches (Carter & Woodroffe 1994).

As field data reflect a range of local variables (textural, inherited factors and dynamics), their generic applicability remains to be tested (Cooper & Pilkey 2004a). A range of analytical techniques (visual, statistical, mathematical) are used to test for relationships between sedimentary/morphological factors and dynamics (e.g. Clarke & Eliot 1988). Unfortunately, the results of field studies are often viewed as universal and empirical relationships are applied elsewhere without due consideration of their limitations (Cooper & Pilkey 2004b). Faced with the difficulty of relating empirical measurements of coastal change to measurements of forcing factors, ever more elaborate approaches have been developed. These include neural networks (Chen et al. 1990), which seek patterns in non-linear systems including

observed sequences of coastal morphological measurement. Such approaches are often constrained by lack of sufficiently long-term data sets to render the approach statistically valid. Constraints on these approaches relate to the representativeness of the data collected and its spatial coverage. Most importantly, the effect of storms is often missed.

8.6.2.2 Historical change and geoindicators

A range of indicators of coastal morphological change is available for most coastlines. These are variable in quality and temporal/spatial coverage. Using long data sets of beach change and storminess, Bryant (1988) ascribed changes in high-tide morphological state to a combination of rainfall, storminess, air circulation and sea-level change. The Dutch beach nourishment approach (Verhagen 1992) involves initially obtaining and analysing a decadal record of weekly beach profiles on the beach to be nourished. The assumption is made that the nourished beach will behave more or less like the natural beach. In the field, a range of indicators exist including sediment accumulation against groynes, rates of inlet movement, etc. (Bush et al. 1996). On many beaches, the presence of groynes, jetties, fishing piers, erosion debris, seawalls and other engineering projects provides geoindicators. Constraints on the historical approach relate to the availability of data, the extent to which coastal behaviour is captured by a series of snapshots, and whether accompanying dynamic data exist to aid interpretation of the morphological information.

8.6.2.3 Modelling

Modelling of coastal processes is a widespread approach in contemporary coastal management. Models fall into two categories: (i) research models and (ii) applied or practical models (Davies & Villaret 2002). Differences between the two are discussed by Thieler et al. (2000). Research models serve the purpose of investigating the mechanisms of sediment transport in the coastal zone and can yield important new discoveries regarding the role of interacting waves and currents, bed friction and sediment texture on, for example, nearshore sediment transport. Applied models, in contrast, set out to deliver a quantitative prediction of nearshore sediment transport or geomorphological response. Applied models are deployed in a routine manner that typically involves testing against a set of field data. Differences between the predicted and observed changes are then considered and coefficients are applied to render the two as close as possible for the historical data. This produces a 'calibrated' or 'tuned' model that is then applied to design a structure, to determine its environmental impacts, or to predict future shoreline evolution.

There are several types of coastal applied model in use. The most commonly used process models in the USA for example are SBEACH and GENESIS. Their formulation is public and therefore can be subjected to public scrutiny (and found wanting – Thieler et al. 2000). In Europe, many of the applied coastal models are proprietary software, developed and used in a competitive environment by coastal engineering firms. The models are often sold and used by other consultants. In this environment, where the precise formulation of the models is not public, such models (e.g. LITPACK, UNIBEST) operate as black boxes. In nature the variability between beaches and even within a single beach is so great that the parameterizations on which applied models are based are not transferable between sites. This seldom contemplated fact is somewhat masked by the calibration process, which in fact involves the use of a constant ('fudge factor') to achieve agreement between predicted and actual change. The 'calibrated' model is then used to predict future scenarios.

Modelling is characterized by endless attempts to compare model output with field data (Mulder et al. 2001). This emphasizes on the one hand the paucity of good field data sets, and on the other the wide variety of model types. In spite of efforts to test models against field data, model refinement is fundamentally prevented by the inability to resolve the spatial variability in factors that control sediment transport and the natural temporal variability in these factors. Models themselves thus do not provide any

more reliable predictions of coastal behaviour than other measures. This fact is not commonly appreciated. Even in the comparatively simple case of aeolian sediment transport, aeolian transport modelling has had little success in predicting actual transport rates at even short time-scales (Sherman et al. 1998).

Modelling approaches are also used in longer-term coastal evolution studies. Here, in spite of a lack of scientific validation, the Bruun 'Rule' model (see 8.7) is the most widely applied approach in predicting coastal response to sea-level rise. The shoreline translation model, a two-dimensional model of coastal profile change (Cowell et al. 1995), utilizes the Bruun approach in maintaining a consistent nearshore profile as the coastline responds to sea-level rise.

8.6.2.4 Composite approaches

Understanding the sedimentology of temperate coasts is an imprecise science that is informed by a range of potential data. Each dataset provides only a piece of the necessary information for correct interpretation and it is important that this is considered in applied sedimentology. The fullest understanding of coastal morphodynamic behaviour is therefore likely to be achieved by the compilation of data from a range of sources, which when compared against each other for consistency or opposing trends, enable description of coastal behaviour.

Understanding coastal sedimentology at time-scales useful to humankind requires utilization of a range of data of varying quality and covering variable time periods. A composite approach using historical, model and field observations could provide the most comprehensive assessment of coastal sedimentary behaviour. In such studies 'order of magnitude' or qualitative answers are sought that aid understanding of the processes operative at varying time-scales.

Building on this type of composite approach, a novel initiative in future shoreline prediction has been undertaken in Great Britain. Termed 'Futurecoast' (Cooper & Jay 2002) this approach uses a combination of geomorphological expert opinion (delimitation of cell boundaries), historical change records, wave modelling, morphological measurements (e.g. beach width) and bathymetric changes (sediment budget guide) to assess the likely future behaviour of the coast. Future coastal evolution is assessed for two scenarios: unconstrained (i.e. assuming no defences or new management practices) and managed (i.e. assuming present management practices continue indefinitely). The objective is to describe the 'behaviour' of a beach system in its coastal context and thus describe the likely future 'behaviour' of the same beach.

8.7 FUTURE ISSUES

Two major issues dominate future concerns about temperate coastlines – climate change and sea-level change. In both instances, studies of coastal response to past changes provide information on likely future scenarios. As the human population is now greater than ever, and is increasingly concentrated in the coastal zone, options for dealing with future changes in the shoreline must consider this constraint.

The sedimentary effects of long-term climatic fluctuation have been analysed in several temperate coastal systems. Two distinct climatic phases have been noted globally in the late Holocene – the Holocene Climatic Optimum and the Little Ice Age. The effects of both have been identified in coastal dune sequences, where climatic deterioration (cooler and with increased storminess) is associated with large-scale instability and development of transgressive sand sheets, and climatic amelioration is linked to enhanced vegetation growth and hence stability (Gilbertson et al. 1996; Wilson et al. 2001). At the decadal time-scale, variations in climate are characterized by ENSO (El Niño–Southern Oscillation) and the NAO (North Atlantic Oscillation). Both oscillations have been associated with changes in coastal behaviour at such time-scales. Goy et al. (2003) ascribed emplacement of beach ridges in southern Spain over a 400-year period to fluctuations in the NAO, with emplacement during stormy periods separated by periods of shoreline stability in intervening calm periods.

Sea-level change is a key issue for temperate coastlines, and is widely associated with problems of coastal erosion. A global predominance of rising sea-levels means that this area has received most attention, however, many temperate coasts are experiencing sea-level fall. As sea-level simply mediates the level at which short-term dynamic processes operate, distinguishing the sea-level-related signature of coastal change from other forcing mechanisms discussed above (see section 8.4) is not straightforward. Nonetheless a number of models of coastal response have been proposed based on field observations, analyses of historical change and laboratory studies. An additional source of information lies in stratigraphical studies (typically spanning several millennia). A number of idealized modes of response to sea-level change have been identified. For a rise in sea-level, simple two-dimensional shore-normal models (Fig. 8.18) involve either an erosional response, a rollover response, or *in situ* drowning (overstepping).

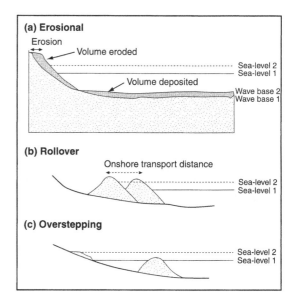

Fig. 8.18 Erosional, rollover and overstepping models of response to sea-level rise. The erosional response (a) involves seaward dispersal of sediment to raise the nearshore profile in tandem with sea-level rise. The coastal migration model (b) envisages retention of a fixed sediment volume that migrates landward via barrier overwash. With no return mechanism, the barrier 'rolls' landward to maintain a fixed volume. In the overstepping mode (c), barriers are unable to respond sufficiently rapidly to sea-level rise and are drowned. (After Carter 1988.)

The erosional response model of sea-level rise is one that envisages maintenance of the nearshore profile by shoreface accretion at the expense of beachface and berm/dune erosion as the beach profile moves upward and landward. This simple model envisages a layer of sediment being preserved on the shoreface as sea-level changes. A numerical form of this response was optimistically termed the Bruun 'Rule' (after the engineer who invented it) by Schwartz (1967), although subsequent studies have shown that, in addition to the lack of field support, it is a flawed concept because of the very restricted conditions in which it is likely to operate and the limited range of processes it considers (List et al. 1997; Pilkey & Cooper 2004).

The rollover response takes place on low barriers where waves overtop the berm, thereby eroding and transporting beachface sediment to the back-barrier area. In a sediment-limited environment this causes the barrier to migrate landward. The migration rates of such systems are dependent largely on the slope of the underlying surface over which they migrate. *In situ* drowning is a phenomenon of barriers that respond more slowly to dynamic forcing than the rate of sea-level rise. Thus the system does not reach equilibrium with a change in sea-level before it is stranded below the level of effective coastal processes. Examples cited include gravel barriers, which have a response time that is slower than sand barriers and are thus left stranded on the shelf as sea-level rises (Carter & Orford 1993). Diagenesis (beachrock and aeolianite formation) may fulfil a similar function and several drowned shorelines on the South African and Australian coasts are preserved as beachrock/aeolianite ridges (Cooper 1991).

Studies of stratigraphical successions that have resulted from sea-level rise have identified different modes of preservation depending on the rate of sea-level rise and the supply of sediment (Belknap et al. 2002; Thom 1984). At the millennial scale, the balance between the rate of sea-level rise and sediment supply are the dominant controls on coastal evolution. At these time-scales, variations in wave dynamics are considered to be masked by the other processes (interestingly, the reverse is true of shorter term observations).

The response of coastal dunes to changing sea-level has received comparatively little attention other than their consideration as part of the beach profile. Dunes are, however, subject to a range of distinctive processes that may vary as sea-level changes. Carter (1991) presented a set of models of potential dune response to sea-level change that incorporated consideration of sediment supply and vegetation cover, two factors that are critical in dune sedimentology (Fig. 8.19).

Changes of beach planform in response to sea-level changes have been much less studied, and are more difficult to relate to sea-level changes as opposed to temporal changes in shoreline dynamics. In the case of gravel barriers, Carter et al. (1987) identified a number of idealized models of planform response mediated largely by sediment supply and antecedent topography. The lack of equivalent studies in sandy environments may reflect the more complex dynamics of those systems. Similarly, the direct relationship tentatively identified by Orford et al. (1995) between gravel barrier retreat and sea-level rise is probably masked in sand systems by the seaward return mechanisms and the range of processes operating in the more dynamically responsive sand systems.

Relative sea-level is falling on some temperate coasts as a result of isostatic uplift, particularly in paraglacial areas, and also as a result of tectonism. Coastal sedimentary responses to falling sea-level are less well studied than rising sea-level. Typical landforms of regressive coasts are beach-ridge plains. These lines of shore-parallel beach-sand ridges are stranded as sea-level falls, isolating earlier formed ridges from the sand source. Several such plains have been attributed to falling relative sea-level (Dominguez et al. 1987; Orford et al. 2003). A series of prograding beach ridges in north-east Ireland were deposited during a fall in sea-level between 6000 and 2000 yr BP (Orford et al. 2003). Firth et al. (1995) described a series of beaches and spits deposited in an embayment in eastern Scotland during falling sea-level as shelf sand was carried onshore

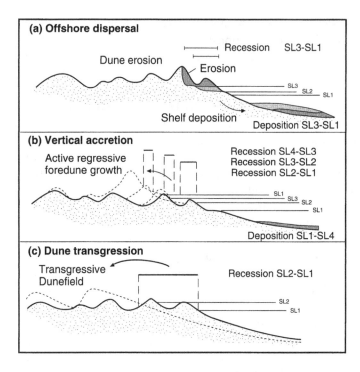

Fig. 8.19 Conceptual models of coastal dune response to rising sea-level. In the erosional response (a) sediment is dispersed offshore. With adequate vegetation, sediment may be transported onshore and trapped in vertically accreting foredunes that accrete as seaward dunes are eroded (b). If vegetation cannot withstand sedimentation rates, a transgressive dunefield may form (c). (After Carter 1991.)

by wave action. Falling sea-level enabled waves to reach sand located progressively further offshore, and increased the accommodation space for sediment accumulation. Littoral deposits, stranded as sea-level falls, may also be subjected to reworking by wind and blown into dunefields. Many of the extensive coastal sand dunes of the north-west Irish coast were emplaced during a fall in sea-level between 5000 and 3000 yr BP (Carter & Wilson 1993; Orford et al. 2003).

The understanding of coastline sedimentary behaviour is of major societal concern because of the growing extent and diversity of human influence on the coast. Global climate change and associated storm patterns, sea-level changes and ecological changes will have an impact on future coastal morphology. At present, understanding of the relationships between forcing factors and coastline response is, at best, limited. This stems in part from the diverse range of controlling factors and in part from heterogeneity in the material factors of the coastline. Many Holocene coastal deposits have been identified that reflect a rapidly changing environmental context (sea-level and climate change). These sequences hold much potential for understand-

ing the likely nature of future coastline changes (e.g. what conditions, of those that determine the different modes of shoreline response to sea-level change discussed above, are likely to occur?). Problems in their interpretation reflect the partial or non-preservation of the full depositional sequence.

A widespread alternative approach in contemporary coastal management practice is the use of applied numerical models of shoreline behaviour. These seek to simulate particular dynamic conditions and to predict future shoreline behaviour. The appeal of such models lies in their normally deterministic output that yields values that can be used in economic planning. The complexities of shoreline behaviour outlined here, however, illustrate that such models do not yield accurate answers because of their inability to take account of all the important factors.

Planning for sea-level rise frequently involves the application of the Bruun concept, which, although it has never been shown to work, has seen application in at least 26 countries since 1995 (Pilkey & Cooper 2004). The application of simple models such as this one demonstrates the need for further investigation of coastline response to sea-level rise in different environmental settings. The identification of the role of sea-level rise is of course difficult to segregate from all of the other forcing factors that drive coastline change at the historical scale.

A key future issue will be to change society's expectation of a precise prediction and to accept a qualitative estimate of future shoreline behaviour based on all available sedimentological information. Interpretations of future shoreline behaviour are best derived from all the evidence available. This may include stratigraphical sequences, historical records of morphological change and of driving forces, observations of contemporary processes and qualitative sedimentary models that identify sediment sources, sinks and shoreline behaviour. The 'Futurecoast' project in England and Wales (Cooper & Jay 2002), which adopts such an approach to future planning of the shoreline, is a significant advance on earlier approaches based largely on application of numerical models.

Future climate change has the potential to alter spatial and temporal patterns of coastal sedimentology. The evidence contained in coastal beach-ridge sequences outlined above shows the importance of variations in climatic conditions at decadal time-scales. Such studies hold much potential for understanding the potential role of future climate change, for example increased periods of storminess. Again, qualitative estimates of future coastal behaviour are likely to be the best that can be achieved.

Finally, the greatest future threat to temperate coastlines is the impact of a growing human population. The potential to have an impact on the coast in many different ways has been outlined above. The extent to which these impacts are understood and mitigated will have probably the most important effect on future coastlines. The human response lies in the realm of politics and economics as much as science, however, better understanding of coastal sedimentary behaviour can inform the decision-making process.

9

Tropical coastal environments: coral reefs and mangroves

Chris Perry

INTRODUCTION

Tropical and subtropical coastlines are characterized by a wide array of sedimentary environments including beaches, dunes, deltas and estuaries, and which are broadly comparable to their temperate counterparts (see Chapters 7 & 8). These lower latitude coastlines are also, however, characterized by two unique sedimentary environments, coral reefs and mangroves, and these, along with their associated sediment substrates, are the focus of this chapter. Although the respective environments are characterized by very different sedimentary processes, sediment production and accumulation in both environments are strongly influenced by biological activity. In coral reefs, the reef structure itself is a product of coral growth, and a high percentage of the sediment substrate typically derives from the breakdown of calcareous skeletal organisms. In mangroves, the trees that colonize these intertidal settings not only contribute abundant organic material to the substrate, but also aid sediment trapping and stabilization. As with most other coastal environments these systems are subject to a high degree of physical reworking. Such natural sediment and shoreline dynamics, combined with, in places, extensive urbanization of the tropical coastal fringe, creates a unique set of management challenges. In many reef and mangrove settings, this is exacerbated by resource extraction and exploitation. From an environmental perspective, anthropogenic-related damage to the biological components of both environments has potential knock-on effects for the physical (including sediment) components. In mangroves, this relates to sediment contamination and the destabilization of sediment substrates following mangrove mortality, and in coral reefs to reduced rates, or modified patterns, of carbonate sediment and framework production. This chapter examines the sources and mechanisms of reef and mangrove sediment accumulation, the response of these sedimentary systems to both natural and anthropogenic change, issues of shoreline management and remediation, and concludes with a review of the likely response of coral reefs and mangroves to future climatic and environmental change.

9.1 NATURE AND DISTRIBUTION OF CORAL REEF AND MANGROVE SEDIMENTARY ENVIRONMENTS

Tropical and subtropical regions are characterized by two distinct sedimentary environments: coral reefs and mangroves. Both are spatially significant and differ from many other sedimentary environments because of the close links between the ecological aspects of the environments and their sedimentology. Tropical coral reefs are, for example, characterized both by the corals themselves and by a diverse associated calcareous fauna (e.g. calcareous algae, molluscs and foraminifera), which, in combination, contribute to the reef structure and the associated sedimentary facies. Similarly, mangrove ecosystems are defined by a characteristic range of floral and faunal associations, which inhabit intertidal sediments and form the basic biological structure of the mangrove. These mangrove trees and shrubs, in turn, promote sediment trapping and may contribute abundant organic material to the mangrove substrate. A fundamental feature of both environments is their diversity in relation to geomorphological settings, their spatial and, in the case of reefs, bathymetric extent, and the sources and rates of sediment production and accumulation. Change within these systems can result from both natural and anthropogenic disturbance. Such changes are capable of altering either the environmental parameters or the substrates on which the biological components depend and, as a result, of modifying sedimentation rates.

9.1.1 Distribution and occurrence of coral reefs

Coral reefs are typically associated with shallow, warm, clear tropical and subtropical marine settings (Fig. 9.1). They have been variously described on the basis of their ecology and geology (see Riegl & Pillar 2000), and are commonly

defined as biologically influenced, wave-resistant build-ups of coral framework and carbonate sediment within tropical and subtropical settings, and which influence sediment deposition in adjacent areas (Longman 1981; Rosen 1990). Coral communities and coral reefs, however, occur in a far wider range of settings, and form a broader range of structures, than can be constrained by such traditional definitions. They occur, for example, in a range of more marginal settings (*sensu* Perry & Larcombe 2003), which include higher latitude (subtropical to warm temperate) environments, along with areas influenced by high turbidity conditions and cool-water upwelling (see section 9.1.3). Although these environments may be characterized by reduced rates of *in situ* calcium carbonate framework and sediment production, and may ultimately produce very different sedimentary deposits, they may still have high coral cover and thus represent important sites of coral-related carbonate production. The coral-related sedimentary environment may thus better be defined as comprising

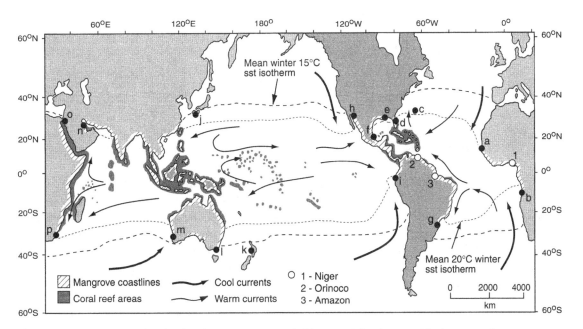

Fig. 9.1 Global distribution of coral reefs and mangroves compared with mean 15 °C and mean 20 °C winter sea-surface temperature isotherms. Latitudinal limits on the distribution of mangroves: a, St Louis, Senegal; b, Lobito, Angola; c, St George, Bermuda; d, St Augustine, Florida; e, Chandeleur Islands, Louisiana; f, Rio Soto La Marina, Mexico; g, Aranangua River, Brazil; h, Puerto de Lobos, Mexico; i, Piura River, Peru; j, Kiire, Kyushu, Japan; k, Raglan Harbour, New Zealand; l, Corner Inlet, Victoria; m, Leschenault Inlet, Australia; n, Qatif, Trucial coast; o, Wadi-Kid, Sinai; p, Kei River, South Africa. (Adapted from Woodroffe & Grindrod 1991; Hubbard 1997.)

Fig. 9.2 Atoll reef showing extensive sheets of carbonate sediment (white areas) accumulated on the leeward side of the reef crest (marked by the area of breaking waves), Courtown Cay Atoll, Caribbean. Field of view is approximately 1 km.

all areas of active coral growth (regardless of wave resistance or framework-building potential) along with the associated carbonate (or mixed carbonate:clastic) sedimentary environment (e.g. lagoons, seagrass beds) (Fig. 9.2).

The primary constructional components of a reef are the hermatypic corals. As a group they are defined by their symbiotic relationship with photosynthetic zooxanthellae algae. These algae provide the corals with additional photosynthetically derived energy, enabling them to thrive within the typically low nutrient waters of the tropics. At a global scale, tropical reef development can be broadly delineated by the mean 20°C sea-surface temperature isotherm (Fig. 9.1). This corresponds to latitudes between about 28°N and 28°S, where reefs currently occupy an estimated area of 255,000 km^2 (Spalding & Grenfell 1997). Within this latitudinal range, coral growth (and the potential for reef development) is highly variable, and is influenced by a range of factors. These include seawater temperature, aragonite

saturation state, salinity, light and nutrient levels (Table 9.1; Fig. 9.3a). Individual coral species can function across wide temperature ranges, but at both higher and lower temperature extremes (Table 9.1) the symbiotic coral–algal relationship breaks down and corals respond by shedding their photosynthetic algae, with consequent impacts for coral growth and calcification (Glynn 1996). Temperature also exerts a fundamental control on calcification because seawater temperature is positively correlated with aragonite saturation state (Buddemeier 1997). This influences calcium carbonate production rates, which decrease with latitude as sea-surface temperatures decrease (Fig. 9.3b & c).

Corals also survive across a range of salinity levels (Table 9.1), but marked reductions in reef-building potential occur in areas subject to either high fluvial discharge (Fig. 9.1) or intense evaporation. Given the dependence of hermatypic corals on photosynthetically derived energy, and in particular the link between photosynthesis and

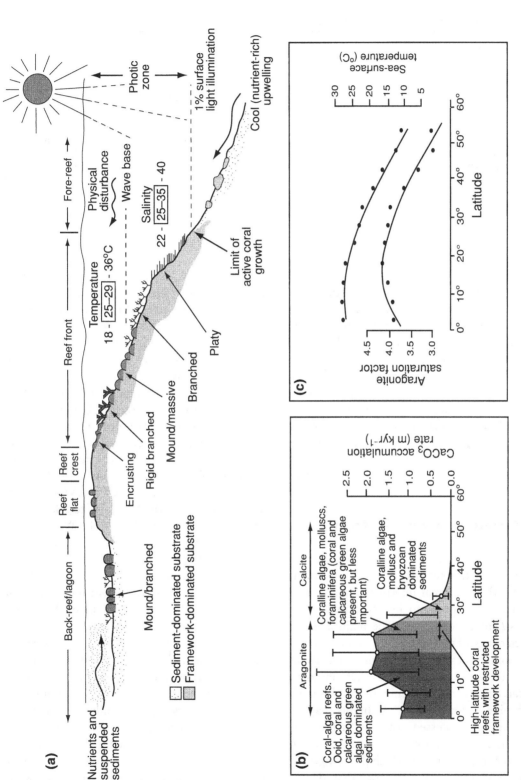

Fig. 9.3 (a) Environmental controls on the development of coral reef communities. (Adapted from James & Bourque 1994.) (b) Latitudinal changes in CaCO₃ accumulation rates and (c) sea-surface temperature and aragonite saturation states. (Adapted from Buddemeier 1997; sediment types after Lees 1975.)

Table 9.1 Marine environmental parameters that influence the distribution of hermatypic corals and of tropical coral reef development. 'Optimal' values for coral growth are shown, along with recorded upper and lower environmental limits. Figures in parentheses are for non-reef building coral communities. (Data from Kleypas et al. 1999.)

Environmental parameter	'Optimal' levels	Environmental limits	
		Lower	Upper
Temperature (°C)*	21.0–29.5	16.0 (13.9)	34.4 (32.1)
Salinity (PSU)†	34.3–35.3	23.3 (20.7)	41.8 (No data)
Nitrate (μmol L^{-1})‡	< 2.0	0.00	3.34 (up to 5.61)
Phosphate (μmol L^{-1})‡	< 0.2	0.00	0.40 (up to 0.54)
Aragonite saturation state (Ω-arag)§	c. 3.83	3.28 (3.06)	No data
Depth of light penetration (m)	c. 50	< 10	c. 90

*Weekly data.
†Monthly average data.
‡Overall averages (1900–1999).
§Overall averages (1972–1978).

coral calcification, light is a key control on coral growth. Light decreases with depth, so that rates of coral growth and calcification also decrease (Huston 1985). The lower limit of hermatypic coral growth is defined as the base of the photic zone (where surface light levels are reduced to 1%; Fig. 9.3a). In clear-water settings this can be as deep as around 90 m (Table 9.1), but occurs at much shallower depths in turbid environments due to reduced light penetration. Finally, elevated nutrient levels (nitrate levels > 2.0 μmol L^{-1}; phosphate levels > 0.20 μmol L^{-1}) may result in reduced rates of coral growth (Tomascik & Sander 1985).

9.1.2 Distribution and occurrence of mangroves

The term mangrove is variously defined and has been used to refer either to the constituent plants of these tropical intertidal forests or to both the community and its associated environment (Tomlinson 1986). The main defining characteristics of mangroves are that they comprise communities of salt tolerant tropical/subtropical trees and shrubs, and as such represent tropical equivalents of temperate salt-marsh communities (Woodroffe 1983). In this chapter, mangroves are discussed not only in the context of the constituent trees and shrubs, but also the sediment substrates on which the mangroves develop and the creek networks that dissect

them (Fig. 9.4). Mangrove ecosystems extend along some 60–75% of tropical and subtropical coastlines (MacGill 1958) and recent mapping estimates suggest a global coverage of around 190,000 km^2 (Spalding et al. 1997).

Opinions vary about the environmental factors limiting mangrove development, with both mean minimum air and sea temperatures having been cited as controls (Woodroffe & Grindrod 1991). Some authors also cite the occurrence of extreme seasonal cold (frost) events (Plaziat 1995). At the global scale, mangrove distributions exhibit a reasonably close correlation with the mean winter 15°C sea-surface isotherm (Woodroffe & Grindrod 1991), which equates to a latitudinal range between about 30°N and 30°S (Fig. 9.1). Actual distributional patterns, however, are variable and reflect local environmental (particularly cool seasonal temperature) constraints, such that (as with coral reefs) their distribution is more restricted on the western coasts of Africa and America (Plaziat 1995; Fig. 9.1).

Such climatic and/or oceanographic constraints are exacerbated by physical constraints, such as local geomorphology, tidal range, seasonal hydrology and substrate availability for colonization (Fig. 9.5). These factors influence mangrove development both at local and regional scales. At the local scale marked physical–chemical gradients across shorelines are determined by the relative importance of tidal against freshwater

Fig. 9.4 Fringing mangrove developed along the shoreline of Inhaca Island, Mozambique. Note the well-developed creek systems that dissect the mangrove. Field of view approximately 0.5 km. (Photograph courtesy of Simon Beavington-Penney.)

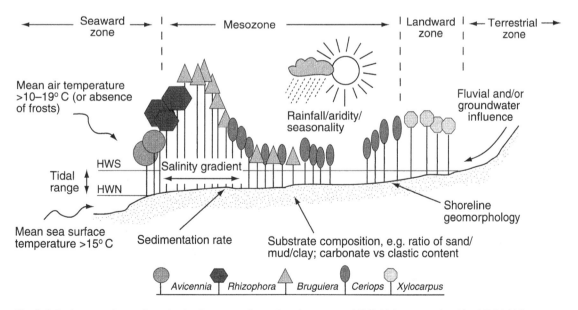

Fig. 9.5 Environmental controls on the development and zonation of mangroves: HWS, high water, spring tides; HWN, high water, neap tides. Temperature ranges are based on data in Woodroffe & Grindrod (1991).

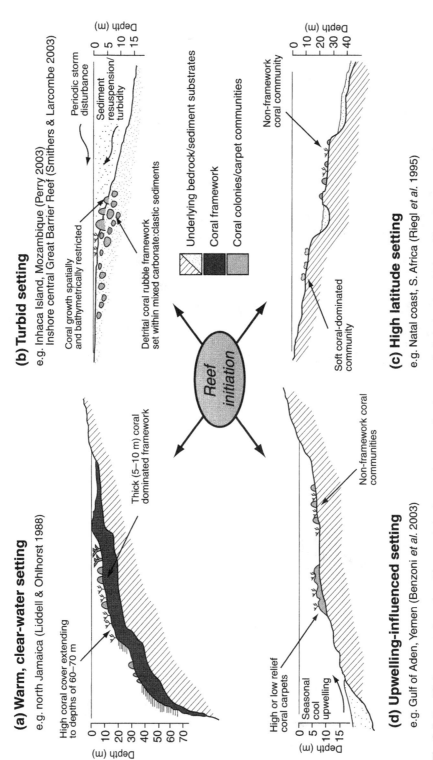

(a) Warm, clear-water setting

e.g. north Jamaica (Liddell & Ohlhorst 1988)

High coral cover extending to depths of 60–70 m

Thick (5–10 m) coral dominated framework

(b) Turbid setting

e.g. Inhaca Island, Mozambique (Perry 2003)
Inshore central Great Barrier Reef (Smithers & Larcombe 2003)

Periodic storm disturbance

Sediment resuspension/ turbidity

Coral growth spatially and bathymetrically restricted

Detrital coral rubble framework set within mixed carbonate:clastic sediments

Underlying bedrock/sediment substrates

Coral framework

Coral colonies/carpet communities

(c) High latitude setting

e.g. Natal coast, S. Africa (Riegl *et al.* 1995)

Non-framework coral community

Soft coral-dominated community

(d) Upwelling-influenced setting

e.g. Gulf of Aden, Yemen (Benzoni *et al.* 2003)

Non-framework coral communities

High or low relief coral carpets

Seasonal cool upwelling

Reef initiation

Depth (m)

Fig. 9.6 Simplified representations of coral framework development in a range of environmentally limited settings. (Adapted from Perry & Larcombe 2003.)

influence, as well as by the seasonality of rainfall (Fig. 9.5). Substrate salinity levels are particularly important and mangrove colonization is enhanced where freshwater inputs, from either river or groundwater sources, dilute salinity levels. At the regional scale, different climatic regimes result in variable fluvial runoff and substrate availability (Tomlinson 1986). Such variable influences result in complex and diverse sediment transport and accumulation processes in different settings, and the development of a diverse range of mangrove sedimentary environments.

9.1.3 Reef types and geomorphology

Reef structures are commonly described in terms of their overall geomorphology and proximity to adjacent landmasses. James & Macintyre (1985) delineate five basic reef types: (i) fringing reefs, (ii) bank-barrier reefs, (iii) barrier reefs, (iv) atolls and (v) patch reefs. The extent of reef development at individual sites is, however, highly variable and reflects local environmental parameters (see section 9.1.1). In warm, clear water and low nutrient environments, reefs can form both spatially and bathymetrically extensive structures. This state is exemplified by the fringing reefs of north Jamaica, where coral growth and framework development has occurred to depths of 80+ m (Liddell & Ohlhorst 1988; Fig. 9.6a) and where framework structures some 5–10+ m thick have developed during Holocene times (Land 1974). In contrast, restricted coral communities, with limited framework development, occur in turbid, nearshore environments. Around Inhaca Island, southern Mozambique, coral communities are restricted by low light levels to depths of < 6 m (Perry 2003) and framework development is replaced by unconsolidated coral rubble, set within a carbonate:clastic sediment matrix (Fig. 9.6b).

Marked variations in calcium carbonate accumulation rates (and reef development) also occur as mean sea-surface temperatures and aragonite saturation state decrease (Fig. 9.3b; Buddemeier 1997). As a result, reef framework development typically decreases towards higher latitude areas as environmental conditions become progressively more marginal for coral survival, and carbonate accumulation rates insufficient for framework construction. This state is illustrated by the reefs along the Natal coast of South Africa (Riegl et al. 1995), where coral communities colonize subtidal bedrock, but framework accumulation does not occur (Fig. 9.6c). Similar reductions in coral growth potential can occur in lower latitude settings (a 'pseudo-high-latitude effect'; Sheppard & Salm 1988) where seasonal upwelling brings cool, nutrient-laden waters to the surface, for example the Gulf of Aden (Fig. 9.6d; Glynn 1993). At these sites, framework development can be both spatially and bathymetrically restricted. Such examples illustrate the diverse range of environmental settings in which coral reefs occur, and which in turn influence patterns of reef sediment accumulation.

9.1.4 Mangrove types and geomorphology

Although mangroves are typified as forming extensive swamps associated with shorelines and estuaries that are accumulating sediment, they actually occupy a diverse range of coastal, offshore (island) and fluvially influenced settings (Thom 1982; Woodroffe 1992). These include:

1 alluvial plains – areas characterized by high fluvial sediment accumulation, throughput or discharge (Fig. 9.7a);

2 tidal plains – areas of high tidal range characterized by strong bi-directional flow patterns (Fig. 9.7b);

3 wave-protected coastlines – mangroves develop along the landward sides of barrier islands and beach ridges (Fig. 9.7c), and the shorelines of protected lagoons (Fig. 9.7d);

4 coastal embayments and drowned valleys (Fig. 9.7e);

5 carbonate-dominated coastal environments – these include subtidal carbonate mudbanks, and substrates associated with intertidal reef flats (Fig. 9.7f).

Although mangroves thus occur in a range of geomorphological settings, they also occur across a range of climatic settings. These span arid, through subtropical to tropical shorelines.

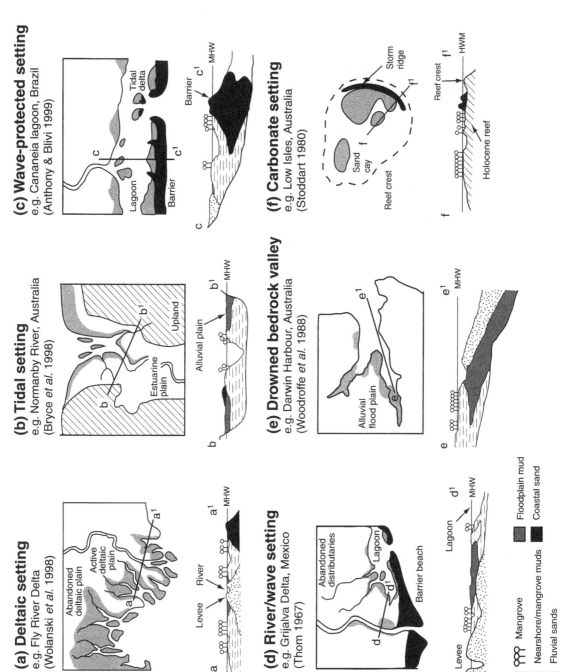

(a) Deltaic setting
e.g. Fly River Delta
(Wolanski *et al.* 1998)

Abandoned deltaic plain

Active deltaic plain

Levee River

MHW

(b) Tidal setting
e.g. Normanby River, Australia
(Bryce *et al.* 1998)

Upland

Estuarine plain

Alluvial plain

MHW

(c) Wave-protected setting
e.g. Cananeia lagoon, Brazil
(Anthony & Blivi 1999)

Tidal delta

Lagoon

Barrier

Barrier

MHW

(d) River/wave setting
e.g. Grijalva Delta, Mexico
(Thom 1967)

Abandoned distributaries

Lagoon

Barrier beach

Lagoon

Levee

MHW

(e) Drowned bedrock valley
e.g. Darwin Harbour, Australia
(Woodroffe *et al.* 1988)

Alluvial flood plain

MHW

(f) Carbonate setting
e.g. Low Isles, Australia
(Stoddart 1980)

Reef crest

Sand cay

Storm ridge

Reef crest

Holocene reef

HWM

Mangrove

Nearshore/mangrove muds

Fluvial sands

Floodplain mud

Coastal sand

Fig. 9.7 Mangrove depositional settings: HWM, high-water mark; MHW, mean high water. (Adapted from Woodroffe 1992.)

These may be characterized by different mangrove species depending upon species tolerance to environmental stress. *Avicennia*, for example, are most resistant to low and high temperatures, and to high soil salinities, whereas *Rhizophora* are resistant to low temperatures (Plaziat 1995). Along arid shorelines, the major environmental gradients relate to salinity, which increases landward due to high evaporation rates. In these environments mangroves thus often form narrow fringes of dwarf *Avicennia* (< 1 m high). In contrast, dense, sprawling forests develop on large areas of deltaic mud and organic-rich substrates along subequatorial shorelines (Smith 1992). In these cases, the dominant *Rhizophora* and *Avicennia* plants may reach heights of 40 m (Plaziat 1995).

9.2 SEDIMENT SOURCES AND SEDIMENT ACCUMULATION PROCESSES

9.2.1 Sources and characteristics of coral reef sediments

Sediments that accumulate on and around coral reefs derive from a range of sources. These include:

1 skeletal sediments – the calcareous remains of reef framework-building and reef-associated organisms;

2 non-skeletal sediments – grains produced by physico-chemical induced carbonate precipitation;

3 allochthonous sediments – grains derived from terrestrial sources and which may be either natural or anthropogenic in origin.

The relative abundance of these sediment contributors varies both within and between environments.

9.2.1.1 Skeletal sediments

Typically the most abundant constituent of reef-related sediments are the skeletal remnants of calcareous reef organisms. These can be subdivided into sediments produced by the breakdown of carbonate framework contributors, and those derived from reef-associated benthic organisms and calcareous algae. Coral reef framework is composed of two main constructional components, the skeletons of hermatypic corals (primary framebuilders) and a diverse array of associated calcareous encrusting faunas (secondary framebuilders). Calcareous encrusters (crustose coralline algae, encrusting forms of bryozoans and foraminifera, and serpulids) produce multiple crusts on the dead surfaces of coral skeletons (Martindale 1992) and help bind and stabilize the reef framework (Rasser & Riegl 2002). Breakdown of these primary and secondary framework contributors (and thus framework-related sediment production) is facilitated by physical and biological activity.

Physical (storm) disturbance results in fragmentation and transport of coral framework, and generation of coral rubble (Fig. 9.8), which in turn can be degraded by physical reworking to produce fine coral sand/silt (Fig. 9.9). However, the release of framework carbonate into sediment results primarily from bioerosion (a term used to describe biological substrate erosion; Neumann 1966). Bioerosion is facilitated by a wide range of reef-associated faunas, including fish and echinoids, and endolithic forms of sponges, bivalves and worms (Fig. 9.8; Hutchings 1986). Framework degradation and sediment production by fish and echinoid species results as a by-product of the search for food. Parrotfish and surgeonfish, for example, have heavily calcified mouthparts and bite off chunks of coral substrate, which is excreted as fine sand (Fig. 9.9; Gygi 1975). Similarly, echinoids such as *Diadema* sp. have heavily calcified feeding apparatus enabling them to remove coral skeleton during feeding (Fig. 9.8). As a by-product, they produce abundant carbonate-rich faecal pellets (Fig. 9.9; Scoffin et al. 1977).

Significant degradation of framework also results from the activities of endolithic boring organisms. These include specific groups of sponges, bivalves and worms (Bromley 1978; Perry 1998a). These organisms, which use either physical and/or chemical processes to excavate tunnels/chambers within dead coral skeleton, produce boreholes > 1 mm in diameter and are

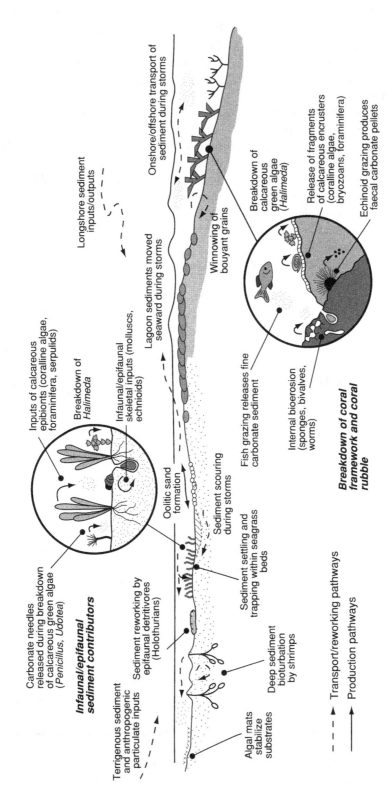

**Infaunal/epifaunal
sediment contributors**

Carbonate needles
released during breakdown
of calcareous green algae
(*Penicillus, Udotea*)

Terrigenous sediment
and anthropogenic
particulate inputs

Sediment reworking by
epifaunal detritivores
(Holothurians)

Algal mats
stabilize
substrates

Deep sediment
bioturbation by shrimps

Inputs of calcareous
epibionts (coralline algae,
foraminifera, serpulids)

Breakdown of
Halimeda

Infaunal/epifaunal
skeletal inputs (molluscs,
echnioids)

Oolitic sand
formation

Sediment scouring
during storms

Sediment settling and
trapping within seagrass
beds

Lagoon sediments moved
seaward during storms

Fish grazing releases fine
carbonate sediment

Internal bioerosion
(sponges, bivalves,
worms)

**Breakdown of coral
framework and coral
rubble**

Longshore sediment
inputs/outputs

Winnowing of
bouyant grains

Onshore/offshore transport of
sediment during storms

Breakdown of
calcareous
green algae
(*Halimeda*)

Release of fragments
of calcareous encrusters
(coralline algae,
bryozoans, foraminifera)

Echinoid grazing produces
faecal carbonate pellets

- - - → Transport/reworking pathways

⟶ Production pathways

Fig. 9.8 Schematic diagram illustrating the main sediment sources and transport pathways within coral-reef-related environments.

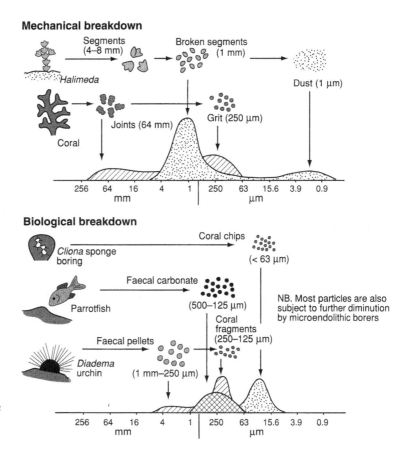

Fig. 9.9 Mechanical and biological breakdown of carbonate sediment contributors and resultant sediment size fractions produced. (Adapted from Scoffin 1987.)

termed macroborers. In addition, a range of microborers, including forms of cyanobacteria, chlorophytes and fungi, facilitate substrate degradation on a microscopic scale (Golubic et al. 1975; Perry 1998b). Although all boring activity results in framework degradation, not all results in sediment production, because varying proportions of the excavated material will be directly dissolved. The most significant sediment producers are the boring sponges. These sponges excavate small fragments of coral skeleton, which are subsequently expelled (Fig. 9.8), resulting in the production of abundant fine-grained sand (Fig. 9.9; Fütterer 1974). Once released into the sediment carbonate grains are subject to further physical, chemical and biological alteration (Perry 2000).

Coral framework is not, however, the only source of reef sediment. Skeletal carbonate sediment is also derived from infaunal and epifaunal organisms as well as a wide array of calcifying algae. Common infaunal and epifaunal sediment contributors include bivalves and gastropods, foraminifera and echinoids (Fig. 9.8; Swinchatt 1965). Skeletal sediments are also contributed by calcareous epiphytic organisms (including crustose coralline algae, foraminifera, serpulids and bryozoans) that encrust seagrass blades and the stems of calcifying algae (such as *Udotea* and *Penicillus*; Fig. 9.8; Land 1970; Nelson & Ginsburg 1986). In addition, carbonate sediment is produced during breakdown of carbonate-secreting algal species. Some, such as *Penicillus* and *Udotea*, release abundant carbonate needles into the sediment (Neumann & Land 1975), whereas others, such as *Halimeda*, produce heavily calcified segments, which often subsequently disintegrate into carbonate needles (Fig. 9.9). Many calcifying algae have high turnover rates, producing several standing crops per year, and are thus significant components of reef sediment budgets (Neumann & Land 1975). Minor additional skeletal, but non-carbonate, sediments derive from diatoms and sponge spicules (both of which are formed from silica). Morphological descriptions of these varied sediment contributors are found in Scoffin (1987).

Production rates by different contributing groups and processes vary markedly between reef environments, thus dictating spatial variations in rates of sediment production and accumulation. At Kailua Bay (Oahu, Hawaiian Islands), both total carbonate sediment production rates and the production rates of specific sediment contributors vary between environments. Nearshore hardgrounds have the highest sediment production rates (0.62 kg m^{-2} yr^{-1}) with production dominated by *Halimeda* (0.25 kg m^{-2} yr^{-1}), bioerosion-derived coral (0.13 kg m^{-2} yr^{-1}) and molluscs (0.18 kg m^{-2} yr^{-1}). In contrast, sediment production rates in reef front sites average 0.41 kg m^{-2} yr^{-1}, of which 95% is derived from coral bioerosion (Harney & Fletcher 2003).

9.2.1.2 Non-skeletal carbonate sediments

In addition to skeletal sediments, non-skeletal grains also occur within reef and reef-related sediments. These include carbonate muds, peloids and ooids (Fig. 9.8). Carbonate silts (grain size < 63 μm) often comprise a volumetrically significant component of lagoon sediments. Much of this material comprises single aragonite crystals derived from the breakdown of carbonate-secreting marine algae, although some may also result from direct precipitation from supersaturated sea waters. Peloids are small, rounded or elliptical grains characterized by a microcrystalline internal structure and have a range of origins (Macintyre 1985). Some represent the calcified remnants of faecal pellets, others fragments of skeletal grains where the internal structure has been modified by microboring. Some peloids are also believed to result from phases of chemical precipitation around a central nucleus. Ooids are also small rounded grains (typically < 1 mm diameter), but exhibit multiple concentric lamellae that form a coating around a central nucleus (often a peloid or skeletal fragment) and result from physico-chemical precipitation of

calcium carbonate under high-energy conditions (Ginsburg 1957).

9.2.1.3 Allochthonous sediments

Although the majority of sediment that accumulates on and around coral reefs is carbonate, there are significant external inputs of sediment in some environments. This is particularly evident in areas where reefs develop close to sources of fluvial sediment discharge (Fig. 9.6). Typically the proportion of terrigenous sediment will decrease with distance from source (e.g. Acker & Stearn 1990), although sediment accumulation within individual coastal environments commonly reflects not only local carbonate and fluvially derived sediment inputs, but also sediment flux into and out of the environment. A detailed sediment budget of a carbonate embayment at Hanalei Bay, Hawaii, influenced by terrigenous sediment input (Calhoun et al. 2002), found that significant proportions of fluvially derived suspended sediments were exported offshore, and the bay was also subject to inputs of carbonate sediment derived from adjacent coastal areas. In this example, the total volume of Holocene carbonate sediment that has accumulated in the bay exceeds estimates of *in situ* Holocene carbonate sedimentation, and the bay is therefore acting as a net sink for sediment produced in adjacent coastal areas. Increasingly associated with terrigenous sediment inputs are a range of dissolved and particulate contaminants linked to anthropogenic (industrial or agricultural) discharges. These include heavy metal and hydrocarbon contaminants and are discussed in section 9.4.4.

9.2.2 Controls on coral reef sediment transport and accumulation

Although local environmental factors influence the composition and abundance of individual sediment contributors, the accumulation of carbonate sediment in reef environments is influenced by a wide range of physical and biogenic processes. These influence sediment transport, reworking, trapping and stabilization.

9.2.2.1 Reef sediment transport

Sediment transport and deposition are determined by two main factors: (i) shear stress and (ii) settling velocity. The former relates to the velocities required to move or entrain sediment particles of a specific size, the latter to the difference between the gravitational and buoyancy forces acting on the particle. The relationship between threshold and settling velocities is relatively well established for quartz grains, which, due to uniform densities, behave in a reasonably predicable fashion dictated by grain size (see Chapter 1). Grain transport and deposition in carbonate sediments, however, are complicated by differences in grain skeletal structure (and hence density) and by grain size, shape and texture. Grains with plate-like morphologies will settle at a slower rate than block or rod-shaped grains and hence such parameters influence grain transport and deposition (Kench & McLean 1996).

These hydraulic controls have been well illustrated in a study of sediment transport in the Cocos (Keeling) Islands (Kench 1997). Carbonate sediment assemblages around the atoll can be related to classes defined by settling velocities and which delineate sediment transport pathways. Essentially two main sediment assemblages are present: one dominated by reef-derived components, and a second produced *in situ* within the lagoon. The former occur within what has been described as an 'active transport zone' in which reef-derived sediments are selectively transported by currents from shallow reef flat areas into the channels entering the central lagoon. Sediments in these areas are dominated by faster settling grains (mainly larger fragments of coral, coralline algae and the foraminifera *Amphistegina* sp.). Slower settling grains (smaller coral grains, *Halimeda* and the foraminifera *Marginopora* sp.) are transported through the reef flat channels and deposited along the lagoon margins (Kench 1997).

9.2.2.2 Reef sediment trapping and stabilization

Although the physical properties of grains dictate sediment entrainment and transport, a range of biogenic components and physico-chemical processes interact to trap and stabilize reef sediments. Particularly important in this respect are marine seagrasses and green algae. The long blades of seagrasses, such as *Thalassia* and *Syringodium*, locally reduce current speeds and promote sediment settling (Scoffin 1970). In the long-term, such processes can lead to the development of carbonate mudbanks (Bosence et al. 1985), although this will depend upon local rates of carbonate mud production (Perry & Beavington-Penney 2005). The dense rhizome (root) networks associated with seagrasses also facilitate substrate stabilization and binding. Similar binding of sediment occurs around the holdfasts of green algae such as *Halimeda*, *Penicillus* and *Udotea* (Scoffin 1970). Organic binding of sediment also occurs in areas where algal-mat communities develop (most commonly in low-energy lagoon settings). Associated with these mats are filamentous algae including the cyanobacteria *Lyngbya* and *Schizothrix*, and the chlorophyte *Enteromorpha* (Scoffin 1970), which promote sediment adhesion and trapping, and in turn enhance substrate stabilization. Binding of substrates also occurs in areas subject to physico-chemical and organically induced carbonate cement precipitation (Scoffin 1987).

9.2.2.3 Reef sediment reworking

In addition to sediment stabilization, significant sediment reworking also occurs, much of which can be attributed to wave and current action (see section 9.2.2.1) and to grain diminution (see section 9.2.1.1). Bioturbation also occurs, however, associated with surface feeders such as holothurians and crabs, and subsurface organisms such as shrimps. Holothurians, for example, are estimated to ingest and excrete up to 250 g of sediment per day. The excreted sediment not only produces a highly homogenized surficial sediment layer, but the ingestion process may also result in chemical grain dissolution (Hammond 1981). Extensive sediment reworking (with volumes of sediment turnover in excess of 11 kg m^{-2} yr^{-1}) may occur associated with infaunal organisms such as the *Callianassa* shrimp (Bradshaw 1997).

Such intense bioturbation results in preferential sediment sorting and aeration of surface sediments, and creates a highly mobile surface layer poorly conducive to colonization (Tudhope & Scoffin 1984). Finer sediments expelled during burrowing are also prone to resuspension and transport (Roberts et al. 1981).

9.2.3 Spatial variations in sediment accumulation

Spatial variations in the abundance and productivity of reef sediment contributors, and the processes of sediment transport and reworking, produce distinct grain assemblages in different reef areas. Across nearshore fringing or bank-barrier reefs, local variations in sediment types are evident between lagoon, reef crest, and shallow and deep reef front environments. Each subenvironment can be delineated on the basis of grain assemblage (i.e. the relative abundance of skeletal grain types) and texture. These reflect not only initial grain inputs, but also grain reworking and transport. Across the narrow (c. 1 km wide) fringing reefs of north Jamaica, clear patterns of sediment accumulation can be identified (Fig. 9.10a), which broadly reflect the abundance of sediment contributing groups on the reef (Boss & Liddell 1987). Lagoon sediments are, for example, characterized by relatively high abundances of *Halimeda*, benthic foraminifera and molluscs, whereas shallow reef-front sediments are dominated by coral, coralline algae and encrusting foraminifera. These patterns are representative of sediment accumulation patterns across many narrow shelf reefs. Distinct patterns of carbonate sediment accumulation are also evident across larger carbonate shelf or platform environments. Reef sediment assemblages (comparable to those described above) occur where reefs are developed along the seaward margins, but significant carbonate production can also occur across the inner platform areas where oolitic sand bodies, seagrass beds or green algal meadows develop (Purdy 1963). Such large-scale patterns of carbonate sediment accumulation are evident across the Great Bahama Bank (Fig. 9.10b), and the land-attached Florida and Great Barrier Reef

shelf settings. Detailed sedimentological descriptions of the two former examples are found in Tucker & Wright (1990).

Although clear patterns of cross-reef or cross-shelf sediment facies therefore can be identified, marked variations in grain assemblages are also evident between reef settings. As outlined in section 9.1.3, the extent of reef development varies markedly between regions and is influenced by spatial and latitudinal changes in marine environmental parameters such as temperature, light penetration and aragonite saturation state. These same limiting factors also influence patterns and rates of reef sediment production and are clearly illustrated in relation to shifts in the types of skeletal and non-skeletal sediment contributors. On a global scale these have been related to latitudinal changes in temperature and salinity (Lees 1975), although related shifts in aragonite saturation state (Buddemeier 1997; Fig. 9.3) are also important. Lees (1975) identified three distinct carbonate grain assemblages that characterize different temperature and salinity zones:

1 a chlorozoan assemblage (characterizing tropical waters) dominated by corals and calcareous green algae;
2 a chloralgal assemblage (characterizing subtropical waters) in which corals and calcareous green algae become progressively less common and the dominant grain types are coralline red algae, molluscs and foraminifera;
3 a foramol association (characterizing warm temperate to cold waters) in which coralline red algae, molluscs and bryozoans dominate.

Transitions between such assemblages therefore occur as environmental conditions become progressively more marginal for coral survival (Halfar et al. 2000).

9.2.4 Sources and characteristics of mangrove sediments

Mangroves colonize a wide range of coastal environments (see section 9.1.4) and the sediments that contribute to mangrove substrates are derived from a range of sources. These can be classified as either:

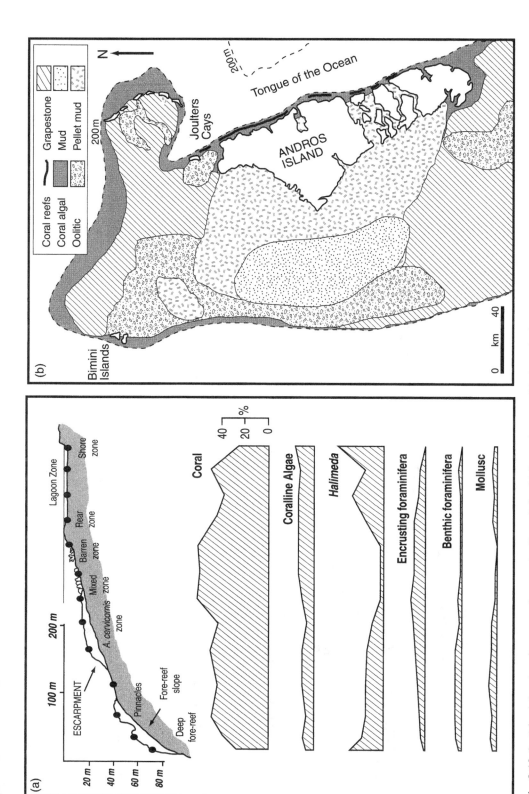

Fig. 9.10 Spatial variations in sediment contributors and sediment facies. (a) Across the Andros Platform, Great Bahama Bank. (Adapted from Purdy 1963.) (b) Across the narrow fringing reefs of north Jamaica. (Adapted from Boss & Liddell 1987.)

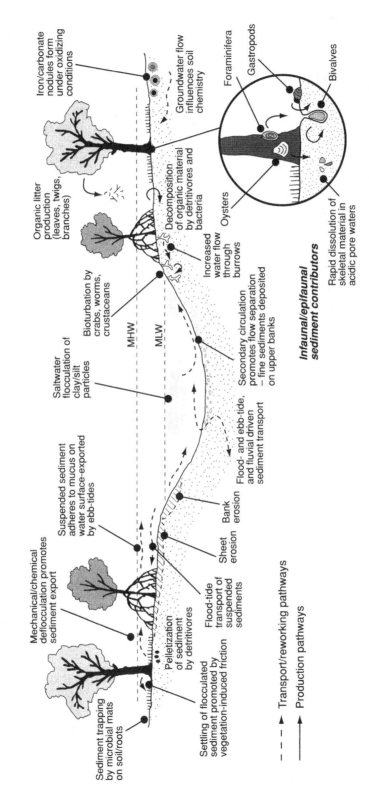

Fig. 9.11 Schematic diagram illustrating the main sediment sources and transport pathways within mangrove environments. MHW, mean high water; MLW, mean low water.

1 allochthonous – sediments derived from outside the mangrove and either terrestrial (often fluvial) or shallow marine in origin;
2 autochthonous – sediments produced *in situ* and which include both organic litter and skeletal material.
Sediment accumulation is influenced by a range of physical and ecological factors that control transport, settling and reworking potential.

9.2.4.1 Allochthonous sediments

A significant proportion of the sediment that accumulates within mangroves derives from either terrestrial (mainly fluvially sourced) or nearshore settings. Such sediment may be moved either as suspended or bedload material (see Chapter 1) and comprises either clastic, carbonate or organic material. The relative importance of sediment sources and of the import/export of sediment varies considerably between settings and depends in part on the relative importance of tidal versus fluvial influence (Fig. 9.11). In the microtidal Richmond River estuary (Australia) 92–99% of the annual suspended sediment load is derived from fluvial inputs, 90% of which enters the estuary during short (2 week) seasonal flood events (Hossain & Eyre 2002). Only 1–2% of estuarine sediment is derived from the continental shelf. In contrast, the Fly River estuary (Papua New Guinea) occurs in a meso/macrotidal setting. Despite high fluvial discharge rates (6000 m^3 s^{-1}), there is a net inflow of suspended sediment from coastal waters which exceeds fluvial discharge by around 10 times (equivalent to 40 t s^{-1}; Wolanski et al. 1998). The seasonality of river flow is also an important control on the net transport of allochthonous sediment in many fluvially influenced settings. The mesotidal Normanby River estuary in northern Australia is, for example, characterized by long dry seasons (8–10 months) with short wet seasonal flow events (Bryce et al. 1998). Although seasonal flood events are significant, with flood-driven sediment transport estimated to range from 6000 to 32,000 t per event, this is insufficient to remove all of the annual sand influx to the estuary (15–30,000 t yr^{-1} of bedload

sands, and 50,000 t yr^{-1} of suspended sediment). As a result, the estuary is dominated by net landward flood-tide sediment transport which accumulates in the upper estuary.

9.2.4.2 Autochthonous sediments

In addition to externally sourced (fluvial or marine) sediment, significant amounts of sediment may also derive from *in situ* sources, the most important of which are organic litter and mangrove-associated skeletal faunas (Fig. 9.11). Most mangroves are characterized by high rates of primary productivity and, as a result, large amounts of organic material, in the form of leaf litter along with decaying roots and branches, accumulate in the sediment (Fig. 9.12). Rates of leaf litter production (a commonly used proxy for organic matter production; Hogarth 1999) range from 5 to 15 t ha yr^{-1}. Production rates are, however, dependent upon both season and setting. At Darwin Harbour (Australia) Woodroffe et al. (1988) measured the highest litter production rates (up to 1400 g m^{-2} yr^{-1}) within tidal creek settings beneath tall (13 m high) *Avicennia* sp. trees. In nearby marginal hinterland settings, production of leaf litter near small (< 2 m high) *Ceriops* sp. was only 300 g m^{-2} yr^{-1}. At all sites, litter production rates were highest during the wet season.

Leaf litter may be broken down either by microbial action or during crab feeding, exported by tidal or river currents, or incorporated into the sediment (Hogarth 1999). In a study from Australia, Robertson et al. (1992) illustrated the spatial variability that occurs within mangroves in terms of reworking of organic matter. In lower intertidal areas, litterfall averages 556 g m^2 yr^{-1}, of which 71% is exported by tidal currents, 28% broken down by crabs and around 1% lost to microbial action. In contrast, within high intertidal settings, only 33% of the 509 g m^{-2} yr^{-1} produced is exported by tides, 34% broken down by crabs and 33% decomposed by microbial activity. In settings with a high tidal range and/or extensive fluvial activity, significant export of organic material may thus occur, with the mangrove acting as a major

Fig. 9.12 Intertidal sediments accumulating in and around the prop roots of *Rhizophora* sp. colonies. Note the abundant leaf and seagrass litter accumulating on the substrate. Inhaca Island, Mozambique.

source for organic matter (Dittmar et al. 2001). In settings characterized by limited tidal exchange and by limited fluvial influence, organic matter is largely retained within the environment and represents a dominant sediment contributor. Wood detritus breaks down much more slowly, but is facilitated by the activities of wood-boring teredinid molluscs.

Another important input into mangrove sediment derives from mangrove-associated shelly faunas. Of these the most abundant are crabs, molluscs and foraminifera (Fig. 9.11). Dominant crab families are the Grapsidae and Ocypodidae and individual species densities can be in the order of 60 m^{-2}. Numerous molluscs, including species of barnacles, oysters, gastropods and bivalves, also occur and are often dominant skeletal contributors (Plaziat 1995), although there are marked variations between environments (Plaziat 1974). Foraminifera are also abundant, although abundance and diversity are strongly influenced by local hydrodynamics, the seasonality of fluvial influence and substrate elevation (Debenay et al. 2002). Although the skeletal remains of molluscs, foraminifera and crabs accumulate with mangrove sediments, much of this material is subject to intense dissolution within acidic porewaters. Although the processes of skeletal modification and degradation in mangroves remain poorly documented, the best preservation is likely in areas where (i) rapid burial in fine-grained (low permeability) sediments occurs, or (ii) either sea water or carbonate muds buffer sediment porewater acidity (Plaziat 1995).

9.2.5 Controls on mangrove sediment transport and accumulation

9.2.5.1 Mangrove sediment transport

The mechanisms and rates of sediment transport within and through mangrove systems are determined by water circulation patterns. These circulation patterns are, in turn, a reflection of both the local hydrology (determined by factors such as freshwater runoff and evapotranspiration) and tidal regime. Mechanisms of sediment transport therefore vary widely depending on the local climate (which may be highly seasonal) and tidal regime (which varies over both diurnal and monthly spring–neap cycles). In addition, very different mechanisms and rates of sediment transport occur in the tidal creeks and on the mangrove flats. The tidal creeks often form long, branched networks and represent the main conduits of water (and sediment movement). In contrast, the mangrove flats are often wide and heavily vegetated and, because of the high frictional effects of the vegetation, represent sites of sediment trapping and accumulation (see section 9.2.5.2). Ratios of creek to flat area are in the order of 2–10 in many mangroves (Wolanski et al. 1992) and thus the mangrove flat represents a significant proportion of the tidal prism (especially during the spring-tide phase). Consequently, the processes of sediment transport in mangroves exhibit marked temporal and spatial variability.

The tidal cycle represents a particularly important influence on sediment transport by dictating current speeds and the magnitude and frequency of tidal inundation. Tidal circulation is the primary cause of water movement through mangrove creeks and there is often a strong asymmetry between ebb- and flood-tide current velocities (see Chapter 1). The ebb-phase is typically shorter and characterized by stronger current speeds. These may be as much as one-third higher than peak flood currents and exceed rates of 1 m s^{-1} (Wolanski et al. 1980). There are also marked differences between the creek networks and the mangrove flats, with current speeds on the flats often less than 0.1 m s^{-1}. As a result, significant variations occur both in suspended and bedload sediment transport (see section 9.2.4.1), as well as in the flux of sediment through tidal creeks and onto the mangrove flats.

An additional influence on sediment transport in the mangrove creeks is the degree of freshwater versus tidal influence, because this dictates the extent of saline incursion. Where there is strong fluvial outflow, complete flushing of saltwater from creeks commonly occurs and results in a net outflow of sediment. In areas of restricted or seasonal outflow, sediment is often retained in upper parts of creek systems (Wolanski et al. 1992). Tidal incursions also influence the areas over which processes such as secondary circulation, saltwater flocculation, baroclinic circulation and tidal pumping occur. These, in turn, influence rates of sediment transport through the tidal creeks (Wolanski et al. 1992). Secondary circulation occurs within creek networks, where marked vertical differences in either salinity or suspended sediment loads create density gradients (Wolanski 1995; Ridd et al. 1998). These are commonly established around meander bends. On the inside of meanders the interface between density layers may be raised resulting in differential transport and accumulation between the upper and lower parts of the channel. Under these conditions, finer sediments may accumulate on the upper, inside parts of the bank, and coarser (bedload) material on the channel floor (Fig. 9.11).

Saline incursions into mangrove creek networks also influence sediment transport through grain flocculation (fine sediment aggregation). In freshwater reaches of creeks, fine sediments rarely flocculate except where organic particles promote grain adhesion and/or suspended sediment concentrations exceed 1 g L^{-1}. As a result, sediment settling velocities are influenced by grain size and density (Wolanski 1995). In saline-influenced creeks (salinity levels > 1), however, saltwater flocculation of fine clay/silt particles occurs (Fig. 9.13), with the metallic and organic coatings on grains promoting grain aggregation. This leads to the formation of large flocs up to 200 μm across. Although both clay and calcareous flocs have a low cohesive strength, the larger grain sizes increase settling velocities, which may be up to 100 times those of unflocculated particles (Furakawa & Wolanski 1996). It is likely that without this process most fine sediment would travel through the mangrove as a 'wash load' (Wolanski 1995).

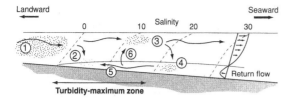

Fig. 9.13 Schematic diagram illustrating the processes of grain flocculation and transport associated with baroclinic circulation in a partially stratified estuary: 1, fluvial transport of unflocculated sediments; 2, saltwater flocculation – large flocs settle out; 3, unflocculated fines remain in suspension; 4, settling of finer grained flocs to the bed; 5, baroclinic circulation transports settled flocs back upstream; 6, disaggregation of reworked flocs and re-entrainment in the water column. The turbidity-maximum zone forms close to the limits of saline intrusion. (Adapted from Wolanski 1995.)

The accumulation of flocculated sediments within mangrove creeks is influenced by two additional factors: (i) baroclinic circulation and (ii) tidal pumping. Baroclinic circulation occurs as a result of the landward movement of denser, saline waters along the bottom of tidal creeks (Fig. 9.13). These currents entrain flocculated sediments as they settle out and return them upstream towards the limit of saline intrusion (Wolanski 1995). This process is strongest in areas with a pronounced salinity gradient, and results in most fine sediment remaining in suspension near the creek floor. Tidal pumping can occur in mangrove creeks where there is an asymmetry between the peak flood and ebb-tidal current velocities. The relative strength of the flood- or ebb-tide current varies depending upon tidal regime and the seasonality of the mangrove system (see section 9.2.4.1), but in sites where tidal currents are stronger during the flood- rather than the ebb-tide phase the result is a net upstream movement of sediment. In combination these two processes form a turbidity maximum zone where fine sediment accumulation is concentrated.

As outlined previously not all sediment that enters mangrove creeks is retained within the creek network and many mangroves are characterized by net sediment export. Sediment export from creek systems will be promoted in systems where the frictional effects of high vegetation cover produce a marked surface-water gradient across the mangrove. This occurs because mangroves reduce flow velocities through the marsh. As a result, flood-tide water levels rise and fall much faster in creek and creek margin areas than on the mangrove flat itself and can create water gradients of up to 1:1000. Where this occurs the tide will be falling at the creek mouth while high tide waters will be ponded on the mangrove flat, producing strong ebb-tide velocities. At Ross Creek, northern Australia, threshold velocities for the fine-grained channel substrates are around 0.4 m s^{-1}, and flood and ebb-tide current velocities are 0.4 and 0.8 m s^{-1} respectively. As a result, bedload transport is more prominent during the ebb-tide phase (Larcombe & Ridd 1995). Under such conditions, sediment will be trapped by vegetation on the mangrove flats during spring-tide phases, but scoured from creek areas and exported (Wolanski et al. 1992).

9.2.5.2 Mangrove sediment trapping and stabilization

Although tidal and fluvial currents exert an important influence on sediment transport, sediment accumulation is strongly influenced by mangrove root type and pneumatophore density, which modifies current velocities and flow regimes (Woodroffe 1992). As currents pass through the dense mangrove root networks on the mangrove flats, the vegetation induces micro-turbulent flow (eddies, jets, stagnation zones), which maintains sediment in suspension. This material is typically transported landward and settles out around slack high tide as flow turbulence reduces (Furukawa & Wolanski 1996). On the mangrove flats this process is restricted to periods of high spring tide, and material is prevented from re-entrainment during the ebb-tide phase by vegetation-induced friction. As a result, tidal currents can act as a 'pump', transporting sediment landward, so that the highest sedimentation rates occur close to the high tide limit. This has been demonstrated at Middle Creek, Australia (Furukawa et al. 1997) where preferential sediment trapping occurs in areas of vegetation-induced flow stagnation and around 80% of suspended sediment is trapped in the mangroves. This equates to 10–12 kg of sediment per metre creek length per spring tide, or a vertical accretion rate of around 0.1 cm yr^{-1}. Krauss et al. (2003) have documented highest sediment settling rates (11.0 mm yr^{-1}) around prop roots, as opposed to pneumatophores (8.3 mm yr^{-1}), although the latter are most effective in terms of sediment retention. Sediment trapping is also aided where bacterial and/or algal mats develop on substrate surfaces. Pelletization of fine sediment by benthic detritivores will also bind sediment and limit sediment entrainment and (re)export.

9.2.5.3 *Mangrove sediment reworking*

Although small-scale effects of physical sediment reworking and transport influence grain entrainment and transport, biological reworking also occurs. Much of this is attributed to the activities of crabs, which rework sediment during feeding and burrow construction. Grapsidae crabs feed primarily on leaf litter, whereas Ocypodidae (fiddler crabs) are detritivores and ingest sand grains in order to remove organic material. This latter group produce large numbers of faecal and pseudofaecal pellets, with sediment turnover rates of up to 0.5 kg m^{-2} day^{-1} (Hogarth 1999). Although this has an important effect on retexturing sediments, the generation of extensive, interconnecting burrow networks also provides a conduit for groundwater flow (Wolanski et al. 1992). Flow velocities in these burrows may be up to 30 mm s^{-1} with 1000–10,000 m^3 of water flow per tidal cycle per square kilometre (Ridd 1996).

9.2.6 Variations in mangrove sediment accumulation

It is now widely accepted that mangroves tend to follow rather than initiate sediment accumulation (Woodroffe 1992), however, once established the mangroves may enhance sediment accumulation by facilitating the trapping of sediment around roots and pneumatophores. A review of recent literature on short-term vertical accretion rates by Ellison (1998) suggests that sediment accretion rates within mangroves are often less than 0.5 cm yr^{-1} (with a maximum of around 1 cm yr^{-1}). Sediment accumulation rates are, however, highly variable both within and between environments, as well as seasonally (Saad et al. 1999). Such differences reflect, in large part, variations in sediment supply and the local processes of sediment transport and reworking. Viewed at a relatively simplistic level, narrow mangrove-fringed shorelines are likely to exert much less influence on sediment budgets than wide mangrove shorelines (Wolanski 1995), although few sufficiently detailed budget studies have been undertaken to enable detailed understanding of the net inputs/outputs of sediment in different mangrove settings.

Different mangrove environments, however, can be identified that have very different sediment substrates and distinct processes of sediment supply and accumulation. The Grijalva–Usumacinta delta, Mexico, for example, forms a complex network of mangrove-fringed lagoons, channels and interdistributary basins (Thom 1967). The delta is developed along a microtidal coastline and is dominated by fluvial inputs (Fig. 9.14a). Fluvial discharge is, however, highly seasonal and during the dry season saline wedges develop within the creeks that may extend up to 30 km inland. Sediment inputs are dominated by fluvially derived inorganic sands, silts and clays, and organic material produced *in situ* from litter-fall. In contrast, the mangroves developed in the vicinity of Coral Creek, north-east Australia are associated with an ebb-tide-dominated estuary, which receives very little fluvial freshwater or sediment input (Fig. 9.14b; Grindrod & Rhodes 1984; Wolanski et al. 1992). As in the previous example, the predominant sediment substrate type across the mangrove is an intertidal organic mud, derived primarily from *in situ* breakdown of organic material. Terrigenous sediments are only important in the upper reaches of creek networks. Mangroves also develop along carbonate-rich shorelines, such as those around the Gulf of St Vincent in southern Australia (Butler et al. 1977). The mangroves develop on mixed carbonate–siliciclastic sediments, although marked variations in sediment content occur across the mangrove (Fig. 9.14c).

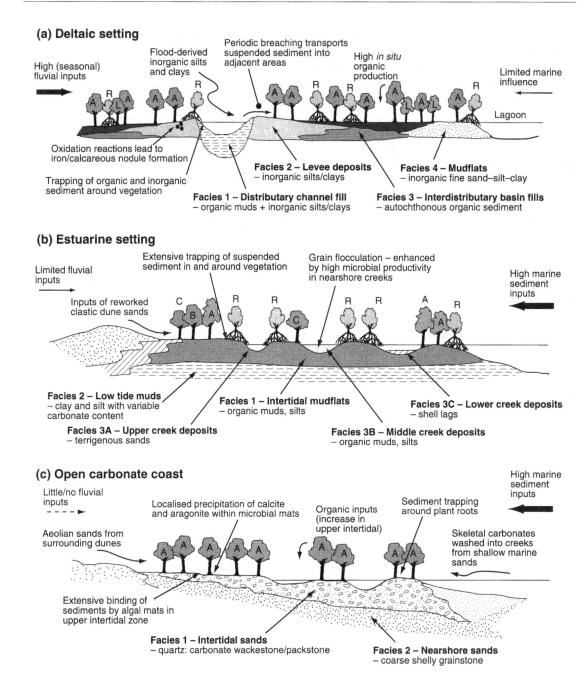

Fig. 9.14 Schematic diagrams illustrating the sediment facies associated with different mangrove settings. (a) Deltaic setting – based on the Grijalva-Usumacinta delta in Tabasco and Campeche, Mexico. Mangrove genera: A, *Avicennia*; R, *Rhizophora*; B, *Bruguiera*; C, *Ceriops*; L, *Languncularia*. (Based on Thom 1967.) (b) Estuarine setting – based on Coral Creek, north-east Australia. (Based on Grindrod & Rhodes 1984.) (c) Open carbonate-dominated shoreline – based on Spencer Gulf and Gulf of St Vincent, South Australia. (Based on Butler et al. 1977.)

9.3 PROCESSES AND IMPACTS OF NATURAL DISTURBANCE EVENTS

Although sediment accumulation within both coral reef and mangrove environments is influenced by a range of physical and biogenic factors, these sedimentary environments are also subject to natural changes in shoreline morphology associated with changing patterns of sediment transport and accumulation. They also respond to larger scale and higher magnitude events such as cyclones and tsunami, and to periodic shifts in oceanographic conditions such as those linked to the El Niño–Southern Oscillation. These events may influence the biological components of the respective environments and thus sediment accumulation.

9.3.1 Natural dynamics of sediment transport and accumulation

Long-term (Holocene) evolutionary trends along coral reef and mangrove-colonized coastlines reflect complex interactions between eustatic and relative sea-level change, and a host of external factors including, climate, marine environmental parameters, sediment supply and shoreline geomorphology. Extensive discussion of these longer term controls is beyond the scope of this section, but in reef environments a range of evolutionary models have been identified, for example, from studies of Holocene reef systems (Neumann & Macintyre 1985). In particular, these illustrate growth response to relative sea-level change. Where reefs initiate at depth, vertical accretion is typical, whereas reefs initiating close to (or reaching) sea-level have tended to prograde laterally. These simple models are often modified by local substrate, tectonic or hydrodynamic conditions (Kennedy & Woodroffe 2002).

Mangrove shorelines have also undergone both progradation (seaward advance) and transgression (shift landward) in response to past sea-level fluctuations (Woodroffe 1992). In general, the post-glacial Holocene sea-level rise resulted in transgression of mangrove shorelines. As sea-levels stabilized during the past 5000 years, areas receiving sufficient sediment supply subsequently

prograded (Wolanski & Chappell 1996). This continues in areas that receive abundant allochthonous sediment inputs. Examples include the Ganges–Brahmaputra delta where seaward accretion rates of 5.5–16 km^2 yr^{-1} have been estimated (Allison et al. 2003), and the Mekong delta where some 62,500 km^2 of deltaic sediment has accumulated over the past 4500 years (Nguyen et al. 2000).

These rapidly prograding mangrove-colonized shorelines represent one end-member in a range of mangrove settings which exhibit variable shoreline dynamics. This reflects not only spatial and temporal variability in rates of sediment supply, but also the unconsolidated nature of the sediment substrates and their susceptibility to nearshore dynamics. Alternating phases of progradation and retreat have, for example, been described along mangrove-fringed coastlines in French Guiana (Froidefrond et al. 1988). This results from longshore sediment transport, but although individual sections of this coast alternate between phases of accretion and erosion, there remains a net balance in the sediment budget along the wider coastal section (Case Study 9.1).

Along the north-east coast of Australia, mangrove-colonized chenier ridge sequences (see section 9.3.2) have shifted from similar phases of relative shoreline stability (punctuated by periodic accretion and erosion) to phases of rapid progradation (Chappell & Grindrod 1984). Prior to 1200 yr BP, coastal evolution was characterized by periods of 'cut-and-recover'. Erosion occurred during chenier ridge migration, but was followed by renewed small-scale progradation (Fig. 9.15). This period was associated with the development of narrow (< 150 m) mangrove fringes colonizing relatively steep shorelines (slope angle 1:200). Since 1200 yr BP, the coastline has undergone rapid progradation. Wide mangrove fringes have developed along lower angle shorelines (1:1000) and promote continued sediment accretion (Fig. 9.15).

In contrast, in north-western Australia there is widespread net erosion of mangrove-colonized tidal flats (Semeniuk 1981). Three types of erosion influence local and regional intertidal geomorphology: cliff erosion, sheet erosion and tidal

Case study 9.1 Cyclical phases of mangrove shoreline accretion and erosion, French Guiana

The nearshore environment along the coast of French Guiana is characterized by the development of highly mobile mangrove-colonized intertidal mudflats. The sediments that accumulate in these mudflats are derived from the Amazon, which discharges an estimated 1200×10^6 t of sediment per year. Around 20% of this sediment is transported north-west alongshore by the Guiana Current (Case Fig. 9.1A), which at its peak (January–April) reaches 5 knots and extends up to 40 km offshore. Sediment is moved both in suspended form ($c. 150 \times 10^6$ m^3 yr^{-1}) and temporarily stored in the mobile mudbanks that develop along the coast ($c. 100 \times 10^6$ m^3 yr^{-1}). This zone of sediment transport thus creates a mobile nearshore setting, with individual sections of the coast alternating between phases of active accretion and erosion.

The distribution of accretionary and erosional sectors is not, however, uniform. Lengths of coastline undergoing active accretion range from 20 to 40 km, are up to 5 km wide at low tide, and comprise on average an estimated 3×10^{12} m^3 of sediment. The mudbanks have a maximum thickness of 10 m and typically exhibit an asymmetric morphology, being steeper on the leeward side. Erosional sectors extend for 8–26 km along the coast and are characterized by numerous smaller embayments (each up to 500 m wide) giving the eroding sectors a 'sawtooth' profile. Mudbank migration and associated degradation of the colonizing mangrove forests occurs primarily as a result of ongoing erosion of the windward (up-current) side of the mudbanks and sediment accretion on the lee side (Case Fig. 9.1A). Migration is attributed primarily to longshore currents and an oblique wave approach. This process, however, may be enhanced in some areas by particularly rapid rates of sediment accretion causing smothering of mangrove pneumatophores and subsequent mortality.

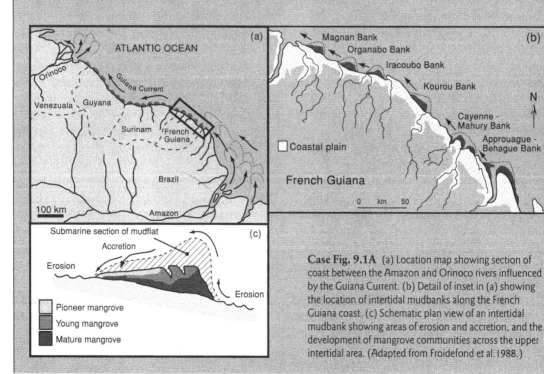

Case Fig. 9.1A (a) Location map showing section of coast between the Amazon and Orinoco rivers influenced by the Guiana Current. (b) Detail of inset in (a) showing the location of intertidal mudbanks along the French Guiana coast. (c) Schematic plan view of an intertidal mudbank showing areas of erosion and accretion, and the development of mangrove communities across the upper intertidal area. (Adapted from Froidefond et al. 1988.)

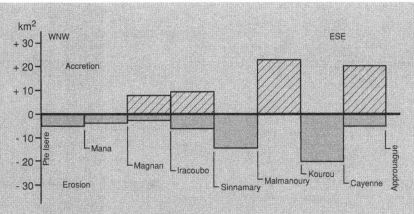

Case Fig. 9.1B Areas of land-gain and loss at different sections along the French Guiana coast between 1979 and 1984. (Adapted from Froidefond et al. 1988.)

Temporal data indicate that sediment mobilization is continual along this coast and that mudbank migration rates average c. 1.2 km yr^{-1}. Different sectors, however, migrate at different rates and displacement rates between 1979 and 1984 ranged from 320 m yr^{-1} (Approuague-Behague bank) to 1220 m yr^{-1} (Organabo bank). Despite these spatial differences, there appears, to be a net balance along the wider coastal sector in terms of the area of annual mudflat gain (around 60 km^2) and erosion (around 58 km^2) (Case Fig. 9.1B). One consequence of this highly mobile intertidal setting is to produce a continually changing mangrove environment. Rapid mangrove colonization follows new substrate accretion along the leeward fringes, with the forests continuing to mature until interrupted by later erosion along the windward margins. Interestingly, this transitional community development is mirrored by changes in sediment accumulation and sediment diagenetic processes, which change with age of the mudbanks. Within the young (and frequently flooded) mangroves, organic matter in the sediment mainly derives from algal-mat communities, and diagenetic processes are dominantly suboxic, leading to rapid degradation of organic matter. As the forests mature, most of the organic matter is derived from higher plant material and diagenetic processes shift to anoxic sulphate-reducing phases, associated with which occurs deposition of pyrite framboids. There is some evidence that these diagenetic characteristics may be preserved beneath areas of newly accreting mudbanks and thus serve as an indicator of previous erosional and accretionary phases.

Relevant reading

Baltzer, F., Allison, M. & Fromard, F. (2004) Material exchange between the continental shelf and mangrove-fringed coasts with special reference to the Amazon–Guianas coast. *Marine Geology* 208, 115–26 (and references therein).

Blasco, F., Saenger, P. & Janodet, E. (1996) Mangroves as indicators of coastal change. *Catena* 27, 167–78.

Debenay, J.P., Guiral, D. & Parra, M. (2002) Ecological factors acting on the microfauna in mangrove swamps. The case of foraminifera assemblages in French Guiana. *Estuarine, Coastal and Shelf Science* 55, 509–33.

Froidefond, J.M., Pujos, M. & Andre, K. (1988) Migration of mud banks and changing coastline in French Guiana. *Marine Geology* 84, 19–30.

Marchand, C., Lallier-Vergès, E. & Baltzer, F. (2003) The composition of sedimentary organic matter in relation to the dynamic features of a mangrove-fringed coast in French Guiana. *Estuarine, Coastal and Shelf Science* 56, 119–30.

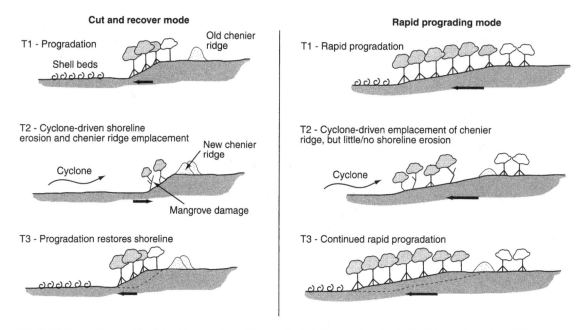

Fig. 9.15 Contrasting models of shoreline evolution at Princess Charlotte Bay, northern Australia. (Adapted from Chappell & Grindrod 1984.)

creek erosion. Cliff erosion is caused by tidal scour and subsequent mass slumping of channel banks and occurs primarily along the seaward fringes of mangroves. Rates of retreat average 2 m yr^{-1}, but in places reach 30–50 m yr^{-1}. Sheet erosion occurs due to sediment desiccation and burrowing and results in vertical stripping of surface sediments, typically removing a few millimetres at a time (Fig. 9.16). This produces highly erodible sediment which is removed during spring tides, and rates of vertical loss average 1–3 cm yr^{-1}. Tidal creek erosion occurs as a result of channel widening and deepening, with lateral erosion of up to 3 m yr^{-1}. Stratigraphical studies indicate that this net erosional regime has been active over the past 5000 years (Semeniuk 1981).

Smaller scale changes in sediment accumulation also occur as a result of local changes in fluvial sediment discharge or nearshore sediment dynamics, and may result in direct erosion or increased sedimentation. At Portuguese Island in southern Mozambique, changes in coastline configuration have caused both direct erosion of mangrove substrates and progressive tidal restriction, leading to a 75% reduction in man-

grove extent between 1958 and 1989 (Hatton & Couto 1992). Conversely, sediment mobilization along mangrove channels in Brunei, Malaysia has caused channel damming and increased mangrove flooding, leading to localized mortality of mangrove species and intertidal fauna (Choy & Booth 1994). A common impact of changing sediment dynamics is modified sediment accumulation rates. Although many mangrove species can tolerate accretion rates of 5–10 mm yr^{-1}, higher rates may result in mangrove mortality due to root smothering (Ellison 1998). Where this occurs, intertidal substrates may become unstable and prone to reworking. Mangrove shorelines are often, therefore, highly dynamic in areas of either high allochthonous sediment input, or where intertidal sediments are subject to frequent mobilization.

In reef environments, rapid geomorphological change, in terms of the position of the reef body and its shoreline geometry, are often less evident. This probably reflects the more rigid structure of many framework-dominated reefs, which require significantly elevated energy regimes, such as those associated with cyclone activity, to facilitate sedimentary responses (although under such

Fig. 9.16 Severe sheet erosion leading to removal of the upper layers of mangrove sediment and the exposure of root networks. Gordon Creek, Australia. (Photograph courtesy of Piers Larcombe.)

conditions major sedimentological changes can occur – see section 9.3.2). Coral communities that occur within some nearshore, turbid settings may, however, be susceptible to episodic sediment mobilization. At Inhaca Island in southern Mozambique, coral communities (but not framework reefs) are patchily developed along the margins of shallow intertidal channels (Perry 2003). These communities undergo episodic mortality due to smothering and burial of coral colonies associated with longshore sediment transport. In turn, new areas of bedrock and coral rubble may be exposed and subsequently colonized (Perry 2005). The natural characteristics of coastal sediment dynamics thus create an ephemeral suite of coral communities.

9.3.2 Physical disturbance (cyclones and tsunami)

Although background or seasonal fluctuations in sediment transport processes exert a clear influence upon reef and mangrove systems,

significant changes also occur in response to low-frequency, high-magnitude events such as cyclones and tsunamis. Tropical cyclones (also termed hurricanes and typhoons, depending upon geographical location) are characterized by cyclonic surface winds formed around centres of low pressure (McGregor & Nieuwolt 1998). These weather systems may be up to 800 km in diameter and are characterized by strong winds (often > 30 m s^{-1}). Impacts on nearshore sediments, and on reef and mangrove communities, result from strong wind-driven currents, high wave heights (5–15 m), and elevated coastal sea-levels.

Physical disturbances to both coral reef and mangrove communities are well documented consequences of cyclones. In coral reefs, wave damage may result in widespread breakage, toppling and abrasion of shallow water (especially branching) corals (Rogers 1993), while high wind speeds and storm-wave surges into intertidal areas cause mangrove uprooting and damage (McCoy

Fig. 9.17 Storm-generated coral rubble ridges. A series of ridges, up to 3 m high, occur at these sites, recording evidence of successive storm events. Triangulos, Campeche Bank, Gulf of Mexico.

et al. 1996). The scale and extent of disturbance are often highly variable even over short spatial scales, but in high-impact areas reduced live coral or mangrove cover may be an immediate consequence of disturbance. Disturbance may be exacerbated in reef environments by the subsequent effects of cyclone-related rainfall, which may induce localized coral bleaching (Van Woesik et al. 1995), and in mangrove settings, increased rainfall may reduce soil salinity levels or cause erosion due to surface runoff.

From a sedimentological perspective, cyclones exert a direct influence on sediment transport, and, over geological time-scales, should be regarded as an important control in tropical sedimentary environments. A detailed review of the geological effects of cyclones within reef environments is provided by Scoffin (1993), who identified a range of erosional and depositional processes. These include both on- and offshore transport of sediment and rubble, much of which is derived from physical breakdown of the coral communities. Onshore transport can

result in the development of rubble cays or ridges (Fig. 9.17; Hayne & Chappell 2001) and sediment deposition in the lee of these features (Fig. 9.18), whereas off-reef transport produces local talus deposits. In shallow water, storm-generated rubble can be a major constituent of reef facies (Blanchon et al. 1997). Storm surges may also result in significant sediment mobilization in shallow water, leading to beach, coral cay and subtidal erosion.

Studies in St Croix provide an insight into the volumes of sediment that may be mobilized during hurricanes. At Salt River Canyon, an estimated 2×10^6 kg of sediment were flushed from the reefs during the passage of Hurricane Hugo in 1989, and at nearby Cane Bay an estimated 336,000 kg of sediment were flushed offshore from a single reef channel (Hubbard 1992). Recent work on the Great Barrier Reef (GBR) shelf demonstrates the extent to which cyclones also influence sedimentation across much wider carbonate shelf systems (Larcombe & Carter 2004; see also Chapter 10). Cyclones, which typically approach

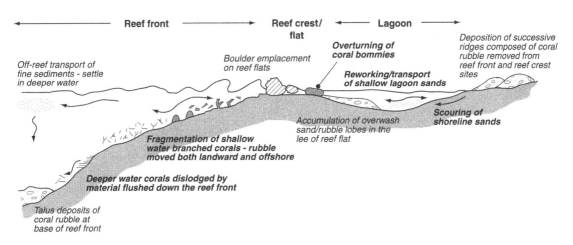

Fig. 9.18 Schematic diagram illustrating the main erosional (bold italics) and depositional (italics) processes associated with cyclone impacts on coral reefs.

the Queensland coast from the north-east, drive along-shelf currents that transport sediment northwards along the shelf, resulting in marked cross-shelf variations in Holocene sediment accumulation (Case Study 9.2).

Along mangrove-colonized shorelines, the main sedimentological impacts of cyclonic disturbance result from shoreline erosion caused by storm surges. In Florida, some 15 m of shoreline erosion occurred following Hurricane Andrew in

---

Case study 9.2 Cyclone-controlled sediment distribution on the Great Barrier Reef shelf (the 'cyclone pump' model)

The Great Barrier Reef (GBR) shelf is the largest modern tropical mixed carbonate–siliciclastic shelf system, extending around 2000 km along the eastern and north-eastern margins of Australia, and varying in width from around 200 km in the south to 50 km in the north. It is characterized by extensive reef development along its seaward margin, but also by sediment deposition on the shelf. In the central regions, the shelf can be partitioned into three distinct shore-parallel sediment zones. The inner shelf (0–22 m depth) is dominated by terrigenous sands and muds (5–10 m thick) and includes the development of intermittent detrital coral build-ups (Smithers & Larcombe 2003). Under fair-weather conditions, sediment transport is dominated by the northwards, along-shelf current-driven movement of sediments resuspended by wave action in nearshore areas, and bedload transport occurs only in the shallow (< 5 m) shoreface zone. The middle-shelf (22–44 m depth) is a zone of sediment starvation and is characterized by only a thin veneer (< 1 m thick) of Holocene shelly, muddy sands. This zone receives little or no terrigenous sediment and there is limited sediment transport under fair-weather conditions. The outer-shelf (40–80 m depth) is dominated by a tract of reef development and by associated carbonate-dominated detrital sediments. In fair weather little or no terrigenous sediment reaches the outer reefs (Case Fig. 9.2a).

Cyclones are a common disturbance on the central GBR (two to three cyclones occur each year at latitude 20°S) and therefore represent a major control on shelf sedimentation. Cyclones influence sediment input to the GBR by:

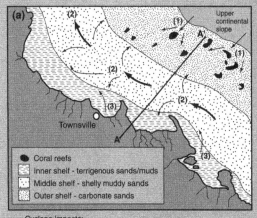

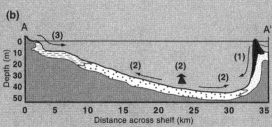

Cyclone impacts:
(1) Breakage of reef framework and mobilization of reef sediment
(2) Cyclone 'pumping' of sediment along and from the middle shelf
(3) Fluvial sediment inputs onto the inner shelf

Case Fig. 9.2 Simplified plan (a) and cross-sectional (b) views showing the effects of cyclones on sediment generation and dispersal across the central Great Barrier Reef. (Adapted from Larcombe & Carter 2004.)

1 causing the breakage of reef framework on the outer shelf, which contributes carbonate sand and gravel on the leeward flanks of individual reefs;
2 generating floods in rivers and discharge of terrigenous sediment to the inner shelf;
3 eroding the sea-floor of parts of the inner and middle-shelf zones (Case Fig. 9.2a & b).
Cyclones also, however, exert a major control on the mobilization and transport of sediment on the shelf. Under cyclonic conditions, fast wind-driven currents (> 60 cm s^{-1}) have been recorded and, in the case of Cyclone Joy (1990), these formed a persistent flow of 9 days duration. Under such conditions, extensive along-shelf transport of sand and gravel occurs, and the cyclones act as a 'pump', transporting large volumes of sediment northwards along the shelf (Larcombe & Carter 2004).

Sedimentary evidence for such transport occurs across much of the inner and middle shelf, in the form of large (up to 2 m high) subaqueous sand and gravel dunes, sediment ribbons and large dune fields. Extensive erosion of the sea-floor also occurs across the middle shelf, which in combination with high rates of cyclone-driven sediment transport may explain the lack of reef development on the middle shelf throughout the Quaternary. The cyclones also redistribute large amounts of the suspended terrigenous sediment that is discharged onto the shelf during storm events, driving this sediment along and onshore, and thus contributing to the development of the terrigenous sediment-dominated inner shelf. In this way, cyclones act both as a control on cross-shelf sediment accumulation and along-shelf sediment transport.

Relevant reading

Gagan, M.K., Chivas, A.R. & Herczeg, A.L. (1990) Shelf-wide erosion, deposition, and suspended sediment transport during Cyclone Winifred, central Great Barrier Reef, Australia. *Journal of Sedimentary Petrology* 60, 456–70.
Larcombe, P. & Carter, R.M. (2004) Cyclone pumping, sediment partioning and the development of the Great Barrier Reef shelf system: a review. *Quaternary Science Reviews* 23, 107–35.
Smithers, S. & Larcombe, P. (2003) Late Holocene initiation and growth of a nearshore turbid-zone coral reef: Paluma Shoals, central Great Barrier Reef, Australia. *Coral Reefs* 22, 499–505.

1992 (Swiadeck 1997), although this is likely to have been mitigated by the presence of nearby protective reef flats, and by the relatively restricted fetch of the approaching waves. Well-documented depositional features of cyclone-influenced tropical shorelines are chenier ridges. These comprise sands and shells winnowed from adjacent mudflats and transported through the mangroves by waves during cyclone-related storm surges. Individual cyclones may result either in the emplacement of chenier ridges or the migration of ridges across chenier plains (Chappell & Grindrod 1984; Woodroffe & Grime 1999). Cyclone-related sedimentation, driven by storm-surges, can also result in widespread mangrove mortality due to root burial (Ellison 1998).

High-energy conditions are not confined to cyclones, with significant physical disturbance also associated with tsunami. The long-wavelength waves associated with tsunami are generated by earthquakes, volcanic eruptions or submarine slides and can move at speeds in excess of 700 km h^{-1} in the open ocean. As water depth decreases nearshore wave height increases dramatically, resulting in significant nearshore damage and sediment and rubble transport (Solomon & Forbes 1999). The high-magnitude (Richter scale 9.0) earthquake-generated Indian Ocean tsunami of 26 December 2004 provided dramatic evidence of both the major geomorphological and sedimentary impacts that these events can have (Fig. 9.19), as well as the high human and socio-economic cost. The impact of this event on coral reef communities appears, as is commonly the case with cyclone damage, to have been spatially very variable, with major coral damage reported at some sites, whereas other areas survived unscathed. At the time of writing a number of post-event geomorphological and geological surveys were being conducted, and these may provide a useful insight into the role (or not) of reefs and mangroves in affording protection to affected shorelines. Common sedimentological indicators of past tsunami events (tsunamites) include the occurrence of large limestone megaclasts on reef flats, along with the occurrence of allochthonous reworked sands and shells (Noormets et al. 2002; van den Bergh et al. 2003).

9.3.3 Large-scale oceanographic changes

In addition to the physical disturbance associated with high-energy events, major changes in nearshore environmental conditions periodically occur due to shifts in oceanographic conditions. The most common example of this is associated with the El Niño–Southern Oscillation (El Niño) which causes major changes in sea-surface temperatures and climate in the Pacific Basin (McGregor & Nieuwolt 1998). During normal (non-El Niño) years, warm water in the equatorial Pacific is driven westwards by prevailing trade winds, resulting in higher and warmer sea-surface conditions in the west. The westward flow is compensated for in the eastern Pacific (i.e. the western seaboard of central America) by upwelling of cool, nutrient-rich waters. These waters limit extensive reef development (see section 9.1.3). During El Niño years, however, the trade winds weaken and warm water sloughs back eastwards, restricting upwelling and bringing warm waters into the nearshore areas of the eastern Pacific. This further influences the coral communities, via coral bleaching and mortality as sea-surface temperatures increase. There are, however, also significant implications for reef carbonate budgets. Studies in Panama (Eakin 1996) documented major reductions in reef carbonate production following the 1982–83 El Niño. Before 1982, reefs around Uva Island were characterized by net carbonate production rates of 0.34 kg m^2 yr^{-1}, but after 1983 a 50% reduction in coral cover, along with continued high rates of bioerosion, led to net reef erosion (average of -0.19 kg m^2 yr^{-1}). Sea-level fluctuations of up to 0.5 m between El Niño years can also result in significant remobilization of nearshore sediments and shoreline erosion (Solomon & Forbes 1999), as changes in climatic conditions are marked by significantly increased rainfall and fluvial runoff.

9.4 PROCESSES AND IMPACTS OF ANTHROPOGENIC DISTURBANCE

Although most reef and mangrove environments are subject to natural physical reworking and

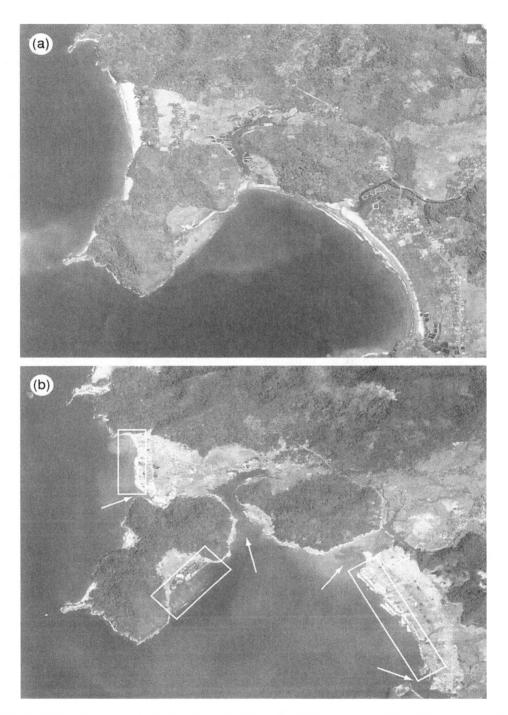

Fig. 9.19 Quickbird satellite imagery showing the effects of the 26 December 2004 tsunami in the vicinity of Gleebruk, south of Banda Aceh, Indonesia. (a) Pre-tsunami image taken on 12 April 2004. (b) Post-tsunami image taken on 2 January 2005. As well as evidence of major damage to fields and villages, there is substantial evidence of major shoreline change, with erosion of beaches (boxed areas) and major changes to river and channel mouths (arrowed). (Imagery courtesy of DigitalGlobe (http://www.digitalglobe.com).)

changes in sediment flux, some significant changes are also linked to anthropogenic disturbance. Many such disturbances may have an immediate impact on the reef or mangrove community, but some may also have longer term sedimentological impacts. In the context of coral reefs, this relates to changes in rates and patterns of carbonate production, in the case of mangroves to the substrate destabilization that can follow community degradation, and in both environments to progressive sediment contamination. A key issue in terms of disturbance relates to the need to place observed environmental or community changes in the context of natural or background community dynamics, i.e. change should not necessarily be regarded as a consequence of anthropogenic disturbance. This section is not intended as a comprehensive review of anthropogenic influences but outlines, via selected examples, some of the main impacts (and uncertainties) of human disturbance in relation to reef and mangrove sedimentology. Useful reviews of potential anthropogenic disturbances to mangroves and coral reefs are provided, respectively, by Ellison & Farnsworth (1996) and Grigg & Dollar (1990).

9.4.1 Impacts of coastline modification and resource extraction

Modification of the coastal fringe as a result of land-use change, construction or resource extraction can exert a major influence on sediment dynamics and coastline stability (see also Chapters 7 & 8). Change may occur either directly due to sediment removal, leading to changes in local sediment budgets, or indirectly as a result of seawall or causeway construction, which may modify local current dynamics and restrict longshore sediment transport (Preu 1989). In reef environments, progressive habitat degradation resulting from coral mining, sand extraction or channel blasting may modify local current pathways, and in mangrove settings, land conversion for agriculture may remove areas from tidal influence and thus alter the area over which tidal influence is exerted. Such changes can lead to modified patterns of sediment transport, erosion and deposition.

In south west Sri Lanka a range of human-related activities have promoted rapid coastal retreat (average retreat 1.1 m yr^{-1}; Preu 1989). Activities include large-scale fluvial sand extraction, which prevents sediment (normally supplied to the beaches) from reaching the coast. In addition, reef degradation resulting from channel blasting and coral mining has led to modified nearshore current dynamics and locally high (up to 4 m yr^{-1}) rates of shoreline retreat. Shoreline erosion rates of 0.5–1 m yr^{-1} have also been reported from the Marshall Islands, where aggregate extraction has reduced sediment supply to the lagoon rim, and causeway and channel constructions have interrupted sediment transport (Xue 2001).

Changes in land-use and the transition of mangrove land for agricultural or development purposes also have an impact on nearshore sediment dynamics. In particular, loss of mangrove swamp has an impact on the spatial extent of tidal influence. This may reduce the tidal prism leading to modified current dynamics, reduced ebb-tide flow and sediment export (see section 9.2.5.1), and thus channel siltation (Wolanski et al. 1992). Land modification for the purposes of shrimp or rice farming also has longer term impacts beyond the immediate loss of mangrove substrate. The sediments in such areas typically become highly saline and acidic, and this often inhibits natural recolonization. This in turn leaves areas of bare, degraded substrate which are more prone to erosion from runoff and waves (Ellison & Farnsworth 1996).

9.4.2 Impacts of increased sediment flux

Although high rates of sediment discharge can be regarded as natural inhibitors or influences on reef and mangrove development (see section 9.1.3), increased or modified rates of sediment influx are a widely cited example of anthropogenic disturbance (Grigg & Dollar 1990). This has been attributed to poor land management practices (which may increase soil erosion and runoff) and to marine dredging and dumping. Two key factors need, however, to be highlighted in relation to sediment inputs into nearshore environments:

sedimentation and turbidity. The former is a measure of downward sediment flux, the latter a measure of suspended sediment load (Te 1997). The two are not, however, directly correlated because high turbidity conditions can occur in areas with relatively low sedimentation rates where wave-driven resuspension rates are high (Thomas et al. 2003). Both factors, however, have a potential impact on coral communities; increased sedimentation leads to smothering and burial, and to an increase in polyp stress, whereas turbidity acts to inhibit light penetration and thus restrict photosynthesis. As a result, sediment influx has been widely perceived as a key threat to coral communities (see Rogers (1990) and references therein).

Recent research, however, has highlighted numerous sites where coral communities (and reefs) not only develop but also persist, under conditions of high turbidity and periodically high sediment flux (Woolfe & Larcombe 1998; Perry 2003; Smithers & Larcombe 2003). These 'reefs' are typically characterized by reduced framework development (although not necessarily reduced coral species diversity) and, as a result, are somewhat distinct from the coral reefs that can develop in clear-water settings. They represent, however, locally important sites of coral community development and bring into question widespread assumptions about the negative effects of sedimentation. Coral reefs occur, for example, at a number of sites along the inshore regions of the Great Barrier Reef, despite high rates of terrigenous sediment influx (Woolfe & Larcombe 1998; Smithers & Larcombe 2003). Furthermore, several studies present evidence suggesting that coral communities have continued to develop under conditions of at least periodically high turbidity and terrigenous sediment input for as much as the past 5000 years (Johnson & Risk 1987; Perry 2005).

Although local increases in the amount of sediment reaching coral reefs (especially in areas where past inputs have been limited) are undoubtedly likely to have a negative impact on corals and to modify reef community structure, too little is known about the occurrence of coral communities under natural conditions

of high turbidity and elevated sedimentation rates to generalize about sediment input, coral response and reef occurrence. Such generalizations are further complicated by the vastly differing regimes of sediment input and sediment composition that characterize different coastal sites. Sediment inputs vary between sites depending upon the frequency (seasonal, episodic), longevity and volumes of terrigenous sediment input, and sedimentation rates and turbidity regimes vary depending upon grain size, degree of sediment resuspension and tidal range. These local variations will influence coral species response to sedimentation in terms of sediment rejection mechanisms and growth strategies. Where sediments do influence coral growth, they are likely to be significant from a sedimentological perspective because rates and patterns of reef carbonate production are directly influenced by the composition of the reef community. Studies in Indonesia have, for example, documented net erosion on a range of reefs subject to high terrigenous sediment and nutrient inputs (see also section 9.4.2) and this has been attributed both to reduced coral cover and increased rates of bioerosion (Edinger et al. 2000).

Despite their association with fine, organic-rich sediments, mangroves are also potentially susceptible to increased sedimentation. Ellison (1998) reviews numerous examples of increased sediment influx relating to dredge spoil dumping, construction-related sedimentation, mining and catchment deforestation. The primary impacts on mangrove communities relate to burial of aerial roots at rates sufficient to restrict soil-gas exchange. This can cause mangrove mortality and thus loss of mangrove cover, and erosion in areas subject to fluvial or marine reworking. Sedimentation rates in excess of 1 cm yr^{-1} are likely to be detrimental, although different species of mangroves exhibit different tolerances to burial (Thampanya et al. 2002).

9.4.3 Impacts of increased nutrient input

The primary source of nutrients into shallow marine and intertidal environments is from anthropogenic sources (e.g. sewage effluent, agricultural runoff), the most important nutrients being nitrogen and phosphorus. In coral reefs, ecological impacts of excess nutrient inputs include suppression of coral growth rates and reproductive capacity (Tomascik & Sander 1985, 1987). In many cases, nutrient inputs are coupled with increased sediment inputs and this complicates isolation of causal mechanisms of disturbance. A possible consequence of nutrification, however, is to increase the competitive advantage of opportunist algal species over corals. This is significant from a sedimentological perspective in relation to changing the composition and abundance of the carbonate-producing community. The potential impacts of such stresses are highlighted in studies from the Florida Keys, where water eutrophication has been cited as a major cause of temporal change in carbonate sedimentation (Case Study 9.3).

Available data indicate that the effects of nutrification may be less significant in mangrove settings, although this may vary in different mangrove environments. Mangroves commonly grow in waterlogged, anaerobic soils, so that anaerobic reactions exert a major influence on phosphorus and nitrogen chemistry. These reactions are strongly influenced by redox potential, which varies depending on the degree of tidal inundation, sediment porosity and organic matter content (Clough et al. 1983). Phosphorus occurs primarily in organic form and, in most mangrove sediments, low redox potentials lead to release of phosphate. Therefore, in systems that receive additional phosphate associated with nutrient inputs, the ability to immobilize phosphorus may be limited. Nitrogen occurs mainly in organic form (primarily ammonium), and any nitrate is rapidly converted by anaerobic bacteria (dentrification) to gaseous nitrogen or nitrous oxide. Most of this occurs in interstitial waters within mangrove sediments and is highly susceptible to leaching by rainwater or during

---

Case study 9.3 Decadal-scale changes in carbonate sediment compositions across the Florida reef tract, USA

Coral reef communities along the Florida reef tract have been subject to a range of both natural and anthropogenic-related disturbance events over the past 40 years. These have included physical damage from hurricanes and extreme cold-water events, as well as the effects of coral disease (White Band Disease, Black Band Disease) and mortality of the key herbivorous echinoid *Diadema antillarum*. In addition, there is evidence that Florida reef-tract waters are becoming increasingly eutrophic as a result of nutrient inputs and infiltration from ground and surface waters (Szmant & Forrester 1996). The consequence of this has been a general decline in reef 'health' as evidenced by widespread reductions in live coral cover and consequent increases in algal biomass.

Comparisons of sedimentary data collected from sites along the Florida reef tract (Case Fig. 9.3) in 1963 with that collected in 1989 indicate differences in the abundance of biogenic sediment constituents that may reflect this progressive decline in coral reef 'health' (Lidz & Hallock 2000). On a reef-wide scale, the major changes in composition have been a doubling in the relative abundance of molluscan grains, and a tripling in abundance of coral fragments. Marked variations occur, however, in different parts of the Keys, reflecting spatial variations in reef vitality. For example, the proportion of coral grains in Upper Keys sediments (a relatively healthy section of the reef) is reduced compared with the large increases that have occurred in the Middle and Lower Keys (where coral communities are in decline) (Case Fig. 9.3b). There is little evidence to suggest that these changes reflect periodic storm/hurricane reworking of sediment, but rather are attributed to increased rates of dead coral substrate erosion by internal

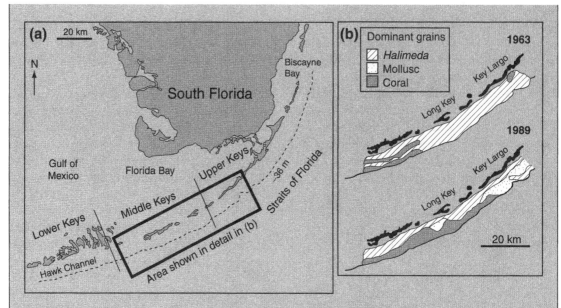

Case Fig. 9.3 (a) Simplified map showing the main geomorphological features of South Florida. (b) Sediment data collected in 1963 and 1989 from the Middle Keys area showing greater abundance of coral within the sediments in 1989. (Adapted from Lidz & Hallock 2000.)

bioeroders (borers). This has, in large part, been linked to increased nutrient levels in nearshore waters as a result of increased sewage discharge. This stimulates phytoplankton production and in turn rates of bioerosion by suspension-feeding boring organisms.

In a more localized study around Key Largo, Cockey et al. (1996) also identified increasing nutrient levels in nearshore waters as a likely cause of temporal changes in the composition of foraminiferal assemblages within reef-related sediments. Samples from 1959 to 1961 indicated a predominance of larger, symbiont-bearing Soritidae foraminifera, which comprised some 50–80% of the species identified. Samples recovered from the same sites in 1991–92, however, showed an overwhelming dominance of smaller, heterotrophic Miliolidae and Rotaliidae. These studies illustrate the close linkages that exist between the reef community and production of carbonate sediment. Major changes in the abundance of either carbonate producers or degraders can result in detectable shifts in the abundance of carbonate sediment constituents over decadal time-scales. These studies illustrate the potential of sedimentary data as geo-indicators of environmental change.

Relevant reading

Cockey, E., Hallock, P. & Lidz, B.H. (1996) Decadal-scale changes in benthic foraminiferal assemblages off Key Largo, Florida. *Coral Reefs* 15, 237–48.
Lidz, B.H. & Hallock, P. (2000) Sedimentary petrology of a declining reef ecosystem, Florida reef tract (U.S.A). *Journal of Coastal Research* 16, 675–97.
Perry, C.T. (1996) The rapid response of reef sediments to changes in community structure: implications for time-averaging and sediment accumulation. *Journal of Sedimentary Research* 66, 459–67.
Szmant, A.M. & Forrester, A. (1996) Water column and sediment nitrogen and phosphorous distribution patterns in the Florida Keys, USA. *Coral Reefs* 15, 21–42.

tidal inundation. Such processes may be enhanced by the high levels of particulate organic matter that are often associated with sewage or agricultural discharges and may speed up anaerobic reactions, increase rates of dentrification and have an impact on mangrove growth. Although mangroves may actually benefit from increased nutrient inputs (and exhibit increased productivity), nutrient enrichment of soils may have a more negative impact by altering soil chemistry and increasing the anaerobic nature of the sediment (Field 1998).

9.4.4 Impacts of contaminant input (oil, chemicals, metals)

Common contaminant inputs into nearshore and intertidal environments include a range of heavy metals, as well as organochlorine pesticides and oil residues. These may be derived from industrial, urban and agricultural discharges, or from contaminant dumping and refuse sites. Contaminant transfer may occur either in suspended or dissolved form, with widespread distribution in tidally influenced settings being enhanced by diffusive flow mechanisms (Wolanski et al. 1992). Transfer through sediments can occur due to seepage and runoff (Clark 1998). Contaminant accumulation is aided, particularly in mangroves, by the fine-grained nature and high organic content of the sediments (Harbison 1986; see also Chapter 7). Mangrove sediments act both as physical traps for fine particulate and transported contaminants, and as chemical traps where metal sulphides precipitate from solution. Such precipitation is driven by a range of bacterial sulphate reduction reactions in the anaerobic sediments. Although there is the potential for precipitated metals to be remobilized into surficial sediments and overlying waters (Harbison 1986), precipitation of metal sulphide is environmentally significant as a mechanism of immobilizing metal contaminants.

Gradients of metal contamination within sediments are common through mangrove swamps (Soto-Jiménez & Páez-Osuna 2001), with the highest metal concentrations tending to occur in areas rich in fine particulate and organic matter.

The extent to which metal enrichment of sediments has a direct impact on the mangrove community is not clear. Some studies report uptake of metals within mangrove roots, often to levels that exceed surrounding sediments (e.g. MacFarlane et al. 2003) but these do not appear to have an impact on mangrove growth. In contrast, uptake within leaves appears low (Machado et al. 2002), but in the context of metal export from mangroves this is significant given the high amounts of leaf detritus that are exported from some mangrove systems (see section 9.2.4.2).

Reef sediments also have potential to accumulate metal and chemical contaminants, and there is evidence of marked spatial variations related to sediment grain size. On the inshore Great Barrier Reef higher metal concentrations occur, for example, within fine-grained, clay-rich harbour sediments compared with coarse-grained carbonate–quartz sands (Esslemont 2000). At these sites, remobilization and transport of fine-grained, metal-rich sediments onto adjacent coral reef areas has occurred periodically due to harbour dredging. Incorporation of metals into coral skeletons may occur via (i) absorption of dissolved metals by coral tissue or during feeding, (ii) particulate trapping in cavities or (iii) direct deposition on coral skeletons through polyp damage (Fallon et al. 2002). Such uptake has documented ecological impacts upon coral fertilization and coral larvae success (Reichett-Brushett & Harrison 1999), although long-term effects on skeletal organisms remain poorly documented.

In extreme cases, inputs of contaminants also have potential to cause changes in carbonate sediment assemblages via progressive sediment dilution. In north Jamaica, bauxite dust from a loading terminal built in the mid-1960s has accumulated within sediments in Discovery Bay, a semi-restricted coastal embayment fronted by fringing reefs and previously dominated by carbonate sediments. Recent studies (Perry & Taylor 2004) have shown that bauxite now comprises upwards of 35% of the sediment in the most heavily impacted areas and some 20–25% of the sediment across much of the southern/central area of the bay (Fig. 9.20).

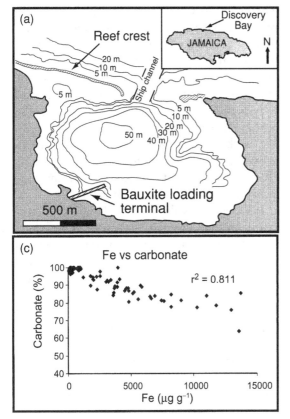

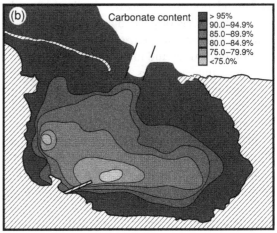

Fig. 9.20 (a) Location of Discovery Bay on the north coast of Jamaica, with the bauxite loading terminal situated in the south-west corner of the bay. (b) Contour map showing variations in carbonate content of sediments. Lowest levels occur near the loading terminal and are closely correlated with levels of iron (c), zinc and manganese. (Adapted from Perry & Taylor 2004.)

Significant ecological impacts on mangroves and corals also can be attributed to oil contamination. In large part, impacts relate to the buoyant nature of spilled oil, which is thus easily transported into intertidal settings by winds and currents. In mangroves, ecological impacts relate primarily to smothering of aerial roots and pneumatophores, leading to oxygen deficiency and rapid mortality of mangrove trees (Levings & Garrity 1994). There also may be impacts on mangrove propagule vitality (Duke & Watkinson 2002). In reef settings, oil may reduce coral growth and modify reproductive rates (Guzmán et al. 1994). The incorporation of oil residues into shallow marine and intertidal sediments is strongly site-dependent and is influenced by the rate of hydrocarbon weathering (Munoz et al. 1997) and the tidal regime, which influences the spatial extent of oil distribution (Lewis 1983). It is clear, however, that mangrove sediments have the potential to act as temporary sinks for oil residues. This is attributed in part to the anaerobic nature of the sediments, which inhibits microbial degradation of hydrocarbons, and to the rapid incorporation of oil into the sediments via burrow networks. Trapping is also aided by the ability of hydrocarbon residues to bond tightly to fine-grained suspended sediments (Ke et al. 2003). The long-term persistence of oil residues in mangrove systems is discussed by Burns et al. (1993), who also illustrate the potential for periodic or pulsed phases of oil residue release as sediments are leached during runoff or storm-wave-related erosion (Case Study 9.4).

9.4.5 Multiple-cause disturbances and changing patterns of sedimentation

Particularly in the case of coral reefs, it is often difficult to isolate the individual factors that induce community decline and shifts in

Case Study 9.4 Long-term persistence of oil residues in mangrove sediments, Panama

In April 1986, rupture of an oil store spilled between 75,000 and 100,000 barrels of medium-weight crude oil, at Bahiá las Minas on the Caribbean coast of Panama (Case Fig. 9.4a). Surveys conducted 2 months after the spill indicated that heavy oiling occurred along some 82 km of coastline between Isla Margarita and Islas Naranjos, which harbours around 16 km^2 of mangroves and 8 km^2 of coral reefs. The only areas along this section of coast to escape heavy oiling were two restricted mangrove lagoons (Case Fig. 9.4a). The oil had a significant negative impact upon the mangrove communities, as well as upon nearshore reefs and seagrass infauna (Jackson et al. 1989). Long-term (> 5 yr) monitoring of mangrove sites along the coast demonstrated:

1 the long-term persistence of residual oils that accumulated within organic-rich mangrove muds;
2 the potential for a range of aromatic hydrocarbon residues to be preserved in the sediments and for intermittent release into the nearshore environment;
3 the impacts on the mangrove communities themselves.

Initial incorporation of oil into the mangrove sediments was rapid. Oil residues were found at depths of 20 cm within 6 months of the spill, and are likely to have migrated into the sediments via diffusive processes, through crab burrows and down dead or decaying mangrove root casts. The oil residues, however, showed evidence, at most sites, of rapid compositional changes resulting from a combination of weathering processes (including evaporation), dissolution, microbial degradation and photochemical decomposition. Undegraded oil residues were found only at a few sites and these are believed to have been preserved within anoxic sediment zones. Despite the apparently rapid compositional changes that occurred, oil residue concentrations remained high within surface sediments 5 years after the spill, and at many sites had increased by an order of magnitude at depths of up to 20 cm (Burns et al. 1994).

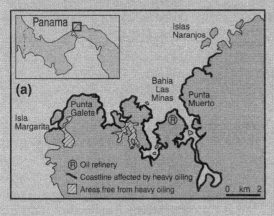
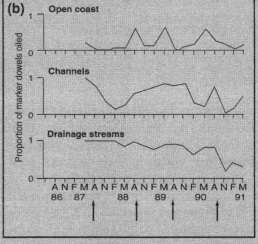

Case Fig. 9.4 (a) Map showing the location of the Bahiá Las Minas oil refinery and the main areas of coast impacted by the April 1986 oil spill. (Adapted from Guzmán et al. 1991.) (b) Plots showing the periodicity of secondary oiling episodes, as indicated by the presence of oil on marker dowels, in different mangrove environments. Arrows denote periods when increased impacts were reported on adjacent coral reef communities (data in Guzmán et al. 1994). These appear to immediately follow each of the secondary oiling episodes. (Adapted from Burns et al. 1993.)

Oiling had a significant impact on the mangrove communities over the 5 years to 1991. On the open coast, area of mangrove was reduced by 13% and root density was reduced by 24%. In channel and lagoon sites, the area of fringing mangroves reduced by 23% (root density decreased by 20%), and in drainage streams, mangrove area reduced by 56.5% (root density declined by 16%) (Levings & Garrity 1994). Release of dissolved and suspended oil residues from mangrove sediments appears to be a relatively frequent occurrence and has led to ongoing, pulsed phases of recontamination long after the initial spill. This secondary oiling occurs following erosion of mangrove substrates and is most likely related to storm activity, periods of increased rainfall leading to sediment washout, or to the cutting and degradation of areas of dead mangrove. This pulsed release of oil from the sediments is evidenced by the occurrence of localized oil slicks (which are most common adjacent to those sites most heavily impacted after the oil spill) and by the re-oiling of artificial marker stakes within the mangroves (Case Fig. 9.4b). This release of oil demonstrates the potential for mangrove sediments to act as long-term storage sites of oil and to periodically release it into the adjacent environments, thus prolonging the time-scales over which oil has an impact on nearshore environments. Time-scales of toxin persistence in mangrove sediments after heavy oiling are estimated to be at least 20 years (Burns et al. 1993).

Relevant reading

Burns, K.A., Garrity, S.D. & Levings, S.C. (1993) How many years until mangrove ecosystems recover from catastrophic oil spills? *Marine Pollution Bulletin* 26, 239–48.

Burns, K.A., Garrity, S.D., Jorissen, D., et al. (1994) The Galeta oil spill: II. Unexpected persistence of oil trapped in mangrove sediments. *Estuarine, Coastal and Shelf Science* 38, 349–64.

Guzmán, H.M., Jackson, J.B.C. & Weil, E. (1991) Short-term ecological consequences of a major oil spill on Panamanian subtidal corals. *Coral Reefs* 10, 1–12.

Guzmán, H.M., Burns, K.A. & Jackson, J.B.C. (1994) Injury, regeneration and growth of Caribbean reef corals after a major oil spill in Panama. *Marine Ecology Progress Series* 105, 231–41.

Jackson, J.B.C., Cubit, J.D., Keller, B.D., et al. (1989) Ecological effects of a major oil spill on Panamanian coastal marine communities. *Science* 243, 37–44.

Levings, S.C. & Garrity, S.D. (1994) Effects of oil spills on fringing red mangroves (*Rhizophora mangle*): losses of mobile species associated with submerged prop roots. *Bulletin of Marine Science* 54, 782–94.

carbonate production. At many sites, reductions in live coral cover and associated increases in macroalgal abundance are more realistically discussed in the context of prolonged and varied disturbances. This is seen particularly clearly on the reefs of north Jamaica, which have undergone major changes in coral community structure over the past 25 years (Liddell & Ohlhorst 1993). Coral cover in shallow reef environments has declined from around 60% to 2–3% at present, and this has been accompanied by a marked increase in the cover of non-calcifying macroalgae. The causes of this decline have been attributed to hurricane disturbance, coral disease and coral bleaching, eutrophication, mortality of the key herbivore *Diadema antillarum*, and a prolonged history of extensive overfishing. The last two factors, in particular, have resulted in the removal of most herbivorous species, and as a result algal species effectively now out-compete corals. Although the impacts on net rates of carbonate production have not yet been quantified (but are likely to be significant), shifts in carbonate sediment production patterns have occurred (Perry 1996).

Similar levels of reef decline to those in Jamaica have also been recorded along the west coast of Barbados, and again reflect a varied

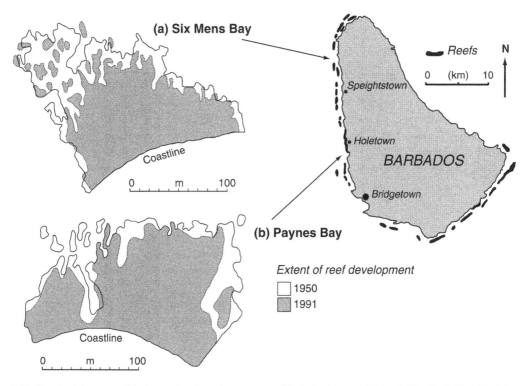

Fig. 9.21 Structural changes on fringing coral reefs on the west coast of Barbados between 1951 and 1991. At Six Mens Bay (a) recession of most of the reef perimeter has occurred, although isolated outliers of live coral remain. This has resulted in a 24% reduction in reef area. At Paynes Bay (b) major loss on the northern flank and in the central region has occurred, leading to a 21% reduction in reef area. At both sites the areas of reef loss now comprise sand and rubble. (Adapted from Lewis 2002.)

history of recent disturbances, including eutrophication and storm damage (Lewis 2002). At these sites, coral cover is reduced but, in addition, recent work has demonstrated large-scale temporal change in the areal extent of reef development. Three main types of structural loss are identified:

1 degradation of spur-and 'groove' structures on the seaward edge of reefs;
2 breaches in the reef front, leading to the development of sand-filled valleys;
3 damage and loss of flank areas.

These sites provide clear evidence for reduced reef development (Fig. 9.21) and changes in carbonate production that can be attributed to recent natural and anthropogenic disturbance events. These types of change, where coral-dominated reefs have shifted to algal-dominated communities, are far from isolated examples, and conform to the 'phase shift' concept of Done (1992). Whether these changes are permanent or whether they represent temporary shifts in community state remains unclear.

9.5 MANAGEMENT AND REMEDIATION OF CORAL REEF AND MANGROVE SEDIMENTS

Coral reef and mangrove environments are susceptible to a range of disturbance events that necessitate increasing attention from a management and remediation perspective. In part, this relates to the effects of anthropogenic-induced disturbance and related ecological degradation, which can in turn lead to destabilization of sediment substrates and increased rates of (shoreline) erosion. Most coastal systems, however, are also subject to shoreline change induced by natural fluctuations in energy inputs (such as those linked to storms, cyclones or tsunami). In these cases, the need for effective management is often driven not so much by the actual event, but as a result of (increasing) human occupation and use of the coastal zone, i.e. urbanization of environments that will naturally respond to changes in nearshore energy levels. Extensive discussion of the ecological aspects of, and problems related to, reef and mangrove management are beyond the scope of this chapter, but useful reviews are provided by Field (1998), Gladstone et al. (1999) and Yap (2000). Rather, this section outlines a number of key management issues relating to shoreline change, coastal sediment budgets and sediment contamination.

9.5.1 Managing natural coastal hazards

Given the naturally dynamic nature of many mangrove and reef-fringed shorelines (see section 9.3.1), understanding the effects of flooding and wave-induced erosion represents an important issue in tropical coastal management. These impacts can be caused both by high-magnitude events, such as cyclones and tsunami, as well as ongoing or periodic processes linked to land subsidence or El Niño-type events. In all cases, resultant changes in nearshore energy levels alter the natural equilibrium of the system and generate a geomorphological response. This most commonly occurs in the form of either shoreline erosion or sediment remobilization and deposition. In essence these sedimentary environments respond to the new (albeit, in many cases, temporary) physical conditions by changing their equilibrium state.

The potential impacts and management implications of such geomorphological changes have been discussed for a range of different South Pacific coastlines (high volcanic islands and atolls) by Solomon & Forbes (1999). In all cases, significant shoreline erosion and flooding occurred, although marked spatial variations occur in the extent of erosion. In some cases these variations related to natural differences in nearshore geomorphology and the extent of reef development (which influence the location and height of breaking waves), whereas in others erosion and flooding were exacerbated by human modification of the nearshore system. These include the removal of nearshore sands during mining, the use of inappropriate

shoreline protection schemes, and the destruction of coastal vegetation. Reclaimed land appears particularly susceptible to erosion. In all cases, economic damage was linked to inappropriate location of roads and buildings in areas that can be expected to be impacted by these types of disturbance.

Natural hazard assessment and consideration of nearshore sediment dynamics therefore form an important aspect of tropical coastal management and an important component of Integrated Coastal Management schemes. Historical records of physical disturbance and shoreline change can provide an insight into the location of high-risk areas and of sites subject to either severe or episodic sediment erosion. Such risks may necessitate the delineation of coastal zones that are deemed inappropriate for subsequent development, although this will present an increasing problem in areas with restricted coastal plains and rapidly increasing populations (Aubanel et al. 1999).

9.5.2 Managing resource extraction and exploitation

Common factors contributing to the magnitude of damage associated with natural disturbance events and, in other cases, direct causes of shoreline erosion (see section 9.4.1), are high levels of sediment extraction (due to mining and dredging), degradation of reef structures (due to coral mining and channel blasting) and removal or loss of intertidal areas. In most cases, subsequent shoreline change (erosion, channel siltation) and a magnification of the effects of episodic high-energy events can be linked to changes in nearshore sediment budgets, current dynamics and artificial modification of the coastal fringe. In addition to limiting or preventing resource extraction, rehabilitating degraded environments is often costly and problematic. The rehabilitation of directly degraded mangrove swamp previously claimed for farming will, for example, necessitate reinstating conditions in relation to topography (and thus tidal inundation), hydrology, sedimentation rates and sediment characteristics. The former may be possible via land regrading and artificial modification of channel networks (Lee et al. 1996). Restoring the physical and chemical properties of mangrove sediments, however, may prove impossible at sites that have been used for shrimp farming or timber harvesting, where extreme acidification and anaerobiosis may have occurred. If devoid of vegetation, these areas may also be prone to rapid soil erosion and removal of organic matter (Field 1998). Under such conditions, natural reseeding or programmes of replanting are unlikely to be successful.

The need for effective regeneration of degraded coral reefs is equally evident both from the perspective of disturbance-related reductions in carbonate production and from the increased rates of adjacent shoreline erosion that can follow reef disturbance (see section 9.4.1). Remedial techniques such as coral transplantation do exist, but as with mangrove replanting, the benefits of this approach are questionable. Edwards & Clark (1998) suggest that natural recovery processes may be far more effective than transplantation, with the exception of areas that are failing to recruit juvenile corals. Even in these cases they suggest a re-emphasis towards transplanting slow-growing, massive corals, which tend to be better natural recruiters than fast-growing branched corals. Massive corals are also less susceptible to the effects of coral bleaching (McClanahan 2000). At best, transplantation is only likely to be effective on a localized scale, so that resources may be better directed at managing or mitigating the causes of degradation (Bellwood et al. 2004).

9.5.3 Managing sediment contaminants

As outlined in section 9.4, nearshore sediments in general, and mangrove sediments in particular, have considerable potential to trap a wide range of contaminants relating to industrial, urban and agricultural discharges. In mangroves, these contaminants, if present in sufficient concentrations, can lead to tree defoliation, seedling mutations, reduced species diversity and mangrove die-back (Ellison & Farnsworth 1996), and in coral reefs to disease, and reduced growth and reproductive potential. In many coastal systems a range of contamin-

ant inputs can often be identified, some being sourced from the coastal fringe, but many linked to activities that occur upstream within coastal catchments. The diverse range of bodies with responsibilities to regulate these different, but linked environments can cause significant management problems.

The central Great Barrier Reef (GBR) provides an interesting perspective on the uncertainties that exist in relation to terrigenous impacts on nearshore marine environments, and the issues of coastal and marine management. There have, for example, been suggestions of increasing sediment runoff and, associated with this, increased nutrient and contaminant (pesticide) levels, linked to agricultural activities in the catchment areas (Haynes & Johnson 2000). Although sediment input rates do appear to have increased, there is conflicting evidence about whether or not this is having a detrimental effect on the marine environment and on the coral reefs in particular, because most fluvially derived sediments accumulate only along nearshore areas of the shelf (Larcombe & Carter 2004 – see Case Study 9.1). The GBR region, however, does highlight the important linkages that exist between coastal environments and river catchments. Management of the GBR is primarily under the jurisdiction of the Great Barrier Reef Marine Park Authority (GBRMPA). Although the GBR itself is widely cited as an example of good marine management practice, the GBRMPA has little influence over managing the catchments that feed onto the shelf. Instead, current management of agricultural activities and hence runoff are under a voluntary code of practice. Given the uncertainties about the actual impacts of terrestrial sediment inputs on the GBR, the effect of changing land-use practices, in terms of the marine environment, seem unclear. The region, however, emphasizes the potential benefits of integrating both coastal and catchment management schemes.

Remediation of contaminated sediments presents an additional challenge and one that is important given the potential for sediments to both store and episodically re-release contaminants (see section 9.4.4). In the case of oil contamination, chemical dispersants traditionally have been used to try and mitigate the effects of oiling. Evidence suggests that although dispersants may be useful in limiting the amount of oil that reaches mangroves, and may limit mangrove tree mortality, they are less effective at preventing oil absorption into mangrove sediments, or subsequent patterns of oil weathering and degradation (Duke et al. 2000). Some success in remediation of oiled sediments has been achieved through the use of bioremediation techniques (Ke et al. 2003). These exploit microorganisms that occur naturally within mangrove sediments, some of which are effective degraders of hydrocarbons. Although rates of bacterial oil degradation may be inhibited under normal anaerobic sediment conditions, rapid degradation rates have been achieved by manipulating bacterial population densities via active sediment aeration and the application of slow-release fertilizers (Ramsay et al. 2000).

9.6 EFFECTS OF GLOBAL CLIMATIC AND ENVIRONMENT CHANGE ON TROPICAL COASTAL SYSTEMS

Many of the predicted changes in global climatic conditions (see Chapter 1) have the potential to influence both coral reef and mangrove environments. Those with direct implications include changes in atmospheric CO_2 concentrations, increased atmospheric and sea-surface temperatures, increased UV radiation, changes to patterns of storm frequency and intensity, and increased sea-level (Ellison & Farnsworth 1996; Wilkinson 1996). Depending upon the magnitude of these changes and the local sedimentary regime, coral reefs and mangroves may exhibit highly variable, and often site-specific, responses. A crucial point to emphasize is that environmental change may have both positive and negative effects upon coral reefs and mangroves.

9.6.1 Impacts of climate change on coral reef and mangrove sedimentary systems

Changes in atmospheric CO_2 levels may have both direct and indirect effects on coral reefs and mangroves. In mangroves, increased CO_2 levels are likely to enhance rates of photosynthesis leading to increased leaf production and mangrove growth (Ellison & Farnsworth 1996), although the positive effects of increased productivity may be restricted locally by sea-level rise (see section 9.6.2). Elevated CO_2 levels may have an impact on coral communities by modifying marine aragonite saturation states and thus reducing rates of coral calcification (Kleypas & Langdon 2002). Recent modelling studies through to 2069 predict that much of the Pacific basin may become marginal (in terms of calcification potential) for corals with respect to aragonite saturation (Guinotte et al. 2003) and would have major implications for reef carbonate production.

Linked to increasing CO_2 levels are probable increases in atmospheric and sea-surface temperatures. In most mangrove settings predicted temperature increases are not likely to be high enough to cause direct mangrove mortality (Ellison & Farnsworth 1996) and, in higher latitude regions, may actually promote expansion of mangrove swamps. Such expansions would be dependent upon local hydrological constraints and, in arid settings, conditions may actually become increasingly marginal for mangrove survival. Higher atmospheric temperatures and increased solar radiation may also cause sediment warming, leading to increased soil respiration, organic matter decomposition, methane (CH_4) and hydrogen sulphide (H_2S) release, and root turnover. This would, in turn, modify the composition and chemistry of mangrove sediments, although the precise impact on mangrove communities is not clear. In contrast, coral reef communities are likely to be influenced significantly by increasing sea-surface temperatures (Chadwick-Furman 1996). From a positive perspective such warming may permit an increase in the latitudinal range of hermatypic corals, and phases of reef 'switch-on' in higher latitude carbonate shelf settings, whereas a negative consequence may be increased coral bleaching. This occurs as a physiological response to stress, and involves coral shedding of the symbiotic algae, leading to reduced coral growth and, if conditions persist, mortality (Glynn 1996). The effects of

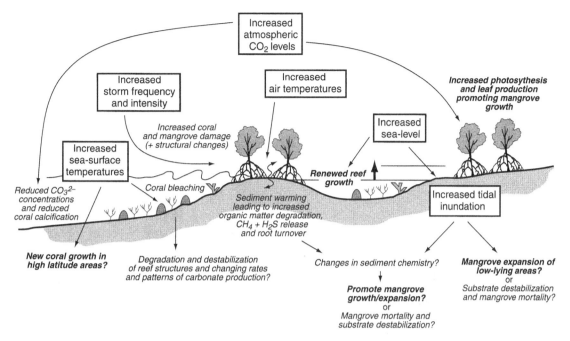

Fig. 9.22 Simplified schematic diagram illustrating some of the key processes associated with climate change (boxed) and both positive (bold italics) and negative (italics) responses within coral reef and mangrove environments. Note that many of the potential responses have questions marks, highlighting either the uncertainties that exist, or the likely site-specific nature of the responses.

such increases were evident most recently on a regional scale following the strong El Niño of 1997–98, which resulted in widespread coral mortality across the Indo-Pacific. Such bleaching-related mortality has major consequences for reef carbonate production, as has been quantified in Panama (Eakin 1996; see section 9.3.3). Temperature-related bleaching events have also been documented in higher latitude settings, where corals are adapted to lower average temperatures (Celliers & Schleyer 2002) and hence the effects of such bleaching events are likely to be globally significant. The rate at which coral communities will recover from these widespread bleaching episodes remains unclear, but may depend upon the ability of corals to adapt to increasing mean sea-surface temperatures.

Shifts towards net erosional regimes on many shallow reefs are thus a potential consequence of increased bleaching. This may lead to destabilization of reef frameworks and reduced structural integrity, although the response of framework accumulation processes (such as bioerosion

and encrustation) to such disturbances remains poorly documented. Any suppression of carbonate production will, however, inhibit the potential for reefs to respond positively to sea-level increases (see section 9.6.2) and increase the frequency of storm-wave overtopping. This, in turn, may result in changes in nearshore sediment dynamics and changes in shoreline morphology (Fig. 9.22). Climate change may also modify storm patterns and intensities, and shift the position (or latitudinal range) of the cyclone/hurricane belt, and thus influence reef and mangrove structure (Smith et al. 1994). In addition to any increase in physical disturbance, more frequent/intense storms may also have an impact on mangroves by increasing rainfall and sediment runoff.

9.6.2 Impacts of sea-level change on coral reef and mangrove sedimentary systems

A major consequence of global climatic change with implications for coral reef and mangrove environments is sea-level change. Future trends (and uncertainties) regarding sea-level are outlined in Chapter 1, but mean sea-level rises of around 50 cm by 2100 are predicted (Wilkinson 1996) and equate to annual increases of around 4 mm yr^{-1}. In the context of published coral growth rate data, which can be as high as 10–12 mm yr^{-1}, these predicted changes thus appear relatively insignificant. Actual reef accretion rates (the rate at which the reef grows vertically) are determined not only by coral growth, however, but also by factors such as the rate of bioerosion and sediment production, and actual rates of net reef accretion are far lower (Stoddart 1990). In addition, reef accretion rates vary spatially across individual reef systems and so different reef subenvironments (reef flat, shallow reef-front, deep reef-front) will have varying potential to maintain their position relative to sea-level (Spencer 1995).

Depending upon the time-scales over which reef accretion trends are viewed, it is possible to reach very different conclusions about the response of reefs to sea-level change. Recent accretion rates suggest that although those areas of reef that are characterized by fast-growing branched corals can accrete at rates that exceed all but the highest sea-level rise estimates (Fig. 9.23a), accretion rates for reefs as a whole (i.e. encompassing lagoon through to deep reef-front environments) fall below even the lower sea-level rise scenarios (Spencer 1995). In contrast, longer term (geological time-scale) data from coring studies indicate that many reef systems have accreted at rates close to or above those predicted over the next century (Fig. 9.23b). Such data, however, include growth that occurred during the early Holocene when sea-levels were rising rapidly (around 5 mm yr^{-1} between 10.5 and 7.7 kyr BP in the Caribbean; Toscano & Macintyre 2003) and it is unclear how modern reefs would respond to such changes. Based on Holocene reef accretion records the threshold rates beyond which reefs are unable to maintain pace with sea-level change lie at around 8–10 mm yr^{-1} (Spencer 1995). One benefit of sea-level rise may be stimulation of reef growth in areas that have reached sea-level and stabilized over the past 5000 years (Wilkinson 1996).

Linked to the response of coral reefs to sea-level rise is the question of how low-lying carbonate islands and rubble cays may respond. These islands comprise semi-consolidated sand and rubble sitting atop reef platforms, and commonly perceived threats include direct erosion (or submergence) and increasing saltwater intrusion into island aquifers (Wilkinson 1996). Recent modelling studies, however, suggest considerable complexity in island response to sea-level rise, related in large part to changes in sediment supply (Kench & Cowell 2002). Reduced sediment supply, combined with increased sea-levels, may result in increased shoreline displacement rates on atolls. Reef flats typically produce little sediment, however, and the most significant displacement rates are caused by changes in the littoral sediment budget. These may occur as the result of a range of additional anthropogenic stressors (see section 9.6.3) and thus magnify the impacts of sea-level change.

The ability of mangrove shorelines to maintain their current positions and geometries in response to sea-level rise appears equally dependent upon the rate of sea-level change and local sediment dynamics, and is likely to be highly site-specific. Stratigraphical studies through Holocene mangroves suggest that many of the large modern mangrove swamps did not exist during the early Holocene (Woodroffe 1990). This is attributed to the rapid rates of sea-level rise (in the order of 10 mm yr^{-1}) and to differences in shoreline geomorphology that prevented sediment accumulation and restricted mangroves to isolated fringing communities. Research indicates that modern mangroves are likely to be eroded if sea-level rises exceed 0.8–0.9 mm yr^{-1} (Ellison & Stoddart 1991), but this will depend upon local sedimentation rates.

Mangroves that currently develop on low lying carbonate islands, which are characterized by autochthonous sedimentation and typically exhibit low accretion rates (< 0.8 mm yr^{-1}), may experience either expansion or contraction depending on local rates of sediment production and accumulation (Ellison & Stoddart 1991). In contrast, both river- and tide-dominated mangroves can potentially receive large amounts of

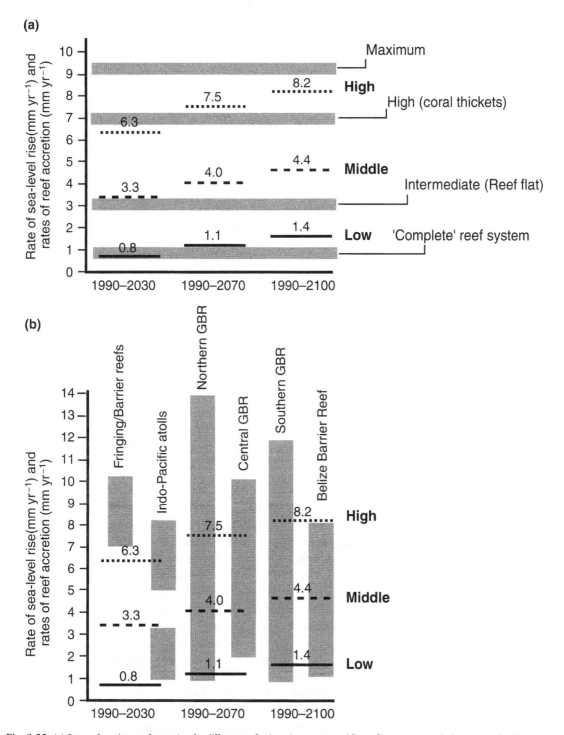

Fig. 9.23 (a) Rates of modern reef accretion for different reef subenvironments and for reef systems as a whole compared with low, medium and high projections of global sea-level rise for three different time periods through to 2100. The different net accretion rates that occur in different reef subenvironments reflect differences in rates of coral growth and coral substrate degradation (see Perry 1999). (b) Rates of Holocene reef growth based on radiocarbon-dated drilled core-sections superimposed above low, medium and high projections of global sea-level rise for three different time periods through to 2100. The often high average rates of accretion during the Holocene reflect growth under different regimes of sea-level rise during the Holocene transgression. Projections based on IS92a emissions scenario. (Adapted from Spencer 1995.)

sediment and so might be expected to respond more positively. These environments, however, are highly variable in form and are characterized by very different sediment dynamics and accretion rates. River-dominated mangroves with high sediment inputs may accrete at rates that are sufficient to keep pace with sea-level rise (Woodroffe 1990). Those associated with smaller river catchments (with lower sediment influx) or areas subject to seasonal sediment influx may undergo landward migration. Tide-dominated mangrove settings also vary significantly in terms of tidal influence and sediment flux and, in such settings, shoreline response will again be highly site specific (Wolanski & Chappell 1996).

Predicting mangrove response to sea-level rise is further complicated by the fact that future sea-level rises will occur not only from conditions of relative stability (Woodroffe 1990), but also across relatively low-lying land (where the spatial extent of sea-level increases is significantly higher). Although this will increase the rate at which tidal inundation and associated changes in sediment characteristics and salinity levels occur (Semeniuk 1994), from a positive perspective it may also facilitate landward migration of mangroves. Indeed, in cases where sediment accumulation rates are high and the seaward margins of the swamps are maintained, the spatial extent of mangrove swamps may actually increase. This, however, will be inhibited in areas where urbanization of the coastal fringe has occurred and under these conditions, regardless of sediment influx, progressive mangrove degradation is likely.

9.6.3 Climatic and sea-level change in relation to increased anthropogenic influence

Although global climate changes and related increases in sea-level are often discussed from the context of increasing ecosystem disturbance, there is an important caveat to such speculation. Predicted changes in temperature, CO_2 concentrations and sea-level through to 2100 are unlikely to approach the scales of change that have occurred in the recent geological past (Wilkinson 1996). For example, major changes in sea-level (of 100+ m) occurred at various times during the Pleistocene. In addition, there is some evidence to suggest that Holocene sea-levels underwent rapid 'jumps' at several points over the past 16,000 years (Toscano & Macintyre 2003) and that rapid rates of sea-level rise (up to 20 mm yr^{-1}) characterized the early Holocene (Wilkinson 1996). Neither appear to have had long-term detrimental effects on reef or mangrove communities, although massive changes in geomorphology and structure have obviously occurred as the world's sedimentary systems migrated across the continental shelves. Similarly, atmospheric CO_2 concentrations and temperatures have fluctuated significantly through the Pleistocene (Wilkinson 1996) without any apparent long-term negative impact on reef and mangrove communities.

Over the coming century, different sedimentary responses may, however, result from the combination of natural variation and anthropogenic influence. Although coral and mangrove communities and their associated sedimentary environments are capable of responding to natural changes in physical and environmental factors, future responses may be influenced by the additional stresses imposed by a range of anthropogenic-related factors. In coral reefs, modified rates and patterns of reef carbonate production may, for example, occur as a result of changes in reef community composition driven by a range of factors including overfishing and pollution. Similarly, in mangrove swamps, mangrove mortality and substrate erosion may occur due to sediment contamination and land-use change. These additional pressures may alter the ways in which reef and mangrove systems respond to climatic and sea-level change. Our understanding of how both reef and mangrove systems are influenced by specific anthropogenic stressors, the mechanisms and rates of recovery and adaptation potential, and the longer term impacts on rates of carbonate production and accretion (in coral reefs) and sediment accumulation (in mangroves) remain poorly understood.

10

Continental shelf environments

Piers Larcombe

10.1 INTRODUCTION

This chapter addresses the environmental sedimentology of continental shelves. 'Continental shelf' is taken to be the sea floor shallower than about 200 m adjacent to continental landmasses, with the outer margin marked by the continental slope (e.g. Whitten & Brooks 1972) (Fig. 10.1). In the long-term, continental shelves receive river sediments, with supply to the shelf modulated by catchment (Milliman 2001; Walling & Fang 2003) and estuarine processes (Dyer 1966, 2000). Some shelves contain major regions of *in situ* sediment production, pre-dominantly calcareous in nature, and shelf sediment dynamics are crucial in influencing the nature and fate of shelf sediments. The nature and morphology of continental shelves are controlled by: (i) the hydraulic regime, and hence sediment transport, (ii) sediment supply and (iii) relative sea-level (Johnson & Baldwin 1996). Other factors, particularly important for carbonate shelves, are climate, biological interactions with the sediments, seawater chemistry and sediment composition. It is increasingly being acknowledged in management regimes that the surface sedimentary systems are vital to marine ecosystems.

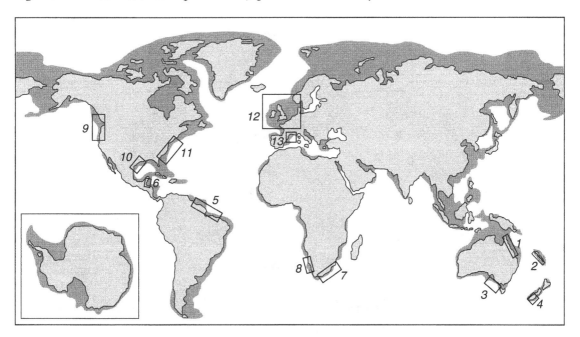

Fig. 10.1 Distribution of modern continental shelves (dark grey). Continental shelves noted in the text are: 1, Great Barrier Reef, Australia; 2, Western New Caledonia; 3, Lacepede shelf, South Australia; 4, South Otago, New Zealand; 5, Amazon; 6, Belize; 7, southeast Africa; 8, southern Namibia; 9, Oregon–Washington; 10, Texas–Louisiana shelf; 11, Mid-Atlantic Bight; 12, UK and western European shelf; 13, north-western Mediterranean. (Modified after Johnson & Baldwin, 1996.)

Direct human influences on shelf sedimentary systems probably can be constrained mainly to the past few hundred years, but such a view requires assessing the nature of environmental change over the longer time-scales of many of the natural driving processes. Further, the sedimentary environmental setting is vital. This chapter thus describes the geoscientific context and sedimentary processes of continental shelves of at least the past few hundred years, noting the influence of human use and management. A deliberate focus on the Australian and north-west European continental shelves introduces the reader to two contrasting but well-described regions. Compared with Chapters 7, 8 & 9, the focus herein is those areas of the shelf where the dominant natural inputs of energy and material are *not* land-influenced, and where environmental management requires understanding marine processes. There is no attempt to be exhaustive, continental shelf settings are also discussed by Walker & James (1992), Wright (1995), Johnson & Baldwin (1996), Wright & Burchette (1996) and Allen (1997).

10.1.1 Nature of shelf sedimentary environments

The Holocene sea-level highstand and the associated relatively deep, modern continental shelves are the geological exception rather than the rule (Shackleton et al. 1990). As a result, modern shelves have significant 'accommodation space' (the total space below sea-level within which it is possible to deposit marine sediment), primary controls on which include relative sea-level and shelf topography. Modern continental shelves are subject to a wide range of physical and biological sedimentary processes, but these are of varied relative significance. Considering physical processes, Swift et al. (1986) concluded that 80% of modern continental shelves are storm dominated, 17% are tidally dominated and only 3% are dominated by oceanic currents. On high-energy, exposed shelves, long-period waves can mobilize sediment even in fair-weather conditions, whereas a complex suite of shelf currents, including waves, wind-driven currents and 'return' currents, can occur during storms (Wright 1995). Thus, a grouping can be made based on the

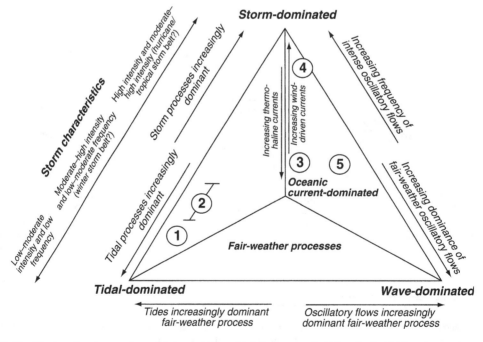

Fig. 10.2 Classification of modern continental shelves based on physical processes: 1, tidal straits; 2, tidal seas; 3, oceanic-current swept shelves; 4, storm-dominated shelves; 5, river mud-dominated shelves. (Simplified after Johnson & Baldwin 1996.)

dominant fair-weather processes and the relative interaction with storm-associated processes (Fig. 10.2), where of primary interest are those processes that are reflected in the nature, distribution and stratigraphy of shelf deposits. Shelf sedimentary processes (Fig. 10.3) operate at a range of physical and temporal scales, so that the relative sedimentary impacts of these processes depend on their frequency, magnitude and recurrence intervals. Sediment transport is a cubic (or quartic) function of flow speed (Soulsby 1983, 1997), which, together with the presence of flow thresholds of transport, means that short periods of fast flows (e.g. related to storms, cyclones or tsunami) may sometimes be more significant in terms of sediment transport than longer periods of slower flow.

10.1.2 Environmental sedimentology of shelf environments

Human influences on the sedimentary regime of continental shelves include, but are not limited to, impacts by fisheries (especially trawling),

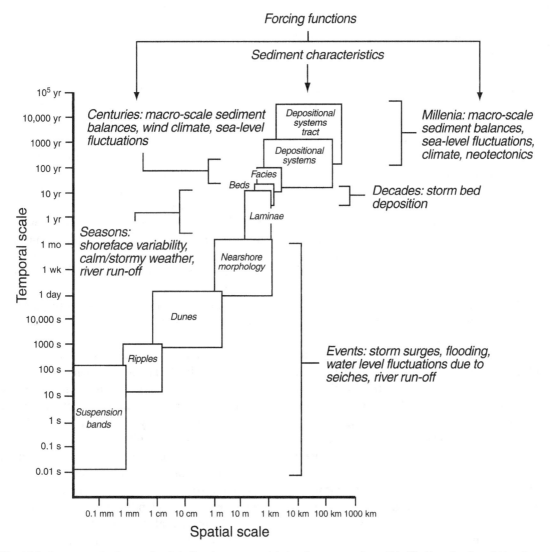

Fig. 10.3 Forces operating in coastal and shelf environments and their sedimentary products. (Modified from Sternberg & Newell 1999; Schwarzer et al. 2003.)

mineral extraction, disposal of dredged material and changed supply of material from rivers. Sediment supply might be either reduced because of damming of large rivers or increased as a result of deforestation in catchments leading to higher sediment yields, or by the canalization of some rivers, which leads to increased bypassing of sediment through the coastal zone. For many types of human impact on the sedimentary environment, the main method of measuring change is restricted to historical surveys, but consideration of the significance of human impacts means that it is necessary to measure change on time-scales of decades to centuries (Berger & Iams 1996; Sarnthein et al. 2002). Of particular interest are those radiometric techniques that allow delineation of chronologies over a few decades (e.g. ^{210}Pb, ^{137}Cs and others). On continental shelves themselves, the best records of past change are generally formed at sites with high rates of sediment accumulation (e.g. the Amazon shelf, Kuehl et al. 1986) or on shelves that have biogenic features with a high preservation potential, from which can be derived a proxy record of environmental change (e.g. Schöne et al. 2004).

Sediments form habitats for marine fauna and flora, and one view of environmental management is that the overall mangement aim should be habitat retention, on the basis that if the habitats themselves are present, then it is likely that the structure and function of the associated ecosystem components will also be retained. Shelf sediments may accumulate contaminants to the extent that the biology is negatively impacted, and these surface accumulations may themselves become sources of pollution, with regard to transport across the shelf and/or bioturbation down into the sediment column (e.g. Kershaw et al. 1988; MacKenzie et al. 1999). Trace metal elements are often of environmental concern, background concentrations of which are controlled by the geological history of a continental shelf and its catchments (e.g. Chapter 1). There are few texts regarding human management of shelf environments, and much information derives from reports of past or ongoing activities. The websites of Government departments

and associated regulatory bodies are increasingly useful. Examples include the US Environmental Protection Agency (EPA), Environment Canada, the UK Department of Environment, Food and Rural Affairs (DEFRA), the Netherlands National Institute for Coastal and Marine Management, the Australian Department of Environment and Heritage, the Great Barrier Reef Marine Park Authority (GBRMPA), the New Zealand Department of Conservation and National Institute for Water and Atmosphere.

10.2 SHELF SETTINGS AND SEDIMENT SOURCES

10.2.1 Shelf settings

Continental shelves receive sediments that have been transported from the terrestrial realm by rivers, are the sites of biogenic sediment production, store sediments in various environments and for various time-scales, and transport some sediments off-shelf down the continental slope towards the abyssal plains. There are two main morphological types of continental shelf (Johnson & Baldwin 1996):
1 pericontinental shelves – which occur on continental margins and equate to those modern continental shelves with the classic division and profile of shoreline, shelf and slope (e.g. the Amazon and Californian shelves);
2 epicontinental shelves – which are partially enclosed seas within continental areas usually with a uniform shallow-dipping ramp profile (e.g. the Gulf of Mexico and the North Sea).

Tectonically, the largest continental shelves are located at passive margins, where they generally develop seaward-thickening accumulations of sediment, supplied by large continental drainage systems. At active plate margins, shelves may develop at the convergent margin itself, upon which shelf areas are relatively small and which may have zones with high sediment accumulation rates, or in foreland basins, where extensive areas of continental shelf may develop and sediment accumulation rates may be high.

Understanding the processes and features of active shelf sedimentary systems involves long time-scales and major changes in sea-level. At the last glacial maximum (LGM), 20,000–18,000 years ago, sea-levels around the globe were 120–130 m below present (Chappell et al. 1996) because large volumes of ocean water were locked up as ice in polar regions. Those physical areas that now form continental shelves were thus subject to subaerial processes and the lowstand coastline was located to seaward of the present continental shelves. Inundation of these areas occurred during the post-glacial transgression, which together with processes operating at the modern sea-level highstand reworked the sub-aerial deposits to various extents. Viewed simply, the post-glacial sea-level rise was rapid until around 8000 years ago, when it slowed to reach its highest level around 6000 years ago. Consequently, many regions of the world's continental shelves have been subject to their present processes for only a few thousand years, and many of the bathymetric, mineralogical and granulometric features of continental shelves and their sediments derive from the actions of past processes that operated at different stages of relative sea-level. Considerable regional variability occurs in relative sea-level, because of the interactions of 'global' sea-level, gravitational variations, tectonics, hydro-isostasy and weather, operating at different temporal and spatial scales.

10.2.2 Sediment sources and characteristics

10.2.2.1 Fluvial/lithogenic material

Chemical and physical weathering of rocks in river catchments leads to sediments being carried by rivers to the sea. Globally, the total load of silt delivered by rivers to the ocean has been estimated to be 13.5×10^9 t yr^{-1} (Milliman & Meade 1983), to which bedload transport adds $1–2 \times 10^9$ t yr^{-1}. The combined rate of sediment supply represents an average denudation of the world's catchments of 50 mm ky^{-1} (Allen 1997), but there is great geographical variation in rates of erosion and transport, controlled by factors including the rate of uplift, climate, rock type, topography and precipitation. Sediment delivery also can be affected by human activities

in catchments and rivers, such as deforestation and conversion to agriculture (e.g. Chapters 2 & 3). There are therefore wide variations in the nature and volume of sediments delivered to the shelves (Chapters 1 & 8). Some shelves receive huge volumes of sediments, particularly from those rivers draining steep tropical catchments with relatively young geology, such as the Indonesian Archipelago (Fig. 10.4). The Amazon River produces an average of 1.2×10^9 t yr^{-1} of sediment, about 90% of which is silt and clay (Dunne et al. 1988), and which represents an average denudation rate of the catchment of 69 mm kyr^{-1}. On relatively dry and old continents, such as Australia, the largest river in terms of sediment delivery is the Burdekin. This delivers 3×10^6 t yr^{-1} of sediment to the continental shelf, the sediment comprising around 75% silt and clay. Sediment yields equate to a catchment denudation rate of 9 mm kyr^{-1} (Belperio 1979; Allen 1997). For some continental shelves that receive relatively little river input, material reworked from the shelf seabed by storms or strong currents may form a significant component of the overall shelf sediment budget.

10.2.2.2 Shelf in situ production

Continental shelves produce significant amounts of 'new' sediment, mainly biogenically, but also locally by inorganic precipitation in some tropical environments. On some shelves and in certain environmental conditions, authigenic minerals are formed, such as the clay mineral glauconite (Bornhold & Giresse 1985; Odin 1988). By far the most significant global contribution is that of biogenic material, which includes skeletal parts of animals and marine plants, of various calcareous and siliceous composition, living and growing within the water column and on the sea bed (Walker & James 1992; Wright & Burchette 1996). Significant production of biogenic sediments needs a suitable regime of salinity, temperature and nutrients, but also good light intensity within the water column, because many primary producers are phototrophic (such as green and red algae) or are mixotrophic (many corals and large benthic

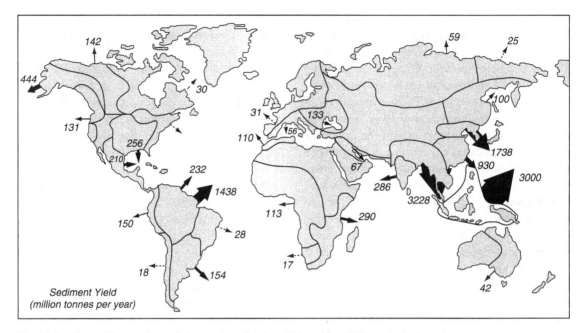

Fig. 10.4 Volume of riverine silt supplied to continental shelves. (Adapted from Milliman & Meade 1993.)

Foraminifera). Other major producers of biogenic sediment are molluscs and bryozoans, which feed on organic matter produced within the water column by phytoplankton, and hence ultimately also rely on light.

Given the above, shelves dominated by biogenic sediment production generally have low rates of terrigenous input, because light intensity is decreased by high rates of sediment input and/or reworking by shelf processes. Globally, major regions of tropical carbonate sedimentation include the seas off Indonesia, northern Western Australia, the Bahamas and the Arabian Gulf (Chapter 9), and there are some areas of carbonate production in cooler waters, most notably the Lacepede shelf, southern Australia (James et al. 1992) but also in Brazil (Gischler & Lomando 1999; Testa & Bosence 1999). There are mechanisms for relatively rapid breakdown of some biogenic sediments, through bioerosion and solution, so that sediment production rates need to be distinguished carefully from eventual sediment accumulation rates (Buddemeier et al. 1974).

Many organisms in these environments are highly environmentally specific, and thus have

potential to form part of assessments of the 'health' of modern ecosystems, or are potential palaeoenvironmental indicators. Notable in this regard are foraminifers and diatoms, but other relevant fossils include dinoflagellate cysts, sponge spicules, pollen, seeds, charcoal and ostracods. There is, however, a general underutilization of microfossil-based techniques in studies of modern environmental sedimentology. As an example, the broad use of diatoms in environmental science is well recognized (Stoermer & Smol 1999) and they are particularly useful in studies of environmental pollution of freshwater lakes and streams. Even in estuarine environments, however, there are relatively few applied studies using diatoms (Sullivan 1999), and there are even fewer studies on continental shelves. Foraminifera are increasingly used in applied environmental studies (Scott et al. 2001), but not yet for open-shelf environments. There appears to be insufficient work specifically designed to determine the tolerances of benthic marine microorganisms to various forms of pollution (Sullivan 1999). The environmental constraints upon many larger organisms are also poorly documented. For corals, there are some generally well-known environmental ranges for their occurrence (e.g. Potts & Jacobs 2003; Chapter 9), but these ranges cover a variety of corals, some of which have variable ways of obtaining energy and some of which are adapted for variable conditions.

10.2.3 Sediment accumulation processes and disturbance events

The wide distribution of modern continental shelves means that, as a group, they experience a range of environmental processes (Fig. 10.3) and thus contain a diverse group of depositional environments. The energy required to transport sediments can be derived from a number of sources. In most cases, the dominant energy supply is either tidal or weather-related, although there is generally a mixture of both. It is also important to consider the relative impact of daily (generally low-energy) versus episodic high-energy phenomena (e.g. hurricanes, tsunami). Finally, there are very well-developed models of across-shelf transport (e.g. Fig. 10.5), but it is now acknowledged that, for most shelves, the along-shelf component of sediment transport is greater than that across-shelf, especially where the coastline (or shallow bathymetry) is relatively straight (e.g. Great Barrier Reef shelf, Otago shelf, Texas–Louisiana shelf, this chapter).

10.2.3.1 Tide-influenced shelves

Tidal forces affect the whole globe, but their influence on the world's oceans is heavily modified by the bathymetry of the continental shelves and by the shape of the coastlines. Tides may occur daily (diurnal) or twice-daily (semi-diurnal): for most places around the world, the semi-diurnal tide is dominant. The dynamics of many continental shelves are dominated by tides, especially where tidal ranges are high (Chapter 1), but even in areas of low tidal range, tidal currents can be strong locally, for example in the Torres Strait, Australia (Harris 1989, 1991) and the Hayasui Strait, Japan (Mogi 1979). On most shelves, the flood and ebb current directions are generally opposed, but as they change in speed they also change direction, so that, overall, the tide describes an open ellipse, with a net direction and magnitude (Stride 1982). Should the threshold for bed sediment transport be exceeded by one or both of the tidal currents, net bed sediment transport results. Over long periods, the result is a pattern of regional bed-sediment-transport pathways, which begin at bed sediment 'parting zones', where they tend to be associated with lag surfaces, and end at bed sediment 'convergences', where there is an accumulation of sandy sediment. On the UK shelf, individual transport

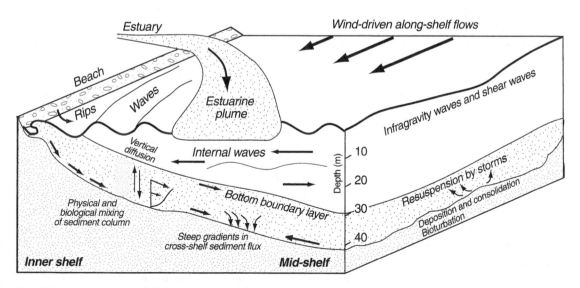

Fig. 10.5 Processes of cross-shelf transport. (Adapted from Nittrouer & Wright 1994.)

pathways are up to 550 km long and 170 km wide (Belderson et al. 1982). The suite of sedimentary facies and bedforms present along such paths is well-defined on tidally dominated shelves and is related both to the peak speed of the tidal current and to sediment availability.

Volumetrically, there is generally little practical impact that humans can have on these sediment transport pathways. A small transport pathway of 30 km in length, 5 km wide and 0.5 m average thickness would contain around 130×10^6 t of sediment (assuming porosity of 30% and density of 2.5), which is, for example, around five times the total weight (25×10^6 t) of aggregate extracted from UK waters each year. Although apparently reassuring, this is an oversimplified view because sediment transport rates along the pathways are generally poorly known, and thefefore there are difficulties in assessing a sustainable rate of removal of material. If the rate of removal exceeds the rate of resupply, or the deposits are relict, there is the potential to create a long-term change in bed habitats.

10.2.3.2 Storm-influenced shelves

Other reviews have noted that most (perhaps 80%) of the world's shelves are storm-dominated (e.g. Johnson & Baldwin 1996; Allen 1997). The effects of both waves and wind-driven currents are relevant. Low-pressure weather systems over the sea generate waves with a range of periods and heights, which may influence shelf sedimentation directly. Sedimentation on some shelves, however, can be dominated by swell waves generated by storms in adjacent sea areas. In deep water, waves travel at speeds relative to their wavelength,

$$C = \sqrt{(g\lambda/2\pi)}$$

i.e. wave speed = $\sqrt{[(\text{gravity} \times \text{wavelength})/(2 \times \pi)]}$

so that long waves travel fastest away from the centre of a storm. Long waves also carry the most energy and transfer that energy deepest into the water column, so that they are able to mobilize sediment on relatively deep areas of continental shelves. As well as generating waves,

wind blowing over the sea surface drags water along with it, forming wind-driven currents. These currents are therefore fastest at, and often limited to, the sea-surface, but where the wind is of sufficient strength, duration and fetch, the whole depth of the shelf water mass may be influenced. A general rule-of-thumb is that the speed of the surface wind-driven current is up to c. 3% of the wind speed (as measured 10 m above the sea surface). Therefore, during storms, wind-driven currents can be of significant magnitude (see also Case Study 9.2).

It is useful to distinguish storm-driven fast unidirectional flows combined with waves, compared with the effects of long-period waves acting alone. In this context, storms include hurricanes, typhoons and cyclones, which directly influence the shelf with wind-driven currents, waves and land runoff. The Caribbean region experiences an average of seven hurricanes each year, some of which enter the Gulf of Mexico. The Texas–Louisiana shelf has an even slope of around 1:1000, with the 80 m contour located c. 80 km offshore. In September 1961, Hurricane Carla, category 5, took 3 days to pass 400 km north-north-west across the shelf, producing a storm sand bed that extended for > 200 km along the shelf. The bed was preserved mostly within the 30 m contour, was thickest, at up to 9 cm, within the 20 m contour and was recognizable down to the 50 m depth contour. At the surface, the bed decreased in grain size offshore. The storm sand bed fined upwards, from a sharp basal surface, through planar laminations to shallowly inclined laminations up to a gradational upper surface. Close to shore, the upper surface was sharp and truncated (Snedden & Nummedal 1991). These sedimentary data are consistent with numerical models of shelf flow, which predicted that the wind regime would have produced:

1 a shoreward flow to the north-west at the surface;

2 a fast shelf-parallel flow to the south-west of > 1.5 m s^{-1} in depths around 20 m;

3 an oblique offshore return flow to the south-south-west or south near the bed (Forristall et al. 1977; Keen & Slingerland 1993).

Further, the oscillatory action of waves would have dominated in shallow water. Hence, fast flows would have mobilized the sands and silts of the inner shelf south-west along the shelf, and the oblique offshore currents would have transported sediments offshore, with finer sediments deposited in deeper water.

The seawards-fining and seawards-thinning storm bed described above, however, is not a universal feature. Shelves of different morphologies and sedimentary regimes can have distinctively different storm sediment dynamics on physical scales of tens to hundreds of kilometres. For the Great Barrier Reef (GBR) shelf, the storm bed coarsens and thickens seawards (Gagan et al. 1988, 1990; Larcombe & Carter 2004). Further, and also in marked contrast with the Texas–Louisiana shelf, the GBR storm bed is best preserved on the inner shelf, where it is buried beneath the depth of subsequent bioturbation by immediate post-storm sedimentation. Storm-bed preservation is a function of cyclone recurrence interval (Case Study 10.1), storm-bed thickness and the rate and depth of between-cyclone bioturbation (Gagan et al. 1988, 1990). In passing, it has long been recognized that the frequent passage of storms is not incompatible with the presence of abundant and apparently delicate benthic organisms on and in shelf sediments (e.g. Vaughan et al. 1987).

Case study 10.1 Time-scales of cyclone-driven sedimentary processes on the northern Great Barrier Reef shelf

Like most regions of the Great Barrier Reef shelf, the shelf off Cairns (latitude 16°50′S) can be divided into three sedimentary regions(Case Fig. 10.1A), which in this region consist of:
1 the inner-shelf sediment prism (at depths of 0–20 m and out to 10–15 km from the coast) formed of bioturbated muddy sand;
2 the middle shelf (depths of 20–40 m), narrow at this point at only 12 km wide, which is mostly formed of bioturbated mixed quartz and calcareous gravelly muddy sands;
3 the outer-shelf reef complex, formed of a series of steep-sided patch reefs between which are shelly calcareous sandy gravels at depths of 40–80 m.

The climate is strongly seasonal, with a summer-monsoonal climate, reflected in the shelf hydrodynamic regime. In summer months (November–March), the weather is punctuated by occasional cyclones, during which shelf waters are influenced by the discharge of muddy river plumes and by wind-driven strong northward along-shelf flows (see also Case Study 9.2).

Major sediment transport events on the shelf are thus generally controlled by cyclones (Case Table 10.1), the most severe (and least frequent) of which generate most sediment transport and may rework the combined results of all intervening (and less severe) cyclones, so that the resulting sedimentary record is dominated by a few cyclone beds. The stratigraphical record of cyclones is varied in nature, patchy and spasmodic, and mostly only available for the period since *c.* 5.5 kyr BP, the mid-Holocene sea-level highstand. Beach ridges on the Cairns coastal plain have an average interval of *c.* 280 years between the emplacement of successive ridges (Jones 1985). Equivalent figures for chenier ridges to the north and south are 177–280 years (Nott & Hayne 2001; Case Fig. 10.1B).

On the Cairns inner shelf, nearshore sediment cores display one or more sharp-based, fining upward, shell hash and sand to mud beds, interpreted as representing storm deposits (Carter et al. 2002). Radiocarbon dates from shells indicate ages of 3100, 2980 and 2830 yr BP for the three beds, indicating > 100-year periods (120 yr and 150 yr) between successive major cyclones. On the inner shelf south of Cairns, modern storm beds (Cyclone Winifred, 1986,

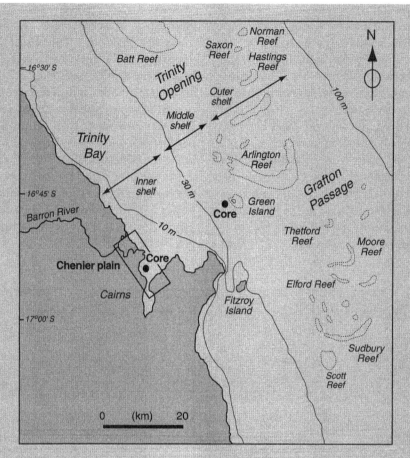

Case Fig. 10.1A Bathymetry of the shelf near Cairns, showing inner shelf, location of cheniers, core sites on inner and mid-shelf, and the outer shelf reef complex.

Case Table 10.1 The main sedimentary effects of cyclones across the main shelf environments of the GBR. (Sources: Hopley 1982; Gagan et al. 1988; Nott & Hayne 2001; Larcombe & Carter 2004.)

Shelf sedimentary environment	Main sedimentary features	Indicated average storm recurrence interval (yr)	Sediment accumulation or erosion?
Coastline or island perimeter	Beach erosion, formation of chenier ridges at muddy coastlines or beach ridges at sandy coastlines	177–280 (cheniers)	Variable
Inner shelf	Unmix sediment, transport it along shelf, creating large dunes of shelly gravel and a graded bed, receive sediment from rivers and from erosion on the middle shelf	120–150 (shellbeds)	Net accumulation
Middle shelf	As above plus formation of sand/gravel ribbons	360–7600 (shellbeds at 40 m depth)	Net erosion. Sediments exported both landward and seaward
Outer shelf	As middle shelf		Net sediment accumulation

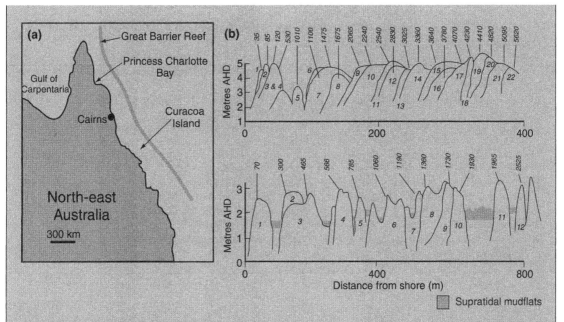

Case Fig. 10.1B Dated chenier-ridge sequences after Nott & Hayne (2001). Study sites and storm deposit data: (a) location map of study sites; (b) stratigraphical relationship of storm deposit/ridges on Curacoa Island (top) and Princess Charlotte Bay (bottom). Successive storm deposits are numbered accordingly. Mean reservoir-corrected radiocarbon age (in yr BP) for each ridge is shown above traces. Note progressive increase in age with distance inland. AHD is Australian Height Datum.

category 3) comprise a moderately well sorted, graded bed of terrigenous sand–mud, locally with a basal shell lag, and 5–20 cm in thickness (Gagan et al. 1988), which was remixed by bioturbation within 12 months. The c. 4 m thick Holocene inner-shelf sediment prism at Cairns thus represents the stratigraphical equivalent of about 40 category 3 cyclones, implying a recurrence interval of c. 140 yr.

On the Cairns middle shelf, a series of five radiocarbon dates from a core located in 40 m of water near Green Island, shows several storm beds in the past 1800 years preserved at an average interval of 360 years, and in nearby cores average intervals are up to 600 years, indicating the generally erosive nature of the middle shelf 'cyclone corridor'. Thus, stratigraphical information from shelf cores and chenier ridges indicates an average return interval for major cyclones (category 3 or higher) of between 150 and 300 yr. Although recurrence intervals and magnitudes will be different, similar processes and products are likely to occur on other tropical storm-influenced shelves, and perhaps other 'rimmed' shelves such as Belize and New Caledonia (section 10.2.3.6), although there are relatively few data on such issues to date. The long recurrence intervals and high magnitudes of such disturbances are important considerations for managers of shelf environments, such as the Great Barrier Reef Marine Park Authority, who in their zoning scheme define allowable human use of different parts of the shelf (section 10.5.2.1).

Relevant reading

Carter, R.M., Larcombe, P., Liu, K., et al. (2002) *The Environmental Sedimentology of Trinity Bay, far North Queensland*. Final Report, James Cook University and Cairns Port Authority, 97 pp.

Gagan, M.K., Johnson, D.P. & Carter, R.M. (1988) The Cyclone Winifred storm bed, central Great Barrier Reef shelf, Australia. *Journal of Sedimentary Petrology* 58, 845–56.

Gagan, M.K., Chivas, A.R. & Herczeg, A.L. (1990) Shelf-wide erosion, deposition, and suspended sediment transport during Cyclone Winifred, central Great Barrier Reef, Australia. *Journal of Sedimentary Petrology* 60, 456–70.

Forristall, G.Z., Hamilton, R.C. & Cardone, V.J. (1977) Continental shelf currents in Tropical Storm Delia: observations and theory. *Journal of Physical Oceanography* 7, 532–46.

Hopley, D. (1982) *The Geomorphology of the Great Barrier Reef: Quaternary Development of Coral Reefs.* Wiley-Interscience, New York.

Hubbard, D.K. (1992) Hurricane-induced sediment transport in open-shelf tropical systems – an example from St. Croix, U.S. Virgin Islands. *Journal of Sedimentary Petrology* 62, 949–60.

Jones, M.R. (1985) *Quaternary Geology and Coastline Evolution of Trinity Bay, North Queensland.* Publication 386, Geological Survey of Queensland, Brisbane, 27 pp.

Larcombe, P. & Carter, R.M. (2004) Cyclone pumping, sediment partitioning and the development of the Great Barrier Reef shelf system: a review. *Quaternary Science Reviews* 23, 107–35.

Nott, J. & Hayne, M. (2001) High frequency of 'super-cyclones' along the Great Barrier Reef over the past 5000 years. *Nature* 413, 508–12.

10.2.3.3 Wave-influenced shelves

The Southern Ocean generates numerous powerful storms, and shelves to its north are subject to the impact of swell waves. On the south-facing Lacepede shelf, South Australia, sedimentary facies are generally arranged in zones parallel to the coast and mud is deposited only below depths of 140 m (James et al. 1992). The sedimentary regime of the South Otago shelf (Fig. 10.6) located on the south-east of South Island, New Zealand, can be characterized as being dominated by the action of swell waves plus storms (Carter et al. 1985; Carter & Carter 1986). Here, swell waves typically mobilize the sandy sea-bed down to depths of 30 m, but a combination of the regional Southland Current, locally strong tides, plus storms, is required to mobilize fine sand to depths of c. 75 m. The grain size of the modern terrigenous inner-shelf sand wedge, which commenced deposition in the past 6500 yr, fines seawards and northwards, with higher mud and biogenic content at its seawards edge. A complex mineralogy and suite of heavy minerals, combined with textural trends and its overall morphology clearly indicate net northward sediment transport. On the inner shelf, the northern front of the sand wedge has advanced northwards during the Holocene highstand at an average rate of 30 m yr^{-1}, and the sand is formed into large dunes in places. To seawards, at depths beneath c. 60 m, lies a gravel and sand facies related to early Holocene lower sea-level, the southern sandy portion of which is being slowly reworked by modern storm-driven processes into sand ribbons, aligned along the shelf.

10.2.3.4 River-influenced shelves

Some continental shelves have their oceanographic and sedimentary regimes heavily influenced by river input, in the form of large volumes of freshwater and/or very high rates of sediment input (e.g. the Yellow Sea–East China Sea, Amazon). With such large systems, riverine, estuarine and shelf processes merge across the shelf and, with such muddy systems, fluid mud processes close to the sea bed become important (Kineke et al. 1996) along with factors such as sediment oxygen concentration, organic content and bioturbation (Johnson & Baldwin 1996).

The Amazon shelf receives c. 1.2×10^9 t yr^{-1} of land-derived sediment from the Amazon River (Milliman & Meade 1983; Dunne et al. 1988; Allen 1997), which is distributed more than 100 km across and 400 km northwards along the shelf, and down to depths of 70 m. Sediment

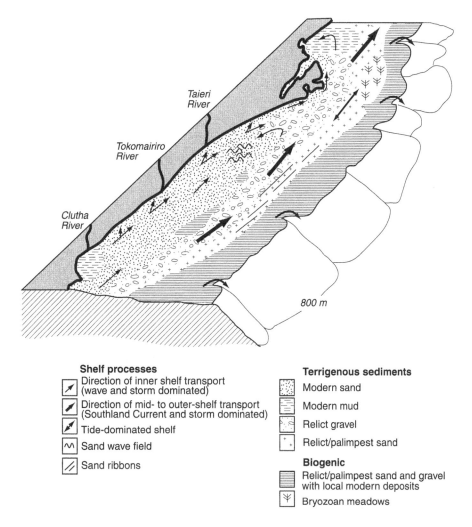

Fig. 10.6 The sedimentary facies and transport regimes of the South Otago shelf. (Adapted from Carter et al. 1985.)

Shelf processes
- Direction of inner shelf transport (wave and storm dominated)
- Direction of mid- to outer-shelf transport (Southland Current and storm dominated)
- Tide-dominated shelf
- Sand wave field
- Sand ribbons

Terrigenous sediments
- Modern sand
- Modern mud
- Relict gravel
- Relict/palimpest sand

Biogenic
- Relict/palimpest sand and gravel with local modern deposits
- Bryozoan meadows

dispersal on this highly dynamic shelf is influenced by the inertia of the river flow, the along-shelf Guiana Current, which flows at 35–75 cm s^{-1} to the north-west, waves driven by trade-winds and by locally strong tidal currents (see also Case Study 9.1). The sediments are sands at river mouths, and further along the transport path are clayey silts, then interbedded silts and sands, and sandy silts, which form the top of the delta and occupy the zone of highest oceanographic energy. Immediately to seawards, in depths of 40–60 m, is a zone of faintly laminated mud, where the highest sediment accumulation rates are located, and the most abundant benthic biology. Over time-scales of 100 years or so, these accumulation rates may reach 10 mm yr^{-1}

(Fig. 10.7). Down-core sulphur concentrations and ^{210}Pb profiles indicate catastrophic accumulation events of up to 5 m yr^{-1}, as well as episodes of high rates of reworking and sediment removal (Aller & Blair 1996; Allison et al. 1996; Sommerfield et al. 1996; see also Kineke et al. 1996). At their seaward edge, these Amazon-derived sediments become mottled and bioturbated, and beyond is a zone of relict fine sands. Organic-rich laminae may occur widely across the shelf, derived from either seasonal river input and/or plankton blooms. A large area of the shelf sediments (31,000 km^2) is gas-enriched (Figueiredo et al. 1996), with the gas formed from breakdown of organic matter derived from marine and terrestrial sources.

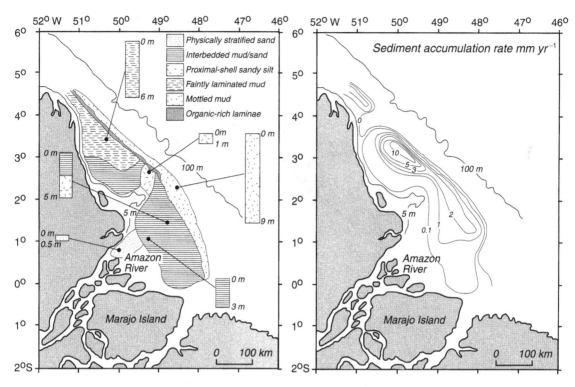

Fig. 10.7 Sedimentary facies and accumulation rates on the continental shelf off the Amazon River. (Adapted from Kuehl et al. 1986.)

10.2.3.5 Oceanic current-influenced shelves

Although sedimentation on many shelves is influenced to some degree by oceanic currents, such as on the Amazon, Newfoundland (Barrie & Collins 1989) and outer Saharan shelves (Newton et al. 1973), several cases occur where the intruding oceanic current (generally a western boundary current) is a major contributor to sediment dispersal, especially on the outer shelf, and in places on the mid-shelf (Case Study 10.2). The long NW–SE–orientated shelf of north-east South America is influenced by the Guiana Current, which flows at 35–75 cm s^{-1} to the north-west, and which supplements strong tidal currents and waves. To the extreme east is the north-east Brazilian shelf, which is relatively starved of land-derived sediments, but which contains a suite of coast-parallel zones of large-scale carbonate and siliciclastic bedforms, driven by oceanic and wind-driven flows, with superposed smaller scale bedforms generated by waves and tides (Testa & Bosence 1999).

10.2.3.6 Wind-driven-current influenced shelves

Except during storms, wind-driven currents are generally only weak, but they may be sedimentologically significant because they may carry fine sediment placed into suspension by tides or fair-weather waves. On the Great Barrier Reef shelf, during fair-weather conditions, trade-winds blow from the south-east, along the shelf. Shelf sediment transport results primarily from wave-induced resuspension, combined with north-ward-directed wind-driven currents and coastal longshore drift (Belperio 1983, 1988; Larcombe et al. 1995; Orpin et al. 1999). Suspended sediment concentrations of 10–100 mg L^{-1}, caused by resuspension of sea-bed mud, occur in a 2–10 km wide coastal belt, which drifts north at rates of up to 15 cm s^{-1} under the influence of

Case study 10.2 The south-east Australian and south-east African continental margins: shelf sedimentation and dispersal influenced by intruding oceanic currents

Off southern Queensland, Australia, shelf sedimentary facies include inner-shelf quartz sands (0–50 m depth), mid-shelf mixed carbonate–quartz sands (40–80 m depth) and, at a range of depths, but mostly on the outer shelf (depths > 40 m), pebble- to cobble-sized rhodoliths (Harris et al. 1996). Rhodoliths are a type of coated grain, with a laminated internal structure related to episodes of growth of coralline algae. With repeated movements, the laminations formed are concentric, and the threshold for movement of 50% of the rhodoliths in this area is 45–80 cm s^{-1}, depending on their size. Current-meter data indicates that the southward-flowing East Australian Current (EAC) episodically intrudes onto the shelf, moderating the tidal flows and inducing near-bed current speeds of up to 130 cm s^{-1}, even at depths of over 70 m (Case Fig. 10.2A). For depths of c. 80 m, only the largest storm waves can create oscillatory currents at the bed of over 60 cm s^{-1}, and tidal currents alone are relatively weak (< 20 cm s^{-1}). The EAC adds up to

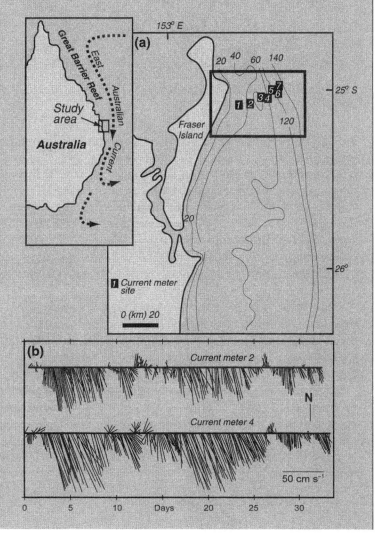

Case Fig. 10.2A (a) The outer part of the southern Queensland shelf and associated currents, and current-meter moorings located off northern Fraser Island. (b) Hourly average current speeds and direction on the outer part of the southern Queensland shelf. Note the residual currents to the south-south-east, indicating the presence of the East Australian Current. (Adapted from Harris et al. 1996.)

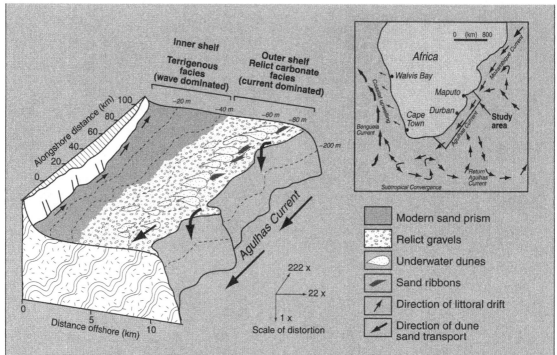

Case Fig. 10.2B Location of Agulhas shelf with regional ocean current patterns and detail of the shelf-parallel sedimentary facies. (Adapted from Flemming 1980, 1981.)

55 cm s^{-1} to the speed of near-bed currents, initiating movement of the rhodoliths and thus forming a major control upon the presence of this facies on this shelf.

A particularly well-documented example of shelf sediment transport dominated by an intruding oceanic current is the south-east African shelf (Flemming 1978, 1980, 1981, 1988) (Case Fig. 10.2B). Here, the western boundary current is the Agulhas Current, which flows to the south-south-west, parallel to the south-east African coast, at speeds up to 2.5 m s^{-1}. Such speeds easily entrain sand and gravel on the sea bed and induce the development of large sedimentary bedforms such as large and very large transverse dunes (*sensu* Ashley et al. 1990) in symmetrical, asymmetrical and 'cat-back' forms (i.e. those where the crest is reversed in orientation compared with the overall bedform).

There is also a severe swell-wave regime, with regular large waves of long period. Given the microtidal regime (i.e. maximum tidal ranges < 2 m), the impact of the Agulhas Current is the dominant mechanism driving sediment transport pathways. At major offsets in the continental margin, the Agulhas Current overshoots the edge of the margin, and, at some distance downstream, the flow 'reattaches' to the margin (Darbyshire 1972). Flow divides at these reattachment zones, so that some sediments to its north are transported back towards the offset in a major eddy, and to its south are transported along the margin. As might be expected, these zones tend to be relatively coarse-grained, ranging from fine sand up to gravel lag pavements. Long-term fluctuations in the current lead to changes in the location of the bedload parting zones of 10–100 km, so that in places a complex suite of sedimentary bedforms occurs which indicates that bedload sediment transport reverses its direction. The suite of sedimentary facies and bedforms on the shelf is very similar to that observed on many tidally dominated shelves, such as the UK shelf (section 10.2.3.1).

Relevant reading

Ashley, G.M., Dalrymple, R.W., Elliott, T., et al. (1990) Classification of large-scale subaqueous bedforms: a new look at an old problem. *Journal of Sedimentary Petrology* **60**, 160–72.

Belderson, R.H., Johnson, M.A. & Kenyon, N.H. (1982) Bedforms. In: *Offshore Tidal Sands* (Ed. A.H. Stride), pp. 27–57. Chapman and Hall, London.

Darbyshire, J. (1972) The effect of bottom topography on the Agulhas Current. *Pure and Applied Geophysics* **101**, 208–20.

Flemming, B.W. (1978) Underwater sand dunes along the southeast African continental margin – observations and implications. *Marine Geology* **26**, 177–98.

Flemming, B.W. (1980) Sand transport and bedform patterns on the continental shelf between Durban and Port Elizabeth (southeast African continental margin. *Sedimentary Geology* **26**, 179–205.

Flemming, B.W. (1981) Factors controlling shelf sediment dispersal along the southeast African continental margin. *Marine Geology* **42**, 259–77.

Flemming, B.W. (1988) Pseudo-tidal sedimentation in a non-tidal shelf environment (southeast African continental margin). In: *Tide-influenced Sedimentary Environments and Facies* (Eds P.L. De Boer, A. van Gelder & S.D. Nio), pp. 167–80. Reidel, Dordrecht.

Harris, P.T., Tsuji, Y., Marshall, J.F., et al. (1996) Sand and rhodolith-gravel entrainment on the mid- to outer-shelf under a western boundary current: Fraser Island continental shelf, eastern Australia. *Marine Geology* **129**, 313–30.

the wind-driven, northward, along-shelf current. Where the coastline is relatively straight, the muddy sediment is transferred northward along the shelf, but where it is indented, fine sediments become trapped in inner-shelf embayments behind headlands or sand spits.

Trade winds also dominate sedimentation on the western New Caledonian shelf, in the south-west Pacific Ocean. Here, the shelf lagoon is 100 km in length, averages *c.* 18 m in depth, and has a funnel shape, narrowing from 40 km wide in the south-east to about 5 km in the north-west, constrained to seawards by a series of long barrier reefs. The maximum tidal range is *c.* 1.7 m, and maximum tidal currents are *c.* 0.2 m s^{-1}. The prevailing winds in the south-west lagoon are moderate to strong south-east trade winds (*c.* 18 knots) that blow for more than 200 days a year (November–May), and which drive shelf currents north-west along the lagoon (Douillet et al. 2001). Wind-driven currents dominate the oceanography when wind speeds are > 2 m s^{-1} (Douillet 1998). Although data are relatively few, resuspension and resedimentation can account for more than 80% of total sedimentation (Clavier et al. 1995), and wind-driven currents assist the shelf to remain relatively free from muddy

sediments, confining the accumulation of muddy material to embayments in the complex coast-line (Debenay 1987).

10.2.3.7 Additional shelf sedimentary processes

As noted above, sedimentary processes on muddy shelves are influenced by factors such as dissolved oxygen concentration, sediment organic matter and bioturbation. On other shelves, especially where carbonate producers are the primary sediment source, various oceanographic factors associated with the water column (nutrient supply, water clarity, water temperature) are all important factors (Reading 1996). Regarding sedimentary 'disturbance', natural hydrocarbon seeps can be important on some continental shelves, because pockmarks produced by escaping fluids can be 0.5 to 20 m deep and 1 m to 1 km in diameter (Hovland & Judd 1988). Pockmarks represent material released from the sea bed and may contain an aggregation of biogenic material in the centre (Dando 2001). Finally, in the light of the (magnitude 9.0) earthquake of 26 December 2004 beneath the sea bed off north-west Indonesia, which created a destructive tsunami that killed more than 150,000 people, it is worth noting that the Pacific Ocean has an average of one to two destructive tsunamis each year; i.e. they are geologically common. Although many continental shelves are clearly influenced by tsunami, there are few existing data from modern shelves on their flows or their deposits.

10.2.3.8 Inferring sedimentary processes from sedimentary facies

Where direct measurements of shelf sedimentary processes do not exist, the nature of long-term shelf sedimentation can be inferred from the characteristics of the deposits, using aspects including the spatial distribution of sedimentary facies and bedforms, gradients of sediment texture and composition, and the shallow stratigraphy. It is of course very helpful to date the sediments and hence infer the timing of the main sedimentary events. Long-term patterns of sediment dispersal sometimes can be demonstrated from regional variations in sedimentation. On the central inner-shelf of the Great Barrier Reef, Holocene sediment accumulation rates decrease with distance from the main source, the Burdekin River, which discharges sediment onto the shelf at an average rate of $3–9 \times 10^6$ t yr^{-1} (Neil et al. 2002). Holocene sediment accumulation rates decrease along the regional shelf-parallel sediment transport pathway, from $c.$ 0.7 to < 0.1 mm yr^{-1} (Woolfe & Larcombe 1998) as a function of the regional trapping ability of a set of north-facing coastal embayments. On the larger scale of the entire central and northern GBR shelf, the texture and composition of terrigenous sediments along the 10 m depth contour, although complex, also indicate general net northward shelf sediment dispersal (Lambeck & Woolfe 2000).

Distance along a transport path sometimes can be reflected in terms of two sedimentary gradients:
1 minerals of low resistance to abrasion and breakage (e.g. carbonate grains) give way to more resistant minerals (e.g. heavy minerals);
2 grains become less angular and more rounded. Concepts of compositional and textural maturity thus result (Pettijohn et al. 1987), with immature sediments tending to be close to source. Care is

needed applying such concepts to modern shelf sedimentary processes, because shelf sediments can reflect, to varying degrees, sedimentary processes that are no longer active, and there can be a variety of grain sources.

Increasingly, routine use of laser-diffraction and other high-resolution techniques indicates that many shelf sediments have polymodal grain-size distributions. Size modes, once identified, can reveal dispersal patterns. For example, on the Californian shelf, the distribution of the dominant size modes shows that Columbia River sediments are transported north-north-west across the shelf, and decrease in size from fine sand to coarse silt (Fig. 10.8). Very fine sands are held close to the coast and accumulate at rates below $c.$ 1.4 mm yr^{-1}, whereas coarse silts dominate along a well-defined axis towards the north-north-west, representing the main transport path of suspended sediment. Accumulation rates decrease along the main transport path from about 7 mm yr^{-1} to 3 mm yr^{-1}. The sediment transport rate of volcanic ash across the shelf has been calculated,

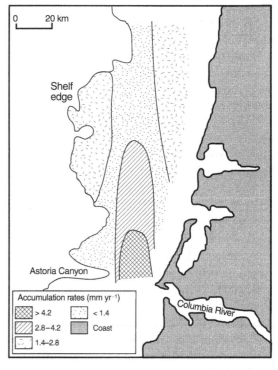

Fig. 10.8 The Oregon–Washington shelf: accumulation rates of sediments. (Adapted from Nittrouer et al. 1979.)

using the time of arrival of material derived from the Mount St Helen's eruption of May 1980 (Chapter 2), which had been carried to the shelf by the river. Only 17 months after the eruption, the ash had been transported to the north-north-west, at least 125 km along the shelf and 40 km across it (Ridge & Carson 1987).

Polymodal shelf sediments aid the delineation of shelf sedimentary facies and infer characteristics of the sediment environments. Woolfe et al. (2000) analysed the silt and sand fractions of 300 samples from the central GBR shelf (Fig. 10.9) to show that the muddy sediments supplied by the Herbert River (Group 2) either are trapped in sheltered environments close to the coast or are deposited below the 10 m depth contour on open coasts. Sediment accumulation rates measured using ^{210}Pb and ^{137}Cs profiles from core samples are $0.7–12.3$ kg m^2 yr^{-1} (mean c. 4 kg m^2 yr^{-1}) (Brunskill et al. 2002). These muds are separated from the coastline by an erosional (or at least highly mobile) sandy subtidal zone (Group 1), and to seawards, much of the shelf is occupied by a muddy, medium to coarse-grained terrigenous or calcareous sand (Group 3) of very low accumulation rate (generally < 0.1 kg m^2 yr^{-1}).

Clearly, on many shelves, characterizing sediments by mean grain size is a simplification. Various computer programs are used by industry to predict sediment transport on continental shelves, some of which use a single grain size. Although such a simple approach can sometimes fit the purpose, predictive models of shelf sediment transport are increasingly being demanded in assessments of human impacts, whether regarding marine engineering projects or regional studies of pollution dispersal and accumulation, so that more complex models need designing and testing. Porter-Smith et al. (2004) have combined information on shelf sediment texture with hydrodynamic models to delineate regions of the Australian continental shelf, based on sediment transport processes.

10.2.3.9 Measuring sedimentary processes

Increasingly, studies on shelves use sea-bed mapping techniques (high-resolution swath-bathymetry and digital side-scan sonar) to produce three-dimensional digital terrain models of the sea bed, which show sediment distribution and bedforms. Repeated surveys can show net changes in bed elevation, and are now standard industry practice in assessing sedimentary impacts of offshore activities. Grabs and corers remain essential elements of field studies (Aller et al. 2004), and nowadays include refined corers that sample the water–sediment interface undisturbed. Accumulation rates of sediments can be made in muds using various radio-tracers (e.g. Brunskill et al. 2002).

Measurements of currents and waves are now generally of very high quality, for example using acoustic devices such as Acoustic Doppler Current Profilers (ADCPs) deployed from buoys, vessels or at the sea bed. Instrument packages deployed at the sea bed can allow simultaneous measurements of near-bed sedimentary processes over periods of days to weeks (e.g. Cacchione et al. 1999). Sediment traps have limited use in shelf studies, because of strong horizontal transport and resuspension events (e.g. Topçu & Brockman 2001; Thomas & Ridd 2004). At the bed itself, time-series measurement of sediment accumulation or erosion is possible at a single point (e.g. Larcombe et al. 1995; Thomas & Ridd 2004). Direct measures of bedform migration are possible through use of video cameras, or sector-scanning sonar. Acoustic techniques also exist to assess sediment transport, especially of sand (Thorne & Hanes 2002). Optical devices now exist for measuring grain-size distributions *in situ* (Gartner et al. 2001; see also Van Walree et al. (2005) for progress on acoustic techniques), and although real-time field measurements of sediment transport rates are increasingly common, they generally remain scientific and investigative techniques rather than being applicable tools. The algorithms applied to satellite images are improving (Binding et al. 2003), but are not yet sufficiently robust to be generally applicable to sedimentary studies. Most shelf sediment transport occurs near the sea-bed, where information from satellites is weakest, especially when sediment transport rates are high.

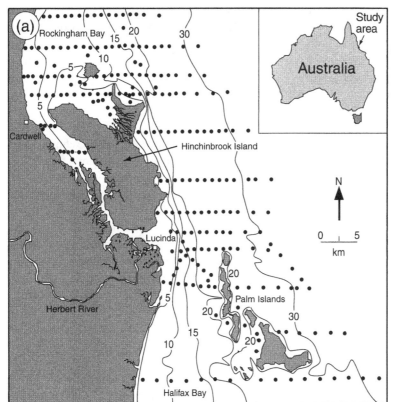

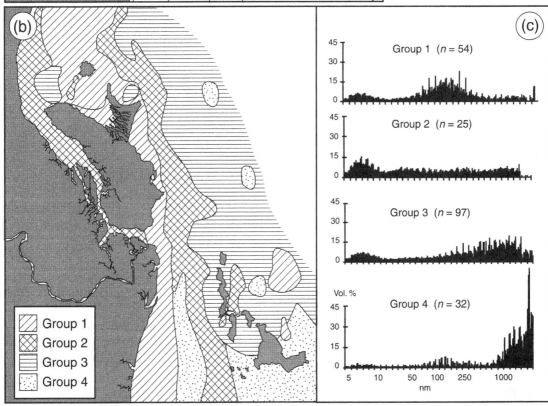

Fig. 10.9 (a) The shelf off the Herbert River delta, central Great Barrier Reef shelf, showing bathymetry (contours in metres) and sample locations (dots). (b) The four delineated sediment groups off the Herbert River delta and (c) their distribution. (Adapted from Woolfe et al. 2000.)

10.3 ANTHROPOGENIC ACTIVITIES – PROCESSES AND IMPACTS

10.3.1 Anthropogenic inputs into shelf sediments

There is a broad range of sources of material inputs into shelf waters, which includes rivers, the atmosphere, maritime transport operations (e.g. dredge material disposal), shipwrecks, offshore hydrocarbon production facilities, smelters and electricity generation plants. Most significant volumes of anthropogenic inputs into the marine environment occur within estuaries or at the coastline, so few inputs occur beyond the influence of coastal processes. There are relatively few inputs of volumetric significance and hence immediate sedimentary consequence.

10.3.1.1 Offshore disposal of dredged material

Most dredging is performed in rivers, estuaries or coastal waters for navigational purposes, for underlying reasons of defence, trade, coastal defence and/or recreation. Offshore, the main sedimentary consequences of the emplacement of dredged material onto the sea bed are that sea-bed habitats may be altered and that contaminants may be introduced offshore. The websites of the London Convention and OSPAR (Convention for the Protection of the Marine Environment of the North East Atlantic) provide background information on the issue of disposal. Information on the total amount of dredged material disposed of at sea is incomplete, but for Europe in 2003 it totalled 63×10^6 t dw (dry weight). Vivian & Murray (2002) have estimated that, world-wide, up to

about 1×10^9 t of dredged material might be disposed of at sea.

10.3.1.2 Offshore dumping of sewage sludge

The dumping of sewage sludge at sea remains in use for some countries around the world, although sewage sludge dumping off the east coast of the USA ended in 1992, and in northeast Atlantic waters, it has been prohibited since 1999. Contamination of the food web can occur by metals, organic and biosynthetic compounds, but in sedimentary terms, the organic content of emplaced material is high, breakdown of which can lower oxygen concentrations in marine sediments and, in some cases, the overlying water as well. There is increasing effort put towards reprocessing sewage sludge on land. In some countries, pellets are being produced for fertilizer, as fuel for industrial power plants and for use in charcoal production.

10.3.1.3 Offshore hydrocarbon production facilities

Hydrocarbon extraction on continental shelves can involve the discharge of contaminated sediments onto the sea floor. In the North Sea, where contamination levels are typically very low in sediments (Table 10.1), between 1 and 1.5×10^6 t of oil-based drill mud and rock cuttings are estimated to lie on the North Sea, derived from the rapid fall-out of discharged drill cuttings onto the sea-bed below oil and gas production platforms. Such discharges are now prohibited, but in parts of the southern North Sea, past discharges resulted in numerous large, discrete, cuttings piles (Daan & Mulder 1996; CEFAS

Table 10.1 Summary of contaminant levels typically found in surface sediments from the North Sea (DTI 2001).

Location	THC ($\mu g\,g^{-1}$)	PAH ($\mu g\,g^{-1}$)	PCB ($\mu g\,kg^{-1}$)	Ni ($\mu g\,g$)	Cu ($\mu g\,g^{-1}$)	Zn ($\mu g\,g^{-1}$)	Cd ($\mu g\,g^{-1}$)	Hg ($\mu g\,g^{-1}$)
Oil and gas installations	10–450	0.02–74.7	–	17.79	17.45	129.74	0.85	0.36
Estuaries	–	0.2–28	6.8–19.1	–	–	–	–	–
Coast	–	–	2	–	–	–	–	–
Offshore	17–120	0.2–2.7	< 1	9.5	3.96	20.87	0.43	0.16

THC, total hydrocarbon; PAH, polycyclic aromatic hydrocarbon; PCB, polychlorinated biphenyls.

2001; Department of Trade and Industry 2001), which are localized hotspots of contamination by hydrocarbons and a range of other compounds. Contamination tends to be limited to within a distance of about 500 m from most offshore platforms, but in some areas hydrocarbon-contaminated sediments extend to 2–8 km from the platforms (OSPAR Commission 2000). In other areas of the North Sea, dispersal has occurred, preventing the formation of such cuttings piles. Effects on benthic communities have been observed out to 3–5 km, although since cuttings discharges were prohibited, the zones of impact have decreased.

10.3.2 Anthropogenic disturbances on shelf sedimentation

10.3.2.1 Dammed rivers

The practice of building dams on rivers to secure water supplies for humans or for hydroelectric power plants has the general effect of decreasing sediment supply down rivers, particularly of bedload, with potential effects on the coastal and nearshore zones (see also Case Study 8.3). On the open shelf there is a decreasing relative effect with distance down the transport path, because shelf processes influence the river-borne sediment supply. For the South Otago shelf, New Zealand (section 10.2.3.3), over the past 9600 years, sediment input to the shelf has been around $2.1 \times 10^6 \, \mathrm{t \, yr^{-1}}$ of bedload, of which around half accumulates in the shelf-sand wedge and half is transported northwards to adjacent shelf regions (Carter 1986). Virtually all the $2.3 \times 10^6 \, \mathrm{t \, yr^{-1}}$ of suspended load received by the shelf is transported northwards along the shelf or offshore to the adjacent continental slope. On one of the three main supplying catchments, the Roxburgh Dam built in 1961 on the Clutha River traps $0.6 \times 10^6 \, \mathrm{t \, yr^{-1}}$ of bedload, which is around 30% of the total bedload supply to the entire South Otago shelf – this is likely to have affected sediment transport and/or the development of the Holocene inner-shelf sediment body. It is unknown whether the coast-shelf sediment regime has adjusted to the

1961 dam, and the variable effect of oceanographic processes is unknown, given the relatively high input of land-derived sediment. Regarding the potential effects of dams on the South Otago shelf-coast system, Carter (1986) considers that the key is whether the sediment being supplied by a river has historically (i) become entrained within the inner-shelf transport system, in which case there would be further impacts, or (ii) has tended to bypass the nearshore zone, in which case the effect would be to reduce sediment transport on the mid-shelf and reduce sand supply to the adjacent shelf and coastline to the north.

10.3.2.2 Trawling

Globally, trawling may be the most intensive of anthropogenic disturbances to the sea bed. An estimated area equivalent to all the world's continental shelves is trawled every 2 years (Watling & Norse 1998), and in the North Sea, total trawling effort is equivalent to the whole North Sea being trawled at least seven times a year. On the regional and local scale, however, trawling is very patchy. As an example, the Dutch beam-trawl fleet visits some areas of the North Sea over 400 times per year but other areas not at all (Rijnsdorp et al. 1998; Trimmer et al. 2005), and there is also significant within-year variation in overall trawling effort. Trawling may mix and resuspend surface sediments (those down to depths of a few centimetres), can disturb sediment down to a decimetre or more, can release nutrients into the water column and can influence the benthic biology (Jennings & Kaiser 1998; Kaiser & De Groot 2000), particularly surface dwellers. A changed benthic biology can alter sediment biogeochemistry because of changes in bioturbation and bio-irrigation. At some heavily trawled sites in the North Sea, biogeochemical processes in the upper layers of sediment, both oxic and anoxic, may be unaffected by trawling in the long-term (Trimmer et al. 2005), but in underlying anoxic sediments, mineralization via sulphate reduction may be stimulated by the extra disturbance, at least in areas where tidal energy is weak.

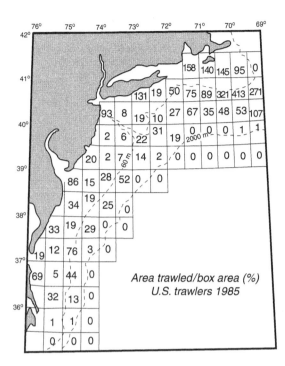

Fig. 10.10 Distribution of trawling by USA vessels over the mid-Atlantic Bight. The number in each box is the relative area of sea bed trawled (i.e. [cumulative trawled area/box area] × 100). Data are for 1985. (Adapted from Churchill 1989.)

Trawling activity is intense on the mid-Atlantic Bight. Churchill (1989) studied the sedimentary impacts of trawling there, using regional data on trawling activity, field observations and measurements of the changes in suspended sediment concentration (SSC) at 100–120 m depth. Overall, trawling effort decreases with depth and distance across the shelf (Fig. 10.10), but trawling is concentrated in some areas, such as the 'Mud Patch', south of Cape Cod, where some areas are trawled completely over three or four times per year. Seasonal variations in natural resuspension and in trawling activity mean that trawling is a significant generator of suspended sediment on the shelf, except during winter and early spring. Once resuspended, material tends to move downslope off the shelf, and Churchill (1989) calculated that 70,000 t yr^{-1} of sediment is lost downslope from the 'Mud Patch', a loss equivalent to 0.2 mm yr^{-1}. This is a significant rate compared with the estimated rate of shelf sediment accumulation of 0.2–0.3 mm yr^{-1} (calculated by

Bothner et al. 1981a,b). Consequently, in places, trawling appears to be preventing shelf sediment accumulation, as well as increasing rates of sediment transfer off the continental shelf.

Further attempts to quantify the sedimentary effects of trawling have been performed by Palanques et al. (2001) on the unfished muddy shelf of the north-west Mediterranean. The work included study of the physical characteristics of the bed sediments at 30–40 m depth, where there is a natural nepheloid layer of SSC 2–3 mg L^{-1}, which extends upwards from the bed for 3–8 m. In calm conditions, use of an Otter trawl with a 4-m-wide line rope, towed along lines 2700 m long, increased the SSC in the nepheloid layer to 5 mg L^{-1}. The impacts remain easily detectable 5 days after trawling, when the total weight of suspended sediment in the area was 350 t, nearly three times the original 120 t. Analysis of the silt content of sediments in core samples indicated that 2–3 cm of material had been eroded by the trawl, and that 10% of this disturbed sediment remained in the water column 4–5 days after trawling.

10.3.2.3 Aggregate dredging

Extraction of marine aggregates from continental shelves is largely based on the need to obtain deposits of a particular range of grain sizes (sands and gravels), primarily for building and construction purposes, but also for beach nourishment and other minor uses. The volume of marine aggregates extracted from the continental shelves around Europe was 40 × 10^6 m^3 yr^{-1} (for 1992–97), with the UK and The Netherlands by far the major extractors (Table 10.2). For the UK, an average of 25 × 10^6 t yr^{-1} of sand and gravel is extracted from the shelf, representing about 20% of all sands and gravels used in the UK. Licenced extraction areas are concentrated on the southern and eastern shelf, close to the economic activity of southern England, although it is worth noting that an area less than 12% of the total is actually used each year (Table 10.3).

Typically, aggregate dredging is performed by trailer suction hopper dredgers that operate while underway, leading to the production of

Country	Total × 10^3 (m³)	Average × 10^3 (m³ yr⁻¹)
Belgium	11,000	1833
Denmark	30,500	5083
France	13,200	2200
Germany	17,000	2833
Iceland	No information	No information
Ireland	Zero abstraction	Zero abstraction
The Netherlands	104,200	17,366
Norway	710	118
Portugal	No information	No information
Spain	No information	No information
Sweden	Zero abstraction	Zero abstraction
UK	81,600	13,600

Table 10.2 Volumes of marine sand and gravel extracted from the western European shelf for the period 1992–97 (OSPAR Commission 2000).

Table 10.3 Summary of UK extraction of terrestrial and marine aggregates (EMU 2004).

	Total area licensed (sand and gravel)	Total area worked annually	Percentage area worked annually	Volume extracted annually (England and Wales)	Maximum size of site	Average size of site
Terrestrial	270 km²	150 km²	55	70 Mt yr⁻¹	9.45 km²	< 1 km²
Marine	1300 km²	150 km²	11.5	23 Mt yr⁻¹		

shallow linear furrows 1–3 m wide and 0.2–0.3 m deep (Kenny & Rees 1994, 1996). Other extraction methods result in saucer-shaped depressions up to 200 m in diameter and 8–10 m deep, with slopes of *c.* 5° (Wenban-Smith 2002; EMU 2004). Together with the direct physical impacts on the sea bed, typically 5% of dredged material is returned to the sea bed during the dredging process, but this may increase to 60% when screening is used to increase the proportion of gravel in the cargo. There are thus also impacts from the settling of sediments through the water column back to the sea bed.

The impacts of aggregate dredging are dependent upon the nature of the substrate being dredged. In physical terms, impacts include:

1 The formation of dredge tracks or depressions on the sea bed (Fig. 10.11).
2 Disruption of the surficial sediments, leading to release of sands and finer sediments from the matrix and from underlying deposits.
3 Generation of sediment plumes at the sea-bed by the action of the drag head, and plumes at the sea surface which then settle to the bed.

4 Changed grain-size characteristics of the bed sediments (McCauley et al. 1977), which might be an increase in gravel content through the exposure of coarser sediments (Kenny et al. 1998), an increase in the proportion of fine sands (Desprez 2000; van Dalfsen et al. 2000), loss of silt- and clay-sized grains (Boyd et al. 2003, 2004) and organic material (EMU 2004).
5 Temporary changes in the sediment transport regime at and near the bed. Typically, this includes the introduction of a local 'slug' of sandy sediment onto the bed, derived from settling of sediments rejected by the screening process on board the dredge and from sands released from the sea bed. Initially, these sands are generally deposited within a few hundred metres of the dredge site. Silts, clays and organic matter are distributed more widely, and all are then liable to redistribution by waves and currents.

It has long been recognized that the sedimentary environment (composition, texture, structure and stability) is a major control upon the benthic biological assemblage (Holme 1961, 1966; Holme & Wilson 1985). Studies of impacts

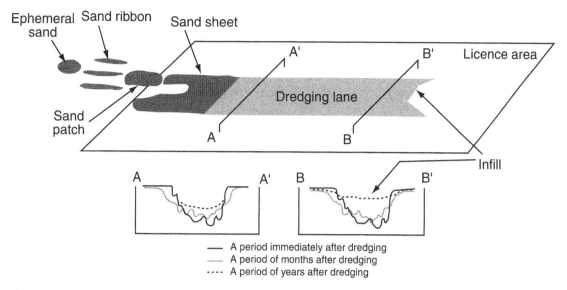

Fig. 10.11 Cartoon of main sedimentary consequences of dredging. Hypothetical distribution of major sedimentary bedforms around an individual dredging lane. Net bed sediment transport is towards the left. A–A′ and B–B′ are sections across the dredging lane.

of marine aggregate dredging on the sea bed have focused on the benthic biological communities, although the primary sedimentary factors are increasingly being recognized. In the UK, various 'biotopes' are defined to help assess the nature of impacts, 'recovery' and mediation, these being (i) shallow-water mobile sands, (ii) shallow-water stable gravels with transient sands and (iii) stable gravels (EMU 2004). Biological 'recovery' requires the sediments to revert to pre-dredge or otherwise stable conditions, and tends to be more rapid in mobile sediments such as sands in shallow waters (few months to 4 years). In contrast, in stable areas (usually gravelly and deep), 'recovery' of the fauna is much slower (up to > 15 yr) because of the presence of long-lived species (EMU 2004). There is great variation in the nature and rates of faunal change 'recovery', however, and meaningful assessments of recovery really can be made only on a site-specific basis.

10.3.2.4 Marine mining

Marine mining differs from aggregate extraction because mining exploits marine sediments specifically for their mineral content rather than their texture. Such deposits fall into the category of 'placer' deposits, which contain economical

quantities of valuable minerals. Those minerals containing gold, diamonds, platinum and tin are the most important. Such minerals tend to be highly resistant to water and abrasion, and of high density (> 2.9). Physical concentration of such minerals usually takes place as a result of transport, and on continental shelves these minerals tend to be associated with deposits of high-energy environments that operated for extended periods.

Diamonds occur in sediments of the western shelf of southern Africa. Exploration licences have been granted and active prospecting is taking place in shelf areas off Namibia and South Africa, at depths down to 200–500 m. Off southern Namibia, diamond-bearing deposits are associated with Pleistocene gravels that developed at the LGM coastline, near the modern 130 m depth contour (Rogers & Li 2002). These beach gravels are now buried beneath a thin Holocene sequence of sediments (< 30 cm thick) associated with the Holocene progradation of the Orange River delta. Overall, the thin Holocene sediment package thus has a characteristic fining upward sequence representing the post-glacial landward migration of sedimentary facies, overlain by a coarsening upward progradational marine deltaic sequence, although bioturbation

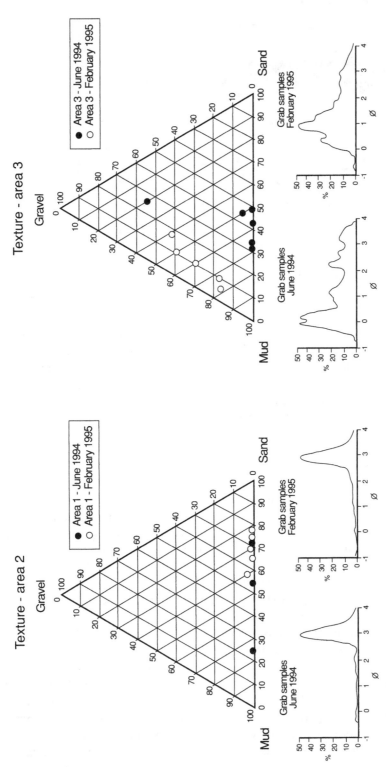

Fig. 10.12 Sediment texture and representative grain-size distributions of the sand fractions of sediments from the middle shelf off Namibia: (left) unmined sediments (unimodal very fine sands); (right) mined sediments (polymodal sands). (Adapted from Rogers & Li 2002.)

has mixed the thin sequence. Although sea-bed sediments deeper than 40 m are generally too deep for common wave disturbance, and thus contain silts and clays, sediments deeper than 100 m are able to be mobilized, either by oscillatory currents driven from long-period swell waves, or by wind-driven currents. Over large areas of the shelf, storms resuspend fine sediments and may create transient nepheloid layers near the bed. On the middle shelf, the natural very fine sands are thus very well sorted and unimodal in nature.

On board the vessel, the mining process involves sorting of sediments into cobbles, pebbles and tailings (sand, silt and clay), which fractions are then discarded over the side via a conveyor belt or a tailings pipe, so that mining activity tends to increase the patchiness of the sea bed. Off southern Namibia, in mined areas, the well-sorted sands are replaced by poorly sorted and heterogeneous sediments, because of the addition of gravels and coarser sands derived from the mined buried lowstand sediments (Fig. 10.12). In areas of the inner shelf, in depths of < 40 m, storms have a relatively great impact in transporting gravel, compared with mining. Although immediate impacts tend to be local, the area mined is expanding rapidly, and 'conservation corridors' are being proposed between mining lanes to preserve the Holocene sequence in some areas, and to provide refuges for benthic organisms to improve recolonization of the substrate after the cessation of mining. Such zones are also proposed between aggregate dredging lanes (Boyd et al. 2004).

10.3.2.5 Offshore windfarms

For many countries, a major driver for the development of Offshore Renewable Energy Developments (OREDs) is the attempt to reduce CO_2 emissions in response to the Kyoto Protocol and to diversify energy supply. The main active OREDs are offshore windfarms, which comprise an array of large individual wind turbines, each comprising an impellor (typically of three or four blades) located on top of a tower connected to a monopile attached to the sea bed.

Between these units runs a network of cables, usually buried beneath sediment. The UK is a leader in the use of this technology, and late in 2005, electricity was already being supplied to the UK National Grid from three offshore windfarms. The first sites each occupy $c.$ 10 km^2 of sea bed and have up to 30 turbines per site, but this is due to increase greatly in the near future to as many as 200–300 turbines per site.

There are potential sedimentary effects during the construction, operation and decommissioning phases of offshore windfarms (Rees, in Judd et al. 2003). In terms of physical sedimentary effects, applications for windfarm developments are considered on a site-by-site basis, requiring a review of the local and regional coastal processes and assessment of whether the structures might change sediment transport patterns, rates and pathways. Monitoring programmes are undertaken to test the predictions made by the Environmental Impact Statements. Such work builds upon basic knowledge of sedimentary impacts of flows around structures (e.g. Allen 1984), and includes accurate swath-bathymetric surveys (Fig. 10.13), repeated at intervals to determine sediment transport associated with bedform migration, and overall changes in bed elevation. Of major interest are the processes of scour around the monopiles. Observation and modelling appear to indicate that, even in areas of highly dynamic sediment transport, the volume of material removed by scour around individual monopiles is insignificant compared with that associated with bedform migration across the whole windfarm site. Further, spacing between monopiles of 300–400 m appears to be sufficient to produce no combined significant sedimentary impacts (Fig. 10.13). This accords with the 'rule of thumb' of little impact upon net flow at 6–10 obstacle diameters downstream. Hence, although there is a range of research aimed at improving understanding of the environmental impacts of windfarms, including a wide range of ecological effects (Gill 2005), there are unlikely to be major sedimentary consequences. The Netherlands' Government has chosen to wait for results from test sites before committing to significant development of offshore windfarms.

Fig. 10.13 Swath-bathymetric image of a nearshore sand bank in the North Sea. The dark oval depressions mark the location of individual monopiles, which are 4.2 m in diameter and 350 m apart, around which have developed scours up to 5 m deep and 50 m in diameter. There are no scours or raised areas are continuous between adjacent monopiles and there is continuity of the crests of large subaqueous dunes across the sandbank and near the monopiles. Thus in this case, the primary physical evidence indicates that the monopiles appear to act individually on the sea-bed rather than in combination. (Image courtesy of EoN (Coventry), Andrews (Great Yarmouth) and CEFAS.)

10.4 SHELF SEDIMENTARY SYSTEMS AND MANAGEMENT

10.4.1 National and international measures

There is a wide range of international and industry measures aimed at protecting the marine environment. Some of the international agreements, conventions and legislation apply to offshore activities, some of which have relevance to shelf sedimentary systems. Especially notable is OSPAR, the Convention for the Protection of the Marine Environment of the North East Atlantic (www.ospar.org). OSPAR requires the application of the precautionary principle, the polluter-pays principle, best-available techniques and best environmental practice, including the use of clean technology. Three annexes to the Convention relate to the prevention and elimination of pollution from land-based sources, by dumping or incineration and from offshore sources, and two others relate to assessing the quality of the marine environment, and protection and conservation of ecosystems and biological diversity.

Other conventions are also relevant in a sedimentary context, such as the United Nations Convention on Biodiversity. As part of implementing this convention, countries are required, for example, to prepare Biodiversity Action Plans, which involves some mapping of their continental shelves (see e.g. http://www.jncc.gov.uk).

10.4.2 Types of impacts on shelf habitats

Shelf sedimentary systems form the physical structure for shelf habitats. This structure can be impacted by:

1 Physical loss, for example by removal or activities that may influence the sedimentary regime and hence change the sea bed.

2 Physical damage, for example by dredging, bottom trawling and extraction. Some habitats may be more resilient and recover faster than others, but extensive physical damage may lead to loss of the original habitat.

3 Non-toxic contamination, such as by enrichment by nutrients or organic matter. Physical

disturbance by fishing may cause changes in nutrient cycling in some marine ecosystems (Duplisea et al. 2001), which may be important in avoiding eutrophication or anoxia in or near the sea-bed.

4 Biological disturbance, e.g. introduction of non-native species, although of the 53 non-native species recorded in British waters, most are causing concern in estuaries and coastal environments, rather than on the open shelf.

5 Climate change, whereby long-term changes in environmental conditions (temperature, salinity, etc.) may result in changes in the distribution and nature of some marine species assemblages. Although not necessarily negative, a changed sea-bed flora or fauna may lead to changes in sea-bed stability or biogeochemical cycling. Where biogenic sediments are significant contributors to present sediment budgets, or the degree and nature of bioturbation are changed, there is the potential for some change to sedimentary conditions at the sea bed.

In terms of inputs of sedimentary material to the shelf system, issues fall into three categories:

1 volume – i.e. the presence of material that may cause a problem to navigation or which may tend to cause bathymetric changes deemed undesirable;

2 composition – especially contamination, where the sediment chemistry or biology may harm the receiving environment (including sediment organic content);

3 texture – which may be sufficiently different from the naturally occurring material to be deemed undesirable.

10.4.3 Environmental remediation

A working definition of remediation is '*the action taken at a site subjected to anthropogenic disturbance to restore or enhance its ecological value*', of which the main categories are:

1 Physical removal of material (dredging), which, as noted elsewhere, is mostly an issue in coastal and estuarine environments.

2 Physical isolation of the material (containment), which may include capping of material *in situ* (Palermo et al. 1998). Capping is the physical covering of disposed material with other (cleaner or coarser) material in order to limit its release and its exchange with the wider marine environment. There is an ongoing debate about the long-term effectiveness of capping, particularly in deep water and in high-energy environments. Capping can be viewed as changing the problem rather than solving it, but can fulfil short- and medium-term needs.

3 Chemical treatment of the sediment.

Amongst others, remediation covers the following management options.

1 Non-intervention, whereby natural processes are allowed to proceed without human intervention (as opposed to, for example, the exclusion of activities such as trawling from occurring nearby).

2 'Restoration', whereby the ecosystem is returned to the condition or state that would exist had no dredging occurred. In practice, this is rarely achievable. Athough it might be considered relatively easy to define a general goal such as 'restoration', the complexity and dynamic nature of marine systems mean that there are genuine difficulties in establishing specific objectives that will be scientifically credible and measurable.

3 Rehabilitation, where some of the original ecological features are replaced by different ones.

4 Habitat enhancement or creation, where the original ecosystem is replaced by another, either at the site of impact or elsewhere.

The need for remediation may depend upon Government policy, or be assessed on a case-by-case basis. Clearly, the sedimentary dynamics of the receiving environment must be well understood, so allowing an assessment to be made of the technology used for material emplacement and the likely and acceptable levels of natural disturbance and dispersal. In terms of sediment dynamics, receiving environments tend to fall between two end members:

1 low-energy environments, quiescent and with relative long-term stability;

2 high-energy environments, dynamic and potentially dispersive.

It is not always clear whether natural retention or dispersal is the best environmental outcome.

10.4.4 'Beneficial use' of dredged material

Dredged material can be a valuable resource, but at present, in the UK, only *c*. 1% of material dredged for disposal at sea is currently reused in the marine environment. This may be higher in other countries but data are not easily available. In many countries, research is being undertaken to identify ways of increasing the percentage of dredged material used in beneficial use schemes. Use of such material on the shelf is minimal, and includes shoreline protection by depositing material on the inner shelf (Small et al. 1997). Most other beneficial use schemes are associated with the coastline, including habitat restoration, the maintenance of coastal sediment cells and the (re-)creation of beach or saltmarsh habitats (e.g. the DECODE project, http://www.cefas.co.uk/decode/use.htm). The joint website of the U.S. EPA and U.S. Army Corps of Engineers lists details of a number of valuable case studies (http://el.erdc.usace.army.mil/dots/budm/).

10.4.5 Managing impacts

Around the world there are generally four main impacts to manage. Regarding material inputs to continental shelves, the disposal of dredge spoil is the main issue, and regarding sea-bed disturbance, the main issues relate to trawling, aggregate dredging and marine mining. Dredge spoil and aggregate dredging are considered below.

10.4.5.1 Dredge spoil disposal

In the USA, the U.S. Army Corps of Engineers (USACE) and the U.S. Environmental Protection Agency (EPA) have statutory responsibilities for the management of dredged material placement in ocean, inland and nearshore waters. When considering open-water placement of dredged material, the potential for water column and benthic effects related to sediment contamination is to be evaluated, and management options considered that aim to reduce the release of contaminants to the water column

during placement and/or subsequent isolation of the material from benthic organisms. Such options include operational modifications, use of subaqueous discharge points, diffusers, subaqueous lateral confinement of material, or capping of contaminated material (Francingues et al. 1985; USACE/EPA 1992). Currently, the broad research programme Dredging Operations and Environmental Research (DOER 2005) is underway, conducting research designed to provide dredging project managers with technology to improve cost-effective operation, evaluation of risks associated with management alternatives and environmental compliance.

Canada has a range of measures in place to protect its marine environment, under the Canadian Environmental Protection Act, 1999 (CEPA), and to meet international commitments. Environment Canada conducts representative monitoring at sites of disposal at sea, and in 2003, monitoring activities were conducted at 17 disposal sites (London Convention 2005; www.ec.gc.ca/seadisposal/) involving assessments of the physical, chemical and biological characteristics. Permits for disposal include the generation of 'impact hypotheses', which form the basis of subsequent monitoring. Physical monitoring involves the collection of relevant geological information for determining the area of deposition, delineating the disposal-site boundaries, studying the accumulation of dredged material within the area of deposition and documenting evidence of sediment transport from the disposal site. Associated biological and chemical assessments are undertaken, the monitoring design for which takes into account the site's size and dispersal characteristics. Major sites of disposal ($> 100,000$ m^3 yr^{-1} of dredged material) are monitored at least every 5 years, and monitoring of other sites is based on volume, proximity to sensitive areas, or level of public concern. In common with many monitoring programmes, Canada's programme depends on funds gained from the collection of fees from permittees. In the UK, annual Aquatic Environment Monitoring Reports (AEMR) include sections on marine monitoring activities (available from www.cefas.co.uk).

10.4.5.2 Aggregate dredging sites

Risk assessment and management are based broadly on maximizing the potential for post-cessation change back to pre-existing conditions. The basic techniques include:

1 minimizing the area of impact and dredging intensities, which determines the starting point for 'recovery';

2 maximizing time-gaps between successive dredging events (c. 6 years can produce near re-establishment, but it depends upon the sediment-ary dynamics), gaps between individual dredging furrows and the depth of individual furrows;

3 using a site-specific assessment of impact and recovery, based on 'types' of gravel biotic assemblages;

4 leaving a thickness of at least 0.5 m of com-parable coarse substrates (Boyd et al. 2004).

For UK waters, a valuable management tool is the Electronic Monitoring System (EMS), which automatically records the geographical location and duration of dredging and dis-posal activity on board each dredging vessel (Fig. 10.14). Such data form a valuable set that can be used to inform risk assessments of planned, active and ceased dredging activities. For one bank off southern England, EMS data showed that dredging intensities in some areas reached up to 5–15 hours per year per 100 m^2 box (Boyd et al. 2004). Note that fishing trawlers over a certain size on the European UK continental shelf are also subject to EMS.

A range of remediation activities related to aggregate dredging are undertaken in UK waters, including minimizing the total area licensed/permitted for dredging, and carefully locating any new dredging areas with regard to the findings of an Environmental Impact Assessment. Further, dredging practices are adopted that minimize impact and operators are required to monitor the environmental impacts of their activities. Finally, dredging operations are controlled through the use of conditions attached to the dredging licence. As a result, basic strategies for remediation for aggregate dredging are developing rapidly (e.g. EMU 2004). Guiding principles include drawing upon the best scientific advice possible to estab-lish (i) the need for remediation, (ii) the general goals and specific objectives of remediation and,

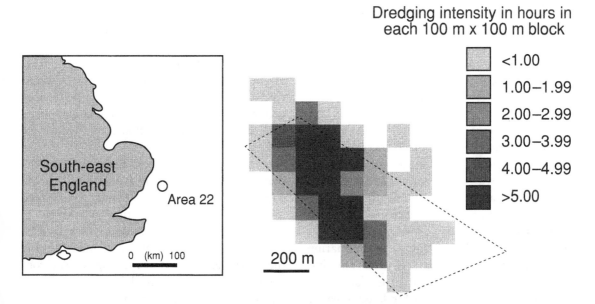

Dredging intensity in hours in each 100 m x 100 m block

	<1.00
	1.00–1.99
	2.00–2.99
	3.00–3.99
	4.00–4.99
	>5.00

Fig. 10.14 Dredging intensity in hours per 10,000 m^2 block for a site off Felixstowe (Area 422), south-east England for the year 1995. Dotted line indicates boundaries of disposal ground. (Adapted from Boyd et al. 2003.)

importantly, (iii) the criteria by which performance will be measured.

10.4.6 Contaminants and sediment quality guidelines

As part of an assessment of environmental impact, many countries have developed Sediment Quality Guidelines, which are used to inform decisions taken about sediment management (Chapter 1; Wenning et al. 2005). A number of legislative drivers require consideration of the potential impact of contaminated sediment on aquatic environments, including, in Europe, the Habitats Directive and the Water Framework Directive. The guidelines are used to assess the quality of material that may be dredged (mostly from harbours and coastal waters) and the quality of sediments (e.g. dredge spoil) placed into the shelf environment. In assessing dredged material and its suitability for disposal to sea, defined concentrations of contaminants ('Action Levels') are not generally used as a simple pass/fail test, but rather are used as part of a 'weight of evidence' approach (CEFAS 2003). This reflects recent weight-of-evidence approaches to environmental management of sediments, whereby multiple lines of evidence concerning ecological assessment are used as an aid to decision making (Chapman 1986, 1996; Burton 2002). One relatively well-developed approach to setting sediment quality criteria is that of Long et al. (1995) and Long & MacDonald (1998) in the USA. A statistical approach matches biological and chemical data from laboratory and field studies in North America, and from modelling work, and the technique has since been developed to set sediment quality guidelines in a number of countries, notably the USA, Canada and Hong Kong.

In the UK, applications for the dredging and disposal of contaminated material must consider alternative options, including placing the material in landfill, 'capping' it in the marine environment and forming a new reclamation site beneath which the material is placed (i.e. an engineered new site). Much related research has been performed by the U.S. Army Corp of Engineers (DOER 2005). On occasion, high costs can lead to a further option, of doing nothing. In forming action levels, the philosophical approach taken is to use sediment chemistry to provide an estimate of background metal concentrations for Action Level 1 (Table 10.4). The derivation of Action Level 2 concentrations has been guided by ecotoxicological information from the literature, applied to concentrations in samples sent in support of applications. Limited ecotoxicological data are available, and most such work has been done in North America (Burton 2002). Work is ongoing to derive background 'reference concentrations' for England and Wales, so the concentrations given here are likely to be revised. In Europe, under OSPAR, there is no consistent approach taken; for example, countries analyse different size fractions, some countries may digest their samples only partially and others digest samples fully using hydrogen fluoride. Although this largely relates to coastal and estuarine sites, it is indicative of some of the problems involved, which could be applicable to shelf environments.

There are also concentrations that are used to assess sediment recovered in post-disposal monitoring programmes, including those for metals (e.g. Rowlatt et al. 2002) and for organic compounds. Environment Canada has a series of lower 'Action Levels' (London Convention 2005) for sediments collected during monitoring programmes, which form part of an integrated assessment procedure (Table 10.5). If sediments are below the lower action levels for contaminants and pass all biological tests, no further action is required. If levels of contaminants or biological test results demonstrate a cause for concern, however, then:

1 compliance is verified with the terms of the permits issued since the site was last monitored;
2 potential sources of pollutants are checked and further site characterization undertaken.
Cursory benthic community surveys can be used as a general sediment quality indicator, but the overall assessment of the disposal site considers all information available from physical, chemical and biological monitoring.

Within Europe, international co-operation on such issues is not finalized, but the indication is that monitoring of disposal sites should aim

Table 10.4 Action levels (mg kg^{-1} dry weight) for metals in sediments in the USA, Canada, Hong Kong, ICES Working Group on Marine Sediments (ICES WGMS) and for England and Wales. (ICES, International Council for Exploration of the Sea; ERL, Effects Range Low; TEL, Threshold Effects Level; ERM, Effect Range Median; PEL, Probable Effect Level; ISQV, Interim Sediment Quality Value). (From CEFAS 2003.)

Metal	Ecotoxicological standards			Chemistry (mainly)	
	NOAA guidelines	Canada	Hong Kong	ICES WGMS range of values	England and Wales existing guidelines (approximate value)
	ERL	*TEL*	*ISQV-low*		*Action Level 1*
As	8.2	7.24	8.2	20–80	20
Cd	1.2	0.676	1.5	0.5–2.5	0.4
Cr	81	52.3	80	60–300	40
Cu	34	18.7	65	20–150	40
Hg	0.15	0.13	0.28	0.1–1	0.3
Ni	20.9	15.9	40	37–130	20
Pb	46.7	30.2	75	30–120	50
Zn	150	124	200	160–700	130
	ERM	*PEL*	*ISQV-high*		*Action Level 2*
As	70	41.6	70	50–1000	50–100
Cd	9.6	4.21	9.6	2.4–12.5	5
Cr	370	160	370	180–5000	400
Cu	270	108	270	90–1500	400
Hg	0.71	0.7	1	0.8–5	3
Ni	51.6	42.8	–	45–1500	200
Pb	218	112	218	100–1500	500
Zn	410	271	410	500–10000	800

to determine temporal trends rather than test for compliance, and that the aim is for no deterioration in sediment quality. There are other approaches being assessed in some countries, such as the 'Added Risk' approach, where background concentrations are determined and a toxicological quotient is added to derive a maximum permissible concentration. For those contaminated sediments deemed unsuitable for conventional disposal, they may be confined, contained, treated or simply not dredged.

10.4.7 Design of research and monitoring programmes

There is a very wide range of advice documents available (mostly on the web) regarding various human activities on the shelf that have sedimentary impacts, and there are a suite of general themes that should be considered in designing research and monitoring for shelf sedimentary systems. It is clearly important to monitor at frequencies and with accuracies that will reveal

information on the relevant shelf sedimentary processes. Given the complexities of monitoring sediment transport processes, this can be expensive. Increasingly, real-time or near real-time data are broadly available on the main oceanographic parameters, through outputs from regional monitoring programmes (e.g. in the USA, European waters and the Great Barrier Reef shelf). It is also sensible to obtain data across spatial scales sufficiently broad to assess the driving forces

Table 10.5 Canadian Environmental Protection Agency (CEPA) Lower Action Levels (from Disposal at Sea Regulations) for selected chemicals in sediments (in mg kg^{-1} dry weight) (London Convention 2005).

Chemical	Current level
Cadmium	0.6
Mercury	0.75
Total PCBs	0.1
Total PAHs	2.5

PCB, polychlorinated biphenyls; PAH, polycyclic aromatic hydrocarbon.

of the relevant sedimentary processes. Maps of the sea-bed facies are essential background information and also provide the basis for the classification of sea-bed habitats. Clearly, it is desirable that programmes should be available for scientific scrutiny, so that the outcomes are transparent and scientifically defensible. It is increasingly important that information and data are rapidly disseminated. Finally, the importance of performance criteria cannot be overemphasized – there may be little point in undertaking work unless progress towards the objectives can be measured and the effectiveness of management actions can be assessed.

There are good recent examples of shelf mapping projects. Extensive mapping work is being undertaken in Australia (www.oceans.gov.au/auscan), which was designed to assist the ecosystem-based approach to environmental management (section 10.4.8) by improving understanding of the links between the physical and biological components. Work in the Irish Sea is developing a physical basis for the conservation of marine habitats (The Joint Nature Conservation Committee 2004). The Irish National Sea-bed Survey is also underway; using techniques including multibeam and single-beam echo sounders, sub-bottom profiling, gravity, magnetics, with ground-truthing performed using box core samples (http://www.gsisea-bed.ie). A pan European project (Mapping European Sea-bed Habitats (MESH) aims to produce sea-bed habitat maps for much of the northwest European shelf, and will improve the input of such data into environmental management within national regulatory frameworks (http://www.searchmesh.net). Complementary regional studies of the composition of shelf sediments (e.g. Stevenson 2001) are also useful, especially when dealing with issues of disposal at sea.

10.4.8 Taking an ecosystem-based approach

Marine systems are naturally dynamic and display natural variability, and the human pressures on them vary with different patterns of human activity. Protection of marine environments needs to be flexible and adaptable. Although

by no means universal, there is increasingly an 'ecosystem' approach taken to environmental management of many ecosystems, including those on continental shelves (e.g. European Marine Strategy 2005). Essentially, such an approach means that, rather than viewing a human impact and its management in isolation, account is taken of the range of components of the system in question, the relevant processes and interactions between components, the environmental factors that influence the system and those factors that might be affected by anthropogenic intervention. For example, except for aggregates or marine minerals, the distribution of shelf sediments and their associated processes is not generally seen as having an inherent value. Rather, sediments are important as a substrate for benthic organisms and as a component of the broader marine ecosystem. Thus, assessment and monitoring of sediments are often performed with regard to their role in forming and maintaining marine habitats.

The ecosystem-based approach requires an understanding of the overall sedimentary physical system. Most continental shelves are huge physical systems, so that there is little practical difference most human activities can make to the systems as a whole, at least in terms of sediment budgets. Consequently, local and subregional scales tend to be most relevant, with biological, chemical and social factors of importance. Mapping the sea-bed is a high priority for the future.

10.5 FUTURE ISSUES

10.5.1 Challenges of managing the impacts of human activities on shelf sedimentation

Study, assessment and management of the impacts of human activities on the sediments of continental shelf environments represent a significant challenge. The reasons include the many physical processes that influence the shelf sedimentary environment (Figs 10.2 & 10.3), ranging from turbulence (scales of seconds and centimetres) through to major sea-level cycles (scales of c. 100,000 yr and c. 100 m). These

Fig. 10.15 Graphic illustrating 22 factors that can influence the decisions taken by UK Government ministers. Note that 'science and technology' forms only one factor. (Used with permission of McNeil Robertson Development.)

directly influence chemical and biological processes and operate on a variety of spatial and temporal scales, which may be hard to sample effectively. In addition, where two or more human activities coincide to influence the sediments, the combined impact may be a complex function of the nature of the individual impacts (e.g. trawling in an area of aggregate dredging). There are also practical difficulties and great expense involved in obtaining information about the shelf sea-bed and its sedimentary processes.

Further, management regimes need to be based on the best available science, so science knowledge needs to be transformed effectively into policy and management actions to prevent environmental damage. Very real constraints exist on our ability to do this, not least because of the difficulty of proving cause and effect, and of dealing with the uncertainties involved. It is also noteworthy that science knowledge is only one of a broad range of factors that can be taken into account when forming policy (Fig. 10.15)

10.5.2 Sedimentary impacts of sea-bed disturbance

The main current issues of anthropogenic sea-bed disturbance were examined in above sec-

tions, some of which are likely to be highly relevant in the future.

10.5.2.1 Trawling

Many countries' economies and populations rely heavily upon trawling on their continental shelves, for example, in 2001, the Great Barrier Reef (GBR) shelf was worth AU\$4.2 million to the Australian economy, and another AU\$0.3 million through recreational and commercial fishing (Productivity Commission 2003). On some shelves, trawling is already heavily regulated, primarily to protect fish-stocks or allow their recovery, but with the added advantage of moderating impacts upon the sea bed. Moves towards enforcing marine zonation of various types are also occurring. Perhaps the most advanced in this process is the GBR shelf, where the wish to minimize potential human impacts has led to the shelf being divided up into a series of complex biophysical zones, within which restrictions apply to various human activities such as anchoring, line fishing and trawling. Such zonation is driven by perceived ecosystem consequences rather than sedimentary ones *per se*, and there remains a need for fundamental research relating the shelf sedimentary environment to the ecology.

10.5.2.2 Aggregate dredging

World-wide, the demand for marine aggregates is largely a function of construction activity. For the UK, most marine sand and gravel is currently extracted from the east and southeast coasts of England, but new discoveries in the eastern English Channel, near the median line with France, are planned for exploitation. The UK's past and predicted requirement is around 25×10^6 t yr^{-1} until 2016, of which these new deposits could provide > 50% (DEFRA 2005). Licence conditions include commitments towards environmental protection and monitoring, so that trends in the sedimentary impacts of marine aggregate extraction will be supported by increasing volumes of relevant sedimentary data.

A Regional Environmental Assessment (REA) of the eastern English Channel has focused upon those sea areas that may be impacted by dredging. The REA includes predictions of dredging-related physical and biotic changes, especially related to recolonization of the bed, but the evidence base for such predictions requires development. The findings of field-based studies differ, ranging from minimal effects of disturbance following cessation of dredging (Robinson et al. 2005) to significant changes in community structure which persist over many years (Desprez 2000). Although some of these differences probably reflect the different sediment types and their associated different physical dynamics, composition and biogeochemical cycling, links between post-dredging ecological change and physical processes need to be strengthened in order to allow better transfer of conclusions to other areas.

10.5.2.3 Offshore renewable energy developments (OREDs)

For countries with continental shelves, the renewable energy resource can include tides, currents, waves and offshore wind. The development of offshore windfarms is now well underway in some countries (section 10.3.2.5.). In December 2003, the UK Government announced a commitment to generating 10% of the UK electricity needs from renewable sources by 2010 and 20% by 2020. Most of this is likely to be met by a major increase in the number and size of offshore windfarms. By late 2005, there were 15 projects awarded and in the early stages of planning, which will amount to a generation capacity of 7.2 GW, potentially contributing electricity to more than 4 million households. Other types of development are increasingly under consideration, including underwater tidal turbines and floating offshore windfarms (Pelc & Fujita 2002; Gill 2005). With offshore structures, the likely impacts upon the shelf sedimentary system are generally local in nature. Some offshore structures, especially those covering extensive areas such as windfarms, cause local disturbance, especially from scour around foundations and the associated impacts upon

local habitats (Rees, in Judd et al. 2003), yet may also be viewed as having positive effects if their presence reduces or halts trawling in the area (Elliott & Cutts 2004).

Extraction of wave energy is also a rapidly expanding field, and the UK is one of the richest nations in terms of potential for wave energy (Pelc & Fujita 2002). Extraction of wave energy reduces the energy reaching the coast, and whether this effect is perceived as beneficial or detrimental depends on the specific coastline. There are no significant impacts likely on shelf sediments, however, unless there is an important active link between the coastal and shelf sediments. Tidal flows are also a huge potential energy source. World-wide, tidal power plants could theoretically generate 500–1000 terraWatt hours per year (TWh yr^{-1}, terra $= 10^{12}$) and western Europe 100 TWh yr^{-1}. Theoretically the UK could generate 50 TWh yr^{-1}, 125 times the current UK use (0.4 TWh yr^{-1}, www.dti.gov.uk) although economic constraints mean that only a fraction of this energy is likely to be exploitable (Pelc & Fujita 2002). Tidal turbines are arrays of impellors located above the bed, generally in locations where tidal current speeds exceed $c.$ 2 m s^{-1}, and appear to have few anticipated sedimentary impacts.

10.5.3 Sedimentary impacts of climate change

For continental shelves, it is probably true that basic knowledge of natural sedimentary changes on time-scales of decades is mostly insufficient to allow anthropogenic changes to be confidently identified and their significance assessed. Studies of marine sediments and fossils can help address whether recently measured shelf sedimentary changes have occurred before. Although the details of cause and effect may often be imperfectly understood, better links are possible between the understanding of the natural driving forces and the tools used for management (Hardman-Mountford et al. 2005; Larcombe et al. 2005; Rogers & Greenaway 2005). With regard to shelf sediments, the main impacts of climate change include results of changes in sea-level and ocean circulation.

10.5.3.1 Impacts of sea-level change

Over the next few decades, changes in mean sea-level (of up to a few decimetres) may affect some parts of the world's continental shelves. For example, the vertical extent of many coral reefs is presently limited by sea-level, the reefs having caught up with sea-level during the late Holocene (Larcombe & Woolfe 1999). As a result, a rise in relative sea-level may increase coral cover on reef flats and thus increase rates of carbonate production. In deeper shelf areas, changed sea-level will have little impact over such time-scales, because here marine processes dominate sedimentation, and changed oceanic circulation patterns, with consequent changes in temperature and storminess, may be more important.

10.5.3.2 Impacts of ocean current change

Changes in shelf sedimentation directly related to varying ocean currents have been noted above (section 10.2.3.5 and Case Study 10.2). Ocean currents may influence shelf sedimentation in other ways, through their introduction of water of varying temperature and salinity, and the consequent potential for altered shelf sediment production. That changes in biogenic sediment production can occur as a result of oceanographic changes is established on time-scales of centuries and millennia (e.g. Scourse et al. 2002), but equivalent data are lacking on shorter time-scales in shelf environments. On some shelves, there are recognized changes in shelf ecology related to variations in inputs of oceanic waters (e.g. Beaugrand 2004), which might be expected to change primary production and the supply of organic matter to the sea bed.

10.5.3.3 Impacts of weather changes (floods, winds, storms)

Shelf sedimentation is likely to be changed if the relative magnitudes between parameters of supply, translation and removal are altered. Thus, if major changes in flood frequency occur, there is the potential for significantly changed sediment supply. Reservoir construction is probably the most important influence on sediment supply, but in many datasets it is difficult to distinguish the influence of climate change from that of other changes in catchment conditions (Walling & Fang 2003). Where shelf sediment distribution depends upon waves and/or wind-driven currents (e.g. the GBR shelf), changed seasonal wind patterns have the potential to change the distribution of shelf sediments. For storm-dominated shelves, changed frequency, magnitude and preferred storm tracks might influence the distribution of shelf sediments. Overall, such oceanographic shifts and associated ecological changes are increasingly acknowledged, but the consequences on the sedimentology are not understood.

10.5.4 Research needs and gaps

For continental shelves in general, and perhaps especially for temperate shelves, there is a lack of detailed information on past shelf 'weather and climates', i.e. the short-, medium- and long-term variation in shelf processes. Knowledge is lacking on the dynamics of coarse sediments, and the role of such sediments in shelf processes (e.g. C and N cycling). Significant limitations in understanding exist regarding the rates of various processes (see Becker et al. 2001), including sediment transport itself and related biogeochemical processes. It is still difficult to make real-time observations of the nature of suspended particulates (Sternberg & Newell 1999). Regarding contamination, research is needed on so-called 'combined effects', whereby the detrimental effects of a contaminant are altered if other chemicals or environmental conditions occur. This is a very difficult issue because of the rapid increase in the number of new chemicals.

Although geophysical survey devices (e.g. swath-mapping) are improving our knowledge of the spatial distribution of various sedimentary facies and biotopes, there remains a lack of basic understanding of the links between shelf sedimentary processes and benthic organisms, especially for sandy and gravelly sediments. Field-based work is needed to document these links. Overall, we can anticipate increasing anthropogenic influences upon some parts of shelf sedimentary systems, although remaining overall at a low level for most shelves. Case studies are required that describe the sedimentary consequences of human intervention over sufficiently long time-scales to allow the design of reliable and scientifically credible indicators of change. For some areas and/or types of human activity, environmental monitoring data of good quality will be available, including time-series data. Two fundamental and linked questions remain:

How do we *distinguish* anthropogenic impacts from natural variation, and assuming that we can do this, how do we then ascribe a *significance* to any differences? These questions are the focus of much ongoing and future research, in the scientific, management and socio-economic spheres.

ACKNOWLEDGEMENT

This work was funded by the UK Department for Environment, Food and Rural Affairs contract numbers A1225 and E3203.

References

Aalto, R., Maurice-Bourgoin, L., Dunne, T., et al. (2003) Episodic sediment accumulation on Amazonian flood plains influenced by El Niño/Southern Oscillation. *Nature* **425**, 493–7.

Acker, K.L. & Stearn, C.W. (1990) Carbonate–siliciclastic facies transition and reef growth on the northeast coast of Barbados, West Indies. *Journal of Sedimentary Petrology* **60**, 18–25.

Ackers, P. & Thompson, G. (1987) Reservoir sedimentation and the influence of flushing. In: *Sediment Transport in Gravel Bed Rivers* (Eds C.R. Thornes, J.C. Bathhurst & R.D. Hey), pp. 845–68. Wiley, Chichester.

Adam, P. (1990) *Saltmarsh Ecology*. Cambridge University Press, Cambridge, 461 pp.

Alam, I.A. (1993) The 1991 Gulf war oil spill: lessons from the past and a warning for the future. *Marine Pollution Bulletin* **27**, 357–60.

Alexander, D. (1998) *Natural Disasters*. UCL Press, London, 632 pp.

Allen, J.R.L. (1964) Studies in fluviatile sedimentation: six cyclothems from the Lower Old Red Sandstone, Anglo-Welsh Basin. *Sedimentology* **3**, 163–98.

Allen, J.R.L. (1970) A quantitative model of grain size and sedimentary structures in lateral deposits. *Geological Journal* **7**, 129–46.

Allen, J.R.L. (1970) *Physical Processes of Sedimentation*. Allen and Unwin, London, 248 pp.

Allen, J.R.L. (1984) *Sedimentary Structures, their Character and Physical Basis*. Elsevier, Amsterdam.

Allen, J.R.L. (1985) *Physical Processes of Sedimentation*. Allen and Unwin, London.

Allen, J.R.L. (1997) Simulation models of salt marsh morphodynamics: some implications for high intertidal sediment couplets related to sea level change. *Sedimentary Geology* **113**, 211–23.

Allen, J.R.L. (2000) Morphodynamics of Holocene salt marshes: a review sketch from the Atlantic and southern North Sea coasts of Europe. *Quaternary Science Reviews* **19**, 1155–231.

Allen, P. (1997) *Earth Surface Processes*. Blackwell Science, Oxford, 404 pp.

Aller, R.A. & Blair, N.E. (1996) Sulfur diagenesis and burial on the Amazon shelf: major control by physical sedimentation processes. *Geomarine Letters* **16**, 3–10.

Aller, R.C., Hannides, A., Heilbrun, C., et al. (2004) Coupling of early diagenetic processes and sedimentary dynamics in tropical shelf environments: the Gulf of Papua deltaic complex. *Continental Shelf Research* **24**, 2455–86.

Allison, M.A., Nittrouer, C.A., Faria, L.E.C. Jr., et al. (1996) Sources and sinks of sediment to the Amazon margin; the Amapa coast. *Geomarine Letter* **16**, 36–40.

Allison, M.A., Khan, S.R., Goodbred, S.L.Jr., et al. (2003) Stratigraphic evolution of the late Holocene Ganges-Brahmaputra lower delta plain. *Sedimentary Geology* **155**, 317–42.

Allott, R.W., Hewitt, C.N. & Kelly, M.R. (1990) The environmental half-lives and mean residence times of contaminants in dust for an urban environment: Barrow-in-Furness. *The Science of the Total Environment* **93**, 403–10.

Alsharhan, A.S. & Kendall, C.G.st.C. (2003) Holocene coastal carbonates and evaporites of the southern Arabian Gulf and their ancient analogues. *Earth-Science Reviews* **61**, 191–243.

Ambers, R.K.R. (2001) Using the sediment record in a western Oregon flood-control reservoir to assess the influence of storm history and logging on sediment yield. *Journal of Hydrology* **244**, 181–200.

Ambio (1976) Special issue on acid rain. *Ambio* **5**(5–6).

Ambio (1990) Special issue on marine eutrophication. *Ambio* **19**, 102–76.

Amos, K.J., Alexander, J., Horn, A., et al. (2004) Supply limited sediment transport in a high-discharge event of the tropical Burdekin River, North Queensland, Australia. *Sedimentology* **51**, 145–62.

Andrews, J.E., Pedley, H.M. & Dennis, P.F. (1994) Stable isotope record of palaeoclimatic change in British Holocene tufa. *The Holocene* **4**, 349–55.

Andrews, J.E., Brimblecombe, P., Jickles, T.D., et al. (1996) *An Introduction to Environmental Change*. Blackwell, Oxford, 209 pp.

Ankley, G.T., Di Toro, D.M., Hansen, D.J., et al. (1996) Technical basis and proposal for deriving sediment quality criteria for metals. *Environmental Toxicology and Chemistry* **15**, 2056–66.

Anthony, E.J. & Blivi, A.B. (1999) Morphosedimentary evolution of a delta-sourced, drift-aligned sand barrier-lagoon complex, western Bight of Benin. *Marine Geology* **158**, 161–76.

Anthony, E.J. & Julian, M. (1999) Source-to-sink sediment transfers, environmental engineering and hazard mitigation in the steep Var River catchment, French Riviera, southeastern France. *Geomorphology* **31**, 337–54.

Anthony, E.J. & Orford, J.D. (2002) Between wave- and tide-dominated coasts: the middle ground revisited. *Journal of Coastal Research, Special Issue* **36**, 8–15.

Antoine, P. (1993) The Somme valley terrace system (northern France); a model of river response to Quaternary climatic variations since 800,000 B.P. *Terra Nova* **6**, 453–64.

Apitz, S. & White, S. (2003) A conceptual framework for river-basin-scale sediment management. *Journal of Soils and Sediments* **3**, 132–8.

Appleby, P.G. & Oldfield, F. (1979) Letter on the history of lead pollution in Lake Michigan. *Environmental Science and Technology* **13**, 478–80.

Arizona Department of Transportation (1975) *Soil Erosion and Dust Control on Arizona Highways*, 4 Vols. Arizona Department of Transportation, Phoenix.

Arnaud-Fassetta, G. (2003) River channel changes in the Rhome Delta (France) since the end of the Little Ice Age: geomorphological adjustment to hydroclimatic change and natural resource management. *Catena* **51**, 141–72.

Ashley, R.M., Fraser, A., Burrows, R., et al. (2000) The management of sediment in combined sewers. *Urban Water* **2**, 263–75.

Aubanel, A., Marguet, N., Colombani, J.M., et al. (1999) Modifications of the shoreline in the Society Islands (French Polynesia). *Ocean and Coastal Management* **42**, 419–38.

Aulitzky, H. (1994) Hazard mapping and zoning in Austria: methods and legal implications. *Mountain Research and Development* **14**(4), 307–13.

Axtmann, E.V. & Luoma, S.N. (1991) Large-scale distribution of metal contamination in fine-grained sediments of the Clark Fork River, Montana, USA. *Applied Geochemistry* **6**, 75–88.

Bailey, R.G., Zoltan, S. & Wiken, E.B. (1985) Ecological regionalization in Canada and the United States. *Geoforum* **16**, 265–75.

Baker, V.R. (1994) Geomorphological understanding of floods. *Geomorphology* **10**, 139–56.

Baker, V.R., Kochel, R.C., Patton, P.C., et al. (1983) Paleohydrological analysis of Holocene flood slack-water sediments. In *International Conference on Fluvial Sedimentology, Glasgow, 1981*, pp. 229–39. Blackwell Scientific, Oxford.

Ballantyne, C.K. (2002) Paraglacial geomorphology. *Quaternary Science Reviews* **21**, 1935–2017.

Baltzer, F., Allison, M. & Fromard, F. (2004) Material exchange between the continental shelf and mangrove-fringed coasts with special reference to the Amazon-Guianas coast. *Marine Geology* **208**, 115–26.

Barnhardt, W.A. & Kelley, J.T. (1995) The accumulation of carbonate sediments on the inner shelf of Maine: a modern consequence of glaciation and sea-level change. *Journal of Sedimentary Research* **65**, 195–208.

Barrell, J. (1912) Criteria for recognition of ancient delta deposits. *Geological Society of America Bulletin* **23**, 377–446.

Barrie, J.V. & Collins, M.B. (1989) Sediment transport on the shelf margin of the Grand Banks of Newfoundland. *Atlantic Geology* **25**, 173–9.

Barry, R.G. (1992) *Mountain Weather and Climate*, 2nd edn. Routledge, London.

Barsch, D. & Caine, N. (1984) The nature of mountain geomorphology. *Mountain Research and Development* **4**(4), 287–98.

Bates, C.C. (1953) Rational theory of delta formation. *Bulletin of the American Association of Petroleum Geologists* **37**, 2119–62.

Bates, R.L. & Jackson, J.A. (1980) *Glossary of Geology*. American Geological Institute, 2nd edn. Falls Church, Virginia.

Battarbee, R.W. (2000) Palaeolimnological approaches to climate change, with special regard to the biological record. *Quaternary Science Reviews* **19**, 107–24.

Baudo, R., Giesy, J.P. & Muntau, H. (Eds) (1990) *Sediments: Chemistry and Toxicity of In-Place Pollutants*. Lewis Publishers, Chelsea, MI, 405 pp.

Beaugrand, G. (2004) The North Sea regime shift: evidence, causes, mechanisms and consequences. *Progress in Oceanography* **60**, 245–62.

Beaumont, P. (1989) *Drylands. Environmental Management and Development*. Routledge, London.

Becker, G.A., Brockmann, U.H., Damm, P., et al. (2001) Instruments and observational methods. In: *Berichte aus dem Zentrum für Meeres- und Klimaforschung – Reihe Z: Interdisziplinäre Zentrumsberichte. Nr. 3. The Changing North Sea: Knowledge, Speculation and New Challenges. Synthesis and New Conception of North Sea Research (SYCON)* (Eds J. Sündermann, S. Beddig, I. Kröncke, G. Radach & K.-H. Schlünzen), pp. 212–38. Zentrum für Meeres- und Klimaforschung der Universität Hamburg (available at http://www.rrz.uni-hamburg.de/SYKON/).

Beckwith, P.R., Ellis, J.B. & Revitt, D.M. (1986) Heavy metal and magnetic relationships for urban source sediments. *Physics of the Earth and Planetary Interiors* **42**, 67–75.

Beets, D.J. & van der Spek, A.J.F. (2000) The Holocene evolution of the barrier and back barrier basins of Belgium and the Netherlands as a function of the late Weichselian morphology, relative sea level rise and sediment supply. *Geologie en Mijnbouw* **79**, 3–16.

Belderson, R.H., Johnson, M.A. & Kenyon, N.H. (1982) Bedforms. In: *Offshore Tidal Sands* (Ed. A.H. Stride), pp. 27–57. Chapman and Hall, London.

Belknap, D.F., Kelley, J.T. & Gontz, A.M. (2002) Evolution of the glaciated shelf and coast of the Gulf of Maine, U.S.A. *Journal of Coastal Research* (Special Issue) **36**, 37–55.

Bellucci, L.G., Frignani, M., Paolucci, D., et al. (2002) Distribution of heavy metals in sediments of the Venice Lagoon: the role of the industrial area. *The Science of the Total Environment* **295**, 35–49.

Bellwood, D.R., Hughes, T.P., Folke, C., et al. (2004) Confronting the coral reef crisis. *Nature* **429**, 827–33.

Belperio, A.P. (1979) The combined use of wash load and the bed material load rating curves for the calculation of total load: an example from the Burdekin River, Australia. *Catena* **6**, 317–29.

Belperio, A.P. (1983) Late Quaternary terrigenous sedimentation in the Great Barrier Reef lagoon. In: *Proceedings of the Great Barrier Reef Conference* (Eds J.T. Baker, R.M. Carter, P.W. Sammario & K.P. Stark), pp. 71–6. James Cook University, Townsville.

Belperio, A.P. (1988) Terrigenous and carbonate sedimentation in the Great Barrier Reef province. In: *Carbonate–Clastic Transitions* (Eds L.J. Doyle & H.H. Roberts), pp. 143–74. Developments in Sedimentology 42, Elsevier, Amsterdam.

Benedict, J.B. (1970) Downslope soil movement in a Colorado alpine region. Rates, processes and climatic significance. *Arctic and Alpine Research* **2**(3), 165–226.

Bengtsson, Å. (1991) *Effects of pollutants on vertebrae and spinal columns in the fourhorn sculpin (Myoxocephalus quadricornis L.) in the Gulf of Bothnia.* Thesis, Umeå Univeristy, Sweden.

Benito, G., Baker, V.R. & Gregory, K.J. (1998) *Palaeohydrology and Environmental Change.* Wiley, Chichester.

Benjamin, M.M. & Leckie, J.O. (1982) Multiple-site adsorption of Cd, Cu, Zn and Pb on amorphous iron oxyhydroxide. *Journal of Colloid and Interface Science* **79**, 209–21.

Benzoni, F., Bianchi, C.N. & Morri, C. (2003) Coral communities of the northwestern Gulf of Aden (Yemen): variation in framework building related to environmental factors and biotic conditions. *Coral Reefs* **22**, 475–84.

Berendsen, H.J.A. & Stouthamer, E. (2000) Weichselian and Holocene palaeogeography of the Rhine-meuse delta, the Netherlands. *Palaeogeography, Palaeoclimatology, Palaeoclimatology* **161**, 311–35.

Berger, A.R. & Iams, W.J. (1996) *Geoindicators: Assessing Rapid Changes in Earth Systems.* Balkema Press, Rotterdam, 466 pp.

Berner, R.A. (1980) *Early Diagenesis. A Theoretical Approach.* Princeton Series on Geochemistry, Princeton University Press, New Jersey, 241 pp.

Bernier, P., Guidi, J.B. & Bottcher, M.E. (1996) Coastal progradation and very early diagenesis of ultramafic sands as a result of rubble discharge from asbestos excavations (northern Corsica, western Mediterranean). *Marine Geology* **144**, 163–75.

Beveridge, J.P. & Culbertson, J.K. (1964) Hyper-concentrations of suspended sediment. *Proceedings of the American Society of Civil Engineers, Journal of the Hydraulics Division* **90** (HY6), 117–28.

Bickerton, M.A. (1995) *Macroinvertebrates and the Determination of In-river Flow Need.* Report to the National Rivers Authority, Anglian Region, Peterborough.

Biggins, P.D.E. & Harrison, R.M. (1986) Chemical speciation of lead compounds in street dusts. *Environmental Science and Technology* **14**, 336–9.

Binding, C.E., Bowers, D.G. & Mitchelson-Jacob, E.G. (2003) An algorithm for the retrieval of suspended sediment concentrations in the Irish Sea from SeaWiFS ocean colour satellite imagery. International Journal of Remote Sensing **24**, 3791–806.

Bird, E.C.F. & Fabbri, P. (1993) Geomorphological and historical changes on the Argentina delta, Lingurian coast, Italy. *GeoJournal* **29**, 428–39.

Bird, E.C.F. (1985) *Coastline Changes: a Global Review.* Wiley, Chichester.

Bird, E.C.F. (1993) *Submerging coasts: the Effects of Rising Sea Level on Coastal Environments.* Wiley, Chichester.

Bird, E.C.F. (1996) *Beach Management.* Wiley, Chichester.

Bird, E.C.F. (1998) *The Coasts of Cornwall.* Alexander Press, Fowey.

Bird, E.C.F. (2000) *Coastal Geomorphology: an Introduction.* Wiley Chichester.

Bird, J.F. (2000) The impact of mining wastes on the rivers draining into Georges Bay, north-east Tasmania. In: *River Management: the Australian Experience* (Eds S. Brizga & B.L. Finlayson), pp. 151–72. Wiley, Chichester.

Birkeland, P.W., Shroba, R.R., Burns, S.F., et al. (2003) Integrating soils and geomorphology in mountains – an example from the Front Range of Colorado. *Geomorphology* **55**, 329–44.

Björnsson, H. (2003) Subglacial lakes and jökulhlaups in Iceland. *Global and Planetary Change* **35**(3–4), 255–71.

Blackwell, M.S.A., Hogan, D.V. & Maltby, E. (2004) The short-term impact of managed realignment on soil environmental variables and hydrology. *Estuarine, Coastal and Shelf Science* **59**(4), 687–701.

Blanchon, P., Jones, B. & Kalbfleisch, W. (1997) Anatomy of a fringing reef around Grand Cayman: storm rubble, not coral framework. *Journal of Sedimentary Research* **67**, 1–16.

Bloesch, J. & Burns, N.M. (1980) A critical review of sedimentation trap technique. *Schweizerische Zeitschrift fur Hydrologie* **42**, 15–55.

Bloesch, J. & Uehlinger, U. (1986) Horizontal sedimentation differences in a eutrophic Swiss lake. *Limnology Oceanography* **31**, 1094–109.

Boorman, L. & Hazeldon, J. (1995) Salt marsh creation and management for coastal defence. In: *Directions in European Coastal Management* (Eds M.G. Healey & J.P. Doody), pp. 175–83. Samara Publishing, Cardigan.

Bornhold, B.D. & Giresse, P. (1985) Glauconitic sediments on the continental shelf off Vancouver, British Columbia, Canada. *Journal of Sedimentary Petrology* 55, 653–64.

Bosence, D.W.J. (1980) Sedimentary facies, production rates and facies models for recent coralline algal gravels, Co. Galway, Ireland. *Geological Journal* 15, 91–111.

Bosence, D.W.J., Rowlands, R.J. & Quine, M.L. (1985) Sedimentology and budget of a Recent carbonate mound, Florida Keys. *Sedimentology* 32, 317–43.

Boss, S.K. & Liddell, D.W. (1987) Patterns of sediment composition of Jamaican fringing reef facies. *Sedimentology* 34, 77–87.

Bothner, M.H., Parmenter, C.M. & Milliman, J.D. (1981a) Temporal and spatial variations in suspended matter in continental shelf and slope waters off the north-eastern United States. *Estuarine, Coastal, and Shelf Science* 13, 213–34.

Bothner, M.H., Spiker, E.C., Johnson, P.P., et al. (1981b) Geochemical evidence for modern sediment accumulation on the continental shelf off south New England, *Journal of Sedimentary Petrology* 51, 281–92.

Bottino, G.M., Chiarle, M., Joly, A., et al. (2002) Modelling rock avalanches and their relation to permafrost degradation in glacier environments. *Permafrost and Periglacial Processes* 13, 283–8.

Boult, S., Collins, D.N., White, K.N., et al. (1994) Metal transport in stream polluted by acid mine drainage; the Afon Goch, Anglesey, UK. *Environmental Pollution* 84, 279–84.

Bowen, H.J.M. (1966) *Trace Elements in Biochemistry*. Academic Press, London, 241 pp.

Boyd, N.A., Taylor, K.G. & Boult, S. (1999) Early diagenetic controls on contaminant geochemistry and mobility in urban water bodies: Salford Quays, Manchester. *Environmental Geochemistry and Health* 22, 317–22.

Boyd, S.E., Limpenny, D.S., Rees, H.L., et al. (2003) Preliminary observations of the effects of dredging intensity on the re-colonisation of dredged sediments off the southeast coast of England (Area 222). *Estuarine, Coastal and Shelf Science* 56, 1–15.

Boyd, S.E., Cooper, K.M., Limpenny, D.S., et al. (2004) *Assessment of the Re-habilitation of the Seabed Following Marine Aggregate Dredging*. Science Series Technical Report 121, Centre for Environment, Fisheries and Aquaculture Science, Lowestoft, 154 pp. (available online at www.cefas.co.uk).

Boyden, C.R., Aston, S.R. & Thornton, I. (1979) Tidal and seasonal variations of trace elements in two Cornish estuaries. *Estuarine Coastal and Marine Science* 9, 303–17.

Bradbury, J.P. (1997) A diatom record of climate and hydrology for the past 200 ka from Owens Lake, California with comparison to other great basin records. *Quaternary Science Reviews* 16, 203–19.

Bradley, S.B. & Cox, J.J. (1990) The significance of the floodplain to the cycling of metals in the river Derwent catchment, UK. *Science of the Total Environment* 97/98, 441–54.

Bradshaw, C. (1997) Bioturbation of reefal sediments by crustaceans in Phuket, Thailand. *Proceedings of Eighth International Coral Reef Symposium*, Vol. 2, pp. 1801–6. Smithsonian Tropical Research Institute, Bilbao, Republic of Panama.

Breed, C.S., McCauley, J.F., Whitney, M.I., et al. (1997) Wind erosion in drylands. In: *Arid Zone Geomorphology: Process, Form and Change in Drylands*, 2nd edn (Ed. D.S.G. Thomas), pp. 437–64. Wiley, Chichester.

Bridge, J.S. (2003) *Rivers and Floodplains. Forms, Processes, and Sedimentary Record*. Blackwell Science, Oxford.

Bridgland, D.R. (2000) River terrace systems in north-west Europe: an archive of environmental change, uplift, and early human occupation. *Quaternary Science Reviews* 19, 1293–303.

Bridgland, D.R. & Allen, P. (1996) A revised model for terrace formation and its significance for the early Middle Pleistocene terrace aggradations of north-east Essex, England. In: *The Early Middle Pleistocene in Europe* (Ed. C. Turner), pp. 121–34. Balkema, Rotterdam.

Bridgland, D.R. & Maddy, D. (2002) Global correlation of long Quaternary fluvial sequences: a review of baseline knowledge and possible methods and criteria for establishing a database. *Netherlands Journal of Geosciences* 81, 265–81.

Brinkhurst, R.O. (1974) *The Benthos of Lakes*. Macmillan, London, 189 pp.

Brinkmann, R. & Tobin, G.A. (2003) *Urban Sediment Removal: the Science, Policy, and Management of Street Sweeping*. Kluwer Academic Press, Dordrecht, 168 pp.

Bristow, C.S. (1996) Reconstructing channel morphology from sedimentary sequences. In: *Advances in Fluvial Dynamics and Stratigraphy* (Eds P.A. Carling & M.R. Dawson), pp. 351–71. Wiley, Chichester.

Brizga, S.O. & Finlayson, B.L. (1994) Interactions between upland catchment and lowland rivers: an applied Australian case study. *Geomorphology* 9(3), 189–201.

Bromberg, S. (1990) Identifying ecological indicators: an environmental monitoring and assessment program. *Journal of Air and Waste Management Association* 40, 976–87.

Bromhead, J.C. & Beckwith, P. (1994) Environmental dredging on the Birmingham canals: water quality and sediment treatment. *Journal of the Institution of Water and Environmental Management* 8, 350–9.

Bromley, R.G. (1978) Bioerosion of Bermuda reefs. *Palaeogeography, Palaeoclimatology, Palaeoecology* **23**, 169–97.

Brookes, A. (1992) Recovery and restoration of some engineered British river channels. In: *River Conservation and Management* (Eds P. Boon, P. Calow & G.E. Petts), pp. 337–53. Wiley, Chichester.

Brown, A.C. & McLachlan, A. (2002) Sandy shore ecosystems and the threats facing them: some predictions for the year 2025. *Environmental Conservation* **29**, 62–77.

Brown, A.G. (1997) *Alluvial Geoarchaeology. Floodplain Archaeology and Environmental Change.* Cambridge Manuals in Archaeology, Cambridge University Press, Cambridge.

Brown, A.G. (1998) Fluvial evidence of the Medieval warm period and the late Medieval climatic deterioration in Europe. In: *Palaeohydrology and Environmental Change* (Eds G. Benito, V.R. Baker & K.J. Gregory), pp. 43–52. Wiley, Chichester.

Brunet, R.C. & Astin, K.B. (2000) A 12-month sediment and nutrient budget in a floodplain reach of the River Adour, southwest France. *Regulated Rivers – Research and Management* **16**, 267–77.

Brunsden, D. & Moore, R. (1999) Engineering geomorphology on the coast: lessons from west Dorset. *Geomorphology* **31**, 391–409.

Brunskill, G.J., Zagorskis, I. & Pfitzner, J. (2002) Carbon burial rates in sediments and a carbon mass balance for the Herbert River region of the Great Barrier Reef continental shelf, north Queensland, Australia. *Estuarine, Coastal and Shelf Science* **54**, 677–700.

Bruun, P. (1962) Sea level rise as a cause of shore erosion. *Journal of Waterways and Harbours Division, American Society of Civil Engineers* **88**, 117–30.

Bryan, G.W. & Langstone, W.J. (1992) Bioavailability, accumulation and effects of heavy metals in sediments with special reference to United Kingdom estuaries – a review. *Environmental Pollution* **76**(2), 89–131.

Bryant, E.A. (1988) The effect of storms on Stanwell Park, N.S.W. beach position, 1943–1980. *Marine Geology* **79**, 171–87.

Bryce, S., Larcombe, P. & Ridd, P.V. (1998) The relative importance of landward-directed tidal sediment transport versus freshwater flood events in the Normanby River estuary, Cape York Peninsula, Australia. *Marine Geology* **149**, 55–78.

Buddemeier, R.W. (1997) Making light work of adaptation. *Nature* **388**, 229–30.

Buddemeier, R.W., Maragos, J.E. & Knutson, D.W. (1974) Radiographic studies of reef coral skeletons: rates and patterns of coral growth. *Journal of Experimental Marine Biology and Ecology* **14**, 179–200.

Burd, F. (1995) *Managed Retreat: a Practical Guide.* English Nature, Peterborough.

Burger, R.L., Fulthorpe, C.S. & Austin, J.A. (2001) Late Pleistocene channel incisions in the southern Eel River Basin, northern California: implications for tectonic vs. eustatic influences on shelf sedimentation patterns. *Marine Geology* **177**, 317–30.

Burkham, D.E. (1972) *Channel Changes of the Gila River in Safford Valley, Arizona.* Professional Paper 655-J, US Geological Survey, Denver, CO.

Burns, K.A., Garrity, S.D. & Levings, S.C. (1993) How many years until mangrove ecosystems recover from catastrophic oil spills? *Marine Pollution Bulletin* **26**, 239–48.

Burton, G.A. (2002) Sediment quality criteria in use around the world. *Limnology* **3**, 65–75.

Burton, Jr., G.A. (Ed.) (1992) *Sediment Toxicity Assessment.* Lewis, Boca Raton, 457 pp.

Bush, D., Pilkey, O.H. & Neal, W.J. (1996) *Living by the Rules of the Sea.* Duke University Press, Durham, NC.

Butcher, S.S., Charlson, R.J., Orians, G.H. & Wolfe, G.V. (Eds) (1992) *Global Biogeochemical Cycles.* Academic Press, London, 379 pp.

Butler, A.J., Depers, A.M., McKillup, S.C., et al. (1977) Distribution and sediments of mangrove forests in south Australia. *Transactions of Royal Society of Southern Australia* **101**, 35–44.

Butler, D.R., Walsh, S.J. & Malanson, G.P. (2003) Introduction to the special issue: mountain geomorphology – integrating earth systems. *Geomorphology* **55**, 1–4.

Butterfield, G.R. (1991) Grain transport rates in steady and unsteady turbulent airflows. *Acta Mechanica* **Supplement 1**, 97–122.

Cacchione, D.A., Wiberg, P.L., Lynch, J., et al. (1999) Estimates of suspended-sediment flux and bedform activity on the inner portion of the Eel continental shelf. *Marine Geology* **154**, 83–97.

Caine, N. (1974) The geomorphic processes of the alpine environment. In: *Arctic and Alpine Environments* (Eds J.D. Ives & R.G. Barry), pp. 721–48. Methuen, London.

Caine, N. (1984) Elevational contrasts in conmtemporary geomorphic activity in the Colorado Front Range. *Studia Geomorphologica Carpatho-Balcanica* **18**, 5–31.

Cairns, Jr., J. & Pratt, J.R. (1987) Ecotoxicological effect indices: a rapidly evolving system. *Water Science and Technology* **19**, 1–12.

Cairns, Jr., J. (1981) *Testing for Effects of Chemicals in Ecosystems.* National Academy Press, Washington, DC, 103 pp.

Calhoun, R.S., Fletcher, C.H. & Harney, J.N. (2002) A budget of marine and terrigenous sediments, Hanalei Bay, Kauai, Hawaiian Islands. *Sedimentary Geology* **150**, 61–87.

Cameron, W.M. & Pritchard, D.W. (1963) Estuaries. In: *The Sea* (Ed. M.N. Hill), pp. 306–24. Wiley, New York.

Carannante, G., Esteban, M., Milliman, J.D., et al. (1988) Carbonate lithofacies as paleolatitude indicators: problems and limitations. *Sedimentary Geology* 60, 333–46.

Carr, A.P. (1969) Size grading along a pebble beach; Chesil Beach, England. *Journal of Sedimentary Petrology* 39, 297–311.

Carter, J., Owens, P.N., Walling, D.E., et al. (2003) Fingerprinting suspended sediment sources in a large urban river system. *The Science of the Total Environment* 314–16, 513–34.

Carter, L. (1986) A budget for modern-Holocene sediment on the South Otago continental shelf. *New Zealand Journal of Marine and Freshwater Research* 20, 665–76.

Carter, L. & Carter, R.M. (1986) Holocene evolution of the nearshore sand wedge, South Otago continental shelf, New Zealand. *New Zealand Journal of Geology and Geophysics* 29, 413–24.

Carter, R.M., Carter, L., Williams, J.J., et al. (1985) Modern and relict sedimentation on the South Otago continental shelf, New Zealand. *New Zealand Oceanographic Institute Memoir* 93, 43 pp.

Carter, R.W.G. (1988) *Coastal Environments*. Academic Press, London.

Carter, R.W.G. (1988) *Coastal Environments*. Cambridge University Press, Cambridge.

Carter, R.W.G. (1991) Near-future sea level impacts on coastal dune landscapes. *Landscape Ecology* 6, 29–39.

Carter, R.W.G. & Orford, J.D. (1993) The morphodynamics of coarse clastic beaches and barriers: a short- and long term perspective. *Journal of Coastal Research* 15, 158–79.

Carter, R.W.G. & Wilson, P. (1993) Aeolian processes and deposits in northwest Ireland. In: *The Dynamics and Environmental Context of Aeolian Sedimentary Systems* (Ed. K. Pye), pp. 173–90. Special Publication No. 72, Geological Society Publishing House, Bath.

Carter, R.W.G. & Woodroffe, C.D. (1994) *Coastal Evolution*. Cambridge University Press, Cambridge.

Carter, R.W.G., Orford, J.D., Forbes, D.L., et al. (1987) Gravel barriers headlands and lagoons: an evolutionary model. In: *Coastal Sediments '87, Proceedings of a Specialty Conference on Advances in Understanding of Coastal Sediment Processes* (Ed. N.C. Kraus), pp. 1776–92. American Society of Civil Engineers, Reston, VA.

Carter, R.W.G., Eastwood, D.A. & Bradshaw, P. (1991) Small-scale sediment removal from beaches in Northern Ireland: environmental impact, community perception and conservation management. *Aquatic Conservation: Marine and Freshwater Ecosystems* 2, 95–113.

CEFAS (2001) *Contaminant Status of The North Sea*. Technical Report TR_004, produced for Strategic Environmental Assessment – SEA2. UK Department of Trade and Industry, Strategic Environmental Assessment – SEA2, August 2001, 101 pp.

CEFAS (2003) *The Use of Action Levels in the Assessment of Dredged Material Placement at Sea and in Estuarine Areas under the Food and Environmental Protection Act (II)*. Report for the Department for Environment, Food and Rural Affairs, Project Code AE0258 (available from www.defra.gov.uk)

Celliers, L. & Schleyer, M.H. (2002) Coral bleaching on high-latitude marginal reefs at Sodwana Bay, South Africa. *Marine Pollution Bulletin* 44, 1380–7.

Chadwick-Furman, N.E. (1996) Reef coral diversity and global change. *Global Change Biology* 2, 559–68.

Chapman, P.M. (1986) Sediment quality criteria from a sediment quality triad: an example. *Environmental Toxicology and Chemistry* 5, 957–64.

Chapman, P.M. (1996) Presentation and interpretation of Sediment Quality Triad data. *Ecotoxicology* 5, 327–39.

Chappell, J. & Grindrod, J. (1984) Chenier plain formation in northern Australia. In: *Coastal Geomorphology in Australia* (Ed. B.G. Thom), pp. 197–231. Academic Press, Sydney.

Chappell, J., Omura, A., Esat, T., et al. (1996) Reconciliation of late Quaternary sea levels derived from coral terraces at Huon Peninsula with deep sea oxygen isotope records. *Earth and Planetary Science Letters* 141, 227–36.

Charlesworth, S.M. & Foster, I.D.L. (1999) Sediment budgets and metal fluxes in two contrasting urban lakes in Coventry, UK. *Applied Geography* 19, 199–210.

Charlesworth, S.M. & Lees, J.A. (1999) The distribution of heavy metals in deposited urban dusts and sediments, Coventry, England. *Environmental Geochemistry and Health* 21, 97–115.

Charlesworth, S.M. & Lees, J.A. (2001) The application of some mineral magnetic measurements and heavy metal analysis for characterising fine sediments in an urban catchment, Coventry, UK. *Journal of Applied Geophysics* 48, 113–25.

Charlesworth, S.M., Everett, M., McCarthy, R., et al. (2003a) A comparative study of heavy metal distribution in deposited street dusts in a large and small urban area: Birmingham and Coventry, West Midlands, UK. *Environment International* 29, 563–73.

Charlesworth, S.M., Harker, E. & Rickard, S. (2003b) Review of sustainable drainage systems (SuDS): a soft option for hard drainage questions? *Geography* 88, 99–107.

Chatres, C.J., Greene, R.S., Ford, G.W., et al. (1985) The effects of gypsum on macroporosity and crusting of two duplex soils. *Australian Journal of Soil Research* 23, 467–79.

Chen, S., Billings, S.A. & Grant, P.M. (1990) Non-linear system identification using neural networks. *International Journal of Control* 51, 1191–214.

Chen, Y.X., Zhu, G.W., Tian, G.M., et al. (2002) Phytoxicity of dredged sediment from urban canal as land application. *Environmental Pollution* 117, 233–41.

Chesnokov, A.V., Govorun, A.P., Linnik, V.G., et al. (2000) ^{137}Cs contamination of the Techa flood plain near the village of Muslumovo. *Journal of Environmental Radioactivity* 50, 179–91.

Chmura, G.L., Coffrey, A. & Crago, R. (2001) Variation in surface sediment deposition on salt marshes in the Bay of Fundy. *Journal of Coastal Research* 17(1), 221–7.

Choy, S.C. & Booth, W.E. (1994) Prolonged inundation and ecological changes in an *Avicennia* mangrove: implications for conservation and management. *Hydrobiologia* 285, 237–47.

Church, M. & Ryder, J.M. (1972) Paraglacial sedimentation: a consideration of fluvial processes conditioned by glaciation. *Geological Society of America Bulletin* 83, 3059–71.

Church, M. & Slaymaker, O. (1989) Disequilibrium of Holocene sediment yield in glaciated British Columbia. *Nature* 337, 452–54.

Churchill, J.H. (1989) The effect of commercial trawling on sediment resuspension and transport over the middle Atlantic Bight continental shelf. *Continental Shelf Research* 9, 841–64.

Clark, C.F., Smith, P.G., Nielson, G., et al. (2000) Chemical characterisation and legal classification of sludges from roads sweepings. *Journal of the Chartered Institution of Water and Environmental Management* 14, 99–102.

Clark, M.W. (1998) Management implications of metal transfer pathways from refuse tip to mangrove sediments. *The Science of the Total Environment* 222, 17–34.

Clark, R.B. (1992) *Marine Pollution*, 3rd edn. Clarendon Press, Oxford.

Clarke, D.J. & Eliot, I.G. (1988) Low frequency changes of sediment volume on the beachface at Warilla beach, New South Wales, 1975–1985. *Marine Geology* 79, 189–211.

Clavier, J., Chardy, P. & Chevillon, C. (1995) Sedimentation of particulate matter in the south-west lagoon of New Caledonia. *Estuarine, Shelf and Coastal Science* 40, 281–94.

Cleary, W.J., Hosier, P.E. & Wells, J. (1976) Genesis and significance of marsh islands in southeastern North Carolina lagoons. *Journal of Sedimentary Petrology* 49, 703–10.

Clements, T., Merriam, R.H., Stone, R.O., et al. (1957) *A Study of Desert Conditions.* Technical Report EP-53, Headquarters Quartermaster Research and Development Command, US Army Environmental Protection Research Division, Natck, MA.

Clough, B.F., Boto, K.G. & Attiwell, P.M. (1983) Mangroves and sewage: a re-evaluation. In: *Tasks for Vegetation Science*, Vol. 8 (Ed. H.J. Teas), pp. 151–61. Dr W. Junk Publishers, The Hague.

Cochran, J.K., Moran, S.B., Fisher, N.S., et al. (2000) Sources and transport of anthropogenic radionuclides in the Ob River system, Siberia. *Earth and Planetary Science Letters* 179, 125–37.

Cohen, J.E., C. Small, C., Mellinger, A., et al. (1997) Estimates of coastal populations *Science* 278, 1211–12.

Cole, K.L. & Mayer, L. (1982) Use of packrat middens to determine rates of cliff retreat in the Eastern Grand Canyon, Arizona. *Geology* 10, 597–9.

Coleman, J.M. (1988) Dynamic changes and processes in the Mississippi river delta. *Geological Society of America Bulletin* 100, 999–1015.

Coleman, M.L. (1985) Geochemistry of diagenetic non-silicate minerals: kinetic considerations. *Philosophical Transactions of the Royal Society of London Series A* 315, 39–56.

Collins, A.L. & Walling, D.E. (2002) Selecting fingerprint properties for discriminating potential suspended sediment sources in river basins. *Journal of Hydrology* 261, 218–44.

Collins, A.L., Walling, D.E. & Leeds, G.J.L. (1997) Source type ascription for fluvial suspended sediment based on a quantitative composite fingerprinting technique. *Catena* 29, 1–17.

Collins, B.D. & Dunne, T. (1986) Erosion of tephra from the 1980 eruption of Mount St Helens. *Geological Society of America Bulletin* 97, 896–905.

Collins, M. & Evans, G. (1986) The influence of fluvial sediment supply on coastal erosion in west and central Africa. *Journal of Shoreline Management* 2, 5–12.

Collinson, J.D. (1986) Alluvial sediments. In: *Sedimentary Environments and Facies*, 2nd edn (Ed. H.G. Reading), pp. 20–62. Blackwell Scientific, Oxford.

Colman, S.M., King, J.W., Jones, G.A., et al. (2000) Holocene and recent sediment accumulation rates in southern Lake Michigan. *Quaternary Science Reviews* 19, 1563–80.

Coltorti, M. (1997) Human impact in the Holocene fluvial and coastal evolution of the Marche region, Italy. *Catena* 30, 311–35.

Cooke, R.U. & Doornkamp, J.C. (1990) *Geomorphology in Environmental Management: a New Introduction*, 2nd edn. Clarendon Press, Oxford.

Cooke, R.U., Brunsden, D., Doornkamp, J.C., et al. (Eds) (1982) *Urban Geomorphology in Drylands.* Oxford University Press, Oxford, 324 pp.

Cooke, R.U., Warren, A. & Goudie, A.S. (1993) *Desert Geomorphology.* University College London Press, London, 526 pp.

Cooper, D.C. & Morse, J.W. (1998) Extractability of metal sulphide minerals in acidic solutions: applications to environmental studies of trace metal contamination within anoxic sediments. *Environmental Science and Technolology* 32, 1076–8.

Cooper, J.A.G. (1991) Beachrock formation in low latitudes: implications for coastal evolutionary models. *Marine Geology* 98, 145–54.

Cooper, J.A.G. (1993) Sedimentary processes in the river-dominated Mvoti estuary, South Africa. *Geomorphology* 9, 271–300.

Cooper, J.A.G. (2001a) Geomorphology of tide-dominated and river-dominated, barred microtidal estuaries: a contrast. *Journal of Coastal Research Special Issue* **34**, 428–36.

Cooper, J.A.G. (2001b) Gemorphological variability among microtidal estuaries from the wave-dominated South African coast. *Geomorphology* **40**, 99–122.

Cooper, J.A.G. & Navas, F. (2004) Natural bathymetric change as a control on century-scale shoreline behavior. *Geology* **32**, 513–16.

Cooper, J.A.G. & Pilkey, O.H. (2004a) Alternatives to the mathematical modelling of beaches. *Journal of Coastal Research* **20**, 641–4.

Cooper, J.A.G. & Pilkey, O.H. (2004b) Longshore drift: trapped in an expected universe. *Journal of Sedimentary Research* **74**, 1–15.

Cooper, J.A.G., Wright, C.I. & Mason, T.R. (1999) Geomorphology and sedimentology of South African estuaries. In: *Estuaries of Southern Africa* (Eds B.R. Allanson & D. Baird), pp. 5–25. Cambridge University Press, Cambridge.

Cooper, N.J. & Jay, H. (2002) Predictions of large-scale coastal tendency: development and application of a qualitative behaviour-based methodology. *Journal of Coastal Research, Special Issue* **36**, 173–81.

Cordery, I., Pilgrim, D.H. & Doran, D.G. (1983) Some hydrological characteristics of arid western New South Wales. In: *Hydrology and Water Resources Symposium 1983, Institute of Australian Engineers*, National Conference Publication 83/13, pp. 287–92.

Costa, J.E. (1987) Hydraulics and basin morphometry of the largest flash floods in the conterminous United States. *Journal of Hydrology* **93**, 313–38.

Costa, J.E. (1988) Rheologic, geomorphic, and sedimentologic differentiation of water floods, hyperconcentrated flows and debris flows. In: *Flood Geomorphology* (Eds V.R. Baker, R.C. Kochel & P.C. Patton), pp. 113–22. Wiley, New York.

Coulthard, T.J. & Macklin, M.G. (2001) How sensitive are river systems to climate and land-use changes? A model-based evaluation. *Journal of Quaternary Science* **16**, 347–51.

Coulthard, T.J., Kirkby, M.J. & Macklin, M.G. (2000) Modelling geomorphic response to environmental change in an upland catchment. *Hydrological Processes* **14**, 2031–45.

Cousset, P. & Meunier, M. (1996) Recognition, classification and mechanical description of debris flows. *Earth Science Reviews* **40**, 209–27.

Cowell, P.J., Roy, P.S. & Jones, R.A. (1995) Simulation of large-scale coastal behaviour using a morphological behaviour model. *Marine Geology* **126**, 45–61.

Cronin, S.J., Neall, V.E., Lecointre, J.A., et al. (1997) Changes in Whangauhu river lahar characteristics during the 1995 eruption sequence, Ruapehu volcano, New Zealand. *Journal of Volcanology and Geothermal Research* **76**, 47–61.

Crooks, S., Schutten, J., Sheern, G.D., et al. (2002) Drainage and elevation as factors in the restoration of salt marshes in Britain. *Restoration Ecology* **10**(3), 591–602.

Cummings, K.W. (1973) Trophic relations in aquatic insects. *Annual Reviews in Entomology* **18**, 183–206.

Cundy, A.B., Croudace, I.W., Thomson, J., et al. (1997) Reliability of salt marshes as 'geochemical recorders' of pollution input: a case study from contrasting estuaries in southern England. *Environmental Science and Technology* **31**, 1093–101.

Cunha, P.P., Martins, A.A., Daveau, S., et al. (2005) Tectonic control of the Tejo river fluvial incision during the late Cenozoic, in Ródão – central Portugal (Atlantic Iberian border). *Geomorphology* **64**, 271–98.

Daan, R. & Mulder, M. (1996) On the short-term and long-term impact of drilling activities in the Dutch sector of the North Sea. – *ICES Journal of Marine Science* **53**, 1036–44.

Dando, P.R. (2001) A review of pockmarks in the UK part of the North Sea, with particular respect to their biology. Technical Report 001 – Pockmarks. Strategic Environmental Assessment – SEA2. UK Department of Trade and Industry, 22 pp.

Darmody, R.G. & Foss, J.E. (1979) Soil-landscape relations of the tidal marshlands of Maryland. *Journal of the Soil Science Society of America.* **43**, 534–41.

Davidson, N.C., d'A Laffoley, D., Doody, J.P., et al. (1991) *Nature Conservation and Estuaries in Great Britain.* Nature Conservancy Council, Peterborough.

Davies, A.G. & Villaret, C. (2002) Prediction of sand transport rates by waves and currents in the coastal zone. *Continental Shelf Research* **22**, 2725–37.

Davies, B.E. & Lewin, J. (1974) Chronosequences in alluvial soils with special reference to historical pollution in Cardiganshire, Wales. *Environmental Pollution* **6**, 49–57.

Davies, J.H. (1964) A morphogenetic approach to world shorelines. *Zeitscrift fur Geomorphologie* **8**, 127–42.

Davies, J.L. (1980) *Geographical Variation in Coastal Development.* Oliver and Boyd, Edinburgh.

Davies, T.R. (1997) Long-term management of facilities on an active alluvial fan – Waiho River fan, Westland, New Zealand. *Journal of Hydrology (New Zealand)* **36**, 127–45.

Davies, T.R. & McSaveney, M.J. (2001) Anthropogenic fanhead aggradation, Waiho River, Westland, New Zealand. In: *Gravel-bed Rivers V* (Ed. M.P. Mosley), pp. 531–53. The Caxton Press, Christchurch.

Davies, T.R., McSaveney, M.J. & Clarkson, P.J. (2003) Anthropic aggradation of the Waiho River, Westland, New Zealand: microscale modelling. *Earth Surface Processes and Landfroms* **28**, 209–18.

Davies, T.R.H. (1991) Research of fluvial processes in mountains – a change of emphasis. In: *Fluvial Hydraulics of Mountain Regions* (Eds A. Armanini & G. Di Silvio), pp. 251–66. Springer-Verlag, Berlin.

Davis, J.A. & Leckie, J.O. (1978) Effects of adsorbec complexing ligands on trace metal uptake by hydrous oxides. *American Chemical Society* **12**, 1309–15.

Davis, R.A. & Hayes, M.O. (1984) What is a wave-dominated coast? *Marine Geology* **60**, 313–29.

Davis, R.A. Jr., Welty, A.T., Borrego, J., et al. (2000) Rio Tinto estuary (Spain): 5000 years of pollution. *Environmental Geology* **39**, 1107–16.

Dawson, A.G. (1994) Geomorphological processes associated with tsunami run-up and backwash. *Geomorphology* **10**, 83–94.

De Miguel, E., Llamas, J.F., Chacon, E., et al. (1997) Origin and patterns of distribution of trace elements in street dust: unleaded petrol and urban lead. *Atmospheric Environment* **31**, 2733–40.

De Ruig, J.H.M. & Hillen, R. (1997) Developments in Dutch coastline management: conclusions from the second governmental coastal report. *Journal of Coastal Conservation* **3**, 203–10.

De Sylva, D. (1986) Increased storms and estuarine salinity and other ecological impacts of the greenhouse effect. In: *Effects of Changes in Stratospheric Ozone and Global Climate*, Vol. 4 (Ed. J.G. Titus), pp. 153–94. U.S. Environmental Protection Agency, Washington.

Dean, R.G. (1974) Compatibility of borrow material for beach fills. *Proceedings of the 14th Coastal Engineering Conference*, New York, American Society of Civil Engineers, pp. 1319–33.

Debenay, J.-P. (1987) Sedimentology in the southwestern lagoon of New Caledonia, SW Pacific. *Journal of Coastal Research* **3**, 77–91.

Debenay, J.P., Guiral, D. & Parra, M. (2002) Ecological factors acting on the microfauna in mangrove swamps. The case of foraminiferal assemblages in French Guiana. *Estuarine, Coastal and Shelf Science* **55**, 509–33.

Dedkov, A.P. & Moszherin, V.I. (1992) Erosion and sediment yield in mountain regions of the world. In: *Erosion, Flows and Environment in Mountain Regions. Proceedings of the Chengdo Symposium, 1992* (Eds D.E. Walling, T.R. Davies & B. Hasholt), pp. 29–36. IAHS Publication 209, International Association of Hydrological Sciences, Wallingford.

DEFRA (2005) *Charting Progress: an Integrated Assessment of the State of UK Seas*. Department for Environment, Food and Rural Affairs, London, 130 pp. (www.defra.gov.uk).

Degens, E.T., Kempe, S. & Richey, J.E. (1991) Summary: Biogeochemistry of major world rivers. In: *Biogeochemistry of Major World Rivers*, SCOPE 42 (Eds E.T. Degens, S. Kempe & J.E. Richey), pp. 323–47. Wiley, Chichester.

DeLaune, R.D., Smith, C.J., Patrick, W.H., et al. (1994) Effect of oil on salt marsh biota: methods for restoration. *Environmental Pollution, Series A.* **36**, 207–27.

Dennis, I.A., Macklin, M.G., Coulthard, T.J., et al. (2003) The impact of the October–November 2000 floods on contaminant metal dispersal in the River Swale catchment, North Yorkshire, UK. *Hydrological Processes* **17**, 1641–57.

Department of Trade and Industry (2001) *Strategic Environmental Assessment of the Mature Areas of the Offshore North Sea (SEA 2)*. Consultation Document. September 2001, 235 pp. (available from www.habitats-directive.org).

Descroix, L. & Gautier, E. (2002) Water erosion in the southern French alps: climatic and human mechanisms. *Catena* **50**, 53–85.

Desprez, M. (2000) Physical and biological impact of marine aggregate extraction along the French coast of the Eastern English Channel: short- and long-term post-dredging restoration. *ICES Journal of Marine Science* **57**, 1428–38.

Dietrich, W.E., Kirchner, J.W., Ikeda, H., et al. (1989) Sediment supply and the development of the coarse surface layer in gravel-bedded rivers. *Nature* **340**, 215–17.

Diskin, M.H. & Lane, L.J. (1972) A basinwide stochastic model of ephemeral stream runoff on southeastern Arizona. *International Association Scientific Hydrologists Bulletin* **17**, 61–76.

Dittmar, T., Lara, R.J. & Kattner, G. (2001) River or mangrove? Tracing major organic matter sources in tropical Brazilian coastal waters. *Marine Chemistry* **73**, 253–71.

Dodd, J., Large, D.J., Fortey, N.J., et al. (2003) Geochemistry and petrography of phosphorus in urban canal bed sediment. *Applied Geochemistry* **18**, 259–67.

DOER (2005) Dredging Operations and Environmental Research, U.S. Army Corps of Engineers, Vicksburg, Mississippi (http://el.erdc.usace.army.mil/dots/doer/).

Dominguez, J.M.L., Martin, L. & Bittencourt, A.C.S.P. (1987) Sea-level history and Quaternary evolution of river mouth-associated beach-ridge plains along the east-southeast Brazilian coast: a summary. In: *Sea-level Fluctuation and Coastal Evolution* (Eds D. Nummedal, O.H. Pilkey & J.D. Howards), pp. 115–27. Special Publication 41, Society of Economic Paleontologists and Mineralogists, Tulsa, OK.

Done, T.J. (1992) Phase shifts in coral reef communities and their ecological significance. *Hydrobiologia* **247**, 121–32.

Doornkamp, J.C. & Ibrahim, H.A.M. (1990) Salt weathering. *Progress in Physical Geography* **X**, 335–48.

Douillet, P. (1998) Tidal dynamics of the south-west lagoon of New Caledonia: observations and 2D numerical modelling. *Oceanologica Acta* **21**, 69–79.

Douillet, P., Ouillon, S. & Cordier, E.A. (2001) A numerical model for fine suspended sediment transport in the south-west lagoon of New Caledonia. *Coral Reefs* **20**, 361–72.

Dregne, H.E. (1976) *Soils of Arid Regions*. Elsevier, Amsterdam, 237 pp.

Dregne, H.E. (1983) *Desertification of Arid Lands*. Harwood Academic, Chur, Switzerland.

Driedger, C.L. (1988) *Geology in action – jökulhlaups on Mount Rainier*. Water Fact Sheet, Open File Report 88-459, US Geological Survey, Denver, CO.

Driedger, C.L. & Fountain, A.G. (1989) Glacier outburst floods at Mount Rainier, Washington State, U.S.A. *Annals of Glaciology* **13**, 51–5.

Droppo, I.G., Irvine, K.N., Murphy, T.P., et al. (1998) Fractionated metals in street dust of a mixed land-use sewershed, Hamilton, Ontario. In: *Hydrology in a Changing Environment*, Vol III (Eds H. Wheater & C. Kirby), pp. 383–94. Wiley, New York.

Dubko, N.V. (1985) *The Labile and Stable Organic Matter*. Ecological System of Naroch Lakes, Minsk, pp. 233–7. (In Russian.)

Duggan, M.J. & Williams, S. (1977) Lead-in-dust in city streets. *The Science of the Total Environment* **7**, 91–7.

Duke, N.C. & Watkinson, A.J. (2002) Chlorophyll-deficient propagules of *Avicennia marina* and apparent longer term deterioration of mangrove fitness in oil-polluted sediments. *Marine Pollution Bulletin* **44**, 1269–76.

Duke, N.C., Burns, K.A., Swannell, R.P.J., et al. (2000) Dispersant use and a bioremediation strategy as alternate means of reducing impacts of large oil spills on mangroves: the Gladstone field trial. *Marine Pollution Bulletin* **41**, 403–12.

Dunkerley, D.L. & Brown, K.J. (1997) Desert Soils. In: *Arid Zone Geomorphology: Process, Form and Change in Drylands*, 2nd edn (Ed. D.S.G. Thomas), pp. 55–68. Wiley, Chichester.

Dunne, T., Bren, D., Mertes, L.A.K., et al. (1988) Exchanges of sediment between the flood plain and channel of the Amazon River in Brazil. *Geological Society of America Bulletin* **110**, 450–67.

Duplisea, D.E., Jennings, S., Malcolm, S.J., et al. (2001) Modelling potential impacts of bottom trawl fisheries on soft sediment biogeochemistry in the North Sea. *Geochemical Transactions* **14**, 1–6.

Dye, K.R. (1998) The typology of intertidal mudflats. In: *Sedimentary Processes in the Intertidal Zone* (Eds K.S. Black, D.H. Paterson & A. Cramp), pp. 11–24. Special Publication 139, Geological Society Publishing House, Bath.

Dyer, K.R. (1986) *Coastal and Estuarine Sediment Dynamics*. Wiley Interscience. New York, 342 pp.

Dyer, K.R. (1997) *Estuaries: a Physical Introduction*, 2nd edn. Wiley, Chichester.

Dyer, K.R. (2000) *Estuaries, A Physical Introduction*, 2nd edn. Wiley.

Dyrynda, P. (1996) Barrages within estuaries – ecological lessons from the Tawe development. *Marine Update 25*. World Wildlife Fund, Godalming.

Eakin, C.M. (1996) Where have all the carbonates gone? A model comparison of calcium carbonate budgets before and after the 1982–1983 El Niño at Uva Island in the eastern Pacific. *Coral Reefs* **15**, 109–19.

Eberhardt, E., Willenberg, H., Loew, S., et al. (2001) Active rockslides in Switzerland – understanding mechanisms and processes. In: *International Conference on Landslides – Causes, Impacts and Countermeasures* (Eds M. Kü, H.H. Einstein, E. Krauter, H. Klapperich & R. Pöttler) pp. 25–34. 17–21 June, Davos, Switzerland, VGE, Essen.

Eden, D.E. & Page, M.J. (1998) Paleoclimatic implications of a storm erosion record from late Holocence lake sediments, North Island, New Zealand. *Palaeogeography, Palaeoclimatology, Palaeoecology* **139**, 37–58.

Edinger, E.N., Limmon, G.V. & Jompa, J. (2000) Normal coral growth rates on dying reefs: are coral growth rates good indictors of reef health. *Marine Pollution Bulletin* **40**, 404–25.

Edwards, A.J. & Clark, S. (1998) Coral transplantation: a useful management tool or misguided meddling? *Marine Pollution Bulletin* **37**, 474–87.

Ehrlich, P.R. & Ehrlich, A.H. (1970) *Population, Resources, Environment*, 2nd edn. Freeman, San Francisco, 383 pp.

Einstein, H.A. (1950) *The Bed-load Function for Sediment Transportation in Open-channel Flows*. Technical Bulletin 1026, U.S. Department of Agriculture and Soil Conservation Service, Washington, DC.

Eisbacher, G.H. & Clague, J.J. (1984) *Destructive Mass Movements in High Mountains: Hazard and Management*. Paper 84-16, Geological Survey of Canada, Vancouver, British Columbia.

Elliott, M. & Cutts, N.D. (2004) Marine habitats: loss and gain, mitigation and compensation. *Marine Pollution Bulletin* **49**, 671–74.

Ellison, A.M. & Farnsworth, E.J. (1996) Anthropogenic disturbance of Caribbean mangrove ecosystems: past impacts, present trends, and future predictions. *Biotropica* **28**, 549–65.

Ellison, J.C. & Stoddart, D.R. (1991) Mangrove ecosystem collapse during predicted sea-level rise: Holocene analogues and implications. *Journal of Coastal Research* **7**, 151–65.

Ellison, J.C. (1998) Impacts of sediment burial on mangroves. *Marine Pollution Bulletin* **37**, 420–6.

Emergy, K.O. & Kuhn, G.G. (1982) Sea cliffs: their processes, profiles and classification. *Geological Society of America Bulletin* **93**, 644–54.

Emmerson, R.H.C., Birkett, J.W., Scromshaw, M., et al. (2000) Solid phase partitioning of metals in managed retreat soils: field changes over the first year of tidal inundation. *The Science of the Total Environment* **254**, 75–92.

EMU (2004) *Marine Aggregate Site Restoration and Enhancement: a Strategic Feasibility and Policy Review*.

Report to the British Marine Aggregate Producers Association, The Crown Estate and English Nature, 125 pp. (available from www.bmapa.org.uk).

Environment Agency UK (2004) *Foresight Future Flooding Report*. http://www.environment-agency.gov.uk/subjects/flood/763964/763974/?version=1&lang=_e (accessed 25 June 2004).

Esslemont, G. (2000) Heavy metals in seawater, marine sediments and corals from the Townsville section, Great Barrier Reef Marine Park, Queensland. *Marine Chemistry* 71, 215–31.

Eswaran, H., Reich, P. & Beinroth, F. (2001) Global desertification tension zones. In: *Sustaining the Global Farm* (Eds D.E. Stott, R.H. Mohtar & G.C. Steinhardt), pp. 24–8. Selected Papers, 10th International Soil Convention Organization Conference, 24–29 May 1999, Purdue. National Soil Erosion Research Laboratory, Agriculture Research Service, US Department of Agriculture. Purdue University, West Lafayette, Indiana.

European Marine Strategy (2005) *Proposal for a Directive of the European Parliament and of the Council establishing a Framework for Community Action in the field of Marine Environmental Policy (Marine Strategy Directive)*. 2005/0211 (COD), European Union (available at http://europa.eu.int).

Evans, J.K., Gottgens, J.F., Gill, W.M. & Mackey, S.D. (2000) Sediment yields controlled by intrabasinal storage and sediment conveyance over the interval 1984–1994. *Journal of Soil and Water Conservation* 55, 264–70.

Evans, M. & Slaymaker, O. (2004) Spatial and temporal variability of sediment delivery from alpine lake basins, Cathedral Provincial Park, southern British Columbia. *Geomorphology* 61, 209–24.

Evenari, M., Shanan, L. & Tadmor, N. (1982) *The Negev: the Challenge of a Desert*, 2nd edn. Harvard University Press, Cambridge, MA, 437 pp.

Eyre, B. & McConchie, D. (1993) Implications of sedimentological studies for environmental pollution assessment and management: example from fluvial systems in north Queensland and western Australia. *Sedimentary Geology* 85, 235–52.

Fallon, S.J., White, J.C. & McCulloch, M.T. (2002) *Porites* corals as recorders of mining and environmental impacts: Misima Island, Papua New Guinea. *Geochimica et Cosmochimica Acta* 66, 45–62.

Farago, M.E., Kavanagh, P., Blanks, R., et al. (1998) Platinum concentrations in urban road dust and soil, and in blood and urine in the United Kingdom. *Analyst* 123, 451–4.

Farella, N., Lucotte, M., Louchouarn, P., et al. (2001) Deforestation modifying terrestrial organic transport in the Rio Tapajos, Brazilian Amazon. *Organic Geochemistry* 32, 1443–58.

Farmer, A. (1997) *Managing Environmental Pollution*. Routledge, London, 246 pp.

Farmer, J.F. & Lyon, T.D.B. (1977) Lead in Glasgow street dirt and soil. *Science of the Total Environment* 8, 89–93.

Farquarson, F.A.K., Meigh, J.R. & Sutcliffe, J.V. (1992) Regional flood frequency analysis in arid and semi-arid areas. *Journal of Hydrology* 138, 487–501.

Fergusson, J.E. & Kim, N.D. (1991) Trace elements in street and house dusts: sources and speciation. *The Science of the Total Environment* 100, 125–50.

Field, C.D. (1998) Rehabilitation of mangrove ecosystems: an overview. *Marine Pollution Bulletin* 37, 383–92.

Figueiredo, A.G. Jr., Nittrouer, C.A., Faria, L.E.C. Jr., et al. (1996) Gas-charged sediments in the Amazon submarine delta. *Geomarine Letters* 16, 31–5.

Filipek, L.H., Nordstrom, D.K. & Ficklin, W.H. (1987) Interaction of acid-mine drainage with waters and sediments of West Squaw Creek in the West Shasta mining district, California. *Environmental Science and Technology* 21, 388–96.

Firth, C.R., Smith, D.E., Hansom, J.D., et al. (1995) Holocene spit development on a regressive shoreline, Dornoch Firth, Scotland. *Marine Geology* 124, 203–14.

Foley, M.G. (1978) Scour and fill in steep, sand-bed ephemeral streams. *Geological Society of America Bulletin* 89, 559–70.

Fookes, P.G., Sweeney, H., Manby, C.N.D., et al. (1985) Geological and geotechnical engineering aspects of low cost roads in mountainous terrain. *Engineering Geology* 21, 1–152.

Ford, T.D. & Pedley, H.M. (1996) A review of tufa and travertine deposits of the world. *Earth-Science Reviews* 41, 117–75.

Forristall, G.Z., Hamilton, R.C. & Cardone, V.J. (1977) Continental shelf currents in Tropical Storm Delia: observations and theory. *Journal of Physical Oceanography* 7, 532–46.

Förstner, U. & Salomons, W. (1981) *Trace Metal Analysis in Polluted Sediments*. Publication No. 248, Delft Hydraulics Laboratory, pp. 1–13.

Förstner, U. & Müller, G. (1974) *Schwermetalle in Flüssen und Seen*. Springer-Verlag, Berlin, 225 pp.

Förstner, U. & Wittmann, G.T.W. (1981) *Metal Pollution in the Aquatic Environment*. Springer-Verlag, Berlin, 486 pp.

Foster, I.D.L. & Walling, D.E. (1994) Using reservoir deposits to reconstruct changing sediment yields and sources in the catchment of the Old mill Reservoir, South Devon, UK, over the past 50 years. *Hydrological Sciences Journal* 39, 347–68.

Fournier, F. (1960) *Climat et Érosion: La Relation Entre l'Érosion du Sol par l'Eau et les Précipitations Atmosphériques*. Presses Universitaires France, Paris.

Francingues, N.R., Palermo, M.R., Lee, C.R., et al. (1985) *Management Strategy for Disposal of Dredged Material: Contaminant Testing and Controls*. Miscellaneous

Paper D-85-1, U.S. Army Engineer Waterways Experiment Station, Vicksburg, MS.

Fraser, R.J. (1993) Removing contaminated sediments from the coastal environment: the New Bedford Harbour project example. *Coastal Management* **21**, 155–62.

French, P.W. (1990) *Coal dust: a marker pollutant in the Severn Estuary and Bristol Channel.* Unpublished PhD thesis. Postgraduate Research Institute for Sedimentology, University of Reading.

French, P.W. (1996) Long-term variability of copper, lead and zinc in salt marsh sediments of the Severn Estuary, UK. *Mangroves and Salt Marshes* **1**(1), 59–68.

French, P.W. (1997) *Coastal and Estuarine Management.* Routledge, London.

French, P.W. (2001) *Coastal Defences: Processes, Problems and Solutions.* Routledge, London.

Frey, R.W. & Basan, P.B. (1978) Coastal salt marshes. In: *Coastal Sedimentary Environments* (Ed. R.A. Davis), pp. 101–69. Springer-Verlag, New York.

Friedman, G.M. & Sanders, J.E. (1978) *Principles of Sedimentology.* Wiley, New York, 792 pp.

Froelich, P.N., Klinkhammer, G.P., Bender, M.L., et al. (1979) Early oxidation of organic matter in pelagic sediments of the eastern equatorial Atlantic: suboxic diagenesis. *Geochimica et Cosmochimica Acta* **43**, 1075–90.

Froidefrond, J.M., Pujos, M. & Andre, X. (1988) Migration of mud banks and changing coastline in French Guiana. *Marine Geology* **84**, 19–30.

Frostick, L.E. & Reid, I. (Eds) (1987) *Desert Sediments: Ancient and Modern.* Geological Society of London Special Publication 35. Blackwell Scientific, Oxford, 401 pp.

Frostick, L.E., Reid, I. & Layman, J.T. (1983) Changing size distribution of suspended sediment in arid-zone flash floods. In: *Modern and Ancient Fluvial Systems* (Eds J.D. Collinson & J. Lewin), pp. 97–106. Special Publication 6, International Association of Sedimentologists. Blackwell Scientific, Oxford.

Fryberger, S.G. & Goudie, A.S. (1981) Arid Geomorphology. *Progress in Physical Geography* **5**, 420–8.

Fryberger, S.G., Schenk, C.J. & Krystinik, L.F. (1988) Stokes surfaces and the effects of near surface groundwater-table on aeolian deposition. *Sedimentology* **35**, 21–41.

Furakawa, K. & Wolanski, E. (1996) Sedimentation in mangrove forests. *Mangroves and Salt Marshes* **1**, 3–10.

Furukawa, K., Wolanski, E. & Mueller, H. (1997) Currents and sediment transport in mangrove forests. *Estuarine, Coastal and Shelf Science* **44**, 301–10.

Fütterer, D.K. (1974) Significance of the boring sponge *Cliona* for the origin of fine grained material of carbonate sediments. *Journal of Sedimentary Petrology* **44**, 79–84.

Gaeuman, D., Schmidt, J.C. & Wilcock, P.R. (2005) Complex channel responses to changes in stream flow and sediment supply on the lower Duschesne River, Utah. *Geomorphology* **64**, 185–206.

Gagan, M.K., Chivas, A.R. & Herczeg, A.L. (1990) Shelf-wide erosion, deposition, and suspended sediment transport during Cyclone Winifred, central Great Barrier Reef, Australia. *Journal of Sedimentary Petrology* **60**, 456–70.

Gagan, M.K., Johnson, D.P. & Carter, R.M. (1988) The Cyclone Winifred storm bed, central Great Barrier Reef shelf, Australia. *Journal of Sedimentary Petrology* **58**, 845–56.

Gaines, R.A. & Maynord, S.T. (2001) Microscale loose-bed hydraulic models. *Journal of Hydraulic Engineering* **May**, 335–8.

Gartner, J.W., Cheng, R.T., Wang, P-F., et al. (2001) Laboratory and field evaluations of the LISST-100 instrument for suspended particle size determinations. *Marine Geology* **175**, 199–219.

Gerard, C. & Chocat, B. (1998) An aid for the diagnosis of sewerage networks: analysis and modelling of the links between network's physical structure and the risk of sediment build-up. *Water Science and Technology* **39**, 185–92.

German, J. & Svensson, G. (2002) Metal content and particle size distribution of street sediments and street sweeping waste. *Water Science and Technology* **46**, 191–8.

Gerrard, A.J. (1990) *Mountain Environments.* Belhaven Press, London.

GESAMP. (1982) *The Review of the Health of the Oceans.* UNESCO, Paris.

Gibbs, R.J. (1973) Mechanisms of trace metal transport in rivers. *Science* **180**, 71–3.

Gibbs, R.J. (1977) Transport phases of transition metals in the Amazon and Yukon Rivers. *Geological Society of America Bulletin* **88**, 829–43.

Gilbertson, D.D. (1981) The impact of past and present land use on a major coastal barrier system. *Applied Geography* **1**, 97–119.

Gilbertson, D.D., Grattan, J.P. & Schwenniger, J.-L. (1996) The Quaternary geology of the coasts of the islands of the southern Outer Hebrides: a stratigraphic survey of the Holocene coastal dune and machair sequences. In: *The Environment of the Outer Hebrides of Scotland: the last 14000 years* (Eds D.D. Gilbertson, M. Kent & J.P. Grattan), pp. 59–71. Sheffield Academic Press, Sheffield.

Gilfillan, E.S., Maher, N.P., Krejsa, C.M., et al. (1995) Use of remote sensing to document changes in marsh vegetation following the *Amoco Cadiz* oil spill (Brittany, France, 1978). *Marine Pollution Bulletin* **30**, 780–7.

Gill, A.B. (2005) Offshore renewable energy: ecological implications of generating electricity in the coastal zone. *Journal of Applied Ecology* **42**, 605–15.

Gilvear, D.J. (1999) Fluvial geomorphology and river engineering: future role utilizing a fluvial hydrosystems framework. *Geomorphology* **31**, 229–45.

Ginsburg, R.N. (1957) Early diagenesis and lithification of shallow-water carbonate sediments in south Florida. In: *Regional Aspects of Carbonate Deposition* (Eds R.J. LeBlanc & J.G. Breeding), pp. 80–100. Special Publication 5, Society of Economic Paleontologists and Mineralogists, Tulsa, OK.

Gischler, E. & Lomando, A.J. (1999) Recent sedimentary facies of isolated carbonate platforms, Belize-Yucatan system, Central America. *Journal of Sedimentary Research* **69**, 747–63.

Gjessing, E.T. (1976) *Physical and Chemical Characteristics of Aquatic Humus*. Ann Arbor Science Publication, 120 pp.

Glade, T. (2003) Landslide occurrence as a response to land use change: a review of evidence from New Zealand. *Catena* **51**, 297–314.

Gladstone, W., Tawfiq, N., Nasr, D., et al. (1999) Sustainable use of renewable resources and conservation in the Red Sea and Gulf of Aden: issues, needs and strategic actions. *Ocean & Coastal Management* **42**, 671–97.

Glennie, K.W. (1987) Desert sedimentary environments, past and present – a summary. *Sedimentary Geology* **50**, 135–66.

Glynn, P.W. (1993) Monsoonal upwelling and episodic *Acanthaster* predation as probable controls of coral reef distribution and community structure in Oman, Indian Ocean. *Atoll Research Bulletin* **379**, 1–66.

Glynn, P.W. (1996) Coral reef bleaching: facts, hypotheses and implications. *Global Change Biology* **2**, 495–509.

Goldstone, M.E., Kirk, P.W.W. & Lester, J.N. (1990) The behaviour of heavy metals during wastewater treatment. II. Lead, nickel and zinc. *Science of the Total Environment* **95**, 253–70.

Golterman, H.L., Sly, P.G. & Thomas, R.L. (1983) *Study of the Relationship between Water Quality Sand Sediment Transport*. Technical. Paper No. 26, Unesco, Paris, 231 pp.

Golubic, S., Perkins, R.D. & Lukas, K.J. (1975) Boring microorganisms and microborings in carbonate substrates. In: *The Study of Trace Fossils* (Ed. R.W. Frey), pp. 229–59. Springer-Verlag, Berlin.

Goodbred, S.L.Jr. (2003) Response of the Ganges dispersal system to climate change: a source-to-sink view since the last interstade. *Sedimentary Geology* **162**, 83–104.

Goodwin, T.H., Young, A.R., Holmes, M.G.R., et al. (2003) The temporal and spatial variability of sediment transport and yields within the Bradford Beck catchment, West Yorkshire, *The Science of the Total Environment* **314–16**, 475–94.

Gore, J.A. (1985) *The Restoration of Rivers and Streams. Theories and Experience*. Ann Arbor Science (Butterworth), Stoneham, MA.

Gottofrey, J. (1990) *The disposition of cadmium, nickel, mercury and methylmercury in fish and effects of lipophilic metal chelation*. Thesis, SLU, Uppsala.

Götz, A. & Zimmermann, M. (1993) The 1991 rock slides in Randa: causes and consequences. *Landslide News* **7**(3), 22–5.

Goudie, A. (Ed.) (1981) *Geomorphological Techniques*. Allen and Unwin, London, 395 pp.

Goudie, A.S. (1977) Sodium sulphate weathering and the disintegration of Mehenjo-Daro, Pakistan. *Earth Surface Processes* **2**, 75–86.

Goudie, A.S. (1984) Salt efflorescence and salt weathering in the Jurza Valley, Karakoram Mountains, Pakistan. In: *The International Karakoram Project* (2 Vols) (Ed. K.J. Miller), pp. 607–15. Cambridge University Press, Cambridge.

Goudie, A.S. (1989) Wind erosion in deserts. *Proceedings of the Geologists Association* **100**, 89–92.

Goudie, A.S. (1997) Weathering processes. In: *Arid Zone Geomorphology: Process, Form and Change in Drylands*, 2nd edn (Ed. D.S.G. Thomas), pp. 25–39. Wiley, Chichester.

Goudie, A.S. & Wells, G.L. (1995) The nature, distribution and formation of pans in arid zones. *Earth Science Reviews* **38**, 1–69.

Goudie, A.S. & Wilkinson, J. (1978) *The Warm Desert Environment*. Cambridge University Press, Cambridge, 88 pp.

Goy, J.L., Zazo, C. & Dabrio, C.J. (2003) A beach-ridge progradation complex reflecting periodical sea-level and climate variability during the Holocene (Gulf of Almeria, Western Mediterranean). *Geomorphology* **50**, 251–68.

Graham, J. (1988) Collection and analysis of field data. In: *Techniques in Sedimentology* (Ed. M.E. Tucker), pp. 5–62. Blackwell Science, Oxford.

Grigg, R.W. & Dollar, S.J. (1990) Natural and anthropogenic disturbance on coral reefs. In: *Ecosystems of the World 25, Coral Reefs* (Ed. Z. Dubinsky), pp. 439–52. Elsevier, Amsterdam.

Grimalt, J.O., Ferrer, M. & Macpherson, E. (1999) The mine tailings accident in Aznalcollar. *Science of the Total Environment* **242**, 3–11.

Grindrod, J. & Rhodes, E.G. (1984) Holocene sea-level history of a tropical estuary: Missionary Bay, north Queensland. In: *Coastal Geomorphology in Australia* (Ed. B.G. Thom), pp. 151–78. Academic Press, Sydney.

Gromaire, M.C., Garnaud, S.S., Saard, M., et al. (2001) Contribution of different sources to the pollution of wet weather flow in combined sewers. *Water Research* **35**, 521–33.

Gromaire-Mertz, M.C., Garnaud, S.S., Gonzalez, A., et al. (1999) Characteristics of urban runoff pollution in Paris, *Water Science and Technology* **39**, 1–8.

Grove, A.T. (1977) The geography of semi-arid lands. *Philosophical Transactions of the Royal Society of London*, Series B, **278**, 457–75.

Guinotte, J., Buddemeier, R.W. & Kleypas, J. (2003) Future coral reef habitat marginality: temporal and spatial effects of climate change in the Pacific basin. *Coral Reefs* **22**, 551–8.

Gupta, A. & Ahmad, R. (1999) Geomorphology and the urban tropics: building an interface between research and usage. *Geomorphology* **31**, 133–49.

Gustafsson, Ö. & Gschwend, P.M. (1997) Aquatic colloids: Concepts, definitions and current challenges. *Limnology Oceanography* **42**, 519–28.

Guzmán, H.M., Burns, K.A. & Jackson, J.B.C. (1994) Injury, regeneration and growth of Caribbean reef corals after a major oil spill in Panama. *Marine Ecology Progress Series* **105**, 231–41.

Gygi, R.A. (1975) *Sparisoma viride* (Bonnaterre), the Spotlight Parrotfish, a major sediment producer on coral reefs of Bermuda. *Ecologiae Geologiae Helvetiae* **68**, 327–59.

Haas, J.N., Richoz, I., Tinner, W., et al. (1998) Synchronous Holocene climatic oscillations recorded on the Swiss Plateau and at timberline in the Alps. *The Holocene* **8**, 301–9.

Hadley Centre (2004) Hadley Centre for Climate Prediction and Research. http://www.metoffice.com/research/hadleycentre/models/modeldata.html (accessed 29 July 2004).

Haeberli, W. (1995) Climate change impacts on glaciers and permafrost. In: *Potential Ecological Impacts of Climate Change in the Alps and Fennoscandian Mountains* (Eds A. Guisan, J.I. Holten, R. Spichiger & L. Tessiers), pp. 97–103. Conservatoire et Jardin Botaniques, Geneva.

Haeberli, W., Rickenmann, D., Zimmermann, M., et al. (1990) Investigation of 1987 debris flows in the Swiss Alps: general concept and geophysical soundings. In: *Hydrology in Mountainous Regions. II – Artificial Reservoirs; Water and Slopes* (Eds R.O. Sinniger & M. Monbaron), pp. 303–10. IAHS Publication 194, Proceedings of two Symposia held at Lausanne, August. International Association of Hydrological Sciences Press, Wallingford.

Haeberli, W., Kääb, A., Paul, F., et al. (2002) A surge-type movement at Ghiacciaio del belevedere and a developing slope instability in the east face of Monte Rosa, Macugna, Italian Alps. *Norwegian Journal of Geography* **56**, 104–11.

Håkanson, L. (1977) The influence of wind, fetch, and water depth on the distribution of sediments in Lake Vänern, Sweden. *Canadian Journal of Earth Science* **14**, 397–412.

Håkanson, L. (1980) An ecological risk index for aquatic pollution control – a sedimentological approach. *Water Research* **14**, 995–1001.

Håkanson, L. (1995) Models to predict organic content of lake sediments. *Ecological Modelling* **82**, 233–45.

Håkanson, L. (1999) *Water Pollution – Methods and Criteria to Rank, Model and Remediate Chemical Threats to Aquatic Ecosystems.* Backhuys Publishers, Leiden, 299 pp.

Håkanson, L. (2004) *Lakes – Form and Function.* The Blackburn Press, New Jersey.

Håkanson, L. & Boulion, V. (2002) *The Lake Foodweb – Modelling Predation and Abiotic/Biotic Interactions.* Backhuys Publishers, Leiden, 344 pp.

Håkanson, L. & Jansson, M. (1983) *Principles of Lake Sedimentology.* Springer-Verlag, Berlin, 316 pp.

Håkanson, L. & Peters, R.H. (1995) *Predictive Limnology. Methods for Predictive Modelling.* SPB Academic Publishing, Amsterdam, 464 pp.

Håkanson, L., Borg, H. & Uhrberg, R. (1990) Reliability of analyses of Hg, Fe, Ca, K, P, pH, alkalinity, conductivity, hardness and colour from lakes. *International Reviews in Hydrobiology* **75**, 79–94.

Håkanson, L., Parparov, A. & Hambright, K.D. (2000) Modelling the impact of water level fluctuations on water quality (suspended particulate matter) in Lake Kinneret, Israel. *Ecological Modelling* **128**, 101–25.

Halfar, J., Godinez-Orta, L. & Ingle, J.C. (2000) Microfacies analysis of Recent carbonate environments in the southern Gulf of California, Mexico – a model for warm-temperate to subtropical carbonate formation. *Palaios* **15**, 323–42.

Hamilton, R.S., Revitt, D.M., Warren, R.S. (1984) Levels and physico-chemical association of Cd, Cu, Pb and Zn in road sediments. *The Science of the Total Environment* **33**, 59–74.

Hamilton, S.K. (1999) Potential effects of a major navigation project (Paraguay-Paraná Hidrovía) on inundation in the Pantanal floodplains. *Regulated Rivers – Research and Management* **15**, 289–99.

Hammarlund, D., Björck, S., Buchardt, B., et al. (2003) Rapid hydrological changes during the Holocene revealed by stable isotope records of lacustrine carbonates from Lake Igelsjön, southern Sweden. *Quaternary Science Reviews* **22**, 353–70.

Hammond, L.S. (1981) An analysis of grain size modification in biogenic carbonate sediments by deposit-feeding holothurians and echinoids (Echinodermata). *Limnology and Oceanography* **26**, 898–906.

Hampel, R. (1980) Geschiebeberechnung für Gefahrenzonenpläne in Wildbachgebieten. *Interpraevent* **4**, 83–92.

Hansen, K. (1961) Lake types and lake sediments. *Verhandlungen der Internationalen Vereinigung der Limnologie* **14**, 285–90.

Harbison, P. (1986) Mangrove muds – a sink and a source for trace metals. *Marine Pollution Bulletin* **17**, 246–50.

Harbor, J. & Warburton, J. (1993) Relative rates of glacial and nonglacial erosion in Alpine environments. *Arctic and Alpine Research* **25**(1), 1–7.

Hardisty, J. & Whitehouse, R.J.S. (1988) Evidence for a new sand transport process from experiments on Saharan dunes. *Nature* **332**, 532–4.

Hardman-Mountford, N.J., Allen, J.I., Frost, M.T., et al. (2005) Diagnostic monitoring of a changing environment: an alternative UK perspective. *Marine Pollution Bulletin* 50, 1463–71.

Harney, J.N. & Fletcher, C.H. (2003) A budget of carbonate framework and sediment production, Kailua Bay, Oahu, Hawaiian Islands. *Journal of Sedimentary Research* 73, 856–68.

Harris, P.T. (1989) Sandwave movement under tidal and wind-driven currents in a shallow marine environment: Adolphus Channel, northeastern Australia. *Continental Shelf Research* 9, 981–1002.

Harris, P.T. (1991) Reversal of subtidal dune asymmetries caused by seasonally reversing wind-driven currents in Torres Strait, northeastern Australia. *Continental Shelf Research* 11, 655–62.

Harris, P.T., Tsuji, Y., Marshall, J.F., et al. (1996) Sand and rhodolith-gravel entrainment on the mid- to outer-shelf under a western boundary current: Fraser Island continental shelf, eastern Australia. *Marine Geology* 129, 313–30.

Harrison, R.M. (1979) Toxic metals in street and household dusts. *The Science of the Total Environment* 11, 89–97.

Harrison, R.M., Yin, J. & Mark, D. (2001) Studies of the coarse particle (2.5–10 μm) component in UK urban atmospheres. *Atmospheric Environment* 35, 3667–79.

Hartley, A.J. (2003) Andean uplift and climate change. *Journal of the Geological Society* 160, 7–10.

Hartley, A.J. & May, G. (1998) Miocene gypcretes from the Calama Basin, northern Chile. *Sedimentology* 45, 351–64.

Hartley, A.J., Chong, G., Houston, J., et al. (2005) 150 million years of climatic stability: evidence from the Atacama Desert, northern Chile. *Journal of the Geological Society, London* 162, 421–4.

Harvey, A.M. (1984) Debris flow and fluvial deposits in Spanish Quaternary alluvial fans: implications for fan morphology. In: *Sedimentology of Gravels and Conglomerates* (Eds E.H. Koster & R. Steel), pp. 123–32. Memoir 10, Canadian Society of Petroleum Geologists, Calgary, Alberta.

Harvey, A.M. (1987) Patterns of Quaternary aggradational and dissectional landform development in the Almería region, southeast Spain: a dry region tectonically active landscape. *Die Erde* 118, 193–215.

Harvey, A.M. (1990) Factors influencing Quaternary alluvial fan development in southeast Spain. In: *Alluvial Fans: a Field Approach* (Eds A.H. Rachocki & M. Church), pp. 247–69. Wiley, Chichester.

Harvey, A.M. (1997) The occurrence and role of arid zone alluvial fans. In: *Arid Zone Geomorphology: Process, Form and Change in Drylands*, 2nd edn (Ed. D.S.G. Thomas), pp. 136–58. Wiley, Chichester.

Harvey, A.M., Foster, G., Hannam, J., et al. (2003) The Tabernas alluvial fan and lake system, southeast Spain: applications of mineral magnetic and pedogenic iron oxide analyses towards clarifying the Quaternary sediment sequences. *Geomorphology* 50, 151–71.

Haslett, S.K. (2000) *Coastal Systems*. Routledge, London.

Haslett, S.K., Cundy, A.B., Davies, C.F.C., et al. (2003) Salt marsh sedimentation over the past *c.* 120 years along the west Cotentin Coast of Normandy (France): Relationship to sea-level rise and sediment supply. *Journal of Coastal Research* 19(3), 609–20.

Hassan, M.A. (1990) Observations of desert flood bores. *Earth Surface Processes and Landforms* 15, 481–5.

Hassan, M.A. & Church, M. (1992) The movement of individual grains on the streambed. In: *Dynamics of Gravel-bed Rivers* (Eds P. Billi, R.D. Hey, C.R. Throne & P. Tacconi), pp. 159–73. Wiley, Chichester.

Hassan, M.A. & Klein, M. (2002) Fluvial adjustment of the Lower Jordan River to a drop in the Dead Sea level. *Geomorphology* 45, 21–33.

Hatton, J.C. & Couto, A.L. (1992) The effect of coastline change on mangrove community structure, Portuguese Island, Mozambique. *Hydrobiologia* 247, 49–57.

Hayes, M.O. (1979) Barrier island morphology as a function of wave and tide regime. In: *Barrier Islands from the Gulf of St. Lawrence to the Gulf of Mexico* (Ed. S.P. Leatherman), pp. 1–29. Academic Press, New York.

Hayes, M.O. (1991) Geomorphology and sedimentation patterns in tidal inlets – a review. *Coastal Sediments '91, Proceedings of a Special Conference on Quantitative Approaches to Coastal Sediment Processes*, Vol. II (Eds N.C. Kraus, K.J. Gingerich & D.L. Kriebel). pp. 1343–55. American Society of Civil Engineers Proceedings, Reston, VA.

Hayes, S.K., Montgomery, D.R. & Newhall, C.G. (2002) Fluvial sediment transport and deposition following the 1991 eruption of Mount Pinatubo. *Geomorphology* 45, 211–24.

Hayne, M. & Chappell, J. (2001) Cyclone frequency during the last 5000 years at Curacoa Island, north Queensland, Australia. *Palaeogeography, Palaeoclimatology, Palaeoecology* 168, 207–19.

Haynes, D. & Johnson, J.E. (2000) Organochlorine, heavy metal and polyaromatic hydrocarbon pollutant concentrations in the Great Barrier Reef (Australia) environment: a review. *Marine Pollution Bulletin* 41, 267–78.

Haynes, F.N. & Coulson M.G. (1982) The decline of *Spartina* in Langstone Harbour, Hampshire. *Hampshire Field Club and Archaeological Society* 38, 5–18.

Hearn, G.J. (1995) Landslide and erosion hazard mapping at Ok-Tedi copper mine, Papua-New-Guinea. *Quaterly Journal of Engineering Geology* 28, 47–60.

Heathcote, R.L. (1983) *The Arid Lands: their Use and Abuse*. Longman, London, 323 pp.

Hegge, B.J., Eliot, I.G. & Hsu, J. (1996) Sheltered sandy beaches. *Journal of Coastal Research* 12, 748–60.

Heim, A. (1882) Der Bergsturz von Elm. *Deutsche Geologische Gesellschaft für Zeitschrift* **34**, 74–115.

Helgen, S.O. & Moore, J.N. (1996) Natural background determination and impact quantification in trace metal-contaminated river systems. *Environmental Science and Technology* **30**, 129–35.

Hellden, U. (1988) Desertification monitoring: is the desert encroaching? *Desertification Control Bulletin* **17**, 8–12.

Hem, J.E. (1972) Chemistry and occurrence of cadmium and zinc in surface water and groundwater. *Water Research* **8**, 661–79.

Hettler, J., Irion, G. & Lehmann, B. (1997) Environmental impact of mining waste disposal on a tropical lowland river system: a case study on the Ok Tedi mine, Papua New Guinea. *Mineralium Deposita* **32**, 280–91.

Hewitt, K. (2004) Geomorphic Hazards in Mountain Environments. In: *Mountain Geomorphology* (Eds P.N. Owens & O. Slaymaker), pp. 187–218. Arnold, London.

Hindson, R.A. & Andrade, C. (1999) Sedimentation and hydrodynamic processes associated with the tsunami generated by the 1755 Lisbon earthquake. *Quaternary International* **56**, 27–38.

Hinrichsen, D. (1998) *Coastal Waters of the World: Trends, Threats, and Strategies.* Island Press, Washington, DC, 420 pp.

Hjalmarson, H.W. (1984) Flash flood in Tanque Verde Creek, Tucson, Arizona. *Journal of Hydraulic Engineering, American Society of Civil Engineers* **110**, 1841–52.

Hjulström, F. (1935) Studies on the morphological activity of rivers as illustrated by the River Fyris. *Bulletin of the Geological Institute of Uppsala* **25**, 221–527.

Hogarth, P.J. (1999) *The Biology of Mangroves.* Oxford University Press, Oxford, pp. 228.

Holme, N.A. (1961) The bottom fauna of the English Channel. *Journal of the Marine Biological Association of the United Kingdom* **41**, 397–461.

Holme, N.A. (1966) The bottom fauna of the English Channel. Part II. *Journal of the Marine Biological Association of the United Kingdom* **46**, 401–93.

Holme, N.A. & Wilson, J.B. (1985) Faunas associated with longitudinal furrows and sand ribbons in a tide-swept area in the English Channel. *Journal of the Marine Biological Association of the United Kingdom* **65**, 1051–72.

Hooke, R.L. (2000) Towards a uniform theory of clastic sediment yield in fluvial systems. *Geological Society of America Bulletin* **112**, 1778–86.

Hopke, P.K., Lamb, R.E. & Natusch, F.S. (1980) Multi-elemental characterization of urban roadway dust. *Environmental Science and Technology* **14**, 164–72.

Horowitz, A.J. (1991) *A Primer on Sediment Trace-element Chemistry.* Lewis Publishers, Michigan.

Horowitz, A.J. & Elrick, K.A. (1987) The relation of stream sediment surface area, grain size and composition to trace element chemistry. *Applied Geochemistry* **2**, 437–51.

Hosiokangas, J., Ruuskanen, J. & Pekkanen, J. (1999) Effects of soil dust episodes and mixed fuel sources on source apportionment of PM10 particles in Kuopio, Finland. *Atmospheric Environment* **33**, 3821–9.

Hossain, S. & Eyre, B. (2002) Suspended sediment exchange through the sub-tropical Richmond River Estuary, Australia: a balance approach. *Estuarine, Coastal and Shelf Science* **55**, 579–86.

Houghton, J.T., Callander, B.A. & Varney, S.K. (1992) *Climate Change.* Cambridge University Press, Cambridge, pp. 69–95.

Hovland, M. & Judd, A.G. (1988) *Seabed Pockmarks and Seepages.* Graham and Trotman, London.

Hsu, K.H. (1975) Catastrophic debris streams (sturzstroms) generated by rockfalls. *Geological Society of America Bulletin* **86**, 129–40.

Hubbard, D.K. (1992) Hurricane-induced sediment transport in open-shelf tropical systems – an example from St. Croix, U.S. Virgin Islands. *Journal of Sedimentary Petrology* **62**, 946–60.

Hubbard, D.K. (1997) Reefs as dynamic systems. In: *Life and Death of Coral Reefs* (Ed. C. Birkeland), pp. 43–67. Chapman and Hall, London.

Hudson-Edwards, K.A., Macklin, M.G., Curtis, C.D., et al. (1996) Processes of formation and distribution of Pb-, Zn-, Cd- and Cu-bearing minerals in the Tyne basin, NE England: implications for metal-contaminated river systems. *Environmental Science and Technology* **30**, 72–80.

Hudson-Edwards, K.A., Macklin, M.G., Curtis, C.D., et al. (1998) Chemical remobilisation of contaminant metals within floodplain sediments in an incising river system: implications for dating and chemostratigraphy. *Earth Surface Processes and Landforms* **23**, 671–84.

Hudson-Edwards, K.A., Macklin, M.G. & Taylor, M.P. (1999a) 2000 years of sediment-borne heavy metal storage in the Yorkshire Ouse basin, NE England. *Hydrological Processes* **13**, 1087–102.

Hudson-Edwards, K.A., Schell, C. & Macklin, M.G. (1999b) Mineralogy and geochemistry of alluvium contaminated by metal mining in the Rio Tinto area, southwest Spain. *Applied Geochemistry* **14**, 55–70.

Hudson-Edwards, K.A., Macklin, M.G., Miller, J.R. & Lechler, P.J. (2001) Sources, distribution and storage of heavy metals in the Río Pilcomayo, Bolivia. *Journal of Geochemical Exploration* **72**, 229–50.

Hudson-Edwards, K.A., Macklin, M.G., Jamieson, H.E., et al. (2003) The impact of tailings dam spills and clean-up operations on sediment and water quality in river systems: the Ríos Agrio-Guadiamar, Aznalcóllar, Spain. *Applied Geochemistry* **18**, 221–39.

Hulme, M. (1992) Rainfall changes in Africa: 1931–60 to 1961–90. *International Journal of Climatology* **12**, 685–99.

Huston, M. (1985) Variation in coral growth rates with depth at Discovery Bay, Jamaica. *Coral Reefs* **4**, 19–25.

Hutchings, P.A. (1986) Biological destruction of coral reefs. *Coral Reefs* **4**, 239–52.

Hutchinson, G.E. (1957) *A Treatise on Limnology*, Vol. 1. *Geography, Physics and Chemistry*. Wiley, New York, 1016 pp.

Hutchinson, G.E. (1967) *A Treatise on Limnology*. Vol. 2. *Introduction to Lake Biology and the Limno-plankton*. Wiley, New York, 1115 pp.

Hutchinson, J.N. (1988) General Report: Morphological and geotechnical parameters of landslides in relation to geology and hydrogeology. In: *Proceedings of the 5th International Symposium on Landslides*, Vol. 1 (Ed. C. Bonnard), pp. 3–35. Balkema, Rotterdam.

Hvitved-Jacobson, T., Yousef, Y.A., Wanielista, M.P., et al. (1984) Fate of phosphorus and nitrogen in ponds receiving highway runoff. *The Science of the Total Environment* **33**, 259–70.

Inbar, M., Tamir, M. & Wittenberg, L. (1998) Runoff and erosion processes after a forest fire in Mount Carmel, a Mediterranean area. *Geomorphology* **24**, 17–33.

Ingersoll, C.G., Haverland, P.S., Brunson, E.L., et al. (1996) Calculation of sediment effect concentrations. *Journal of Great Lakes Research* **22**, 602–23.

Inman, D.L. & Jenkins, S.A. (1999) Climate change and the episodicity of sediment flux of small Californian rivers. *Journal of Geology* **107**, 251–70.

Inouchi, Y., Kinugasa, Y., Kumon, F., et al. (1996) Turbidites as records of intense paleoearthquakes in Lake Biwa, Japan. *Sedimentary Geology* **104**, 117–25.

Ives, J.D. & Messerli, B. (1989) *The Himalayan Dilemma – Reconciling Development and Conservation*. Routledge, London.

Jäckli, H. (1957) Gegenwartsgeologie des undnerischen Rheingebietes-ein Beitrag zur exogenen Dynamik alpiner Gebirgslandschaften. *Beitraege zur Geologie der Schweiz, Geotechnische Serie* **36**.

James, A. (1999) Time and persistence of alluvium: River engineering, fluvial geomorphology, and mining sediment in California. *Geomorphology* **31**, 265–90.

James, C.S. (1985) Sediment transfer to overbank sections. *Journal of Hydraulic Research* **23**, 435–52.

James, L.A. (1989) Sustained storage and transport of hydraulic gold mining sediment in the Bear River, California. *Annals of the Association of American Geographers* **79**, 570–92.

James, N.P. & Bourque, P.A. (1992) Reefs and mounds. In: *Facies Models: Response to Sea Level Change* (Eds R.G. Walker & N.P. James), pp. 323–47. Geological Association of Canada, St John's, Newfoundland.

James, N.P. & Macintyre, I.G. (1985) *Carbonate Depositional Environments Modern and Ancient*. Colorado School of Mines Quarterly 80.

James, N.P., Bone, Y., Von de Borsch, C.C., et al. (1992) Modern carbonate and terrigenous clastic sediments on a cool water, high energy, mid-latitude shelf: Lacepede, southern Australia. *Sedimentology* **38**, 877–903.

Janda, R.J., Daag, A.S., Delos Reyes, P.J., et al. (1996) Assessment and response to lahar hazard around Mount Pinatubo, 1991 to 1993. In: *Fire and Mud, Eruptions and Lahars of Mount Pinatubo, Philippines* (Eds C.G. Newhall & R.S. Punongbayan), pp. 107–39. PHIVOLCS Press, Quezon City, and University of Washington Press, Seattle.

Jansen, J.M.L. & Painter, R.B. (1974) Predicting sediment yield from climate and topography. *Journal of Hydrology* **21**, 371–80.

Jansson, M.B. (1988) A global survey of sediment yield. *Geografiska Annaler* **70A**(1–2), 81–98.

Jennings, S. & Kaiser, M.J. (1998) The effects of fishing on marine ecosystems. *Advances in Marine Biology* **34**, 203–352.

Johnson, D.P. & Risk, M.J. (1987) Fringing reef growth on a terrigenous mud foundation, Fantome Island, central Great Barrier reef, Australia. *Sedimentology* **34**, 275–87.

Johnson, H.D. & Baldwin, C.T. (1996) Shallow clastic seas. In: *Sedimentary Environments: Processes, Facies and Stratigraphy*, 3rd edn (Ed. H.G. Reading), pp. 232–80. Blackwell Science, Oxford.

Johnsson, M.J., Stallard, R.F. & Lundberg, N. (1991) Controls on the composition of fluvial sands from a tropical weathering environment – sands of the Orinoco River drainage-basin, Venezuela and Columbia. *Geological Society of America Bulletin* **103**, 1622–47.

Jones, B.F. & Bowser, C.J. (1978) The mineralogy and related chemistry of lake sediments. In: *Lakes: Chemistry, Geology, Physics* (Ed. A. Lerman), pp. 179–235. Springer-Verlag, Berlin.

Jones, D.K.C. (1992) Landslide hazard assessment in the context of development. In: *Geohazards: Natural and Man-made* (Eds G.J.H. McCall, D.J.C. Laming & S.C. Scott), pp. 117–41. Chapman and Hall, London.

Jones, D.K.C. (1993a) Global warming and geomorphology. *The Geographical Journal* **159**, 124–30.

Jones, D.K.C. (1993b) Slope instability in a warmer Britain. *The Geographical Journal* **159**, 184–95.

Jones, D.K.C., Brunsden, D. & Goudie, A.S. (1983) A preliminary geomorphological assessment of the Karakoram Highway. *Quarterly Journal of Engineering Geology* **16**, 331–55.

Jonsson, P. (1992) *Large-scale changes of contaminants in Baltic Sea sediments during the twentieth century*. Thesis, Uppsala University, Sweden.

Judd, A., Franklin, F. & Faire, S. (Eds) (2003) *OSPAR Workshop – Environmental Assessment of Renewable Energy in the Marine Environment*, Maldon, 17–18 September, 109 pp.

Kääb, A. (2002) Monitoring high-mountain terrain deformation from repeated air and spaceborne optical data: examples using digital aerial imagery and ASTER data. *ISPRS Journal of Photogrammetry and Remote Sensing* 57, 39–52.

Kääb, A., Wessels, R., Haeberli, W., et al. (2003) Rapid ASTER imaging facilitates timely assessment of glacier hazards and disasters. *Eos (Transactions of the American Geophysical Union)* 84, 13,117.

Kaiser, M.J. & De Groot, S.J. (2000) *The Effects of Fishing on Non-target Species and Habitats: Biological, Conservation and Socio-economic issues.* Blackwell Science, Oxford.

Karrickhoff, S.W. (1981) Semi-empirical estimation of sorption of hydrophobic pollutants on natural sediments and soils. *Chemosphere* 10, 833–46.

Karwan, D.L., Allan, J.D. & Bergen, K.M. (2001) Changing near-stream land use and river channel morphology in the Venezuelan Andes. *Journal of the American Water Resources Association* 37, 1579–87.

Kasse, C. (1998) Depositional model for cold-climate tundra rivers. In: *Palaeohydrology and Environmental Change* (Eds G. Benito, V.R. Baker & K.J. Gregory), pp. 83–97. Wiley, Chichester.

Kasse, C., Vandenberghe, J., Van Huissteden, J., Bohncke, S.J.P., et al. (2003) Sensitivity of Weichselian fluvial systems to climate change (Nochten mine, eastern Germany). *Quaternary Science Reviews* 22, 2141–56.

Kato, M., Fukusawa, H. & Yasuda, Y. (2003) Varved lacustrine sediments of Lake Tougou-ike, Western Japan, with reference to Holocene sea-level changes in Japan. *Quarternary International* 105, 33–7.

Ke, L., Wang, W.Q., Wong, T.W.Y., et al. (2003) Removal of pyrene from contaminated sediments by mangrove microcosms. *Chemosphere* 51, 25–34.

Kearney, M.S. & Stevenson, J.C. (1991) Island land loss and marsh vertical accretion rate evidence for historical sea-level changes in Chesapeake Bay. *Journal of Coastal Research* 7, 403–15.

Keen, T.R. & Slingerland, R.L. (1993) Four storm-event beds and the tropical cyclones that produced them: a numerical hindcast. *Journal of Sedimentary Petrology* 63, 218–32.

Kelderman, P., Drossaert, W.M.E., Min, Z., et al. (2000) Pollution assessment of the canal sediments in the city of Delft (the Netherlands). *Water Research* 34, 936–44.

Kemp, A.L.W., Thomas, R.L., Dell, C.I., et al. (1976) Cultural impact of the geochemistry of sediments in Lake Erie. *Journal of the Fisheries Research Board of Canada* (Special Issue) 33, 440–62.

Kench, P.S. (1997) Physical processes in an Indian Ocean atoll. *Coral Reefs* 17, 155–68.

Kench, P.S. & Cowell, P.J. (2002) Variations in sediment production and implications for atoll island stability under rising sea level. *Proceedings of 9th international Coral Reef Symposium*, Bali, Vol. 2, pp. 1181–6. Ministry of Environment (Indonesia), Indonesian Institute of Sciences and International Society for Reef Studies.

Kench, P.S. & McLean, R.F. (1996) Hydraulic characteristics of bioclastic deposits: new possibilities for environmental interpretation using settling velocity fractions. *Sedimentology* 43, 561–70.

Kennedy, D.M. & Woodroffe, C.D. (2002) Fringing reef growth and morphology: a review. *Earth-Science Reviews* 57, 255–77.

Kenny, A.J., Rees, H.L., Greening, J., et al. (1998) The effects of gravel extraction on the macrobenthos at an experimental dredge site off North Norfolk, UK (results 3 years post-dredging). *International Council for the Exploration of the Seas, Council Minutes* 14, 1–7.

Kenny, A.J. & Rees, H.L. (1994) The effects of marine gravel extraction on the macrobenthos: Early post-dredging recolonization. *Marine Pollution Bulletin* 28, 442–7.

Kenny, A.J. & Rees, H.L. (1996) The effects of marine gravel extraction on the macrobenthos: results 2 years post-dredging. *Marine Pollution Bulletin* 32, 615–22.

Keown, M.P., Dardeau, E.A. & Causey, E.M. (1986) Historic trends in the sediment flow regime of the Mississippi River. *Water Resources Research* 22, 1555–64.

Kershaw, P.J., Swift, D.J. & Denoon, D.C. (1988) Evidence of recent sedimentation in the eastern Irish Sea. *Marine Geology* 85, 1–14.

Kim, K.W., Myung, J.H., Ahn, J.S., et al. (1998) Heavy metal contamination in dusts and stream sediments in the Taejon Area, Korea. *Journal of Geochemical Exploration* 64, 409–19.

Kineke, G.C., Sternberg, R.W., Trowbridge, J.H., et al. (1996) Fluid-mud processes on the Amazon continental shelf. *Continental Shelf Research* 16, 667–96.

Kirchner, J.W., Finkel, R.C., Riebe, C.S., et al. (2001) Mountain erosion over 10 yr, 10 k.y., and 10 m.y. time scales. *Geology* 29, 591–4.

Kirkby, M.J., Abrahart, R., McMahon, M.D., Shao, J. & Thornes, J.B. (1998) MEDALUS soil erosion models for global change. *Geomorphology* 24, 35–49.

Kleypas, J. & Langdon, C. (2002) Overview of CO_2-induced changes in seawater chemistry. *Proceedings of 9th International Coral Reef Symposium*, Bali, Vol. 2, pp. 1085–9. Ministry of Environment (Indonesia), Indonesian Institute of Sciences and International Society for Reef Studies.

Kleypas, J.A., McManus, J.W. & Meñez, L.A.B. (1999) Environmental limits to coral reef development: where do we draw the line? *American Zoologist* 39, 146–59.

Knighton, D. (1998) *Fluvial Forms and Processes*. Arnold, London.

Knighton, D. & Nanson, G. (1997) Distinctiveness, diversity and uniqueness in arid zone river systems. In: *Arid Zone Geomorphology: Process, Form and Change in Drylands*, 2nd edn (Ed. D.S.G. Thomas), pp. 185–203. Wiley, Chichester.

Knox, J.C. (1987) Historical valley floor sedimentation in the upper Mississippi valley. *Annals of the Association of American Geographers* 77, 224–44.

Knox, J.C. (1995) Fluvial systems since 20,000 years BP. In: *Global Continental Palaeohydrology* (Eds K.J. Gregory, L. Starkel & V.R. Baker), pp. 87–108. Wiley, Chichester.

Knox, J.C. (2001) Agricultural influence on landscape sensitivity in the Upper Mississippi River Valley. *Catena* 42, 193–224.

Knudson, T. (1991) Mountain lake turns to desert. *Sacramento Bee*, **10 June**.

Komar, P.D. (1998) *Beach and Nearshore Sedimentation*, 2nd edn. Prentice Hall, Upper Saddle River, NJ.

Komatsubara, J. (2004) Fluvial architecture and sequence stratigraphy of the Eocene to Oligocene Iwaki Formation, northeast Japan: channel-fills related to the sea-level change. *Sedimentary Geology* 168, 109–23.

Kondolf, G.M. (1995) Managing bedload sediment in regulated rivers: examples from California, USA. In: *Natural and Anthropogenic Influences in Fluvial Geomorphology* (Eds J.E. Costa, A.J. Miller, K.W. Potter & P.R. Wilcock), pp. 125–213. Monograph 89, Geophysical Union of America, Washington, DC.

Kornman, B.A. & De Deckere, E.M.G.T. (1998) Temporal variation in sediment erodability and suspended sediment dynamics in the Dollard estuary. In: *Sedimentary Processes in the Intertidal Zone* (Eds K.S. Black, D.H. Paterson & A. Cramp), pp. 231–41. Special Publication 139, Geological Society Publishing House, Bath.

Kranck, K. (1973) Flocculation of suspended sediment in the sea. *Nature* 24, 348–50.

Krank, K. (1975) Sediment deposition from flocculated suspensions. *Sedimentology* 22, 111–23.

Kranck, K. (1979) Particle matter grain-size characteristics and flocculation in a partially mixed estuary. *Sedimentology* 28, 107–14.

Kratzer, C.R. (1999) Transport of sediment-bound organochlorine pesticides to the San Joaquin River, California. *Journal of the American Water Resources Association* 35, 957–81.

Krauss, K.W., Allen, J.A. & Cahoon, D.R. (2003) Differential rates of vertical accretion and elevation change among aerial root types in Micronesian mangrove forests. *Estuarine, Coastal and Shelf Science* 56, 251–9.

Krumbein, W.C. (1963) A geological process–response model for analysis of beach phenomena. *Bulletin of the Beach Erosion Board* 17, 1–15.

Kuehl, S., DeMaster, D.J. & Nittrouer, C.A. (1986) Nature of sediment accumulation on the Amazon continental shelf. *Continental Shelf Research* 6, 209–25.

Laban, P. (1978) *Field Measurements on Erosion and Sedimentation in Nepal*. PAO/UNDP/IWM?SP/05, Department of Soil Conservation and Watershed Management, Nepal.

Laban, P. (1979) *Landslide Occurrence in Nepal*. Phewa Tal Project Report no. 8P/13, Integrated Watershed Management Project, Kathmandu.

Lacerda, L.D. & Salomons, W. (1992) Mercurio na Amazonia: uma Bomba Relogio Quimica? *Serie Tecnologia Ambiental* 3, CETEM/CNPq, Rio de Janeiro.

Laird, J.R. & Harvey, M.D. (1986) Complex response of a chaparral drainage basin to fire. In: *Drainage Basin Sediment Delivery* (Ed. R.F. Hadley), pp. 165–83. Publication 159, International Association of Hydrological Sciences, Washington, DC.

Lambeck, A. & Woolfe, K.J. (2000) Composition and textural variability along the 10-metre isobath, Great Barrier Reef – evidence for pervasive northward sediment transport. *Australian Journal of Earth Sciences* 47, 327–35.

Lamprey, H.F. (1975) *Report on the Desert Encroachment Reconnaissance in Northern Sudan, October 21–November 10, 1975*. National Council for Research, Ministry of Agriculture, Food and Resources, Khartoum, 16 pp.

Lancaster, N. (1994) Dune morphology and dynamics. In: *Geomorphology of Desert Environments* (Eds A.D. Abrahams & A.J. Parsons), pp. 474–505. Chapman and Hall, London.

Lancaster, N. (1995) *The Geomorphology of Desert Dunes*. Routledge, London, 211 pp.

Land, L.S. (1970) Carbonate mud: production by epibiont growth on *Thalassia testudinium*. *Journal of Sedimentary Petrology* 42, 179–82.

Land, L.S. (1974) Growth rate of a West Indian (Jamaican) reef. *Proceedings of Second International Coral Reef Symposium*, Vol. 2, pp. 409–12. The Great Barrier Reef Committee, Brisbane.

Langbein, W.B. & Schumm, S.A. (1958) Yield of sediment in relation to mean annual precipitation. *Transactions of the American Geophysical Union* 39, 1076–84.

Langedal, M. (1997a) Dispersion of tailings in the Knabeåna–Kvina drainage basin, Norway, 1: evaluation of overbank sediments as sampling medium for regional geochemical mapping. *Journal of Geochemical Exploration* 58, 157–72.

Langedal, M. (1997b) Dispersion of tailings in the Knabeåna–Kvina drainage basin, Norway, 2: mobility of Cu and Mo in tailings-derived fluvial sediments. *Journal of Geochemical Exploration* 58, 173–83.

Larcombe, P. & Carter, R.M. (2004) Cyclone pumping, sediment partitioning and the development of the Great Barrier Reef shelf system: a review. *Quaternary Science Reviews* 23, 107–35.

Larcombe, P. & Ridd, P.V. (1995) Megaripple dynamics and sediment transport in a mesotidal mangrove creek: implications for palaeoflow reconstructions. *Sedimentology* **42**, 593–606.

Larcombe, P. & Woolfe, K.J. (1999) Terrigenous sediments as influences upon Holocene nearshore coral reefs, central Great Barrier Reef, Australia. *Australian Journal of Earth Sciences* **46**, 141–54.

Larcombe, P., Ridd, P.V., Prytz, A. & Wilson, B. (1995) Factors controlling suspended sediment on inner-shelf coral reefs, Townsville, Australia. *Coral Reefs* **14**, 163–71.

Larcombe, P., Parker, R., Dye, S., et al. (2005) Marine environmental change over decades to millennia – evaluating proxy parameters and their potential use in understanding the state of the marine ecosystem. Final report to Defra.

Large, D.J., Fortey, N.J., Milodowski, A.E., et al. (2002) Petrographic observations of iron, copper and zinc sulfides in freshwater canal sediment. *Journal of Sedimentary Research* **71**, 61–9.

Laronne, J.B. & Reid, I. (1993) Very high bedload sediment transport in desert ephemeral rivers. *Nature* **366**, 148–50.

Laronne, J.B., Reid, I., Yitshak, Y., et al. (1994) The non-layering of gravel streambeds under ephemeral flow regimes. *Journal of Hydrology* **159**, 353–63.

Lau, Y.L., Oliver, B.G. & Kirshnappan, B.G. (1989) Transport of some chlorinated contaminants by the water, suspended sediments, and bed sediments in the St. Clair and Detroit Rivers. *Environmental Toxicology and Chemistry* **8**, 291–301.

Lave, J. & Avouac, J.P. (2001) Fluvial incision and tectonic uplift across the Himalayas of central Nepal. *Journal of Geophysical Research – Solid Earth* **106**, 26,561–91.

Lavee, H. & Poesen, J.W.A. (1991) Overland flow generation and continuity on stone-covered soil surfaces. *Hydrological Processes* **5**, 345–60.

Lavier, L.L., Steckler, M.S. & Brigaud, F. (2001) Climatic and tectonic control on the Cenozoic evolution of the West African margin. *Marine Geology* **178**, 63–80.

Le Houérou, H.N., Bingham, R.L. & Skerbek, W. (1988) Relationship between the variability of primary production and the variability of annual precipitation in world arid lands. *Journal of Arid Environments* **15**, 1–18.

Lecce, S.A. & Pavlowsky, R.T. (2001) Use of mining-contaminated sediment tracers to investigate the timing and rates of historical flood plain sedimentation. *Geomorphology* **38**, 85–108.

Lecoanet, H., Leveque, F. & Ambrosi, J.P. (2003) Combination of magnetic parameters: an efficient way to discriminate soil-contamination sources (south France). *Environmental Pollution* **122**, 229–34.

Lee, S.K., Tan, W.H. & Havanond, S. (1996) Regeneration and colonisation of mangrove on clay-filled reclaimed land in Singapore. *Hydrobiologia* **319**, 23–35.

Leeder, M. (1999) *Sedimentology and Sedimentary Basins: from Turbulence to Tectonics.* Blackwell Science, Oxford.

Leeder, M.R. (1999) *Sedimentology and Sedimentary Basins.* Blackwell Science, Oxford.

Leeks, G.J.L. & Roberts, G. (1987) The effects of forestry on upland streams – with special reference to water quality and sediment transport. In: *Environmental Aspects of Plantation Forestry in Wales* (Eds J.E.G. Good & Institute of Terrestrial Ecology), pp. 9–24. Publication Number 22, Institute of Terrestrial Ecology, Huntingdon.

Lees, A. (1975) Possible influence of salinity and temperature on modern shelf carbonate sedimentation. *Marine Geology* **19**, 159–98.

Legesse, D., Gasse, F., Radakovitch, O., et al. (2002) Environmental changes in a tropical lake (Lake Abiyata, Ethiopia) during recent centuries. *Palaeogeography, Palaeoclimatology, Palaeoecology* **187**, 233–58.

Legleiter, C.J., Lawrence, R.L., Fonstad, M.A., et al. (2003) Fluvial response a decade after wildfire in the northern Yellowstone ecosystem: a spatially explicit analysis. *Geomorphology* **54**, 119–36.

Leigh, D.S. (1997) Mercury-tainted overbank sediment from past gold mining in North Georgia, USA. *Environmental Geology* **30**, 244–51.

Leonard, L., Clayton, T., Dixon, K., et al. (1989) An analysis of replenished beach design parameters on U.S. East Coast barrier islands: *Journal of Coastal Research* **6**, 15–36.

Leopold, L.B., Emmett, W.W. & Myrick, R.M. (1966) *Channel and Hillslope Processes in a Semi-arid Area of New Mexico.* Professional Paper 352-G, US Geological Survey, Denver, CO.

Levings, S.C. & Garrity, S.D. (1994) Effects of oil spills on fringing red mangroves (*Rhizophora mangle*): losses of mobile species associated with submerged prop roots. *Bulletin of Marine Science* **54**, 782–94.

Lewin, J. & Macklin, M.G. (1987) Metal mining and floodplain sedimentation in Britain. In: *International Geomorphology 1986, Part 1* (Ed. V. Gardiner), pp. 1009–27. Wiley, Chichester.

Lewin, J. & Wolfendon, P.J. (1978) The assessment of sediment sources: a field experiment. *Earth Surface Processes* **3**, 171–8.

Lewin, J., Davies, B.E. & Wolfenden, P.J. (1977) Interactions between channel change and historic mining sediments. In: *River Channel Changes* (Ed. R.C. Gregory), pp. 353–67. Wiley, New York.

Lewin, J., Bradley, S.B. & Macklin, M.G. (1983) Historical valley alluviation in mid-Wales. *Geological Journal* **18**, 331–50.

Lewis, J.B. (2002) Evidence from aerial photography of structural loss of coral reefs at Barbados, West Indies. *Coral Reefs* **21**, 49–56.

Lewis, R.R. (1983) Impacts of oil spills on mangrove forests. In: *Tasks for Vegetation Science*, Vol. 8 (Ed. H.J. Teas), pp. 171–83. Dr W. Junk Publishers, The Hague.

Lewis, S.G., Maddy, D. & Scaife, R.G. (2001) The fluvial system response to abrupt climate change during the last cold stage: the Upper Pleistocene River Thames fluvial succession at Ashton Keynes, UK. *Global and Planetary Change* **28**, 341–59.

Lick, W., Lick, J. & Ziegler, C.K. (1992) Flocculation and its effect on the vertical transport of fine-grained sediments. *Hydrobiologia* **235/236**, 1–16.

Liddell, W.D. & Ohlhorst, S.L. (1988) Hard substrata community patterns, 1–120 m, north Jamaica. *Palaios* **3**, 413–23.

Liddell, W.D. & Ohlhorst, S.L. (1993) Ten years of disturbance and change on a Jamaican fringing reef. *Proceedings of Seventh International Coral Reef Symposium*, Vol. 1, pp. 149–55. University of Guam Press, UOG Station, Guam.

Likens, G., Wright, R., Galloway, J., et al. (1979) Acid rain. *Scientific American* **241**, 43–51.

Linton, R.W., Natusch, D.F.S., Solomon, R.L., et al. (1980) Physicochemical characterisation of lead in urban dusts. A microanalytical approach to lead tracing. *Environmental Science and Technology* **4**, 159–64.

List, J.H., Sallenger, A.H., Hansen, M.E., et al. (1997) Accelerated relative sea-level rise and rapid coastal erosion: testing a causal relationship for the Louisiana barrier islands. *Marine Geology* **140**, 347–65.

Livingstone, I. & Warren, A. (1996) *Aeolian Geomorphology: an Introduction*. Longman, London.

London Convention (2005) *Compendium of Monitoring Activities at Disposal at Sea Sites in 2003*. Submitted by Canada. London Convention, Scientific Group, 28th Meeting, 23–27 May 2005, paper LC/SG 28/INF.4. 30 pp.

Long, E.D. & MacDonald, D.D. (1998) Recommended uses of empirically derived, sediment quality guidelines for marine and estuarine ecosystems. *Human and Ecological Risk Assessment* **4**, 1019–93.

Long, E.R., MacDonald, D.D., Smith, S.L., et al. (1995) Incidence of adverse biological effects within ranges of chemical concentrations in marine and estuarine sediments. *Environmental Management* **19**, 81–97.

Longfield, S.A. & Macklin, M.G. (1999) The influence of recent environmental change on flooding and sediment fluxes in the Yorkshire Ouse basin. *Hydrological Processes* **13**, 1051–66.

Longman, M.W. (1981) A process approach to recognizing facies of reef complexes. In: *European Fossil Reef Models* (Ed. D.F. Toomey), pp. 9–40. Special

Publication 30, Society for Economic Paleontologists and Mineralogists, Tulsa, OK.

Louis, H. (1975) *Neugefasstes Höhendiagramm der Erde*, pp. 305–26. Bayerische Akademie der Wissenschaften (Mathematisch-Naturwissenschaftliche Klasse Sitzungsberichte), Munich.

Lovley, D.R. & Anderson, R.T. (2000) Influence of dissimilatory metal reduction on fate of organic and metal contaminants in the subsurface. *Hydrogeology Journal* **8**, 77–88.

Lovley, D.R. (1993) Dissimilatory metal reduction. *Annual Reviews in Microbiology* **47**, 263–90.

Lucchitta, I. (1975) *Application of ERTS Images and Image Processing to Regional Geologic Problems and Geologic Mapping in Northern Arizona: Part IVB, the Shivwitts Plateau*. Technical Report 32-1597, National Aeronautical and Space Administration, Washington, DC, 41–72.

Ludlam, S.D. (1981) Sedimentation rates in Fayetteville Green Lake, New York, U.S.A. *Sedimentology* **28**, 85–96.

Ly, C.K. (1980) The role of the Akosombo Dam on the Volta River in causing coastal erosion in central and eastern Ghana (west Africa). *Marine Geology* **37**, 323–32.

Mabutt, J.A. (1978) The impact of desertification as revealed by mapping. *Environmental Conservation* **5**, 45–56.

Mabutt, J.A. (1985) Desertification of the world's rangelands. *Desertification Control Bulletin* (United Nations Environmet Programme, Nairobi) **12**, 1–11.

Mabutt, J.A. (1986) Desertification in Australia. In: *Arid Land Development and the Combat against Desertification: an Integrated Approach*, pp. 101–12. United Nations Environment Programme, Moscow.

MacFarlane, G.R., Pulkownik, A. & Burchett, M.D. (2003) Accumulation and distribution of heavy metals in the grey mangrove, *Avicennia marina* (Forsk.) Vierh.: biological indication potential. *Environmental Pollution* **123**, 139–51.

MacGill, J.T. (1958) Map of coastal landforms of the world. *Geographical Revue* **48**, 402–5.

Machado, W., Silva-Filho, E.V., Oliveira, R.R., et al. (2002) Trace metal retention in mangrove ecosystems in Guanabara Bay, SE Brazil. *Marine Pollution Bulletin* **44**, 1277–80.

Macintyre, I.G. (1985) Submarine cements – the peloidal question. In: *Carbonate Cements* (Eds N. Schneidermann & P.M. Harris), pp. 109–16. Special Publication 36, Society for Economic Paleontologists and Mineralogists, Tulsa, OK.

Mackay, N. (1990) Understanding the Murray. In: *The Murray* (Eds N. Mackay & D. Eastburn), pp. viii–xix. Murray-Darling Basin Commission, Canberra.

MacKenzie, A.B., Cook, G.T. & McDonald, P. (1999) Radionuclide distributions and particle size associations in Irish Sea surface sediments: implications for actinide dispersion. *Journal of Environmental Radioactivity* **44**, 275–96.

Mackey, D. & Paterson, S. (1982) Fugacity revisited. *Environmental Science and Technology* **16**, 654–60.

Macklin, M.G. (1988) *A Fluvial Geomorphological Based Evaluation of Contamination of the Tyne Basin, North-east England by Sediment-borne Heavy Metals.* Unpublished Report to the Natural Environment Research Council, Swindon, 29 pp.

Macklin, M.G. (1992) Metal contaminated soils and sediment: a geographical perspective. In: *Managing the Human Impact on the Natural Environment: Patterns and Processes* (Ed. M.D. Newson), pp. 172–95. Belhaven Press, London.

Macklin, M.G. (1996) Fluxes and storage of sediment-associated heavy metals in floodplain systems: assessment and river basin management issues at a time of rapid environmental change. In: *Floodplain Processes* (Eds M.G. Anderson, D.E. Walling & P.D. Bates), pp. 441–60. Wiley, Chichester.

Macklin, M.G. & Lewin, J. (1989) Sediment transfer and transformation of an alluvial valley floor: the River South Tyne, Northumbria, UK. *Earth Surface Processes and Landforms* **14**, 232–46.

Macklin, M.G. & Lewin, J. (1993) Holocene river alluviation in Britain. *Zeitschrift für Geomorphologie Supplement-Band* **88**, 109–22.

Macklin, M.G. & Lewin, J. (2003) River sediments, great floods and centennial-scale Holocene climate change. *Journal of Quaternary Science* **18**, 101–5.

Macklin, M.G., Rumsby, B.T. & Newson, M.D. (1992) Historic overbank floods and vertical accretion of fine-grained alluvium in the lower Tyne valley, north east England. In: *Dynamics of Gravel-bed Rivers, Proceedings of the Third International Workshop on Gravel-bed Rivers* (Eds P. Billi, R.D. Hey, P. Tacconi & C. Thorne), pp. 564–80. Wiley, Chichester.

Macklin, M.G., Ridgway, J., Passmore, D.G., et al. (1994) The use of overbank sediment for geochemical mapping and contamination assessment: results from selected English and Welsh floodplains. *Applied Geochemistry* **9**, 689–700.

Macklin, M.G., Brewer, P.A., Balteanu, D., et al. (2003) The longer term fate and environmental significance of contaminant metals released by the January and March 2000 mining tailings dam failures in the Maramure County, upper Tisa Basin, Romania. *Applied Geochemistry* **18**, 241–57.

Macleod, C.L., Scrimshaw, M.D., Emmerson, R.H.C., et al. (1999) Geochemical changes in metal nutrient loading at Orplands Farm managed realignment site, Essex, UK. *Marine Pollution Bulletin* **38**(12), 1115–25.

Madrid, L., Diaz-Barrientos, E. & Madrid, F. (2002) Distribution of heavy metal contents of urban soils in parks of Seville. *Chemosphere* **49**, 1301–8.

Magilligan, F.J., Gomez, B., Mertes, L.A.K., et al. (2002) Geomorphic effectiveness, sandur development, and the pattern of landscape response during jökulhlaups: Skeiðoardrsandur, southeastern Iceland. *Geomorphology* **44**(1–2), 95–113.

Maizels, J. (1997) Jökulhlaup deposits in proglacial areas. *Quaternary Science Reviews* **16**(7), 793–819.

Major, J.J. (2003) Post-eruption hydrology and sediment transport in volcanic river stystems. *Water Resources Impact* **5**(3), 10–15.

Major, J.J., Pierson, T.C., Dinehart, R.L., et al. (2000) Sediment yield following severe volcanic disturbance – a two decade perspective from Mount St Helens. *Geology* **28**, 819–22.

Major, J.J., Scott, W.E., Driedger, C., et al. (2005) *Mount St. Helens Erupts Again: Activity from September 2004 through March 2005.* Fact Sheet FS2005-3036, U.S. Geological Survey, Denver, CO http://vulcan.wr.usgs.gov/Volcanoes/MSH/Publications/FS2005-3036/FS2005-3036.pdf (accessed July 2005).

Mann, K.H. (2000) *Ecology of Coastal Waters.* Blackwell, Oxford.

Marsalek, J. & Marsalek, P.M. (1997) Characteristics of sediments from a stormwater management pond. *Water Science Technology* **36**, 117–22.

Marston, R.A., Girel, J., Pautou, G., et al. (1995) Channel metamorphosis, floodplain disturbance, and vegetation development: Ain River, France. *Geomorphology* **13**, 121–31.

Martindale, W. (1992) Calcified epibionts as palaeoecological tools: examples from the Recent and Pleistocene reefs of Barbados. *Coral Reefs* **11**, 167–77.

Marui, H., Watanabe, N., Sato, O., et al. (1997) Gamahara torrent debris flow on 6 December 1996, Japan – 10 debris flow disaster. *Landslide News* **10**, 4–6.

Marutani, T., Brierley, G.J., Trustrum, N.A., et al. (2001) *Source-to-sink Sedimentary Cascades in Pacific Rim Geosystems.* Matsumoto Sabo Works Office, Ministry of Land, Infrastructure and Transport, Japan.

Massadeh, A.M. & Snook, R.D. (2002) Determination of Pb and Cd in road dusts over the period in which Pb was removed from petrol in the UK. *Journal of Environmental Monitoring* **4**, 567–72.

Masselink, G. & Short, A.D. (1993) The influence of tide range on beach morphodynamics: a conceptual model. *Journal of Coastal Research* **9**, 785–800.

May, J.P. & Tanner, W.F. (1973) The littoral power gradient and shoreline changes. In: *Coastal Geomorphology* (Ed. D.R. Coates), pp. 43–61. State University of New York, Binghampton.

May, R.M. (1978) Human reproduction reconsidered. *Nature* **272**, 491–5.

McAlister, J.J., Smith, B.J. & Neto, J.A.B. (2000) The presence of calcium oxalate dihydrate (weddellite) in street sediments from Niteroi, Brazil and its health implications. *Environmental Geochemistry and Health* **22**, 195–210.

McCarthy, J.J., Canziani, O.F., Leary, N.A., et al. (2001) *Climate Change 2001: Impacts, Adaptation and Vulnerability*. Contribution of Working Group II to the Third Annual Assessment Report of the Intergovernmental Panel on Climate Change. Cambridge University Press, Cambridge.

McCauley, J.E., Parr, R.A. & Hancock, D.R. (1977) Benthic infauna and maintenance dredging: a case study. *Water Research* **11**, 233–42.

McCauley, J.F., Breed, C.S., El-Baz, F., et al. (1979) Pitted and fluted rocks in the western Desert of Egypt – Viking comparisons. *Journal of Geophysical Research* **84**, 8222–32.

McCauley, J.F., Ward, A.W., Breed, C.S., et al. (1977) *Experimental Modelling of Wind Erosion Forms*. Technical Memorandum X-3511, National Aeronautics and Space Administration, Washington, DC, pp. 150–2.

McCave, I.N. (1970) Deposition of fine-grained suspended sediments from tidal currents. *Journal of Geophysics Research* **75**, 4151–9.

McClanahan, T.R. (2000) Bleaching damage and recovery potential of Maldivian coral reefs. *Marine Pollution Bulletin* **40**, 587–97.

McCoy, E.D., Mushinsky, H.R., Johnson, D., et al. (1996) Mangrove damage caused by Hurricane Andrew on the southwestern coast of Florida. *Bulletin of Marine Science* **59**, 1–8.

McFadden, L.D., Wells, S.G. & Jercinovich, M.J. (1987) Influences of aeolian and pedogenic processes on the origin and evolution of desert pavements. *Geology* **15**, 504–8.

McFadden, L.D., Wells, S.G., Dohrenwend, J.C., et al. (1984) Cumulic soil formed in eolian parent materials on flows of the Cima Volcanic field, Mohave Desert. *Geological Society of America Guidebook 14*, Annual Meeting, Reno, NV, pp. 134–49.

McGregor, G.R. & Nieuwolt, S. (1998) *Tropical Climatology*. Wiley, Chichester.

McKee, E.D. (Ed.) (1979) *A Study of Global Sand Seas*. Professional Paper 1052, US Geological Survey, Denver, CO.

McKenna-Neumann, C. & Nickling, W.G. (1989) A theoretical and wind tunnel investigation of the effect of capillary water on the entrainment of soil by wind. *Canadian Journal of Soil Science* **69**, 79–96.

McKnight, D.M. & Bencala, K.E. (1989) Reactive iron transport in an acidic mountain stream in Summit County, Colorado: a hydrologic perspective. *Geochimica et Cosmochimica Acta* **53**, 2225–34.

McLusky, D.S. (1989) *The Estuarine Ecosystem*, 2nd edn. Blackie, Glasgow.

McMahon, T.A. (1979) Hydrological characteristics of arid zones. In: *The Hydrology of Areas of Low Precipitation. Proceedings of a Symposium held during the XVII General Assembly of the IUGG at Canberra, December*, pp. 105–23. IAHS Publication 128, International Association of Hydrological Sciences Press, Wallingford.

McManus, J. (1988) Grain size determination and interpretation. In: *Techniques in Sedimentology* (Ed. M.E. Tucker), pp. 63–85. Blackwell Science, Oxford.

McSaveney, M.J. (1978). Sherman Glacier rock avalanche, Alaska, USA. In: *Rockslides and Avalanches*, Vol. 1, *Natural Disasters* (Ed. B. Voight), pp. 197–58. Elsevier, Amsterdam.

Meade, R.H. & Parker, R.S. (1985) Sediment in rivers of the United States. *U.S. Geological Survey Watersupply Paper* **2275**, 49–60.

Mei-e, R. & Xianmo, Z. (1994) Anthropogenic influences on changes in the sediment load of the Yellow River, China, during the Holocene. *The Holocene* **4**, 314–20.

Meigs, P. (1953) World distribution of arid and semi-arid homoclimates. In: *Arid Zone Hydrology. UNESCO Arid Zone Research Series* **1**, 203–9.

Memon, F.A. & Butler, D. (2002a) Identification and modelling of dry weather processes in gully pots. *Water Research* **36**, 1351–9.

Memon, F.A. & Butler, D. (2002b) Assessment of gully pot management strategies for runoff quality control using a dynamic model. The *Science of the Total Environment* **295**, 115–29.

Merilehto, K., Kenttämies, K. & Kämäri, J. (1988) *Surface Water Acidification in the ECE Region*. Nordic Council of Ministers, NORD 1988:89, 156 pp.

Merrington, G. & Alloway, B.J. (1994) The transfer and fate of Cd, Cu, Pb and Zn from two historic metalliferous mine sites in the U.K. *Applied Geochemistry* **9**, 677–87.

Messerli, B. (1983) Stability and instability of mountain ecosystems: introduction to the workshop. *Mountain Research and Development* **3**(2), 81–94.

Meybeck, M. (1979) Concentrations des eaux fluviales en éléments majeurs et apports en solution aux océans. *Revue de Géologie Dynamique et de Géographie Physique* **21**, 215–46.

Meybeck, M. (2003) Global analysis of river systems: from Earth system controls to Anthropocene syndromes. *Philosophical Transactions of the Royal Society of London, Series B*, Online DOI 10.1098/rstb.2003.1979, 21 pp.

Middleton, N. (1997) Desert Dust. In: *Arid Zone Geomorphology: Process, Form and Change in Drylands*, 2nd edn (Ed. D.S.G. Thomas), pp. 413–36. Wiley, Chichester.

Miguel, A.G., Cass, G.R., Glovsky, M.M., et al. (1999) Allergens in paved road dust and airborne particles. *Environmental Science and Technology* **33**, 4159–68.

Miller, A.J. & Gupta, A. (Eds) (1999) *Varieties of Fluvial Form*. Wiley, Chichester.

Miller, C. (1999) Field measurements of longshore sediment transport during storms. *Coastal Engineering* **36**, 301–21.

Miller, J.R. (1997) The role of fluvial geomorphic processes in the dispersal of heavy metals from mine sites. *Journal of Geochemical Exploration* **58**, 101–18.

Miller, J.R., Lechler, P.J. & Desilets, M. (1998) The role of geomorphic processes in the transport and fate of mercury within the Carson River Basin, west-central Nevada. *Environmental Geology* **33**, 249–62.

Milliman, J.D. (2001) Delivery and fate of fluvial water and sediment to the sea: a marine geologists's view of European rivers. In: *A Marine Science Odyssey into the 21st Century* (Eds J.M. Gili, J.L. Pretus & T.T. Packard). *Scientia Marina* **65**, 121–32.

Milliman, J.D. & Meade, R.H. (1983) World-wide delivery of river sediment to the oceans. *Journal of Geology* **91**, 1–21.

Milliman, J.D. & Syvitski, P.M. (1992) Geomorphic/tectonic control of sediment discharge to the ocean: the importance of small mountainous terrain. *The Journal of Geology* **100**, 525–44.

Milliman, J.D., Broadus, J.M. & Gable, F. (1989) Environmental and economic implications of rising sea level and subsiding deltas: the Nile and Bengal examples. *Ambio* **18**, 340–5.

Millington, A.C. (1989) African soil erosion: nature undone and the limits of technology. *Land degradation and Rehabilitation* **1**, 279–90.

Möeller, I., Spencer, T. & French, J.R. (1996) Wind wave attenuation over salt marsh surfaces: preliminary results from Norfolk, England. *Journal of Coastal Research* **12**, 1009–16.

Mogi, A. (1979) *An Atlas of the Sea Floor around Japan: Aspects of Submarine Geomorphology*. University of Tokyo Press.

Monitor (1986) Acid and acidified waters (in Swedish). Swedish Environmental Protection Agency, Solna, 180 pp.

Moore, J.N., Brook, E.J. & Johns, C. (1987) Grain size partitioning of metals in contaminated, coarse-grained floodplain sediment: Clark Fork River, Montana. *Environmental Geology and Water Science* **14**, 107–15.

Morehead, M.K., Syvitski, J.P. & Hutton, E.W.H. (2001) The link between abrupt climate change and basin stratigraphy: a numerical approach. *Global and Planetary Change* **28**, 107–27.

Morgan, K.P.C. (1994) *Soil Erosion and Conservation*. Longman, London.

Morris, S.E. & Moses, T.A. (1987) Forest fire and the natural soil erosion regime in the Colorado Front Range. *Association of American Geographers* **77**, 245–54.

Morrison, G.M., Revitt, D.M., Ellis, J.B., et al. (1988) Transport mechanisms and processes for metal species in a gully pot system. *Water Research* **22**, 1417–27.

Morrison, G.M., Revitt, D.M. & Ellis, J.B. (1995) The gully pot as a biochemical reactor. *Water Science Technology* **31**, 229–36.

Morton, R.A., Gibeaut, J.C. & Paine, J.G. (1995) Meso-scale transfer of sand during and after storms: implications for prediction of shoreline movement. *Marine Geology* **126**, 161–79.

Mosley, M.P. & Schumm, S.A. (2001) Gravel bed rivers – the view from the hills. In: *Gravel-bed Rivers V* (Ed. M.P. Mosley), pp. 479–505. The Caxton Press, Christchurch.

Motelica-Heino, M., Rauch, S., Morrison, G.M., et al. (2001) Determination of palladium, platinum and rhodium concentrations in urban road sediments by laser ablation-ICP-MS. *Analytica Chimica Acta* **436**, 233–44.

Mount, N.J., Sambrook Smith, G.H. & Stott, T.A. (2005) An assessment of the impact of upland afforestation on lowland river reaches: the Afon Trannon, mid-Wales. *Geomorphology* **64**, 255–69.

Mualem, Y., Assouline, S. & Rohdenburg, H. (1990) Rainfall induced soil seal: a critical review of observation and models. *Catena* **17**, 185–203.

Mulder, J.P.M., van Koningsveld, M., Owen, M.W., et al. (2001) *Guidelines on the Selection of CZM tools*. Report RIKZ/2001.020, Rijkswaterstaat/RIKZ, UK Environment Agency, Swindon.

Muller, A. & Wessels, M. (1999) The flood in the Odra river 1997 – impact of suspended solids on water quality. Acta Hydrochim. *Hydrobiologia* **27**, 316–20.

Munawar, M. & Dave, G. (Eds) (1996) *Development and Progress in Sediment Quality Assessment: Rationale, Challenges, Techniques and Strategies*. SPB Academic Publishing, Amsterdam, 255 pp.

Munoz, D., Guiliano, M., Doumenq, P., et al. (1997) Long term evolution of petroleum biomarkers in mangrove soils (Guadeloupe). *Marine Pollution Bulletin* **34**, 868–74.

Myers, N. (1993) *Gaia: an Atlas of Planet Management*. Anchor and Doubleday, New York.

Nageotte, S.M. & Day, J.P. (1998) Lead concentrations and isotope ratios in street dust determined by electrothermal atomic absorption spectrometry and inductively coupled plasma mass spectrometry. *Analyst* **123**, 59–62.

Nanayama, F., Shigeno, K., Satake, K., et al. (2000) Sedimentary differences between the 1993 Hokkaido-nanse-oki tsunami and the 1959 Miyakojima typhoon at Taisea, southwestern Hokkaido, northern Japan. *Sedimentary Geology* **135**, 255–64.

Nash, D.J., Thomas, D.S.G. & Shaw, P.A.A. (1994) Timescales, environmental change and dryland valley development. In: *Environmental Change in Drylands: Biogeographical and Geomorphological Perspectives* (Eds A.C. Millington & K. Pye), pp. 25–41. Wiley, Chichester.

Neall, V.E. (1976) Lahars as major geological hazards. *Bulletin of the International Association of Engineering Geologists* **14**, 233–40.

Needleman, H.L., Gunnoe, C. & Leviton, A. (1979) Deficits in psychological and classroom performance in children with elevated dentine lead levels. *New England Journal of Medicine* **300**, 689–95.

Neil, D.T., Orpin, A.R., Ridd, P.V., et al. (2002) Sediment yield and impacts from river catchments to the Great Barrier Reef lagoon. *Marine and Freshwater Research* **53**, 733–52.

Nelsen, J.E. & Ginsburg, R.N. (1986) Calcium carbonate production by epibionts on *Thalassia* in Florida Bay. *Journal of Sedimentary Petrology* **56**, 622–8.

Nelson, E.J. & Booth, D.B. (2002) Sediment sources in an urbanizing, mixed land-use watershed. *Journal of Hydrology* **264**, 51–68.

Nelson-Smith, A. (1972) Effects of oil industry on shore life in estuaries. *Proceedings of the Royal Society of London* **180B**, 487–96.

Nemec, W. & Steel, R.J. (1984) Alluvial and coastal conglomerates: their significant features and some comments on gravely mass-flow deposits. In: *Sedimentology of Gravels and Conglomerates* (Eds E.H. Koster & R.J. Steel), pp. 1–31. Memoir 10, Canadian Society of Petroleum Geologists, Calgary.

Neumann, A.C. (1966) Observations on coastal erosion in Bermuda and measurement of the boring rate of the sponge, *Cliona lampa*. *Limnology and Oceanography* **11**, 92–108.

Neumann, A.C. & Land, L.S. (1975) Lime mud deposition and calcareous algae in the Bight of Abaco, Bahamas: a budget. *Journal of Sedimentary Petrology* **45**, 763–86.

Neumann, A.C. & Macintyre, I. (1985) Reef response to sea-level rise: keep-up, catch-up or give-up. *Proceedings of 5th International Coral Reef Symposium*, Vol. 3, pp. 105–10. National Museum of Natural History and the Practical School of Advanced Studies, Tahiti, French Polynesia.

Newson, M. (1992) *Land, Water and Development. River Basin Systems and their Sustainable Management.* Routledge, London.

Newson, M.D. (1980) The erosion of drainage ditches and its effects on bed load yields in Wales: reconnaissance case studies. *Earth Surface Processes and Landforms* **5**, 275–90.

Newton, R.S., Siebold, E. & Werner, F. (1973) Facies distribution patterns on the Spanish Saharan continental shelf mapped with side-scan sonar. *'Meteor' Forschungsergebnisse, Reihe C: Geologie und Geophysik* **15**, 55–77.

Nguyen, V.L., Oanh, T.K. & Tateishi, M. (2000) Late Holocene depositional environments and coastal evolution of the Mekong River Delta, southern Vietnam. *Journal of Asian Earth Sciences* **18**, 427–39.

Nicholls, R.J. (2004) Coastal flooding and wetland loss in the 21st Century: changes under the SRES climate and socio-economic scenarios. *Global Environmental Change* **14**, 69–86.

Nickling, W.G. (1988) The initiation of particle movement by wind. *Sedimentology* **35**, 499–511.

Nickling, W.G. (1994) Aeolian sediment transport and deposition. In: *Sediment Transport and Depositional Processes* (Ed. K. Pye), pp. 293–350. Blackwell, Oxford.

Nittrouer, C.A. & Wright, L.D. (1994) Transport of particles across continental shelves. *Reviews in Geophysics* **32**, 85–113.

Nittrouer, C.A., Sternberg, R.W., Carpenter, R., et al. (1979) The use of ^{210}Pb geochronology as a sedimentological tool: application to the Washington continental shelf. *Marine Geology* **31**, 297–316.

Noormets, R., Felton, E.A. & Crook, K.A.W. (2002) Sedimentology of rocky shorelines: 2 Shoreline megaclasts on the north shore of Oahu, Hawaii – origins and history. *Sedimentary Geology* **150**, 31–45.

Nordstrom, K.F. (2000) *Beaches and Dunes of Developed Coasts.* Cambridge University Press, Cambridge, 383 pp.

Novotny, V. (2003) *Water Quality, Diffuse Pollution and Watershed Management*, 2nd edn. Wiley, New York.

Nürnberg, G.K. (1996) Trophic state of clear and colored, soft- and hardwater lakes with special consideration of nutrients, anoxia, phytoplankton and fish. *Journal of Lake and Reservoir Management* **12**, 432–47.

Nürnberg, G.K. & Shaw, M. (1998) Productivity of clear and humic lakes: nutrients, phytoplankton, bacteria. *Hydrobiologia* **382**, 97–112.

O'Hara, S.L., Street-Perrott, F.A. & Burt, T.P. (1993) Accelerated soils erosion around a Mexican highland lake caused by pre-hispanic agriculture. *Nature* **363**, 48–51.

Oak, H.L. (1984) The boulder beach: a fundamentally distinct sedimentary assemblage. *Annals of the Association of American Geographers* **74**, 71–82.

Oberlander, T.M. (1994). Rock varnish in deserts. In: *Geomorphology of Desert Environments* (Eds A.D. Abrahams & A.J. Parsons), pp. 107–19. Chapman & Hall, London.

Odin, G.S. (1988) Green marine clays: oolitic ironstone facies, verdine facies, glaucony facies, and celadonite-bearing facies: a comparative study. *Developments in Sedimentology 45*. Elsevier, New York, 445 pp.

OECD (1982) *Eutrophication of Waters. Monitoring, Assessment and Control.* Organization of Economic Cooperation and Development, Paris, 154 pp.

OECD (1991) *Environmental Indicators.* Organization of Economic Cooperation and Development, Paris, 77 pp.

Oertel, G.F. (1985) The barrier island system. *Marine Geology* **63**, 1–18.

Old, G.H., Leeks, G.J.L., Packman, J.C., et al. (2003) The impact of a convectional summer rainfall event on river flow and fine sediment transport in a highly urbanised catchment: Bradford, West Yorkshire. *The Science of the Total Environment* **314–16**, 495–512.

Oldfield, F., Hunt, A., Jones, M.D.H., et al. (1985) Magnetic differentiation of atmospheric dusts. *Water, Air and Soil Pollution* 317, 516–18.

O'Neill, R.V., Gardner, R.H., Barnthouse, L.W., et al. (1982) Ecosystem risk analysis: a new methodology. *Environmental Toxicology and Chemistry* 1, 167–77.

Orford, J.D., Carter, R.W.G., McKenna, J., et al. (1995) The relationship between the rate of mesoscale sea-level rise and the retreat rate of swash-aligned gravel-dominated coastal barriers. *Marine Geology* 124, 177–86.

Orford, J.D., Cooper, J.A.G. & McKenna, J. (1999) Mesoscale temporal changes to foredunes on Inch Spit, south-west Ireland. *Zeitschrift fur Geomorphologie N.F.* 43, 439–61.

Orford, J.D., Murdy, J.M. & Wintle, A.G. (2003) Prograded Holocene beach ridges with superimposed dunes in north-east Ireland: mechanisms and time-scales of fine and coarse beach sediment decoupling and deposition. *Marine Geology* 194, 47–64.

Ornstein, P., Jaboyedoff, M. & Rouiller, J.D. (2001) Surveillance géodésique du site de Randa (VS): gestion des measures 1-D et 3-D. *Publication Soc, Suisse Méc Sols Roches* 143, 82–91.

Orpin, A.R., Ridd, P.V. & Stewart, L.K. (1999) Assessment of the relative importance of major sediment transport mechanisms in the central Great Barrier Reef lagoon. *Australian Journal of Earth Sciences* 46, 883–96.

OSPAR Commission (2000) *Quality Status Report 2000, Region II – Greater North Sea*. OSPAR (Oslo/Paris convention (for the Protection of the Marine Environment of the North-East Atlantic)) Commission, London, 136 + xiii pp. (www.ospar.org)

Ostapenia, A.P. (1985) *Ratio between the components of seston*. Ecological system of Naroch lakes. Minsk, pp. 232–3 (in Russian).

Ouchi, S. (1985) Response of alluvial rivers to slow active tectonic movement. *Geological Society of America Bulletin* 96, 504–15.

Owens, P.N. & Slaymaker, O. (Eds) (2004) *Mountain Geomorphology*. Arnold, London, 313 pp.

Owens, P.N. & Walling, D.E. (2002) The phosphorus content of fluvial sediment in rural and industrialized river basins. *Water Research* 36, 685–701.

Owens, P.N., Walling, D.E. & Leeks, G.J.L. (2000) Tracing fluvial suspended sediment sources in the catchment of the River Tweed, Scotland, using composite fingerprints and a numerical mixing model. In: *Tracers in Geomorphology* (Ed. I.D.L. Foster), pp. 291–308. Wiley, Chichester.

Owens, P.N., Walling, D.E., Carton, J., et al. (2001) Downstream changes in the transport and storage of sediment-associated contaminants (P, Cr and PCBs) in agricultural and industrialized drainage basins. *The Science of the Total Environment* 266, 177–86.

Owens, P.N., Apitz, S., Batalla, R., et al. (2004) Sediment management at the river basin scale: synthesis of the SedNet Working Package 2 outcomes. *Journal of Soils and Sediments* 4, 219–22.

Page, M.J., Trustrum, N.A. & Dymond, J.R. (1994) Sediment budget to assess the geomorphic effect of a cyclonic storm, New Zealand. *Geomorphology* 9(3), 169–88.

Page, M.J., Reid, L.M. & Lynn, I.H. (1999) Sediment production from Cyclone Bola landslides, Waipaoa catchment. *Journal of Hydrology (New Zealand)* 38(2), 289–308.

Painter, R.B., Blyth, K., Mosedale, J.C. & Kelly, M. (1974) The effects of afforestation on erosion processes and sediment yields. In: *Effects of Man on the Interface of the Hydrological Cycle with the Physical Environment*, Proceedings of a Symposium held at Paris, September, pp. 62–7. IAHS Publication 113, International Association of Hydrological Sciences Press, Wallingford.

Palanques, A., Guilen, J. & Puig, P. (2001) Impact of bottom trawling on water turbidity and muddy sediment of an unfished continental shelf. *Limnology and Oceanography* 46, 1100–10.

Palermo, M.R., Clausner, J.E., Rollings, M.P., et al. (1998) *Guidance for Sub-aqueous Dredged Material Capping*. Technical Report DOER-1, U.S. Army Engineer Waterways Experiment Station, Vicksburg, MS.

Parkman, R.H., Curtis, C.D., Vaughan, D.J. & Charnock, J.M. (1996) Metal fixation and mobilization in the sediments of Afon Goch Estuary, Dulas Bay, Anglesey. *Applied Geochemistry* 11, 203–10.

Passmore, D.G. & Macklin, M.G. (1994) Provenance of fine-grained alluvium and late Holocene land-use change in the Tyne basin, northern England. *Geomorphology* 9, 127–42.

Pearce, F. (1996) Crumbling away. *New Scientist* 152, 1061–2.

Pearson, T.H. & Rosenberg, R. (1976) A comparative study on the effects on the marine environment of wastes from cellulose industries in Scotland and Sweden. *Ambio* 5, 77–9.

Peart, M.R. & Walling, D.E. (1986) Fingerprinting sediment source: the example of a drainage basin in Devon, UK. In: *Drainage Basin Sediment Delivery*, pp. 41–55. IAHS Publication 159, International Association of Hydrological Sciences Press, Wallingford.

Pedley, H.M. (1990) Classification and environmental models of cool freshwater tufas. *Sedimentary Geology* 68, 143–54.

Pedley, H.M. (1992) Freshwater (phytoherm) reefs: the role of biofilms and their bearing on marine reef cementation. *Sedimentary Geology* 79, 255–74.

Pedley, M. (2000) Ambient temperature freshwater microbial tufas. In: *Microbial Sediments* (Eds R.E. Riding & S.M. Awramik). Springer-Verlag, Berlin.

Pedley, M., Andrews, J. & Ordonez, S. (1996) Does climate control the morphological fabric of freshwater carbonates? A comparative study of Holocene barrage tufas from Spain and Britian. *Palaeogeography, Palaeoeclimatology, Palaeoecology* 121, 239–57.

Peel, R.A. (1970) Landscape sculpture by wind. *21st International Geographical Congress Selected Papers* 1, 99–104.

Pelc, R. & Fujita, R.M. (2002) Renewable energy from the ocean. *Marine Policy* 26, 471–9.

Pennington, W., Cambray, R.S. & Fischer, E.M. (1973) Observations on lake sediments using fallout of Cs-137 as a tracer. *Nature* 242, 324–6.

Perry, C.T. (1996) The rapid response of reef sediments to changes in community structure: implications for time-averaging and sediment accumulation. *Journal of Sedimentary Research* 66, 459–67.

Perry, C.T. (1998a) Macroborers within coral framework at Discovery Bay, north Jamaica: species distribution and abundance and effects on coral preservation. *Coral Reefs* 17, 277–87.

Perry, C.T. (1998b) Grain susceptibility to the effects of microboring: implications for the preservation of skeletal carbonates. *Sedimentology* 45, 39–51.

Perry, C.T. (1999). Reef framework preservation in four contrasting modern reef environments, Discovery Bay, Jamaica. *Journal of Coastal Research* 15, 796–812.

Perry, C.T. (2000) Factors controlling sediment preservation on a north Jamaican fringing reef: a process based approach to microfacies analysis. *Journal of Sedimentary Research* 70, 633–48.

Perry, C.T. (2003) Coral reefs in a high latitude, siliciclastic barrier island setting: reef framework and sediment production at Inhaca Island, southern Mozambique. *Coral Reefs* 22, 485–97.

Perry, C.T. (2005) Structure and development of detrital reef deposits in turbid nearshore environments, Inhaca Island, Mozambique. *Marine Geology* 214, 143–61.

Perry, C.T. & Beavington-Penney, S.J. (2005) Epiphytic calcium carbonate production and facies development from seagrass beds in high latitude, reef-related environments, Inhaca Island, Mozambique. *Sedimentary Geology* 174, 161–76.

Perry, C.T. & Larcombe, P. (2003) Marginal and non-reef building coral environments. *Coral Reefs* 22, 427–32.

Perry, C.T. & Taylor, K.G. (2004) Impacts of bauxite sediment inputs on a carbonate-dominated embayment, Discovery Bay, Jamaica. *Journal of Coastal Research* 20, 1070–9.

Pethick, J. (1984) *An Introduction to Coastal Geomorphology*. Edward Arnold, London.

Petit-Maire, N. & Guo, Z.T. (1998) Mid-Holocene climatic change and Man in the present-day Sahara desert. In: *Quaternary Deserts and Climatic Change* (Eds A.S. Alsharhan, K.W. Glennie, G.L. Whittle & C.G.St.C. Kendall), pp. 351–6. Balkema, Rotterdam.

Pettijohn, F.J., Potter, P.E. & Siever, R. (1987) *Sand and Sandstone*. Springer-Verlag, Berlin, 553 pp.

Petts, G.E. (1984) *Impounded Rivers*. Wiley, Chichester.

Piégay, H., Walling, D.E., Landon, N., et al. (2004) Contemporary changes in sediment yield in an alpine mountain basin due to aforestation (the upper Drôme in France). *Catena* 55, 183–212.

Pierson, T.C. (1988) Hazardous hydrologic consequences of volcanic eruptions and goals for mitigative action: an overview. In: *Hydrology of Disasters Proceedings Technical Conference WNMO, Geneva* (Eds O. Starosolszky & O.M. Melder), pp. 220–36.

Pilkey, O.H. (2003) *Barrier Islands*. Columbia University Press, Columbia, OH.

Pilkey, O.H. & Cooper, J.A.G. (2004) Society and sea level rise. *Science* 303, 1781–2.

Pilkey, O.H. & Dixon, K.L. (1996) *The Corps and the Shore*. Island Press, Washington, DC.

Pilkey, O.H., Neal, W.J., Riggs, S.R., et al. (1998) *The North Carolina Shore and its Barrier Islands*. Duke University Press, Durham, NC.

Pirazzoli, P.A. (1996) *Sea-level Change: the Last 20,000 Years*. Wiley, Chichester.

Pirrie, D., Power, M.R., Wheeler, P.D. & Ball, A.S. (2000) A new occurrence of diagenetic simonkolleite from the Ganel Estuary, Cornwall. *Geoscience in Southwest England* 10, 18–20.

Plafker, G. & Eriksen, G.E. (1978) Nevados Huascaran avalanches, Peru. In: *Rockslides and Avalanches*, Vol. 1, *Natural Disasters* (Ed. B. Voight), pp. 48–55. Elsevier, Amsterdam.

Plaziat, J.C. (1974) Mollusc distribution and its value for recognition of ancient mangroves. *Proceedings of International Symposium on the Biology and Management of Mangroves*, Honolulu, Vol. 2, pp. 456–65.

Plaziat, J.C. (1995) Modern and fossil mangroves and mangals: their climatic and biogeographic variability. In: *Marine Palaeoenvironmental Analysis from Fossils* (Eds D.W.J. Bosence & P.A. Allison), pp. 73–96. Special Publication 83, Geological Society Publishing House, Bath.

Poesen, J. (1986) Surface sealing as influenced by slope angle and position of simulated stones in the top layer of loose sediments. *Earth Surface Processes and Landforms* 11, 1–10.

Porter-Smith, R., Harris, P.T., Andersen, O.B., et al. (2004) Classification of the Australian continental shelf based on predicted sediment threshold exceedance from tidal currents and swell waves. *Marine Geology* 211, 1–20.

Postma, G. (1986) Classification for sediment gravity flow deposits based on flow conditions during sedimentation. *Geology* 14, 291–4.

Potts, D.C. & Jacobs, J.R. (2003) Evolution of reef-building scleractinian corals in turbid environments: a paleo-ecological hypothesis. In: *Proceedings of the*

9th International Coral Reef Symposium, Bali, Vol. 2, pp. 249–54. Ministry of Environment (Indonesia), Indonesian Institute of Sciences and International Society for Reef Studies.

Power, J, McKenna, J., MacLeod, M., et al. (2000) Developing integrated participatory management strategies for Atlantic dune systems in County Donegal, Ireland. *Ambio* **29**, 143–9.

Powers, M.C. (1982) *Comparison Chart for Estimating Roundness and Sphericity*. AGI Data Sheet 18, American Geological Institute, Alexandria, VA.

Preu, C. (1989) Coastal erosion in southwestern Sri Lanka: consequences of human interference. *Malaysian Journal of Tropical Geography* **20**, 30–42.

Price, L.W. (1981) *Mountain and Man: a study of Process and Environment*. University of California Press, Berkeley.

Pringle, A.W. (1995) Erosion of a cyclic saltmarsh in Morecambe Bay, North-West England. *Earth Surface Processes and Landforms* **20**, 387–405.

Productivity Commission (2003) *Industries, Land Use and Water Quality in the Great Barrier Reef Catchment*. Research Report, 415 pp. (available at http://www.pc.gov.au/study/gbr/finalreport/index.html).

Prospero, J.M., Bonatti, E., Schubert, C., et al. (1970) Dust in the Caribbean atmosphere traced to an African dust storm. *Earth and Planetary Science Letters* **9**, 287–93.

Purdy, E.G. (1963) Recent calcium carbonate facies of the Great Bahama Bank. 2. Sedimentary facies. *Journal of Sedimentary Petrology* **71**, 474–97.

Pye, K. & French, P.W. (1993) *Targets for Coastal Habitat Re-creation*. English Nature Science Series No. 13, English Nature, Peterborough.

Pye, K. & Tsoar, H. (1990) *Aeolian Sand and Sand Dunes*. Unwin Hyman, London. 396 pp.

Qu, W. & Kelderman, P. (2001) Heavy metal contents in the Delft canal sediments and suspended solids of the River Rhine: multivariate analysis for source tracing. *Chemosphere* **45**, 919–25.

Quanterra (2003) *Short Guide about Slope Instabilitites between Lausanne and Zermatt (Switzerland)*. (18) Randa. http://www.quanterra.org/guide/guide1_18.htm. Accessed May, 2005.

Radoane, M., Radoane, N. & Dumitriu, D. (2003) Geomorphological evolution of longitudinal river profiles in the Carpathians. *Geomorphology* **50**, 293–306.

Rae, J.E. & Allen, J.R.L. (1993) The significance of organic matter degradation in the interpretation of historical pollution trends in depth profiles of estuarine sediment. *Estuaries* **16**, 678–82.

Raetzo, H., Latelin, O., Bollinger, D., et al. (2002) Hazard assessment in Switzerland – codes of practice for mass movements. *Bulletin of Engineering Geology and the Environment* **61**, 263–8.

Rahn, K.A., Borys, R.D. & Shaw, G.E. (1981) Asian desert dust over Alaska: anatomy of an Arctic haze episode. In: *Desert Dust* (Ed. T.L. Péwé), pp. 37–70. Special Paper 186, Geological Society of America, Boulder, CO.

Ramsay, M.A., Swannells, R.P.J., Shipton, W.A., et al. (2000) Effect of bioremediation on the microbial community in oiled mangrove sediments. *Marine Pollution Bulletin* **41**, 413–19.

Rapp, A. (1960) Recent developments of mountain slopes in Kärkevagge and surroundings, northern Scandanavia. *Geografiska Annaler* **42A**, 71–200.

Rasser, M.W. & Riegl, B. (2002) Holocene coral reef rubble and its binding agents. *Coral Reefs* **21**, 57–72.

Rawat, J.S. & Rawat, M.S. (1994) Accelerated erosion and denudation in the Nani Kosi watershed, Central Himalaya, India. Part I: sediment load. *Mountain Research and Development* **14**, 25–38.

Reading, H.G. (Ed.) (1996) *Sedimentary Environments: Processes, Facies and Stratigraphy*, 3rd edn. Blackwell Scientific, Oxford, 688 pp.

Reddering, J.S.V. (1983) An inlet sequence produced by migration of a small microtidal inlet against longshore drift: the Keurbooms inlet, South Africa. *Sedimentology* **30**, 201–18.

Reeves, C.C., Jr. (1968) *Introduction to Paleolimnology*. Elsevier, Amsterdam, 228 pp.

Reichett-Brushett, A.J. & Harrison, P.L. (1999) The effects of copper on the settlement success of larvae from the scleractinian coral *Acropora tenuis*. *Marine Pollution Bulletin* **41**, 385–91.

Reid, I. & Frostick, L.E. (1987) Flow dynamics and suspended sediment properties in arid zone flash floods. *Hydrological Processes* **1**, 239–53.

Reid, I. & Frostick, L.E. (1994) Fluvial sediment transport and deposition In: *Sediment Transport and Depositional Processes* (Ed. K. Pye), pp. 89–155. Blackwell Science, Oxford.

Reid, I. & Laronne, J.B. (1995) Bedload sediment transport in an ephemeral stream and a comparison with seasonal and perennial counterparts. *Water Resources Research* **31**, 773–81.

Reid, I., Laronne, J.B., Powell, D.M., et al. (1994). Flash floods in desert rivers: studying the unexpected. *Eos (Transactions of the American Geophysical Union)* **75**, 452.

Reisner, M. (1986) *Cadillac Desert: the American West and its disappearing water*. Viking Penguin, London.

Renard, K.G. & Keppel, R.V. (1966) Hydrographs of ephemeral streams in the Southwest. *Proceedings of the American Society of Civil Engineers, Journal of the Hydraulics Division* **92** (HY2), 33–52.

Rhoades, J.D. (1990) Soil salinity – causes and controls. In: *Techniques for Desert Reclamation* (Ed. A.S. Goudie), pp. 109–34. Wiley, Chichester.

Rice, S.P. & Church, M. (1998) Grain size along two gravel-bed rivers: statistical variation, spatial pattern and sedimetary links. *Earth Surface Processes and Landforms* 23, 345–63.

Richard, G.A. (1978) Seasonal and environmental variations in sediment accretion in a Long Island salt marsh. *Estuaries* 1, 29–35.

Richards, K.S. (1982) *Rivers: Form and Process in Alluvial Channels.* London, Methuen.

Rickenmann, D. (1990) Debris flows 1987 in Switzerland: modelling and fluvial sediment transport. In: *Hydrology in Mountainous Regions. II – Artificial Reservoirs; Water and Slopes* (Eds R.O. Sinniger & M. Monbaron), pp. 371–8. IAHS Publication 194, Proceedings of two Symposia held at Lausanne, August. International Association of Hydrological Sciences Press, Wallingford.

Rickenmann, D. & Zimmermann, M. (1993) The 1987 debris flows in Switzerland: Documentation and analysis. *Geomorphology* 8, 175–89.

Ridd, P.V. (1996) Flow through animal burrows in mangrove creeks. *Estuarine, Coastal and Shelf Science* 43, 617–25.

Ridd, P.V., Stieglitz, T. & Larcombe, P. (1998) Density-driven secondary circulation in a tropical mangrove estuary. *Estuarine, Coastal and Shelf Science* 47, 621–32.

Ridge, M.J.H. & Carson, B. (1987) Sediment transport on the Washington continental shelf: estimates of dispersal rates from Mount St. Helen's ash. *Continental Shelf Research* 7, 759–72.

Riegl, B. & Piller, W.E. (2000) Reefs and coral carpets in the northern Red Sea as models for organism-environment feedback in coral communities and its reflection in growth fabrics. In: *Carbonate Platform Systems: Components and Interactions* (Eds E. Insalaco, P.W. Skelton & T.J. Palmer), pp. 71–88. Special Publication 178, Geological Society Publishing House, Bath.

Riegl, B., Schleyer, M.H., Cook, P.J., et al. (1995) Structure of Africa's southernmost coral communities. *Bulletin of Marine Science* 56, 648–63.

Riethmüller, R., Hakvoort, J.H.M., Heineke, M., et al. (1998) Relating erosion shear stress to tidal flat surface colour. In: *Sedimentary Processes in the Intertidal Zone* (Eds K.S. Black, D.H. Paterson & A. Cramp), pp. 283–93. Special Publication 139, Geological Society Publishing House, Bath.

Rijnsdorp, A.D., Bujiis, A.M., Storbeck, F., et al. (1998) Microscale distribution of beam trawl effort in the southern North Sea between 1993 and 1996 in relation to the trawling frequency of the sea bed and the distribution of benthic organisms. *ICES Journal of Marine Science* 55, 403–19.

Ringrose-Voase, A.J., Rhoades, D.W. & Hall, G.F. (1989) Reclamation of a scalded, red duplex soil by waterponding. *Australian Journal of Soil Research* 27, 779–95.

Rivera-Duarte, I. & Flegal, A.R. (1997a) Porewater gradients and diffusive benthic fluxes of Co, Ni, Cu, Zn and Cd in San Francisco Bay. *Croatica Chemica Acta* 70, 389–417.

Rivera-Duarte, I. & Flegal, A.R. (1997b) Pore-water silver concentrations and benthic fluxes from contaminated sediments of San Francisco Bay, California, U.S.A. *Marine Chemistry* 56, 15–26.

Robbins, J.A. (1978) Geochemical and Geophysical Applications of Radioactive Lead. In: *Biogeochemistry of Lead in the Environment* (Ed. J.O. Nriagu), pp. 285–93. Elsevier, Amsterdam.

Roberts, H.H., Wiseman, J. & Suchanek, T.H. (1981) Lagoon sediment transport: the significant effect of *Callianassa* bioturbation. *Proceedings of Fourth International Coral Reef Symposium*, Vol. 1, pp. 459–65. Marine Science Center, University of the Philippines, Manila, Philippines.

Robertson, A.I., Alongi, D.M. & Boto, K.G. (1992) Food chains and carbon fluxes. In: *Tropical Mangrove Ecosystems* (Eds A.I. Robertson & D.M. Alongi), pp. 293–326. Coastal and Estuarine Studies 41, American Geophysical Union, Washington.

Robertson, D.J., Taylor, K.G. & Hoon, S.R. (2003) Geochemical and mineral characterisation of urban sediment particulates, Manchester. UK. *Applied Geochemistry* 18, 269–82.

Robinson, J.E., Newell, R.C., Seiderer, L.J., et al. (2005) Impacts of aggregate dredging on sediment composition and associated benthic fauna at an offshore dredge site in the southern North Sea. *Marine Environmental Research* 60, 51–68.

Robinson, M. & Lambrick, G.H. (1984) Holocene alluviation and hydrology in the Upper Thames basin. *Nature* 308, 809–14.

Rodier, M. (1992) The Rance tidal power station: a quarter of a century. In: *Tidal Power: Trends and Developments* (Eds Institute of Civil Engineers), pp. 301–8. Thomas Telford Press, London.

Roesli, U. & Schindler, C. (1990) Debris flows 1987 in Switzerland: geological and hydrogeological aspects. In: *Hydrology in Mountainous Regions. II – Artificial Reservoirs; Water and Slopes* (Eds R.O. Sinniger & M. Monbaron), pp. 379–86. IAHS Publication 194, Proceedings of two Symposia held at Lausanne, August. International Association of Hydrological Sciences Press, Wallingford.

Rogers, C.S. (1990) Responses of coral reefs and reef organisms to sedimentation. *Marine Ecology Progress Series* 62, 185–202.

Rogers, C.S. (1993) Hurricanes and coral reefs: the intermediate disturbance hypothesis revisited. *Coral Reefs* 12, 127–37.

Rogers, J. & Li, X.C. (2002) Environmental impact of diamond mining on continental shelf sediments off southern Namibia. *Quaternary International* 92, 101–12.

Rogers, S.I. & Greenaway, B. (2005) A UK perspective on the development of marine ecosystem indicators. *Marine Pollution Bulletin* **50**, 9–19.

Rosen, B.R. (1990) Reefs and carbonate build-ups. In: *Palaeobiology – a Synthesis* (Eds D.E.G. Briggs & P.R. Crowther), pp. 341–6. Blackwell Scientific, Oxford.

Rosen, M.R. (1994) The importance of groundwater in playas: a review of playa classifications and the sedimentology and hydrology of playas. In: *Paleoclimate and Basin Evolution of Playa Systems* (Ed. M.R. Rosen), pp. 1–18. Special Paper 289, Geological Society of America, Boulder, CO.

Rostad, C.E., Pereira, W.E. & Leiker, T.J. (1999) Distribution and transport of selected anthropogenic lipophilic organic compounds associated with Mississippi River suspended sediment, 1989–1990. *Archives of Environmental Contamination and Toxicology* **36**, 248–55.

Rowan, J.S., Goodwill, P. & Franks, S.W. (2000) Uncertainty estimation in fingerprinting suspended sediment sources. In: *Tracers in Geomorphology* (Ed. I.D.L. Foster), pp. 279–90. Wiley, Chichester.

Rowlatt, S., Matthiessen, P., Reed, J., et al. (2002) *Review and Recommendations of Methodologies for the Derivation of Sediment Quality Guidelines*. R&D Technical Report P2-082/TR, Environment Agency, Bristol, 113 pp.

Roy, P.S., Cowell, P.J., Ferland, M.A., et al. (1994) Wave-dominated coasts. In: *Coastal Evolution* (Eds R.W.G. Carter & C.D. Woodroffe), pp. 121–86. Cambridge University Press, Cambridge.

Rozan, R.F., Lassman, M.E., Ridge, D.P., et al. (2000) Evidence for iron, copper and zinc complexation as multinuclear sulphide clusters in oxic rivers. *Nature* **406**, 879–82.

Rumsby, B.T. & Macklin, M.G. (1994) Channel and floodplain response to recent abrupt climate change: the Tyne basin, northern England. *Earth Surface Processes and Landforms* **19**, 499–515.

Rust, B.R. (1978) A classification of alluvial channel systems. In: *Fluvial Sedimentology* (Ed. A.D. Miall), pp. 187–98. Memoir 5, Canadian Society of Petroleum Geologists, Calgary.

Rust, B.R. & Waslenchuk, D.G. (1974) The distribution and transport of bed sediments and persistent pollutants in the Ottawa River, Canada. In: *Proceedings of the International Conference on the Transportation of Persistent Chemicals in the Aquatic Ecosystem*. Ottawa, Canada, pp. 1–25.

Ryding, S.O. & Borg, H. (1973) *Sediment-chemical Studies in Lake Lilla Ullevifjärden*. Report 58, National Swedish Environmental Protection Board, NLU, Uppsala, 77 pp. (In Swedish.)

Saad, S., Husain, M.L., Yaacob, R., et al. (1999) Sediment accretion and variability of sedimentological characteristics of a tropical estuarine mangrove: Kemaman, Terengganu, Malaysia. *Mangroves and Salt Marshes* **3**, 51–8.

Saito, Y., Yang, Z. & Hori, K. (2001) The Huanghe (Yellow River) and Changjiang (Yangtze River) deltas: a review on their characteristics, evolution and sediment discharge during the Holocene. *Geomorphology* **41**, 219–31.

Salomons, W. & De Groot, A.J. (1978) Pollution history of trace elements in sediments, as affected by the Rhine River. In: *Environmental Biochemistry*, Vol. 1 (Ed. W.E. Krumbein), pp. 149–62. Ann Arbor Science Publications, Maine.

Salomons, W. & Förstner, U. (1984) *Metals in the Hydrocycle*. Springer-Verlag, Heidelberg, 349 pp.

Sanderson, G. & Elliot, I.G. (1996) Shoreline salients, cuspate forelands and tombolos on the coast of Western Australia. *Journal of Coastal Research* **12**, 761–73.

Santschi, P.H. & Honeyman, B.D. (1991) Radio-isotopes as tracers for the interactions between trace elements, colloids and particles in natural waters. In: *Heavy Metals in the Environment* (Ed. J.-P. Vernet), pp. 229–46. Elsevier, Amsterdam.

Sarnthein, M. & Walger, E. (1974) Der äolische sandstrom aus der W. Sahara zur Atlantikküste. *Geologische Rundschau* **63**, 1065–87.

Sarnthein, M., Kennett, J.P., Allen, J.R.M., et al. & SCOR-IMAGES Working Group 117 (2002) Decadal-to-millennial-scale climate variability – chronology and mechanisms: summary and recommendations. *Quaternary Science Reviews* **10**, 1121–8.

Sarre, R.D. (1989) Aeolian sand drift from the intertidal zone on a temperate beach: potential and actual rates. *Earth Surface Processes and Landforms* **14**, 247–58.

Sarre, R.D. (1990) Evaluation of aeolian sand transport equations using intertidal-zone measurements, Saunton Sands, England – reply. *Sedimentology* **37**, 389–92.

Sartor, J.D. & Gaboury, D.R. (1984) Street sweeping as a water pollution control measure: lessons learned over the past ten years. *The Science of the Total Environment* **33**, 171–83.

Sartor, J.D., Boyd, G.B. & Agardy, F.J. (1974) Water pollution aspects of street surface contaminants. *Journal of Water Pollution Control Federation* **46**, 458–67.

Schell, C., Black, S. & Hudson-Edwards, K.A. (2000) Sediment source characteristics of the Río Tinto, Huelva, south-west Spain. In: *Tracers in Geomorphology* (Ed. I.D.L. Foster), pp. 503–20. Wiley, Chichester.

Schick, A.P. (1970) Desert floods. In: *Symposium of the Results of Research on Representative Experimental Basins*, pp. 478–93. International Association of Scientific Hydrologists, Wallingford.

Schick, A.P. (1988) Hydrologic aspects of floods in extreme arid environments. In: *Flood Geomorphology* (Eds V.R. Baker, R.C. Kochel & P.C. Patton), pp. 189–203. Wiley, New York.

Schindler, C., Cuenod, Y., Eisenlohr, T., et al. (1993) The events of Randa, April 18th and May 9th, 1991 – an uncommon type of rockfall. *Ecologae Geologicae Helvetiae* **86**(3), 643–65.

Schlager, W. (2003) Benthic carbonate factories of the Phanerozoic. *International Journal of Earth Science* **92**, 445–64.

Schlyter, P., Jönsson, P., Nyberg, R., et al. (1993) Geomorphic process studies related to climate change in Kärkevagge, Norhern Sweden – status of research. *Geografiska Annaler* **75A**(1–2), 55–60.

Schmidt, K.H. (1980) Eine neue Methode sur Ermittlung von Stufenrückwanderungsratendargestellt am Beispiel der Black Mesa Schichtstufe, Colorado Plateau, USA. *Zeitschrift für Geomorphologie* **24**, 180–91.

Schmidt, K.H. (1989) The significance of scarp retreat for Cenozoic landform evolution on the Colorado Plateau. *Earth Surface Processes and Landforms* **14**, 93–105.

Schöne, B.R., Freyre Castro, A.D., Feibig, J., et al. (2004) Sea surface water temperatures over the period 1884–1983 reconstructed from oxygen isotope ratios of a bivalve mollusc shell (*Arctica islandica*, southern North Sea). *Palaeogeography, Palaeoclimatology, Palaeoecology* **212**, 215–32.

Schrott, L., Hufschmidt, G., Hankammer, M., et al. (2003) Spatial distribution of sediment storage types and quantification of valley fill deposits in an alpine basin, Reintal, Bavarian Alps, Germany. *Geomorphology* **55**, 45–63.

Schumm, S.A. (1968) *River Adjustment to Altered Hydrologic Regimen – Murrumbidgee River and Palaeochannels, Australia*. Professional Paper 598, US Geological Survey, Denver, CO.

Schumm, S.A. (1969) River metamorphosis. *Journal of the Hydraulic Division, American Society of Civil Engineers* **95** (HY1), 255–73.

Schumm, S.A. & Chorley, R.J. (1966) Talus weathering and scarp recession in the Colorado Plateaus. *Zeitschrift für Geomorphologie*, Supplementband **10**, 11–35.

Schumm, S.A. & Lichty, R.W. (1963) Channel widening and flood-plain construction along Cimarron River in south-western Kansas. *United States Geological Survey Professional Paper* **352D**, 71–88.

Schumm, S.A. & Lichty, R.W. (1965) Time, space and causality in geomorphology. *American Journal of Science* **263**, 110–19.

Schwartz, M. (1967) The Bruun theory of sea-level rise as a cause of shore erosion. *Journal of Geology* **75**, 76–92.

Schwarzer, K., Diesing, M., Larson, M., et al. (2003) Coastline evolution at different time scales: examples from the Pomeranian Bight, southern Baltic sea. *Marine Geology* **194**, 79–101.

Scoffin, T.P. (1970) The trapping and binding of subtidal carbonate sediments by marine vegetation in Bimini lagoon, Bahamas. *Journal of Sedimentary Petrology* **40**, 249–73.

Scoffin, T.P. (1987) *An Introduction to Carbonate Sediments and Rocks*. Blackie, Glasgow.

Scoffin, T.P. (1993) The geological effects of hurricanes on coral reefs and the interpretation of storm deposits. *Coral Reefs* **12**, 203–21.

Scoffin, T.P., Stearn, C.W., Boucher, D., et al. (1977) Calcium carbonate budget of a fringing reef on the west coast of Barbados. *Bulletin of Marine Science* **30**, 475–508.

Scott, D.B., Medioli, F.S., Schafer, C.T. (2001) *Monitoring in coastal environments using Foraminifera and Thecamoebian indicators*. Cambridge University Press, 192 pp.

Scott, K.M. (1988) *Origins, Behaviour, and Sedimentology of Lahars and Lahar-runout Flows in the Toutle–Cowlitz River System*. Professional Paper 1447A, US Geological Survey, Denver, CO, 74 pp.

Scott, K.M., Jana, R.J., de la Cruz, E.G., et al. (1996) Channel and sedimentation responses to large volumes of 1991 volcanic deposits on the east flank of Mount Pinatubo. In: *Fire and Mud, Eruptions and Lahars of Mount Pinatubo, Philippines* (Eds C.G. Newhall & R.S. Punongbayan), pp. 971–88. PHIVOLCS Press, Quezon City, and University of Washington Press, Seattle.

Scourse, J.D., Austin, W.E.N., Long, B.T., et al. (2002) Holocene evolution of seasonal stratification in the Celtic Sea: refined age model, mixing depths and foraminiferal stratigraphy. *Marine Geology* **191**, 119–45.

Scruton, P.C. (1960) Delta building and the deltaic sequence. In: *Recent Sediments. North West Gulf of Mexico* (Eds F.P. Shepard, F.B. Phleger & T.H. van Andel), pp. 82–102. American Association of Petroleum Geologists, Tulsa, OK.

Sear, D.A., Newson, M.D. & Brookes, A. (1995) Sediment-related river maintenance: the role of fluvial geomorphology. *Earth Surface Processes and Landforms* **20**, 629–47.

Sear, D.A., Lee, M.W.E., Oakey, R.J., Carling, P.A. & Collins, M.B. (2000) Coarse sediment tracing technology in littoral and fluvial environments: a review. In: *Tracers in Geomorphology* (Ed. I.D.L. Foster), pp. 21–55. Wiley, Chichester.

Selley, R.C. (1994) *Applied sedimentology*. Academic Press, London.

Semadeni-Davies, A. (2004) Urban water management vs climate change: impacts on cold region waste water inflows. *Climatic Change* **64**, 103–26.

Semeniuk, V. (1981) Long-term erosion of the tidal flats, King Sound, north western Australia. *Marine Geology* **43**, 21–48.

Semeniuk, V. (1994) Predicting the effect of sea-level rise on mangroves in northwestern Australia. *Journal of Coastal Research* **10**, 1050–76.

Serrano-Belles, C. & Leharne, S. (1997) Assessing the potential for lead release from road dusts and soils. *Environmental Geochemistry and Health* **19**, 89–100.

Shackleton, N.J., Berger, A. & Peltier, W.R. (1990) An alternative astronomical calibration of the lower Pleistocene timescale based on ODP Site 677. *Transactions of Royal Society of Edinburgh: Earth Sciences* **81**, 251–61.

Shakesby, R.A., Chafer, C.J., Doerr, S.H., et al. (2003) Fire severity, water repellency characteristics and hydrogeomorphological changes following the Christmas 2001 Sydney forest fires. *Australian Geographer* **34**, 147–75.

Sharon, D. (1974) The spatial pattern of convective rainfall in Sukumaland, Tanzania – a statistical analysis. *Archiv für Meteorologie, Geophysik und Bioklimatologie, Series B* **22**, 201–18.

Shaw, J. (1985) Beach morphodynamics of an Atlantic Coast embayment – Runkerry Strand, Northern Ireland. *Irish Geography* **18**, 50–8.

Shaw, P.A. & Thomas, S.G. (1997) Pans, playas and salt lakes. In: *Arid Zone Geomorphology: Process, Form and Change in Drylands*, 2nd edn (Ed. D.S.G. Thomas), pp. 293–317. Wiley, Chichester.

Shaw, T.L. (1995) Environmental effects of estuary barrages. *Proceedings of the Institute of Civil Engineers: Water, Maritime and Engineering* **112**, 48–59.

Sheppard, C.R.C. & Salm, R.V. (1988) Reef and coral communities of Oman, with a description of a new coral species (Order Scleractinia, Genus *Acanthastrea*). *Journal of Natural History* **22**, 263–79.

Sherman, D. & Bauer, B.B. (1993) Dynamics of beach-dune systems. *Progress in Physical Geography* **17**, 413–47.

Sherman, D.J., Jackson, D.W.T. & Namikas, S. (1998) Wind blown sand on beaches: an evaluation of models. *Geomorphology* **22**, 113–33.

Sherman, D.J., Barron, K.M. & Ellis, J.T. (2002) Retention of beach sands by dams and debris basins in southern California. *Journal of Coastal Research* (Special Issue) **36**, 662–74.

Shi, Z., Hamilton, L.J. & Wolanski, E. (2000) Near-bed currents and suspended sediment transport in saltmarsh canopies. *Journal of Coastal Research* **16**, 909–14.

Shine, J.P., Ika, R. & Ford, T.E. (1998) Relationship between oxygen consumption and sediment-water fluxes of heavy metals in coastal sediments. *Environmental Toxicology and Chemistry* **17**, 2325–7.

Short, A.D. (1999) *Handbook of Beach and Shoreface Morphodynamics*. Wiley, Chichester, 379 pp.

Siebert, L. (1984) Large volcanic debris avalanches: characteristics of source areas, deposits and associated eruptions. *Journal of Volcanology and Geothermal Research* **22**(304), 163–97.

Silvester, R. (1976) Headland defense of coasts. *Coastal Engineering* **1**, 1394–406.

Simmons, I.G. & Innes, J.B. (1996) Prehistoric charcoal in peat profiles at North Gill, North Yorkshire Moores, England. *Journal of Archaeological Science* **23**, 193–7.

Singh, B.P., Pawar, J.S. & Karlupia, S.K. (2004) Dense mineral data from the north-western Himalayan foreland sedimentary rocks and recent river sediments: evaluation of the hinterland. *Journal of Asian Earth Sciences* **23**, 25–35.

Sivan, D., Wdowinski, S., Lambeck, K., et al. (2001) Holocene sea-level changes along the Mediterranean coast of Israel, based on archaeological observations and numerical model. *Palaeogeography, Palaeoclimatology, Palaeoecology* **167**, 101–17.

Slaymaker, O., Souch, C., Menounos, B., et al. (2003) Advances in Holocene mountain geomorphology inspired by sediment budget methodology. *Geomorphology* **55**, 305–16.

Sly, P.G. (1978) Sedimentary processes in lakes. In: *Lakes: Chemistry, Geology, Physics* (Ed. A. Lerman), pp. 65–89. Springer-Verlag, Berlin.

Small, D.L., Payonk, P.M. & Jarrett, J.T. (1997) Beneficial use of dredged material in nearshore placement areas in North Carolina. In: *Proceedings: International Workshop on Dredged Material Beneficial Uses* (Ed. M.C. Landin), p. 152. Unnumbered U.S. Army Corps of Engineers Document, Waterways Experiment Station, Vicksburg, Mississippi.

Smith, B. & Warke, P. (1997) Controls and uncertainty in the weathering environment. In: *Arid Zone Geomorphology: Process, Form and Change in Drylands*, 2nd edn (Ed. D.S.G. Thomas), pp. 41–54. Wiley, Chichester.

Smith, B.A. & Marchand, M. (2001) Coastal erosion and catchment management. *Coastline* **10**(3), 3–6.

Smith, K. (2001) *Environmental Hazards – Assessing Risk and Reducing Disaster*, 3rd edn. Routledge, London.

Smith, N.D. (1974) Sedimentology and bar formation in the Upper Kicking Horse River, a braided outwash stream. *Journal of Geology* **82**, 205–24.

Smith, T.J. (1992) Forest Structure. In: *Tropical Mangrove Ecosystems* (Eds A.I. Robertson & D.M. Alongi), pp. 101–36. Coastal and Estuarine Studies 41, American Geophysical Union, Washington.

Smith, T.J., Robblee, M.B., Wanless, H.R., et al. (1994) Mangroves, hurricanes, and lighting strikes. *BioScience* **44**, 256–62.

Smithers, S. & Larcombe, P. (2003) Late Holocene initiation and growth of a nearshore turbid-zone coral reef: Paluma Shoals, central Great Barrier Reef, Australia. *Coral Reefs* **22**, 499–505.

Smol, J.P. (2002) *Pollution of Lakes and Rivers: a Paleoenvironmental Perspective*. Arnold, London.

Snedden, J.W. & Nummedal, D. (1991) Origin and geometry of storm-deposited sand beds in moderns sediments of the Texas continental shelf. In: *Shelf Sand and Sandstone Bodies* (Eds D.J.P. Swift, G.F. Oertel, R.W. Tillman & J.A. Thorne), pp. 283–308. Special Publication 14, International Association of Sedimentologists. Blackwell Scientific, Oxford.

Södergren, A. (Ed.) (1992) *Bleached Pulp Mill Effluents. Composition, Fate and Effects in the Baltic Sea.* Report 4047, Swedish Environmental Protection Agency, Stockholm, 150 pp.

Södergren, A., Jonsson, P., Bengtsson, B.-E., et al. (1988) *Biological Effects of Bleached Pulp Mill Effluents.* Report 3558, Swedish Environmental Protection, Stockholm, 134 pp.

Soldati, M., Corsini, A. & Pasuto, A. (2004) Landslides and climate change in the Italian Dolomites since the late glacial. *Catena* **55**, 141–61.

Solomon, S.M. & Forbes, D.L. (1999) Coastal hazards and associated management issues on South Pacific Islands. *Ocean & Coastal Management* **42**, 523–54.

Sommerfield, C.K., Nittrouer, C.A. & DeMaster, D.J. (1996) Carbon-isotope systematics on the Amazon shelf. *Geomarine Letters* **16**, 17–23.

Soto-Jiménez, M.F. & Páez-Osuna, F. (2001) Distribution and normalization of heavy metal concentrations in mangrove and lagoonal sediments from Mazatlán Harbor (SE Gulf of California). *Estuarine, Coastal and Shelf Science* **53**, 259–74.

Soulsby, R. (1997) *Dynamics of Marine Sands.* Thomas Telford, London.

Soulsby, R.L. (1983) The bottom boundary layer of shelf seas. In: *Physical Oceanography of Coastal and Shelf Seas* (Ed. B. Johns), pp. 189–266. Elsevier, Amsterdam.

Spalding, M.D. & Grenfell, A.M. (1997) New estimates of global and regional coral reef areas. *Coral Reefs* **16**, 225–30.

Spalding, M.D., Blasco, F., Field, C.D. (1997) *World Mangrove Atlas.* International Society for Mangrove Ecosystems, Okinawa, Japan.

Spedding, N. (2000) Hydrological controls on sediment transport pathways: implications or debris-covered glaciers. In: *Debris-covered Glaciers*, Workshop held at Seattle, September, pp. 133–42. IAHS Publication 264, International Association of Hydrological Sciences Press, Wallingford.

Spencer, T. (1995) Potentialities, uncertainties and complexities in the response of coral reefs to future sea-level rise. *Earth Surface Processes and Landforms* **20**, 49–64.

Speth, J.G. (1994) *Towards an Effective and Operational International Convention on Desertification.* International Negotiating Committee, International Convention on Desertification, United Nations, New York.

Stanley, D.J. & Warne, A.G. (1998) Nile delta in its destruction phase. *Journal of Coastal Research* **14**, 794–825.

Starkel, L. (1983) The reflection of hydrologic changes in the fluvial environment of the temperate zone during the last 15,000 years. In: *Background to Palaeohydrology: a Perspective* (Ed. K.J. Gregory), pp. 213–35. Wiley, Chichester.

Staub, J.R., Among, H.L. & Gastaldo, R.A. (2000) Seasonal sediment transport and deposition in the Rajang River delta, Sarawak, East Malaysia. *Sedimentary Geology* **133**, 249–64.

Sternberg, R.W. & Newell, A.R.M. (1999) Continental shelf sedimentology: scales of investigation define future research opportunities. *Journal of Sea Research* **41**, 55–71.

Stevenson, A.G. (2001) Metal concentrations in marine sediments around Scotland: a baseline for environmental studies. *Continental Shelf Research* **21**, 879–97.

Stigliani, W.M. (Ed.) (1991) *Chemical Time Bombs: Definition, Concepts, and Examples.* Executive Report 16, International Institute of Applied Systems Analysis, Laxenburg, Austria.

Stoddart, D.R. (1980) Mangroves as successional stages, inner reefs of the Great Barrier Reef. *Journal of Biogeography* **7**, 269–84.

Stoddart, D.R. (1990) Coral reefs and islands and predicted sea-level rise. *Progress in Physical Geography* **14**, 521–36.

Stoddart, D.R., Reed, D.G. & French, J.R. (1989) Understanding salt-marsh accretion, Scolt Head Island, Norfolk, England. *Estuaries* **12**, 228–36.

Stoermer, E.F. & Smol, J.P. (1999) *The Diatoms: Applications for the Environmental and Earth Sciences.* Cambridge University Press, Cambridge, 469 pp.

Stokes, G.G. (1851) *Collected Papers*, Vol. III. Cambridge Transactions Vol. IX (see e.g. Lamb, H., 1945. *Hydrodynamics*. Gover Publications, New York, 450 pp.)

Stokes, W.L. (1968) Multiple parallel truncation bedding planes – a feature of wind-deposited sandstone formations. *Journal of Sedimentary Petrography* **38**, 510–15.

Stone, M. & Marsalek, J. (1996) Trace metal composition and speciation in street sediment: Sault Ste, Marie, Canada. *Water, Air and Soil Pollution* **87**, 149–69.

Stow, D.A.V. (1986) Deep clastic seas. In: *Sedimentary Environments and Facies*, 2nd edn (Ed. H.G. Reading), pp. 399–470. Blackwell Science, Oxford.

Stride, A.H. (Ed.) (1982) *Offshore Tidal Sands.* Chapman and Hall, London, pp. 95–125.

Stum, W. (1992) *Chemistry of the Solid–Water Interface* Wiley Interscience, New York.

Sturm, M. (1975) Depositional and erosional sedimentary features in a turbidity current controlled basin (Lake brienz). *IX International Congress of Sedimentology*, Nice, Theme 5, Vol. 5/2.

Sturm, M. & Matter, A. (1978) Turbidites and varves in Lake Brienz (Switzerland): deposition of clastic detritus by density currents. In: *Modern and Ancient Lake Sediments* (Eds A. Matter & M.E. Tucker), pp. 147–68. Special Publication 2, International Association of Sedimentologists. Blackwell Scientific, Oxford.

Sullivan, M.J. (1999) Applied diatom studies in estuaries and shallow coastal waters. In: *The Diatoms: Applications for the Environmental and Earth Sciences* (Eds E.F. Stoermer & J.P. Smol), pp. 334–51. Cambridge University Press, Cambridge.

Summerfield, M.A. (1991) *Global Geomorphology*. Longman, New York, 537 pp.

Summerfield, M.A. & Hulton, N.J. (1994) Natural controls of fluvial denudation rates in major world drainage basins. *Journal of Geophysical Research* 99(B7), 13871–83.

Surian, N. & Rinaldi, M. (2003) Morphological response to river engineering and management in alluvial channels in Italy. *Geomorphology* 50, 307–26.

Sutherland, R.A. (2003) Lead in grain size fractions of road-deposited sediment. *Environmental Pollution* 121, 229–7.

Swennen, R., Van Keer, I. & De Vox, W. (1994) Heavy metal contamination in overbank sediments of the Geul river (East Belgium): its relation to former Pb-Zn mining activities. *Environmental Geology* 24, 12–21.

Swiadeck, J.W. (1997) The impacts of Hurricane Andrew on mangrove coasts in southern Florida: a review. *Journal of Coastal Research* 13, 242–5.

Swift, D.J.P., Han, G. & Vincent, C.E. (1986) Fluid processes and sea-floor responses on a modern storm-dominated shelf: middle-Atlantic shelf of North America. Part 1. The storm current regime. In: *Shelf Sands and Sandstones* (Eds R.J. Knight & J.R. McLean), pp. 99–119. Memoir 11, Canadian Society of Petroleum Geologists, Calgary.

Swinchatt, J.P. (1965) Significance of constituent composition, texture and skeletal breakdown in some Recent carbonate sediments. *Journal of Sedimentary Petrology* 35, 71–90.

Szmant, A.M. & Forrester, A. (1996) Water column and sediment nitrogen and phosphorous distribution patterns in the Florida Keys, USA. *Coral Reefs* 15, 21–42.

Taylor, K.G., Boyd, N.A. & Boult, S. (2003) Sediments, porewaters and diagenesis in an urban water body, Salford, UK: impacts of remediation. *Hydrological Processes* 17, 2049–61.

Te, F.E. (1997) Turbidity and its effect on corals: a model using the extinction coefficient (k) of photosythetic active radiation (PAR). *Proceedings of Eighth International Coral Reef Symposium, Panama* 2, 1899–904.

Teal, J.M., Farrington, J.W., Burns, K.A., et al. (1992) The west Falmouth oil spill after 20 years: fate of fuel oil compounds and effects on animals. *Marine Pollution Bulletin* 24, 607–14.

Teller, J.T. (2001) Formation of large beaches in an area of rapid differential isostatic rebound: the three-outlet control of Lake Agassiz. *Quaternary Science Reviews* 20, 1649–59.

Testa, V. & Bosence, E.J. (1999) Physical and biological controls on the formation of carbonate and siliciclastic bedforms on the north-east Brazilian shelf. *Sedimentology* 46, 279–301.

Thampanya, U., Vermaat, J.E. & Terrados, J. (2002) The effects of increasing sediment accretion on the seedlings of three common Thai mangrove species. *Aquatic Biology* 74, 315–25.

The Joint Nature Conservation Committee (2004) *The Irish Sea Pilot Final Report*. 176 pp. (available at www.jncc.gov.uk).

The Times Atlas of the World (1983) *Comprehensive Edition*, 6th edn. John Bartholomew and Times Books Limited, London.

Thieler, E.R., Pilkey, O.H., Young, R.S., et al. (2000) The use of mathematical models to predict beach behavior for U.S. coastal engineering: a critical review. *Journal of Coastal Research* 16, 48–70.

Thom, B.G. (1967) Mangrove ecology and deltaic geomorphology: Tabasco, Mexico. *Journal of Ecology* 55, 301–43.

Thom, B.G. (1982) Mangrove ecology: a geomorphological perspective. In: *Mangrove Ecosystems in Australia, Structure, Function and Management* (Ed. B.F. Clough), pp. 3–17. Australian National University Press, Canberra.

Thom, B.G. (1984) Transgressive and regressive stratigraphies of coastal sand barriers in southeast Australia. *Marine Geology* 56, 137–58.

Thomas, D.S.G. (1989) Reconstructing ancient arid environments. In: *Arid Zone Geomorphology: Process, Form and Change in Drylands*, 2nd edn (Ed. D.S.G. Thomas), pp. 311–34. Wiley, Chichester.

Thomas, D.S.G. (1997a) Arid environments. In: *Arid Zone Geomorphology: Process, Form and Change in Drylands*, 2nd edn (Ed. D.S.G. Thomas), pp. 3–12. Wiley, Chichester.

Thomas, D.S.G. (1997b) Sand seas and aeolian bedforms. In: *Arid Zone Geomorphology: Process, Form and Change in Drylands*, 2nd edn (Ed. D.S.G. Thomas), pp. 273–412. Wiley, Chichester.

Thomas, D.S.G. (1997c) Science and the desertification debate. *Journal of Arid Environments* 37, 599–608.

Thomas, D.S.G. & Middleton, N.J. (1994) *Desertification: Exploding the Myth*. Wiley. Chichester, 1194 pp.

Thomas, M.G. (1998) Late Quaternary landscape instability in the humid and sub-humid tropics. In: *Palaeohydrology and Environmental Change* (Eds G. Benito, V.R. Baker & K.J. Gregory), pp. 247–58. Wiley, Chichester.

Thomas, R.L., Jacquet, J.M., Kemp, A.L.W., et al. (1976) Surficial sediments in Lake Erie. *Journal of the Fisheries Research Board of Canada* 33, 385–403.

Thomas, S. & Ridd, P.V. (2004) Review of methods to measure short time scale sediment accumulation. *Marine Geology* 207, 95–114.

Thomas, S., Ridd, P.V. & Day, G. (2003) Turbidity regimes over fringing coral reefs near a mining site at Lihir Island, Papua New Guinea. *Marine Pollution Bulletin* **46**, 1006–14.

Thompson, T.A. & Baedke, S.J. (1995) Beach-ridge development in Lake Michigan: shoreline behavior in response to quasi-periodic lake-level events. *Marine Geology* **129**, 163–74.

Thorne, P.D. & Hanes, D.M. (2002) A review of acoustic measurement of small-scale sediment processes. *Continental Shelf Research* **22**, 603–32.

Thorne, R.D., Hey, R.D. & Newson, M.D. (Eds) (1997) *Applied Fluvial Geomorphology for River Engineering*. Wiley, Chichester.

Thornes, J.B. (1974) The rain in Spain. *Geogographical Magazine* **47**, 337–43.

Thornes, J.B. (1977) Channel changes in ephemeral streams: observations, problems and models. In: *River Channel Changes* (Ed. R.C. Gregory), pp. 353–67. Wiley, Chichester.

Thornthwaite, C.W. (1948) An approach towards a rational classification of climate. *Geographical Review* **38**, 55–94.

Thornton, I., Watt, J.M., Davies, D.J.A., et al. (1994) Lead contamination of UK dusts and soils and implications for childhood exposure – an overview of the work of the Environmental Geochemistry Research Group, Imperial College, London, England, 1981–1992. *Environmental Geochemistry and Health* **16**, 113–22.

Thurman, E.M. (1985) *Organic Geochemistry of Natural Waters*. Martinus Nijhoff/Dr W. Junk Publishers, Dordrecht, 487 pp.

Tomascik, T. & Sander, F. (1985) Effects of eutrophication on reef-building corals I. growth rate of the reef-building coral *Montastrea annularis*. *Marine Biology* **87**, 143–55.

Tomascik, T. & Sander, F. (1987) Effects of eutrophication on reef-building corals. III. Reproduction of the reef-building coral *Porites porites*. *Marine Biology* **94**, 77–94.

Tomlinson, P.B. (1986) *The Botany of Mangroves*. Cambridge University Press, Cambridge.

Topçu, D.H. & Brockmann, U. (2001) *Synthesis and New Conception of North Sea Research (SYCON) Working Group 5: Fluxes of Matter*. Zentrum für Meeres – und Klimaforschung der Universität, Hamburg, 167 pp.

Törnqvist, T.E. (1993) Holocene alteration of meandering and anastomising fluvial systems in the Rhine–Meuse delta (central Netherlands) controlled by sea level rise and subsoil erodability. *Journal of Sedimentary Petrology* **63**, 683–93.

Törnqvist, T.E. (1994) Middle and late Holocene avulsion history of the River Rhine (Rhine–Meuse Delta, Netherlands). *Geology* **22**, 711–14.

Toscano, M.A. & Macintyre, I.G. (2003) Corrected western Atlantic sea-level curve for the last 11,000 years based on calibrated 14C dates from *Acropora palmata* framework and intertidal mangrove peats. *Coral Reefs* **22**, 257–70.

Trauth, M.H., Alfonso, R.A. & Haselton, K.R. (2000) Climate change and mass movement in the NW Argentine Andes. *Earth and Planetary Science Letters* **179**, 243–56.

Trefry, J.H. & Presley, B.J. (1976) Heavy metal transport from the Mississippi River to the Gulf of Mexico. In: *Marine Pollution Transfer* (Eds H.L. Windom & R.A. Duce), pp. 39–76. Health Publications, Lexington.

Trenhaile, A.S. (1997) *Coastal Dynamics and Landforms*. Clarendon Press, Oxford.

Trichet, J., Defarge, C., Tribble, J., et al. (2001) Christmas Island lagoonal lakes, models for the deposition of carbonate–evaporite–organic laminated sediments. *Sedimentary Geology* **140**, 177–89.

Trimmer, M., Petersen, J., Sivyer, D.B., et al. (2005) Impacts of long-term benthic trawl disturbance on sediment sorting and biogeochemistry in the southern N. Sea. *Marine Ecology Progress Series* **298**, 79–94.

Trustrum, N.A., Gomez, B., Page, M.J., et al. (1999) Sediment production, storage and output: the relative role of large magnitude events in steepland catchments. *Zeitschrift für Geomorphologie* (N.F. supplement) **115**, 71–86.

Tucker, C.J., Dregne, H.E. & Newcomb, W.W. (1991) Expansion and contraction of the Sahara Desert from 1980 to 1990. *Science* **253**, 299.

Tucker, M.E. & Wright, V.P. (1990) *Carbonate Sedimentology*. Blackwells, Oxford.

Tudhope, A.W. & Scoffin, T.P. (1984) The effects of *Callianassa* bioturbation on the preservation of carbonate grains in Davies reef lagoon, Great Barrier Reef, Australia. *Journal of Sedimentary Petrology* **54**, 1091–6.

Udvardy, M.D.F. (1981) The riddle of dispersal: dispersal theories and how they affect vicariance biogeography. In: *Vicariance Biogeography: a Critique* (Eds G. Nelson & D.E. Rosen), pp. 6–29. Columbia University Press, New York.

UN (1977) Status of desertification in the hot arid regions, climatic aridity index map and experimental world scheme of aridity and drought probability, at a scale of 1:25,000,000. Explanatory note. *UN Conference on Desertification A/CONF.* 74/31, New York.

Underwood, G.J.C. & Smith, D.J. (1998) *In-situ* measurements of exopolymer production by intertidal epipelic diatom-dominated biofilms. In: *Sedimentary Processes in the Intertidal Zone* (Eds K.S. Black, D.H. Paterson & A. Cramp), pp. 125–34. Special Publication 139, Geological Society Publishing House, Bath.

UNEP (1992) *World Atlas of Desertification*. Edward Arnold, Sevenoaks, 69 pp.

UNESCO (1979) *Map of the World Distribution of Arid Regions.* Technical Note 4, Man and the Biosphere Programme, UNESCO, New York.

United Nations (2003) *Water and Sanitation in the World's Cities: Local Action for Global Goals.* Press Release HAB/184, 19 March, UN-HABITAT, Nairobi.

USACE (2003) *Shore Protection Manual.* United States Army Corps of Engineers, Vicksburg, Mississippi. Available at: http://bigfoot.wes.army.mil/cem026.html

USACE/EPA (1992) *Evaluating Environmental Effects of Dredged Material Management Alternatives – a Technical Framework.* EPA842-B-92-008, Environmental Protection Agency, Office of Marine and Estuarine Protection, Washington, DC.

Van Dalfsen, J.A., Essink, K., Toxvig Madsen, H., et al. (2000) Differential response of macrozoobenthos to marine sand extraction in the North Sea and the Western Mediterranean. *ICES Journal of Marine Science* 57, 1439–45.

Van den Bergh, G.D., Boer, W., de Hass, H., et al. (2003) Shallow marine tsunami deposits in Teluk Banten (NW Java, Indonesia), generated by the 1883 Krakatau eruption. *Marine Geology* 197, 13–34.

Van Walree, P.A., Tjegowski, J., Labanc, C., et al. (2005) Acoustic seafloor discrimination with echo shape parameters: a comparison with the ground truth. *Continental Shelf Research* 25, 2273–93.

Van Woesik, R., De Vantier, L.M. & Glazebrook, J.S. (1995) Effects of Cyclone 'Joy' on nearshore coral communities of the Great Barrier Reef. *Marine Ecology Progress Series* 128, 261–70.

Vaughan, N.D., Johnson, T.C., Mearns, D.L., et al. (1987) The impact of Hurricane Diana on the North Carolina continental shelf. *Marine Geology* 76, 169–76.

Veldkamp, A. & Van Dijke, J.J. (1998) Modelling long-term erosion and sedimentation processes in fluvial systems: a case study for the Allier/Loire system In: *Palaeohydrology and Environmental Change* (Eds G. Benito, V.R. Baker & K.J. Gregory), pp. 53–66. Wiley, Chichester.

Verhagen, H. (1992) Method for artificial beach renourishment, *Proceedings of the 23rd International Conference on Coastal Engineering,* Venice, Italy, pp. 2474–85.

Viklander, M. (1998) Particle size distribution and metal content in street sediments. *Journal of Environmental Engineering* 124, 761–66.

Viles, H. & Spencer, T. (1995) *Coastal Problems: Geomorphology, Ecology and Society at the Coast.* Edward Arnold, London.

Viseras, C., Calvache, M.L., Soria, J.M., et al. (2003) Differential features of alluvial fans controlled by tectonic or eustatic accommodation space, examples from the Betic Cordillera, Spain. *Geomorphology* 50, 181–20.

Vivian, C.M.G. & Murray, L.A. (2001) Pollution, solids. In: *Encyclopaedia of Ocean Sciences* (Eds J. Steele, S. Thorpe & K.K. Turekian), pp. 2236–41. Academic Press, London.

Vollenweider, R.A. (1968) *The Scientific Basis of Lake Eutrophication, with Particular Reference to Phosphorus and Nitrogen as Eutrophication Factors.* Technical Report DAS/DSI/68.27, Organization of Economic Cooperation and Development, Paris, 159 pp.

Vollenweider, R.A. (1976) Advances in defining critical loading levels for phosphorus in lake eutrophication. *Istituto Italiano di Idrobiologia* 33, 53–83.

Walder, J.S. & Driedger, C.L. (1995) Frequent outburst floods from South Tahoma Glacier, Mount Rainier, USA: relations to debris flows, meteorological origin and implications for subglacial hydrology. *Journal of Glaciology* 41(137), 1–10.

Walker, R.G. & James, N.P. (Eds) (1992) *Facies Models: Response to Sea-level Change.* Geological Association of Canada, St John's, Newfoundland.

Walker, W.J., McNutt, R.P. & Maslanka, C.K. (1999) The potential contribution of urban runoff to surface sediments of the Passaic River: sources and chemical characteristics. *Chemosphere* 38, 363–77.

Wallin, M., Håkanson, L. & Persson, J. (1992) Load models for nutrients in coastal areas, especially from fish farms (in Swedish with English summary). *Nordiska ministerrådet, Copenhagen* 502, 207 pp.

Walling, D.E. (1987) Rainfall, runoff and erosion of the land: a global view. In: *Energetics of Physical Environment* (Ed. K.J. Gregory), pp. 89–117. Wiley, Chichester.

Walling, D.E. (1990) Linking the field to the river: sediment delivery from agricultural land. In: *Soil Erosion on Agricultural Land* (Eds J. Boardman, I.D.L. Foster & J.A. Dearing), pp. 129–52. Wiley, Chichester.

Walling, D.E. (1999) Linking land use, erosion and sediment yields in river basins. *Hydrobiologia* 410, 223–40.

Walling, D.E. & Fang, D. (2003) Recent trends in the suspended sediment loads of the world's rivers. *Global and Planetary Change* 39, 111–26.

Walling, D.E. & Kleo, A.H.A. (1979) Sediment yields of rivers in areas of low precipitation: a global view. In: *The Hydrology of Areas of Low Precipitation,* Proceedings of the Canberra Symposium, December, pp. 479–93. IAHS Publication 128, International Association of Hydrological Sciences Press, Wallingford.

Walling, D.E. & Webb, B.W. (1983) Patterns of sediment yield. In: *Background to Palaeohydrology: a Perspective* (Ed. K.J. Gregory), pp. 69–100. Wiley, Chichester.

Walling, D.E., Peart, M.R., Oldfield, F., et al. (1979) Suspended sediment sources identified by magnetic measurements. *Nature* 281, 110–13.

Walling, D.E., Woodward, J.C. & Nicholas, A.P. (1993) A multi-parameter approach to fingerprinting suspended sediment sources. In: *Erosion and Sediment Transport Monitoring Programmes* (Eds J. Bogen, D.E. Walling & T. Day), pp. 329–38. IAHS Publication 215, International Association of Hydrological Sciences Press, Wallingford.

Walling, D.E., Owens, P.N. & Leeks, G.J.L. (1999) Fingerprinting suspended sediment sources in the catchment of the River Ouse, Yorkshire, UK. *Hydrological Processes* 13, 955–75.

Walling, D.E., Russell, M.A., Hodgkinson, R.A., et al. (2002) Establishing sediment budgets for two small lowland agricultural catchments in the UK. *Catena* 47, 323–53.

Walling, D.E., Owens, P.N., Carter, J., et al. (2003) Storage of sediment-associated nutrients and contaminants in river channels and floodplain systems. *Applied Geochemistry* 18, 195–220.

Warburton, J. (1990) An alpine proglacial fluvial sediment budget. *Geografiska Annaler* 72A(3–4), 261–72.

Warren, A. (1979) Aeolian processes. In: *Process in Geomorphology* (Eds C. Embleton & J. Thornes), pp. 325–551. Edward Arnold, Sevenoaks.

Warren, N., Allan, I.J., Carter, J.E., et al. (2003) Pesticides and other micro-organic contaminants in freshwater sedimentary environments – a review. *Applied Geochemistry* 18, 159–94.

Wasson, R.J. & Hyde, R. (1983) Factors determining desert dune type. *Nature* 304, 337–9.

Watling, L. & Norse, E.A. (1998) Disturbance of the seabed by mobile fishing gear: a comparison to forest clearcutting. *Conservation Biology* 12, 1180–97.

Watson, A. & Nash, D.J. (1997) Desert crusts and varnishes. In: *Arid Zone Geomorphology: Process, Form and Change in Drylands*, 2nd edn (Ed. D.S.G. Thomas), pp. 69–107. Wiley, Chichester.

Watts, C.D., Naden, P.S., Cooper, D.M., et al. (2003) Application of a regional procedure to assess the risk to fish from high sediment concentrations. *The Science of the Total Environment* 314–16, 551–65.

Webb, R.H., Pringle, P.T., Reneau, S.L., et al. (1988) Monument Creek debris flow, 1984 – implications for formation of rapids on the Colorado River in Grand Canyon National Park. *Geology* 6, 50–4.

Wei, C. & Morrison, G.M. (1994a) Platinum in road dusts and urban river sediments. *The Science of the Total Environment* 147, 169–74.

Wei, C. & Morrison, G.M. (1994b) Platinum analysis and speciation in urban gullypots. *Analytica Chimica Acta* 284, 587–92.

Wells, S.G. & Harvey, A.M. (1987) Sedimentalogic and geomorphic variations in storm-generated alluvial fans, Howgill Fells, northwest England. *Geological Society of America Bulletin* 98, 182–98.

Wenban-Smith, F. (2002) *Palaeolithic and Mesolithic Archaeology on the Sea Bed: Marine Aggregate Dredging and the Historic Environment*. Report to the British Marine Aggregate Producers Association, london, 18 pp. (available from www.bmapa.org.uk).

Wenning, R.J., Batley, G.R., Ingersoll, C.G., et al. (Eds) (2005) *Use of sediment quality guidelines & related tools for the assessment of contaminated sediments*. Society of Environmental Toxicology and Chemistry, 816 pp.

Werritty, A. & Leys, K.F. (2001) The sensitivity of Scottish rivers and upland valley floors to recent environmental change. *Catena* 42, 251–73.

Westaway, R., Maddy, D. & Bridgland, D. (2002) Flow in the lower continental crust as a mechanism for the Quaternary uplift of south-east England: constraints from the Thames terrace record. *Quaternary Science Reviews* 21, 559–603.

Wetzel, D.L. & Van Vleet, E.S. (2003) Persistence of petroleum hydrocarbon contamination in sediments of the canals of Venice, Italy: 1995 and 1998. *Marine Pollution Bulletin* 46, 1015–23.

Wetzel, R.G. (2001) *Limnology and Lake Ecosystems*, 3rd edn. Academic Press, San Diego, 1006 pp.

Whipple, K.X. & Trayler, C.R. (1996) Tectonic control of fan size: the importance of spatially variable subsidence rates. *Basin Research* 8, 351–66.

White, S., García-Ruiz, J.M., Martí, C., et al. (1997) The 1996 Biescas campsite disaster in the Central Spanish Pyrenees, and its temporal and spatial context. *Hydrological Processes* 11, 1797–812.

Whiteley, J.D. & Murray, F. (2003) Anthropogenic platinum group element (Pt, Pd and Rh) concentrations in road dusts and roadside soils from Perth, Western Australia. *The Science of the Total Environment* 317, 121–35.

Whitney, M.I. & Dietrich, R.V. (1973) Ventifact sculpture by wind-blown dust. *Geological Society of America Bulletin* 84, 2561–82.

Whitten, D.G.A. & Brooks, J.R.V. (1972) *The Penguin Dictionary of Geology*. Penguin Books, London, 514 pp.

Whittow, J. (1980) Landslides and avalanches – avalanches. In *Disasters: the Anatomy of Environmental Hazards*, pp. 163–70. Penguin, Harmondsworth.

Wicklund, A. (1990) *Metabolism of cadmium and zinc in fish*. Thesis, Uppsala University, Sweden.

Wiedermann, H.V. (1972) Shell deposits and shell preservation in Quaternary and Tertiary estuarine sediments in Georgia, USA. *Sedimentary Geology* 7: 103–25.

Wiggs, G.F.S., Livingstone, I., Thomas, D.S.G., et al. (1994) The effect of vegetation removal on air flow structure and dune mobility in the southwest Kalahari. *Land Degradation and Rehabilitation* 5, 13–24.

Wilkinson, C.R. (1996) Global change and coral reefs: impacts on reefs, economies and human cultures. *Global Change Biology* 2, 547–58.

Willets, B.B. (1983) Transport by wind of granular materials of different grain shapes and densities. *Sedimentology* **30**, 669–79.

Willets, B.B. & Rice, M.A. (1985) Wind tunnel tracer experiments using dyed sand. In: *Proceedings of International Workshop on the Physics of Blown Sand*, pp. 225–42. Memoir 8, Department of Theoretical Statistics, Aarhus University, Denmark.

Williams, M.A.J. & Balling, R.C. (1995) *Interactions of Desertification and Climate*. Edward Arnold, London, 270 pp.

Williams, M.A.J. (1994) Cenozoic climatic changes in deserts: a synthesis. In: *Geomorphology of Desert Environments* (Eds A.D. Abrahams & A.J. Parsons), pp. 644–70. Chapman and Hall, London.

Wilson, I.G. (1973) Ergs. *Sedimentary Geology* **10**, 77–106.

Wilson, P. & Braley, S.M. (1997) Development and age structure of Holocene coastal sand dunes at Horn Head, near Dunfanaghy, Co. Donegal, Ireland. *The Holocene* **7**, 187–97.

Wilson, P., Orford, J.D., Knight, J., et al. (2001) Late-Holocene (post-4000 years BP) coastal dune development in Northumberland, northeast England. *The Holocene* **11**, 215–29.

Winkler, E.M. & Wilheim, E.J. (1970) Saltburst by hydration pressures in architectural stone in urban atmosphere. *Geological Society of America Bulletin* **81**, 567–72.

Winkler, E.M. (1977) Insolation of rock and stone, a hot item. *Geology* **5**, 188–9.

Wisniak, J. & Garces, I. (2001) The rise and fall of the salitre (sodium nitrate) industry. *Indian Journal of Chemical Technology* **8**, 427–38.

Wolanski, E. (1995) Transport of sediment in mangrove swamps. *Hydrobiologia* **295**, 31–42.

Wolanski, E. & Chappell, J. (1996) The response of tropical Australian estuaries to a sea level rise. Journal of Marine Systems **7**, 267–79.

Wolanski, E., Jones, M. & Bunt, J.S. (1980) Hydrodynamics of a tidal creek-mangrove swamp system. *Australian Journal of Marine and Freshwater Research* **31**, 431–50.

Wolanski, E., Mazda, Y. & Ridd, P. (1992) Mangrove Hydrodynamics. In: *Tropical Mangrove Ecosystems* (Eds A.I. Robertson & D.M. Alongi), pp. 43–62. Coastal and Estuarine Studies 41, American Geophysical Union, Washington.

Wolanski, E., Gibbs, R.J., Spagnol, S., et al. (1998) Inorganic sediment budget in the mangrove-fringed Fly River Delta, Papua New Guinea. *Mangroves and Salt Marshes* **2**, 85–98.

Wolfenden, P.J. & Lewin, J. (1977) Distribution of metal pollutants in floodplain sediments. *Catena* **4**, 309–17.

Wolman, M.G. & Gerson, R. (1978) Relative scales of time and effectiveness of climate in watershed geomorphology. *Earth Surface Processes* **3**, 189–208.

Wood, P.A. (1978) Fine sediment mineralogy of source rocks and suspended sediments, Rother catchment, West Sussex. *Earth Surface Processes* **3**, 255–63.

Woodhouse, W.W. (1978) *Dune Building and Stabilization with Vegetation*. Special Report 3, Coastal Engineering Research Center, U.S. Army Corps Engineers, Springfield, VA.

Woodroffe, C. (1992) Mangrove sediments and geomorphology. In: *Tropical Mangrove Ecosystems* (Eds A.I. Robertson & D.M. Alongi), pp. 7–41. Coastal and Estuarine Studies 41, American Geophysical Union, Washington.

Woodroffe, C.D. (1983) Development of mangrove forests from a geological perspective. In: *Tasks for Vegetation Science 8* (Ed. H.J. Teas), pp. 1–17. Dr W. Junk Publishers, The Hague.

Woodroffe, C.D. (1990) The impact of sea-level rise on mangrove shorelines. *Progress in Physical Geography* **14**, 483–520.

Woodroffe, C.D. (2003) *Coasts: Form, Process and Evolution*. Cambridge University Press, Cambridge.

Woodroffe, C.D. & Grime, D. (1999) Storm impact and evolution of a mangrove-fringed chenier plain, Shoal Bay, Darwin, Australia. *Marine Geology* **159**, 303–21.

Woodroffe, C.D. & Grindrod, J. (1991) Mangrove biogeography: the role of Quaternary environmental and sea-level change. *Journal of Biogeography* **18**, 479–92.

Woodroffe, C.D., Bardsley, K.N., Ward, P.J., et al. (1988) Production of mangrove litter in a macro-tidal embayment, Darwin Harbour, N.T., Australia. *Estuarine, Coastal and Shelf Science* **26**, 581–98.

Woodward, J.C. (1995) Patterns of erosion and suspended sediment yield in Mediterranean river basins. In: *Sediment and Water Quality in River Catchments* (Eds I.D.L. Foster, A.W. Gurnell & B.W. Webb), pp. 365–89. Wiley. Chichester.

Woodward, J.C., Lewin, J. & Macklin, M.G. (1992) Alluvial sediment sequences in a glaciated catchment: the Voidomatis Basin, Northwestern Greece. *Earth Surface Processes and Landforms* **16**, 207–26.

Woolfe, K.J. & Larcombe, P.I. (1998) Terrigenous sediment accumulation as a regional control on the distribution of reef carbonates. In: *Reef and Carbonate Platforms in the Pacific and Indian Oceans* (Eds G.F. Camoin & P.J. Davies), pp. 295–310. Blackwell Science, Oxford.

Woolfe, K.J., Larcombe, P. & Stewart, L.K. (2000) Shelf sediments adjacent to the Herbert River delta, Great Barrier Reef, Australia. *Australian Journal of Earth Sciences* **47**, 301–8.

World Commission on Dams (2000) *Dams and Development: a New Framework for Decision-Making*. The Report of the World Commission on Dams. An Overview, 16 November 2000. http://www.dams.org (accessed 25 June 2004.)

Wright, L.D. & Coleman J.M. (1971) The discharge/wave power climate and the morphology or delta coasts. *Association of American Geographers Proceedings* 3, 186–9.

Wright, L.D. & Coleman, J.M. (1972) River delta morphology: wave climate and the role of the subaqueous profile. *Science* 176, 282–4.

Wright, L.D. & Short, A.D. (1984) Morphodynamic variability of surf zones and beaches: a synthesis. *Marine Geology* 56, 93–118.

Wright, L.D. (1982) Deltas. In: *The Encyclopedia of Beaches and Coastal Environments* (Ed. M. Schwartz), pp. 358–68. Hutchinson-Ross, Stroudsberg, PA.

Wright, L.D. (1995) *Morphodynamics of Inner Continental Shelves*. CRC Press, London, 241 pp.

Wright, V.P. & Burchette, T.P. (1996) Shallow-water carbonate environments. In: *Sedimentary Environments: Processes, Facies and Stratigraphy*, 3rd edn (Ed. H.G. Reading), pp. 325–94. Blackwell Science, Oxford.

Wu, J.S., Holman, R.E. & Dorney, J.R. (1996) Systematic evaluation of pollution removal by urban wet detention ponds. *Journal of Environmental Engineering* 122, 983–8.

Xie, S., Dearing, J., Bloemendal, J., et al. (1999) Association between the organic matter content and magnetic properties in street dust, Liverpool, UK. *The Science of the Total Environment* 241, 205–14.

Xue, C. (2001) Coastal erosion and management of Majuro Atoll, Marshall Islands. *Journal of Coastal Research* 17, 909–18.

Yair, A. (1994) The ambiguous impact of climate change at a desert fringe: northern Negev, Israel. In: *Environmental Change in Drylands: Biogeographical and Geomorphological Perspectives* (Eds A.C. Millington & K. Pye), pp. 199–227. Wiley, Chichester.

Yair, A. & Gerson, R. (1974) Mode and rate of escarpment retreat in an extremely arid environment (Sharm el Sheikh, southern Sinai Peninsula). *Zeitschrift für Geomorphologie*, Supplementband 21, 202–15.

Yair, A. & Lavee, H. (1976) Runoff generative process and runoff yield from arid talus mantled slopes. *Earth Surface Processes* 1, 235–47.

Yan, P., Shi, P., Gao, S., et al. (2002) ^{137}Cs dating of lacustrine sediments and human impacts on Dalian Lake, Qinghai province, China. *Catena* 47, 91–9.

Yap, H.T. (2000) The case for restoration of tropical coastal ecosystems. *Ocean and Coastal Management* 43, 841–51.

Yeats, P.A. & Bewers, J.M. (1982) Discharge of metals from the St. Lawrence River. *Canadian Journal of Earth Sciences* 19, 982–92.

Yechieli, Y. & Wood, W.W. (2002) Hydrogeologic processes in saline systems: playas, sabkhas, and saline lakes. *Earth-Science Reviews* 58, 343–65.

Young, R.A. (1985) Geomorphic evolution of the Colorado Plateau margin in west-central Arizona: a tectonic model to distinguish between the causes of rapid, symmetrical scarp retreat and scarp dissection. In: *Tectonic Geomorphology* (Eds M. Morisawa & J.T. Hack), pp. 261–78. Allen & Unwin, Boston.

Yunker, M.B., Macdonald, R.W., Vingarzan, R., et al. (2002) PAHs in the Fraser River basin: a critical appraisal of PAH ratios as indicators of PAH source and composition. *Organic Geochemistry* 33, 489–515.

Zhu, B.Q., Chen, Y.W. & Peng, J.H. (2001) Lead isotope geochemistry of the urban environment in the Pearl River Delta. *Applied Geochemistry* 16, 409–17.

Zimbelman, J.R., Williams, S.H. & Tchakerian, V.P. (1995) Sand transport paths in the Mojave Desert, southwestern United States. In: *Desert Aeolian Processes* (Ed. V.P. Tchakarian), pp. 101–29. Chapman and Hall, London.

Zimmermann, M. (1990) Debris flows 1987 in Switzerland: geomorphological and meteorological aspects. In: *Hydrology in Mountainous Regions. II – Artificial Reservoirs; Water and Slopes* (Eds R.O. Sinniger & M. Monbaron), pp. 387–93. IAHS Publication 194, Proceedings of two Symposia held at Lausanne, August. International Association of Hydrological Sciences Press, Wallingford.

Zimmermann, M. & Haeberli, W. (1992) Climate change and debris flow activity in high-mountain areas – a case study in the Swiss Alps. *Catena Supplement* 22, 59–72.

Zolitschka, B. (1998) A 14,000 year sediment yield record from western Germany based on annually laminated lake sediments. *Geomorphology* 22, 1–17.

Index

Note: page numbers in **bold** refer to tables, those in *italics* refer to figures and boxes.

The manufacturer's authorised representative in the EU for product safety is Oxford University Press España S.A. of El Parque Empresarial San Fernando de Henares, Avenida de Castilla, 2 – 28830 Madrid (www.oup.es/en or product.safety@oup.com). OUP España S.A. also acts as importer into Spain of products made by the manufacturer.

Printed in the USA/Agawam, MA
January 13, 2025

880951.011